Lecture Notes in Mechanical Engineering

Lecture Notes in Mechanical Engineering (LNME) publishes the latest developments in Mechanical Engineering—quickly, informally and with high quality. Original research or contributions reported in proceedings and post-proceedings represents the core of LNME. Volumes published in LNME embrace all aspects, subfields and new challenges of mechanical engineering.

To submit a proposal or request further information, please contact the Springer Editor of your location:

Europe, USA, Africa: Leontina Di Cecco at Leontina.dicecco@springer.com

China: Ella Zhang at ella.zhang@cn.springernature.com

India, Rest of Asia, Australia, New Zealand: Swati Meherishi at swati.meherishi@springer.com

Topics in the series include:

- Engineering Design
- Machinery and Machine Elements
- Mechanical Structures and Stress Analysis
- Automotive Engineering
- Engine Technology
- Aerospace Technology and Astronautics
- Nanotechnology and Microengineering
- Control, Robotics, Mechatronics
- MEMS
- Theoretical and Applied Mechanics
- Dynamical Systems, Control
- Fluid Mechanics
- Engineering Thermodynamics, Heat and Mass Transfer
- Manufacturing Engineering and Smart Manufacturing
- Precision Engineering, Instrumentation, Measurement
- Materials Engineering
- Tribology and Surface Technology

Indexed by SCOPUS, EI Compendex, and INSPEC.

All books published in the series are evaluated by Web of Science for the Conference Proceedings Citation Index (CPCI).

To submit a proposal for a monograph, please check our Springer Tracts in Mechanical Engineering at https://link.springer.com/bookseries/11693.

Holger Kohl · Günther Seliger · Franz Dietrich · Giampaolo Campana
Editors

Safe and Sustainable Value Creation by Design

Proceedings of the 21st Global Conference on Sustainable Manufacturing (GCSM 2025) September 10–12, 2025, Bologna, Italy, Volume 2

Springer

Editors
Holger Kohl
Institute of Machine Tools and Factory Management
Technische Universität Berlin
Berlin, Germany

Franz Dietrich
Technical University of Berlin
Berlin, Germany

Günther Seliger
Technische Universität Berlin
Berlin, Germany

Giampaolo Campana
Dipartimento di Ingegneria Industriale
University of Bologna
Bologna, Italy

ISSN 2195-4356 ISSN 2195-4364 (electronic)
Lecture Notes in Mechanical Engineering
ISBN 978-3-032-21153-8 ISBN 978-3-032-21154-5 (eBook)
https://doi.org/10.1007/978-3-032-21154-5

This work was supported by GPE Global Production Engineering UG.

This Springer imprint is published by the registered company Springer Nature Switzerland AG
The registered company address is: Gewerbestrasse 11, 6330 Cham, Switzerland

Preface

We are pleased to publish a collection of papers presented at the 21st Global Conference on Sustainable Manufacturing (GCSM), held on September 10–12, 2025, in Bologna, Italy. The conference is annually sponsored by the International Academy for Production Engineering (CIRP), committed to excellence in the creation of sustainable products and processes. GCSM 2024 was jointly organized by the Institute for Machine Tools and Factory Management (IWF) at the TU Berlin and Fraunhofer Institute for Production Systems and Design Technology (IPK) together with the University of Bologna.

The GCSM 2025 brought together more than 160 attendees from over 30 countries providing a global forum of academics, researchers, and specialists from universities, research institutes and industry from across the globe, working on topics related to sustainable manufacturing. A unique feature of the GCSM conference series is its integration of industrial engineering perspectives, sustainable manufacturing applications in emerging and developing countries, as well as education and workforce development for advancing sustainable manufacturing. Plenary keynote speeches by experienced personalities from academics and industry, paper sessions presentations and workshops of student teams from different countries offered new insights and chances for exchange of ideas. The conference featured five keynote speakers who shared recent advances in cutting-edge research and industry practices; these prominent and internationally recognized experts elaborated how technologies in the product, process and system domains can enable sustainable manufacturing.

This volume documents over 120 contributions presented at GCSM 2025 in 24 sessions held over three days. The proceedings are organized according to the conference program, classified into four broad categories as: Sustainable Processes, Sustainable Manufacturing Systems, Sustainable Manufacturing Products, and Crosscutting Topics in Sustainable Manufacturing. The papers cover a variety of topics in these areas related to modelling and simulation of manufacturing processes, product design for sustainability, metrics for sustainability assessment, energy efficiency in manufacturing, strategies and business models, as well as education and workforce development for sustainable manufacturing. All papers published in these proceedings have been single-blind reviewed by experts from the international scientific committee, which are listed in the next section. These reviews resulted in an overall acceptance rate of 60.6%.

Holger Kohl
Franz Dietrich
Günther Seliger
Giampaolo Campana

Organization

International Organizing Committee

Holger Kohl (International Chairman)
Franz Dietrich (Local Chairman)
Günther Seliger (Founding Chairman)
Valentin Eingartner (Organizational Team)
Maxim Mintchev (Organizational Team)
Gaurav Mahajan Anant (Organizational Team)

National Organizing Committee

Giampaolo Campana (National Chairman)
Sebastián Mur (Latin American Advisor)
Barbara Cimatti (Local Organizational Team)
Luca Anzolin (Local Organizational Team)
Mariam Maisuradze (Local Organizational Team)

International Scientific Committee

Ahmed Abu Hanieh	Birzeit University, Palestine
Feri Afrinaldi	Andalas University, Indonesia
Tülin Aktin	İstanbul Kültür Üniversitesi, Turkey
Elita Amrina	Andalas University, Indonesia
Helmi Attia	McGill University, Canada
Fazel Ansari	Technische Universität Wien, Austria
Fazleena Badurdeen	University of Kentucky, USA
Dirk Bähre	Universität des Saarlandes, Germany
Peter Ball	University of York, UK
Thomas Bergs	RWTH Aachen, Germany
Leopoldo De Bernardez	Instituto Tecnológico de Buenos Aires, Argentina
Clemens Biet	Technische Universität Berlin, Germany
Dermot Brabazon	Dublin City University, Ireland
Christopher Brown	Worcester Polytechnic Institute, USA
Erhan Budak	Sabancı Üniversitesi, Turkey
Luiz Cesar Ribeiro Carpinetti	Universidade de São Paulo, Brazil

Felipe Cerdas	Technische Hochschule Würzburg-Schweinfurt, Germany
Jun-Ki Choi	University of Dayton, USA
George Chryssolouris	University of Patras, Greece
Gabriela Corsano	Consejo Nacional Investigaciones Científicas y Técnicas, Argentina
Roy Damgrave	University of Twente, Netherlands
Berend Denkena	Gottfried Wilhelm Leibniz Universität Hannover, Germany
Franz Dietrich	Technische Universität Berlin, Germany
Matthew Doolan	Australian National University, Australia
Klaus Dröder	Technische Universität Braunschweig, Germany
Joost Duflou	Katholieke Universiteit Leuven, Belgium
Arne Glodde	Technische Universität Berlin, Germany
Domingo González	Pontificia Universidad Católica del Perú, Peru
Karl Haapala	Oregon State University, USA
Michael Zwicky Hauschild	Denmarks Teknise Universitet, Denmark
Carlos Hernandez	Universidad Cayetano Heredia, Peru
I. S. Jawahir	University of Kentucky, USA
Roland Jochem	Technische Universität Berlin, Germany
Sami Kara	The University of New South Wales
Takayuki Kataoka	Kindai University, Japan
Yusuf Kaynak	Marmara University, Turkey
Karel Kellens	Katholieke Universiteit Leuven, Belgium
Yuki Kinoshita	SECOM CO., LTD., Japan
Holger Kohl	Technische Universität Berlin, Germany
Philip Koshy	McMaster University, Canada
Daniel Martinez Krahmer	Instituto Nacional de Tecnología Industrial, Argentina
Sze Yin Kwok	Blekinge Institute of Technology, Sweden
Lang Van Thang	IUH, Vietnam
Bert Lauwers	Katholieke Universiteit Leuven, Belgium
Le Doan Quoc Bao	Vietnamese-German University, Vietnam
Hoew Pueh Lee	National University of Singapore, Singapore
Laura V. Lerman	Technische Universität München, Germany
Yongrong Li	SKF Lubrication Systems Germany GmbH, Germany
Makinde Olasumbo	University of Johannesburg, South Africa
Sotiris Makris	LMS - University of Patras, Greece
Raúl Marino	Instituto Tecnológico de Buenos Aires, Argentina
Khaled Medini	École des Mines de Saint-Étienne, France
Shreyes Melkote	Georgia Tech Manufacturing Institute, USA

Robert Mies	Technische Universität Berlin, Germany
Laszlo Monostori	Hungarian Academy of Sciences, Hungary
Sandra Naomi Morioka	Federal University of Paraíba, Brazil
Anke Müller	Hochschule Hof, Germany
Bernd Muschard	Technische Universität Berlin, Germany
Nabil Nasr	Rochester Institute of Technology, USA
Vi Nguyen	Vietnamese-German University, Vietnam
Nguyen Xuan Thanh	Vietnamese-German University, Vietnam
Aldo Roberto Ometto	Universidade de São Paulo, Brazil
Soh Khim Ong	National University of Singapore, Singapore
Tiaan Oosthuizen	Institute for Technology & Society, South Africa
Emanuele Pagone	Cranfield University, UK
Shahin Rahimifard	Loughborough University, UK
Paul Refalo	University of Malta, Malta
Konstantinos Salonitis	Cranfield University, UK
Sebastian Schlund	Technische Universität Wien, Austria
Julia Schwarzkopf	Hochschule für Technik und Wirtschaft Berlin, Germany
Günther Seliger	Technische Universität Berlin, Germany
Luca Settineri	Politecnico Di Torino, Italy
Semih Severengiz	Hochschule Bochum, Germany
Vennan Sibanda	National University of Science and Technology, Zimbabwe
Cecilia Smoglie	Instituto Tecnológico de Buenos Aires, Argentina
Rikard Söderberg	Chalmers University of Technology, Sweden
Rainer Stark	Technische Universität Berlin, Germany
Frank Straube	Technische Universität Berlin, Germany
Nicole Stricker	Hochschule Aalen, Germany
John W. Sutherland	Purdue University, USA
Sebastian Thiede	University of Twente, Netherlands
Olga Timoteo	Universidad Cayetano Heredia, Peru
Tran Le Luu	Vietnamese-German University, Vietnam
Eckart Uhlmann	Technische Universität Berlin, Germany
Thomas Volling	Technische Universität Berlin, Germany
Rok Vrabic	University of Ljubljana, Slovenia
Christian Wacker	Technische Universität Braunschweig, Germany
Lihui Wang	KTH Royal Institute of Technology, Sweden
Rafi Wertheim	Technische Universität Chemnitz, Germany
Shuho Yamada	Toyama Prefectural University, Japan
Tetsuo Yamada	University of Electro-Communications, Japan
Hitomi Yamaguchi	University of Florida, USA
Chris Yuan	Case Western University, USA

Michael Zaeh	Technische Universität München, Germany
Zhang Weimin	Tongji University, China
Fu Zhao	Purdue University, USA

Local Scientific Committee

Giacomo Bergamini	University of Bologna, Italy
Barbara Cimatti	University of Bologna, Italy
Anna Costa	Italian National Research Council (CNR), Italy
Monica Fabrizio	Italian National Research Council (CNR), Italy
Marco Giorgetti	University of Bologna, Italy
Luca Pasquini	University of Bologna, Italy
Luca Settineri	Politecnico di Torino, Italy
Roberto Teti	University of Naples "Federico II", Italy

Contents

Factory Planning

Production Planning and Control

Education and Training

Student Sessions

Factory Planning

Ecologically Interlinked Manufacturing Districts – a Method for Generating Local Synergies to Increase Sustainable Value Creation

Michael Hertwig[1(✉)], Jannik Munderich[2], and Adrian Barwasser[1]

[1] Fraunhofer Institute for Industrial Engineering IAO, Nobelstrasse 12, 70569 Stuttgart, Germany
michael.hertwig@iao.fraunhofer.de

[2] Institute for Human-Factors and Technology Management IAT, University of Stuttgart, Nobelstrasse 12, 70569 Stuttgart, Germany

Abstract. The sustainable transition is being taken seriously by industries worldwide, but progress remains slower than anticipated. The primary challenge lies in the limited scope for holistic improvements due to economic hurdles and isolated decision-making within companies. Creating synergies between different industrial and municipal entities offers a promising solution. However, this complex process is often managed by municipal authorities without adequate tool support. To address this, we propose a methodological approach that simplifies the collection and processing of relevant data. Beyond gaining corporate buy-in, obtaining and analyzing company-specific data is crucial. A software-based tool has been developed to systematically identify, quantify, and visualize synergy potentials across different stakeholders. These insights enable the development of tailored, location-specific scenarios, providing a concrete roadmap for cross-organizational sustainable transformation. The necessary collaborative decision-making process—in which all relevant stakeholders should be involved—is facilitated through an interactive visualization tool. The additional benefits of the application will be outlined based on two illustrative examples.

Keywords: Sustainable Industrial district · urban production · sustainable value creation networks

1 Introduction

The EU Green Deal and its associated transformation trajectories challenge not only companies [1] but also municipalities, as achieving these goals requires collaboration between industry and municipal administrations [2]. For instance, Baden-Württemberg mandates that municipalities with over 20,000 inhabitants conduct a municipal heating plan [3], assessing current heating consumption, building stock, and energy sources to facilitate sustainable transformation [4].

Despite existing political and regulatory frameworks for sustainable production and value creation, these initiatives often fail to interlock effectively for all local stakeholders.

H. Kohl et al. (Eds.): GCSM 2025, LNME, pp. 3–13, 2026.
https://doi.org/10.1007/978-3-032-21154-5_1

Many municipal administrations struggle to bridge the gap due to a lack of human resources and necessary skills.

This article addresses (i) how to methodically support convergence, (ii) the tools that can serve as empowering instruments, and (iii) lessons learned from existing case studies.

2 Related Work in Urban Sustainability and Glocal Production

2.1 Sustainability, Urban Efficiency, and Glocal Production

The discourse on sustainable transformation is increasingly intertwined with urban efficiency and glocal production. Sustainability is understood not only as an ecological responsibility but also as an economic and social imperative in urban contexts [5]. Cities are recognized as key arenas for transformation, serving as hubs for infrastructure, consumption, and innovation [6].

A central concept is urban efficiency, which refers to the resource-saving, multifunctional, and integrated use of urban spaces. The integrated management of energy, water, and waste flows – termed "urban metabolism" – enables a holistic understanding of urban systems [7]. This approach is particularly effective in densely populated industrial areas, where waste and by-products can be repurposed.

The concept of glocal production complements this by linking local production networks with global supply chains. This fosters resilient systems characterized by digital fabrication methods like 3D printing [8]. According to Ståhlbröst and Holst [9], these decentralized systems reduce carbon dioxide emissions and logistical costs while promoting social inclusion and economic resilience.

2.2 Regulatory Frameworks and Cross-Sectoral Collaboration Approaches

The European Union's Green Deal aims to achieve climate neutrality by 2050, requiring significant shifts in industry, energy systems, and urban development [10]. In Germany, the state of Baden-Württemberg became the first to mandate municipal heat planning in 2020, obligating cities and large towns to assess heat demand, building stock, and energy sources to support a sustainable transition [11].

However, the integration of industrial actors into municipal transformation processes remains insufficient. While municipalities are required to develop heat plans, there are often no binding mechanisms for systematically involving industrial companies. This results in isolated actions lacking coordinated strategies between public and private stakeholders. This challenge is further exacerbated by the varying dynamics and unequal resources and capacities among these actors. Fratini et al. [12] emphasize the need for methodologically supported tools that facilitate mapping and assessing local resource flows and stakeholder participation.

Concepts like Industrial Symbiosis and Urban Industrial Ecology highlight the potential of site-based collaboration to leverage synergies in energy, material, and infrastructure use [13]. In industrial clusters with diverse sectors, such cooperation can provide significant ecological and economic benefits [14]. However, implementation is challenging due to the lack of technical, human, and institutional resources among municipal actors [15]. Additionally, identifying and evaluating synergies is often hindered by limited data and insufficient methodological frameworks. Recent research emphasizes the need for practical tools that can be used by diverse stakeholders to support shared transformation strategies in varying urban-industrial contexts [16].

2.3 Digital Tools and Their Limitations

Digital tools have emerged to support material flow analyses and energy balances, with platforms like S.W.A.N.[1] facilitating the identification of industrial symbioses through waste and resource flow analysis [17]. Recently, digital twins have gained popularity, allowing for modeling urban-industrial interactions and optimizing cross-sectoral synergies [18]. However, these tools often fall short in practice, as many are designed for individual firms and lack adaptability for broader municipal applications. Implementation is also hindered by fragmented data infrastructure, insufficient interoperability, and limited user accessibility, especially for smaller municipalities [16, 18]. To promote inclusive cooperation and effective transformations, digital tools must prioritize usability, transparency, and integration.

3 Improvement of Productive Urban Districts

3.1 Supportive Methodology

To systematically identify transformation needs and derive suitable measures for the development of ultra-efficient industrial districts, a structured step-by-step approach was developed (Tab. 1). The challenge lies in engaging diverse stakeholders along a common schedule while ensuring a shared understanding of local priorities and potentials.

The following table summarizes the six key steps of the applied methodology:

To complement this process, a targeted training concept was developed for company sustainability managers and municipal climate protection officers. The goal is to enable these key actors to further explore relevant aspects of transformation and drive the process forward independently where necessary.

In practice-oriented workshops, municipal actors are introduced to a standardized approach for sustainable site transformation. Real-world challenges in implementing local measures are discussed collaboratively. Practical tools and guidelines for identifying and prioritizing actions are presented. A central focus lies in fostering peer exchange and network formation to promote mutual learning. At the enterprise level, individual potential analyses are conducted to identify actionable ultra-efficiency measures tailored to company-specific needs. These are intended to enhance resilience, compliance, and long-term competitiveness under dynamic conditions.

[1] Developed by EU-Interreg - digital **S**olid **W**aste reuse pl**A**tform for Balka**N** (2017–2019).

Tab. 1. Concrete approach to potential identification in the municipal context

Step	Description	Activity
1	Kick-off event and self-assessment questionnaire	Present the objective and proceedings to the main stakeholders (typically major, largest local company, municipal economic advisor, sustainability managers) Initiate the data and information collection for the self-assessment questionnaire (see Fig. 3, Annex)
2	Initial roundtable with local actors and data collection	Contact all potentially involved parties to inform them about initiative (present objective, proceedings, potential individual benefits) Start data collection using a standardized data collection sheet (input and output flows, descriptive information)
3	First assessment of symbiotic potential and interaction analysis	Evaluate the provided data set Derive an interdependency analysis by comparing similar input and outputs (see Fig. 4, Annex) Identify information gap for deep dive analysis
4	On-site data collection and detailed interviews	Perform a detailed data collection on spot by experts, based on information gaps identified Perform semi-structured interviews to understand limitations and location specifics Summarize all data in a comprehensive way to be extended and approved by the stakeholder
5	Identification of suitable inputs/outputs and ultra-efficiency measures; concept drafting	Collect existing and innovative measures for advancement in different action fields Detailed analysis of the interdependencies (see Fig. 5, Annex) Identify relationships between advancement measures and KPI Prioritize of advancement measures
6	Refinement of measures and development of implementation pathway	Quantify the advancement measures Derive potential benefits (measurable in 1 one year, 5 years and 10 years) Visual presentation of data for discussion Creation of scenarios with different effective dates

3.2 Tool-Based Analysis

The tool must enhance the process by effectively illustrating complex stakeholder interactions and clearly visualizing relevant information. Its goal is to promote a shared understanding, facilitating faster discourse convergence. To achieve this, a quantitative model (i) and appropriate visualization (ii) are necessary.

(i) The calculation module assesses the impact of additional development measures using available stakeholder data. Effects are calculated for a specified set of parameters across five action areas, and the modular design accommodates the evaluation of supplementary measures. The tool chain gathers data input in an Excel sheet for each company (see Fig. 6, Annex), allowing an independent organization to merge and import data, thus ensuring confidentiality. A MATLAB script computes both individual and mutual effects of the development measures. The resulting data is exported to an Excel file, with separate tabs for each company, and subsequently transferred to a visualization interface for interactive analysis.

(ii) Achieving a shared understanding of objectives and values is crucial for fostering compromise among diverse organizations seeking collaboration. Research indicates that graphics and illustrations enhance stakeholder convergence, particularly during the collective analysis of quantitative data [19]. Infographics that establish common reference points and highlight changes are particularly effective [20]. In this application case, the local environment is identified as a foundational aspect through stakeholder discussions, with recognizable infrastructural or ecological elements facilitating this identification. Engaging organizations in the presentation allows for the assessment of various developmental impacts. A calculation tool provides essential benchmarks for each entity, promoting balanced discussions on both positive and negative outcomes and enabling the targeted implementation of relevant measures. To ensure independence from specific operating systems, the visualization employs open-source web technology (Fig. 1), allowing content to be presented locally or online and accommodating unique local conditions.

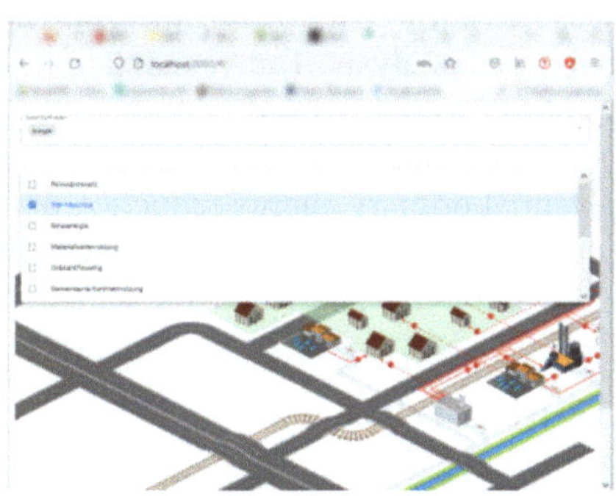

Fig. 1. Web-based visualization of the interdependencies with the possibility to show impacts of measures

4 Case Study

Previous projects have demonstrated a holistic perspective of industrial sites in relation to residential areas, particularly in the context of urban production, where economic interests conflict with the diverse needs of the community, resulting in a complex landscape. The urban production approach aims not only to mitigate negative impacts but also to create positive contributions to the surrounding area, thereby enhancing local tolerance and acceptance while maintaining a focus on economically viable solutions.

In the following sections two exemplary projects (Fig. 2, left Rheinfelden, right Fellbach) are presented to provide an insight into realizable benefits for the local community by empowering cross-organizational collaboration.

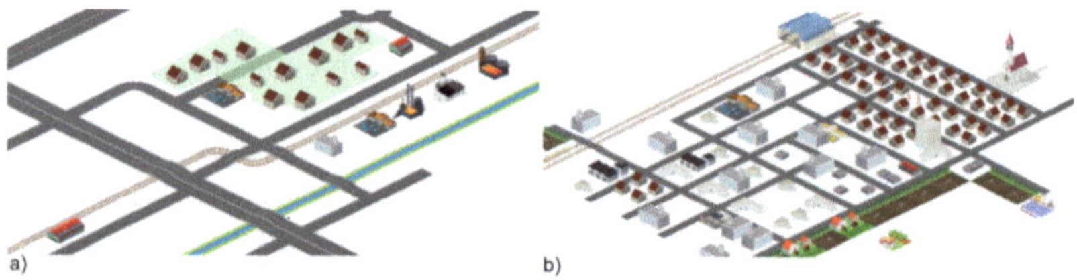

Fig. 2. Visualization of the exemplary project aiming for cross-organizational collaboration

4.1 Industrial District – Rheinfelden

In Rheinfelden (Baden), a town in southern Germany, not only the municipal authorities but also larger companies are driving the transformation. As the location has two industrial areas, that frame the residential areas, the analysis also involved transportation route and reducing fragmentation of functions [21].

Following round tables, data collection and interviews on spot were performed. In total 30 advancement measures across the different action fields have been derived. The detailed analysis of the measures allowed a prioritization on 12 measures that have been quantified by using the tool-based analysis. The visualization was used for the decision workshop where all local entities were involved.

The top measure to be implemented was developing an industrial heat supply network. The reason was a potential primary energy reduction of about 26%. Based on the quantification of all aspects, the projected construction costs will be recovered within six years, by establishing the jointly developed business model.

4.2 Urban Productive District - Fellbach

Fellbach, the second project site, is characterized by small-scale industries and agricultural businesses adjacent to residential and recreational areas. Given the complex stakeholder dynamics and conflicting interests, municipal representatives organized workshops with citizens in addition to corporate discussions. The initiative was presented to all stakeholders, with company data collected directly and household data sourced from the city and consultancy agencies.

Close collaboration with municipal energy suppliers allowed for the alignment of further development measures with the roadmap, focusing on supplementary initiatives. In total, 22 measures were identified.

Based on stakeholder prioritization, two measures were quantified: (i) renewable energy potential and (ii) water reserves from local precipitation.

(i) Solar generation potential from photovoltaics on the 21 largest roofs can cover about one-third of the area's energy demand (6.2 GWh), and potentially up to half with full coverage, excluding specific energy efficiency measures.
(ii) The cost-effectiveness of rainwater collection is debated due to required infrastructure. Farmers use freshwater for irrigation, while companies pay for water charges on sealed areas, allowing potential savings to fund construction. Analysis indicates annual savings of approximately € 1,100 for industries and € 33,000 for farmers. The city proposed developing a lake for water storage and recreation or smaller reservoirs at industrial sites.

5 Conclusion and Future Work

This article presented a method for providing structured support to municipalities undergoing transformation through a holistic assessment. The approach facilitates the collection, evaluation, and use of targeted information, supported by tools. Interaction between representatives of independent organizations aims to create a common understanding, simplifying implementation.

Two case studies demonstrated the method's potential for transformation initiatives, allowing for the identification of added value focused on implementation.

In future work, training formats will be implemented, enabling municipal climate managers to drive transformation more cost-effectively using the approach and IT-based model. Strengthening transformation within companies will also be addressed in training courses for sustainability officers. The training concept and its application results will be prepared for subsequent publication.

Annex

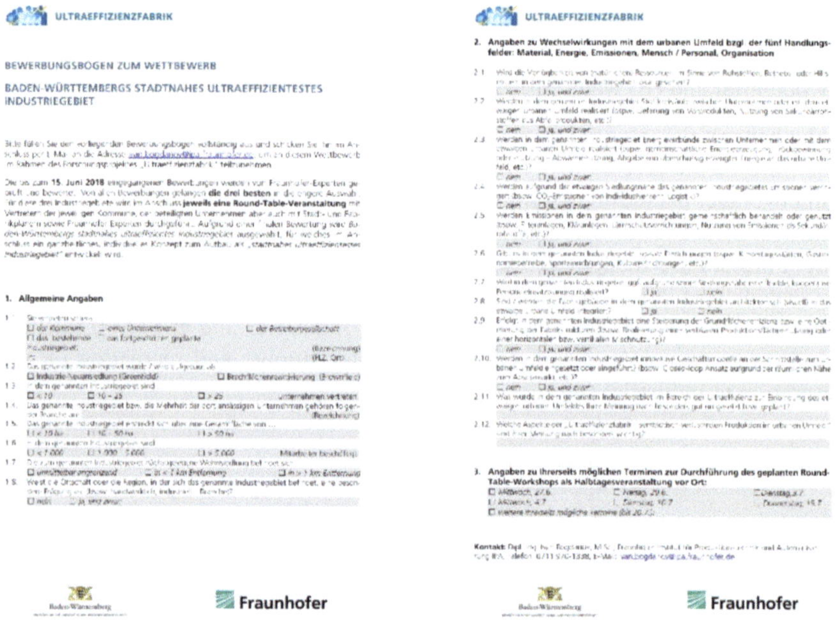

ULTRAEFFIZIENZFABRIK

BEWERBUNGSBOGEN ZUM WETTBEWERB

BADEN-WÜRTTEMBERGS STADTNAHES ULTRAEFFIZIENTESTES INDUSTRIEGEBIET

... **15. Juni 2018** ... **die drei besten** ... **jeweils eine Round-Table-Veranstaltung** ...

1. Allgemeine Angaben

ULTRAEFFIZIENZFABRIK

2. Angaben zu Wechselwirkungen mit dem urbanen Umfeld bzgl. der fünf Handlungsfelder: Material, Energie, Emissionen, Mensch / Personal, Organisation

3. Angaben zu Ihrerseits möglichen Terminen zur Durchführung des geplanten Round-Table-Workshops als Halbtagesveranstaltung vor Ort:

Kontakt: ...

Fraunhofer

Fraunhofer

Fig. 3. Self-assessment questionnaire (in German language)

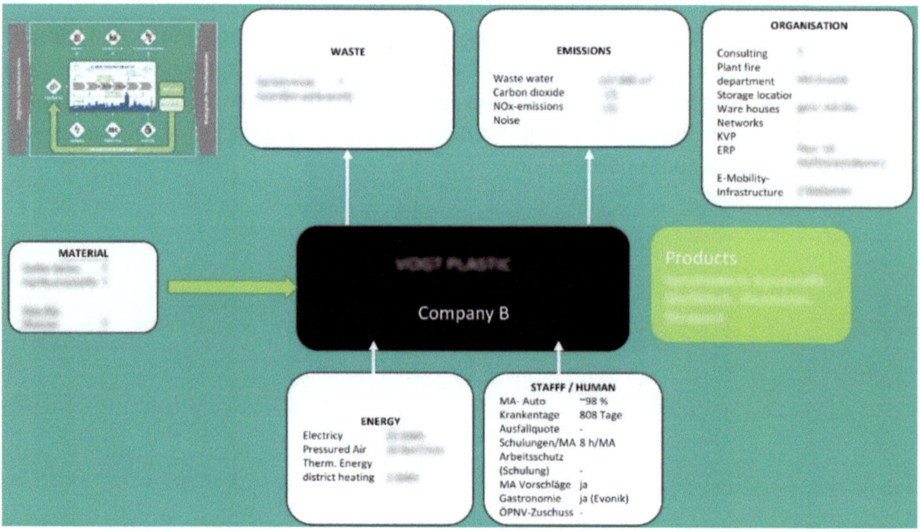

Fig. 4. Data visualization for each stakeholder to identify interdependencies

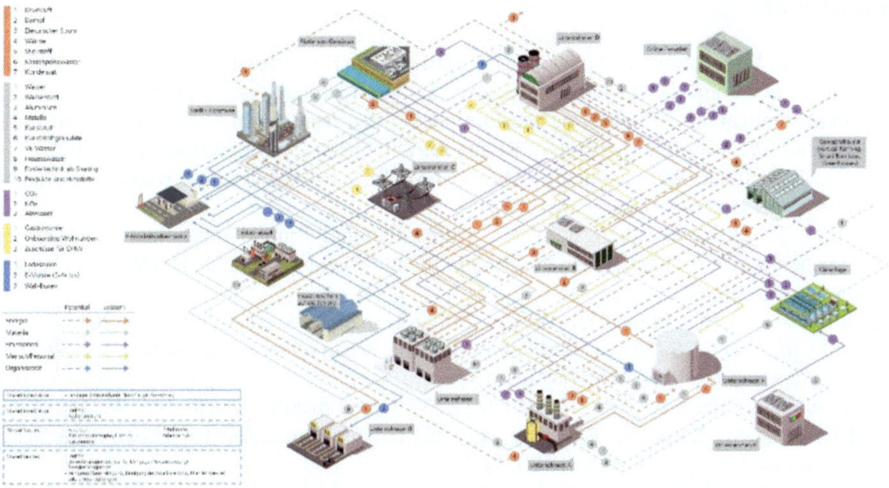

Fig. 5. Interdependency analysis visualized with all potential and existing interrelations

ULTRAEFFIZIENZ FABRIK

Ultraeffizientes Industrie-/Gewerbegebiet - Datenerhebung für die Konzeptentwicklung
Unternehmen MUSTERFIRMA GmbH

Mitarbeiterzahl	500	Branche	Metallverarbeitung	Angaben für das Jahr
Umsatz / Jahr (T€)	3,5 T€	Ansprechpartner	Max Mustermann	2017

	Type	Amount	Source	Sink	Additional information
Material	Steel	70 tons	Supplier	Company	
	Wood				
	Hydrochloric acid	750.000 liters	Company	network partner	34% of the partner company's hydrochloric acid requirements are covered
	Hydrogen	1.870.000 liters	network partner	Company	43% of the company's hydrogen needs are covered
	Packaging	8.500 tons	surrounding area, roughly 2. Mio people	Company	
	Steel chips	15 tons	Company	residual recyclers	
	Steel waste	3 tons	Company	residual recyclers	90 % residual recycler, 10 % workshop for training
Energy	Electricity	1,2 MWh	Electricity provider	Company	
	Thermal energy	70 MWh	District heating supplier	Company	
	Thermal energy	-	Company	Umwelt	Waste heat at 35 °C is not reused
	Thermal energy	20 MWh	Company	City	80% of waste heat at 75 °C is delivered to the city
	Compressed Air	70.000 Nm³	own compressors	Company	
	Compressed Air	10.000 Nm³	Compressors, network partner	Company	12.5% of the company's compressed air requirements are covered
Emission	CO_2	-	Company	Environment	Curing fumes through chimney, filter in use
	NOx-Emission	10 t	Company	Environment	Exhaust gases through chimney, filter technology in use
	Waste Water	350.000 m³	Company	Waste water treatment plant of the city	
	Process water	150.000 m³	Company	Chemical waste water treatment plant	
	Noise	70 dB	Company	Environment	Sound insulation of machines - 25% noise reduction
Human / Staff	employee - car	120 von 500	Home	Company	
	employee - public transport	350 von 500	City	Company	bus connection to city center
	employee - bicycle use	30 von 500			5 of it by e-bike
	illness days	12			2% work accident, 12% commuting accident, 80% illness (flu etc.), 6% physical complaints
	Training and qualification of employees	50 h	Lean Experts / IT Experts / etc.	Employee	
	Gastronomie am Standort	100.000 visitors			
Organisation	Consultancy service	5 Days	Consultancy office	Company	
	external logistic servicee	250.000 €	Logistic provider	Company	
	Anzahl Feuerwehreinsätze	1	Fire brigade city		
	Network / esxchange	4 events	-	-	2 events at the site, 2 events at the city's business development department
	online material supervision	15-Min.-intervalle			
	e-mobility infrastructure	10 cars, 5 charging solutions			Extension by 50% within next 2 Years

Fig. 6. Example for the data collection sheet for quantification of advancement measures

References

1. Institut für Wirtschaftsforschung (ifo): European Green Deal. Bottlenecks bis 2030(73) (2020)
2. Grabbe, H., Potočnik, J., Dixson-Declève, S.: International system change compass. The global implications of achieving the European green deal. Open Society European Policy Institute; SYSTEMIQ; The Club of Rome, Zürich. opensocietyfoundations.org (2022)
3. § 27 Kommunale Wärmeplanung. KlimaG BW. In: Klimaschutz- und Klimawandelanpassungsgesetz Baden-Württemberg (2023)
4. Ortner, S., et al.: Handlungsleitfaden Wärmeplanung. Empfehlungen zur methodischen Vorgehensweise für Kommunen und andere Planungsverantwortliche, Heidelberg, Freiburg, Stuttgart, Berlin (2024)
5. Elmqvist, T.: Urban Planet, 1st edn. Cambridge University Press, Cambridge (2018)
6. Acuto, M., Parnell, S., Seto, K.C.: Building a global urban science. Nat. Sustain. **1**, 2–4 (2018). https://doi.org/10.1038/s41893-017-0013-9
7. Derrible, S., Cheah, L., Arora, M., Yeow, L.W.: Urban metabolism. In: Shi, W., Goodchild, M.F., Batty, M., Kwan, M.-P., Zhang, A. (eds.) Urban informatics. The Urban Book Series, pp. 85–114. Springer, Singapore (2021)
8. Srai, J.S., et al.: Distributed manufacturing: scope, challenges and opportunities. Int. J. Prod. Res. **54**, 6917–6935 (2016). https://doi.org/10.1080/00207543.2016.1192302
9. Ståhlbröst, A., Holst, M.: Reflecting on actions in living lab research. Technol. Innov. Manag. Rev. (2017). https://doi.org/10.22215/timreview/1055
10. Europäische Kommission: Der europäische Grüne Deal (2025). https://commission.europa.eu/strategy-and-policy/priorities-2019-2024/european-green-deal_de. Accessed 25 Apr 2025
11. Ministerium für Umwelt, Klima und Energiewirtschaft Baden-Württemberg: Kommunale Wärmeplanung (2024). https://um.baden-wuerttemberg.de/de/klima-energie/energieeffizienz/kommunen/kommunale-waermeplanung. Accessed 25 Apr 2025
12. Fratini, C.F., Georg, S., Jørgensen, M.S.: Exploring circular economy imaginaries in European cities: a research agenda for the governance of urban sustainability transitions. J. Clean. Prod. **228**, 974–989 (2019). https://doi.org/10.1016/j.jclepro.2019.04.193
13. Chertow, M.R.: Industrial symbiosis: literature and taxonomy. Ann. Rev. Energy Environ. **25**(1), 313–337 (2000). https://doi.org/10.1146/annurev.energy.25.1.313
14. Geng, Y., Sarkis, J., Ulgiati, S., Zhang, P.: Measuring China's circular economy. Science (2013). https://doi.org/10.2307/41942068
15. Fraccascia, L., Yazan, D.M.: The role of online information-sharing platforms on the performance of industrial symbiosis networks. Res. Conservat. Recycl. (2018). https://doi.org/10.1016/j.resconrec.2018.03.009
16. Lombardi, D.R., Laybourn, P.: Redefining industrial symbiosis: crossing academic–practitioner boundaries. J. Ind. Ecol. **16**(1), 28–37 (2012). https://doi.org/10.1111/j.1530-9290.2011.00444.x
17. Papageorgiou, A., Björklund, A., Sinha, R.: Applying material and energy flow analysis to assess urban metabolism in the context of the circular economy. J. Ind. Ecol. **28**(4), 885–900 (2024). https://doi.org/10.1111/jiec.13504
18. de Benedictis, R., et al.: Digital twins for intelligent cities: the case study of Matera. J. Reliable Intell. Environ. **11**, 6 (2025). https://doi.org/10.1007/s40860-025-00245-3
19. Katz, J.: Designing information. In: Human Factors and Common Sense in Information Design. Wiley, Hoboken (2012)
20. Smiciklas, M.: The power of infographics. In: Using Pictures to Communicate and Connect with Your Audience. Que Pub, Indianapolis Ind. (2012)
21. Hertwig, M., Bogdanov, I., Beckett, M., Waltersmann, L., Lentes, J.: Symbiotic loss-free industrial production in ultra-efficient urban industrial parks. Procedia CIRP **98**, 637–642 (2021). https://doi.org/10.1016/j.procir.2021.01.167

Technical and Organisational Readiness Framework for Facilitating Development of Sustainable Manufacturing Processes by Urban and Creative Factories

Fedoua Kasmi[1](✉), Michel Langhammer[2], Anaëlle Hily[1], and Robert Mies[3]

[1] Université de Lorraine, ERPI, 54000 Nancy, France
{fedoua.kasmi,anaelle.hily}@univ-lorraine.fr
[2] Helmut-Schmidt-Universität, Holstenhofweg 85, 22043 Hamburg, Germany
michel.langhammer@hsu-hh.de
[3] Technische Universität Berlin, Pascalstraße 8-9, 10587 Berlin, Germany
robert.mies@tu-berlin.de

Abstract. Urban factories, such as makerspaces, fab labs, and co-working manufacturing spaces, integrate manufacturing into cities by supporting circular, local supply chains, creativity and collaborative production. However, despite their potential, they lack standardised methods to analyse their readiness as manufacturing hubs. Existing models focus on large-scale industry, overlooking the constraints of these spaces within urban settings. This paper presents a technical and organisational readiness framework tailored to collaborative urban factories. The framework draws on literature combining sustainable urban manufacturing, Industry 5.0, and New European Bauhaus principles, as well as insights from workshops and surveys of eight urban factories, three of which are briefly presented as case studies. It identifies key physical, digital, and organisational features essential for enabling effective small-batch production capabilities.

Keywords: Sustainable Urban Manufacturing · Infrastructure · Readiness

1 Introduction and Theoretical Background

Globalisation has driven technological progress but also shifted manual labour, technical knowledge, and innovation away from cities, increasing the environmental impact of mass production. In response, sustainable urban manufacturing reimagines factories to meet economic, environmental, and social goals within compact, community-integrated urban settings. Advances in cleaner, quieter, and smaller-scale technologies now make such integration feasible. This paper introduces a framework to analyse the physical, digital, and organisational capabilities of urban and creative factories, highlighting their role in facilitating sustainable, small-scale manufacturing by small and medium-sized manufacturers and creatives.

H. Kohl et al. (Eds.): GCSM 2025, LNME, pp. 14–21, 2026.
https://doi.org/10.1007/978-3-032-21154-5_2

Within the wider urban manufacturing literature, urban factories or micro-factories have emerged as manifestations of distributed and small-scale manufacturing. The precise definition of an "urban factory" remains rather vague, reflecting the nascent stage of this development and the diversity of its manifestations [1]. Three main characteristics emerge from the literature:

- **Integration into urban areas:** Urban factories are flexible facilities embedded in densely populated zones, contrasting with the traditional separation of manufacturing from residential zones [2]. This integration offers advantages such as proximity to customers, suppliers, and skilled labour, access to infrastructure, creative industries and the capacity for localised product-service systems and on-demand production [3, 4].
- **Digitalisation and smart technologies:** The relatively recent concept of urban micro-factories suggests increasingly adopting Industry 4.0 technologies within urban factories, forming the basis of "urban smart factories" [5, 6]. IoT enables real-time equipment and energy monitoring; reconfigurable setups and cobots support custom, small-batch production [7]. Cloud-based platforms and computer-aided tools (CAx) enable global design with local fabrication. Extended reality (XR) technologies can augment design for manufacturing with connected optics-enabled visualisations of computed 3D spaces and virtual metaphors. Interoperability challenges are eventually addressed through (open) standards, semantic models, and ontologies to reflect seamless data integration and collaboration. Digital product passports (DPP) can enhance traceability and circularity compliance in future [8].
- **Localised circular urban production:** Sustainable urban manufacturing integrates circular economy principles, where urban waste is reused in local production cycles [4]. Urban factories exploit the city's "urban metabolism", using materials from localised sources, urban mines (e.g., end of life, material, e-waste), and industrial by-products [3]. This fosters urban scale synergies for closely involving local stakeholders across the value chain, while waste becomes a resource, encompassing open co-creation, production, sourcing, distribution, and waste management [9].

In sum, urban factories support sustainable manufacturing through localised, circular, and digital processes. They also enable openness, collaboration, and access to shared tools and knowledge serving as platforms for the Do-It-Together (DIT) approach [10, 11]. This model enables communities to co-create customised, sustainable products, combining professional expertise for small-batch manufacturing for commercial activities, creative exploration and citizen participation.

Expanding beyond, Industry 5.0 and the New European Bauhaus (NEB) vision are shaping the future of urban manufacturing in Europe. Industry 5.0 already emphasises human-centricity, sustainability, and resilience, promoting collaborative environments where humans co-create with the support of machines and technological mediation, whilst circularity is embedded in design[1]. The NEB adds an important cultural and aesthetic dimension, advocating the creation of functionally designed spaces that are at their core "beautiful, sustainable, together." Applied to manufacturing, it reimagines factories

[1] See European Commission report on 'Industry 5.0' under URL: https://research-and-innovation.ec.europa.eu/research-area/industrial-research-and-innovation/industry-50_en.

as inclusive, functional, and artistic hubs for promoting creativity, civic engagement, as well as (open) urban innovation. In combination, these frameworks reinforce a directional push towards human-centric and creative urban factories actively imagining and shaping a better future for society.

Collaborative spaces such as makerspaces and fab labs can support this vision as creative, distributed, and market-oriented production sites. These open-access workshops, equipped with digital fabrication tools, enable individuals to prototype and produce locally [4], fostering grassroots innovation through peer production and knowledge sharing [12, 13]. In Europe, they may increasingly reflect NEB values, acting as inclusive places for design, technology, and creativity. Aligned with Industry 5.0, they may help democratise small-scale manufacturing within local ecosystems. However, their production capacity remains underdeveloped in the literature. Hennelly et al. [14] note that most makerspaces focus on education or design rather than full-scale production. Whilst this is insinuated in the promise of fab labs as places to "make almost anything", it can be particularly attributed to limited infrastructure capabilities. These spaces often lack the specialisation, precision, reliability, and scalability required for commercial and artistic manufacturing. Moreover, reliance on grants and memberships undermines financial sustainability [ibid.]. To scale their impact, dedicated infrastructure is essential. Meyer et al. [3] highlight the need for access to reliable energy, water, materials, and functional space. Additionally, Hausleitner et al. [15] stress the importance of effective logistics, supply chain coordination, and waste management for operational sustainability.

Building on the gaps identified in previous literature (particularly the lack of focus on production capabilities within makerspaces and fab labs), this study develops a readiness framework to analyse the capacity of these collaborative spaces for sustainable urban manufacturing. The characteristics highlighted in the literature underscore infrastructure needs: adaptable, modular spaces and multifunctional equipment suited to constrained urban environments (**physical**); robust digital capabilities for managing data, connectivity, and automation in small-scale manufacturing (**digital**); and organisational capacities necessary for effectively governing circular supply chains and facilitating distributed collaboration (**organisational**). These insights form the foundation of the framework's structure, which underpins the readiness analysis in the following sections.

2 Method and Results

This study is part of the LAUDS Factories project (Local Accessible Urban Digital Sustainable Factories), a Horizon Europe initiative aligned with Industry 5.0 and NEB principles. Our research analyses eight LAUDS factories from various European countries (Spain, Germany, France, Denmark, Belgium, Ukraine), encompassing diverse collaborative settings such as makerspaces, fab labs, and third places. The criteria of the infrastructure components for the readiness framework were initially identified through a structured review of relevant literature and refined through a workshop with consortium participants (researchers, managers, and makers). Data collection (Jan-Mar 2025), using 87 questions, combined electronic questionnaires and semi-structured interviews (2–3 h each) offering deeper insight into operational challenges and strategic priorities. Given the exploratory nature of the readiness concept in urban factories contexts and their inherent heterogeneity, qualitative methods were relevant. This allowed context-sensitive refinement, adjustments to the criteria, and enabled detailed examination of infrastructure capacities and capabilities. The data was coded and thematically analysed by three researchers using QualCoder. Coding involves labelling segments of data to identify patterns and themes, allowing structured interpretation of complex qualitative material. Figure 1 visualises the resulting framework as well as the distribution and thematic weight of coded responses mapped to the framework's three infrastructure dimensions, illustrating how qualitative insights align with physical, digital, and organisational readiness criteria. This figure offers an empirical grounding of the framework, highlighting how the themes identified in interviews and open responses correspond with the structure derived from literature. The relative size of each segment indicates the number of coded references. Larger segments reflect greater emphasis in the qualitative responses, highlighting which aspects of infrastructure readiness were most frequently discussed across cases.

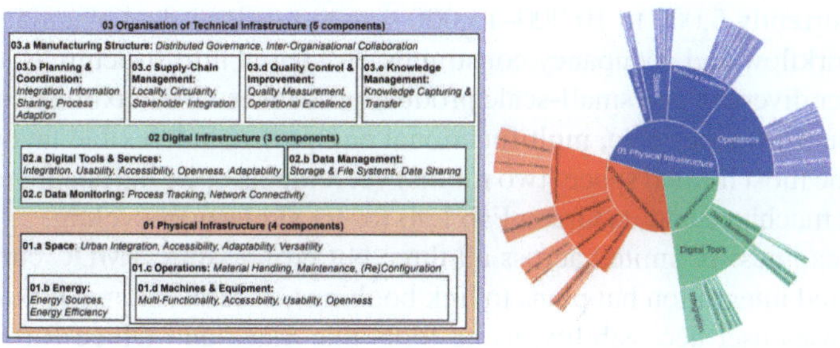

Fig. 1. Framework structure for assessing the readiness of infrastructure capabilities (left) and distribution of coded qualitative data aligned with readiness framework dimensions (right)

The physical infrastructure layer (purple/01) focuses on space use (01.a), energy (01.b), operational flexibility in urban settings (01.c) as well as machines and equipment (01.c). It captures how factories accommodate diverse activities through modular layouts, intuitive machines, open-source hardware, and reconfigurable workstations. Criteria such as accessibility, energy efficiency, and in-house maintenance reflect the unique constraints of city-based production.

The digital infrastructure layer (green/02) is the core of Industry 4.0/5.0, enabling data-driven decision-making and seamless communication. The readiness evaluates the integration and capabilities of digital tools and services (02.a), data management with use of shared repositories and standard file formats for production (02.b), as well as the monitoring of relevant production data to evaluate actions and generate needed datasets for DPP services with considering the network connectivity and interoperability of used hard- and software components (02.c).

The organisational layer (red/03) describes the organisation around the physical and digital infrastructure. The manufacturing structure assesses distributed project management and inter-organisational cooperation (03.a). The integration of planning and coordination of projects with information sharing should enable structuring and flexibility in project management (03.b). Supply chain management with local supply chains and stakeholders' integration focuses on circularity, implementing processes for material sorting, recycling, reuse and waste minimisation (03.c). Quality control for process improvement (03.d) and knowledge management (03.e) are further integral components.

As full analysis is ongoing, this paper presents findings from three case studies reflecting diverse geographic and operational contexts. They are analysed below with a summary table for applying the framework for an initial qualitative analysis of strengths and limitations extracted from coded transcripts of interviews and aggregated for the physical, digital and organisational layers (see Table 1):

The three LAUDS factory case studies reflect distinct models of urban creative factories shaped by their local contexts, stakeholder ecosystems, and production missions. All prioritise small-scale production, creative prototyping, and collaborative innovation, though their infrastructure readiness varies based on specific strategies, constraints, and objectives.

Each space demonstrates the ability to create modular, multifunctional environments suited to diverse user needs. TMDC stands out for its industrial ambition, planning to scale from currently 5,000 to 10,000–15,000 sqm to resolve urban planning constraints as well as workflow and occupancy constraints. Its shared infrastructure is already optimised for user diversity and small-scale production. Maker V10 maximises a more compact space through accessible, multifunctional equipment and flexible layouts. LSCLL operates in the most limited space (two rooms) yet remains open and adaptive, leveraging open-source machines and its Green Fab Lab for local plastic recycling.

Digital readiness is limited across all three but progressing. TMDC currently lacks automation and integration but plans to link bookings, access, and machine data. Maker V10 emphasises user accessibility, using tools like Slack and Office RnD, with early trials in AI-based knowledge capture, though manufacturing data tracking remains limited. LSCLL uses open-source tools (e.g., Fabmanager) for collaboration and machine bookings but lacks a fully integrated digital workflow or energy/data monitoring, as they would like to establish for future DPP compliance.

From an organisational point of view, the three spaces encourage peer production, but with different structures. LSCLL demonstrates strong stakeholder engagement through its living lab model, but is dependent on a central technical manager for production. Maker V10 reflects a decentralised, collaborative approach with informal management. TMDC combines peer dynamics with structured coordination, with the aim of bridging the gap between creative production and commercial scalability.

3 Conclusion

This paper assessed the infrastructural readiness of three urban creative factories TMDC (Barcelona), Maker V10 (Copenhagen), and LSCLL (Nancy) using a structured framework across physical, digital, and organisational infrastructure capabilities. Despite differing contexts, shared challenges include limited workflow space, fragmented digital infrastructure, low automation, and informal planning. Circular practices like material reuse and recycling are present, and shared spaces enhance resource efficiency. However, the absence of real-time monitoring hinders systematic assessment of environmental impact. These findings support prior research suggesting that most of these spaces

Table 1. Overview of strengths and limitations of technical infrastructures for the three cases

Layers	Strengths	Limitations/gaps
Case 1 – MAKER (Denmark): *is a hybrid workshop integrating digital tools (CNC, laser cutters, 3D printers) with traditional craftsmanship. Designed for accessibility, they support small-batch production, material innovation, and custom work, aiming to foster opportunities for entrepreneurship*		
Physical	- Flexible, multi-use space - High equipment reliability - Modular, adaptive layout	- Space limitations hinder scalability - Manual material handling - Undue external dependencies for maintenance of machines
Digital	- Digital tools for core functions (user, access, project management) - QR-linked machine documentation	- Lack of machine usage tracking - Poor integration between digital tools / user systems
Organisa-tional	- Circular practices in material use - Peer-to-peer knowledge sharing	- Informal, ad-hoc coordination - Challenges in capturing tacit knowledge
Case 2 – TMDC (Spain): *is a co-working fabrication space that brings together designers, engineers and manufacturers. It operates a large workshop offering machinery and space with a focus on wood fabrication. They facilitate diverse projects from individual custom work to coordinated professional projects*		
Physical	- Urban location with large production spaces - Modular spatial design - Large-format capability - Machine multifunctionality	- Urban planning constraints - Manual material flow - Machines availability - No energy metrics - Inefficient multitool machines
Digital	- Digital tools for core functions - Newer machines with connectivity - Wireless connectivity is available	- Weak system integration - Users prefer analogue informal direct communication for scheduling - CAD/CAM compatibility issues - Missing operational usage data
Organisa-tional	- Circular practices in material use - Informal knowledge exchange for process improvement - Introductions and basic guidelines for machine usage - Distributed, peer-driven coordination - Cross-organisational collaboration	- Lacks formal project structures - Informal project management - Planning depends on individuals - Tacit knowledge not captured - Uncontrolled material logistics
Case 3 – LSCLL (France): *is an action research platform of Université de Lorraine (Nancy, France), supports early-stage development of systemic innovations aligned with Industry 5.0. Anchored in a territorial network of urban and artistic spaces like L'Octroi Nancy, LSCLL offers a relevant case for exploring urban and creative production dynamics*		
Physical	- Accessible central location - Open, repairable tools - Flexible spatial setup	- Tight limited space - Limited machine versatility - Ad-hoc material flow - No energy monitoring system
Digital	- Open-source and cloud-based software tools (e.g., Fabmanager, Onshape, and GitHub) - Support for distributed design and open data	- Low automation across digital workflows and operations - Challenges in data management, - Nascent stage in real-time data collection
Organisa-tional	- Stakeholder integration with community-driven projects - Process agility, daily scheduling - Circular practices through the Green Fab Lab for eco-design experimentation	- Centralised production around one technical manager - Informal production planning - Lack of formal production quality KPI - Dependency with university regulations (e.g., new equipment…)

remain oriented towards education/learning and design with relatively few evolving into full production capabilities. However, our hypothesis suggests that not all spaces need to become production-intensive. Their value lies in leveraging specific strengths to empower local actors (e.g., artists, creatives, entrepreneurs) who often lack production infrastructure. Notably, the three cases illustrate distinct strategic orientations shaping their readiness: TMDC is evolving into a scalable, semi-industrial hub with ambitions to expand and increase its level of automation; Maker V10 maintains a craft- and arts-oriented, human-centric model; and LSCLL focuses on research and demonstration, embedding itself in local cultural ecosystems with mixed partnerships (public, private, citizen). These diverse approaches highlight the need for development strategies that go beyond infrastructure metrics and consider the intended function and trajectory of each type of space. Future research should examine how such differentiated models contribute to sustainable urban manufacturing by supporting various stages of production and broadening access to local capabilities. Finally, this research constitutes the first step of our methodology for the development of a readiness framework tailored to collaborative urban factories. Due to the diversity of these spaces and the lack of standardised assessment tools, the qualitative method was relevant for exploring their infrastructure. It allowed us to capture context-specific challenges and stakeholder perspectives. However, the method is resource-intensive and reliant on researcher involvement, limiting practical usability. Thus, the next phase will adapt the framework into a self-assessment tool based on maturity models theory. This will support more systematic, autonomous application by factory managers and operators. Future work includes defining of adapted maturity levels, refining assessment scales, and re-evaluating cases to develop recommendations tailored to their strategic paths.

Acknowledgement. This research was created in the frame of the Horizon Europe project LAUDS Factories, funded by the European Union under Grant Agreement No. 101135986.

References

1. Ijassi, W., Evrard, D., Zwolinski, P.: Characterizing urban factories by their value chain: a first step towards more sustainability in production. Procedia CIRP **105**, 290–295 (2022)
2. Herrmann, C., Juraschek, M., Burggräf, P., Kara, S.: Urban production: state of the art and future trends for urban factories. CIRP Ann. **69**(2), 764–787 (2020)
3. Meyer, K., Görgens, S., Juraschek, M., Herrmann, C.: Increasing resilience of material supply by decentral urban factories and secondary raw materials. Front. Manuf. Technol. **3**, 1106965 (2023)
4. Tsui, T., Peck, D., Geldermans, B., van Timmeren, A.: The role of urban manufacturing for a circular economy in cities. Sustainability **13**(1), 1 (2021)
5. Carvalho, N., Chaim, O., Cazarini, E., Gerolamo, M.: Manufacturing in the fourth industrial revolution: a positive prospect in Sustainable Manufacturing. Procedia Manuf. **21**, 671–678 (2018)
6. Sajadieh, S.M.M., Son, Y.H., Noh, S.D.: A conceptual definition and future directions of urban smart factory for sustainable manufacturing. Sustainability **14**(3), 3 (2022)
7. Amjad, M.S., Diaz-Elsayed, N.: Smart and sustainable urban manufacturing for a circular economy. Environ. Dev. Sustain. **26**(12), 31789–31815 (2024)
8. Lopes, C., Barata, J.: Digital product passport: a review and research agenda. Procedia Comput. Sci. **246**, 981–990 (2024)

9. Ijassi, W., Evrard, D., Zwolinski, P.: Proposal of a circularity design approach for urban factories based on local stakeholders' engagement. Procedia CIRP **116**, 13–18 (2023)
10. Dupont, L., Kasmi, F., Pearce, J.M., Ortt, R.J.: Do-it-together and innovation: transforming European industry. J. Innov. Econ. Manag. **40**(1), 1–11 (2023)
11. Redlich, T., Moritz, M., Buxbaum-Conradi, S., Krenz, P., Heubischl, S., Basmer-Birkenfeld, S.V.: OpenLabs - collaborative industrialization with distributed and open source microfactories. Appl. Mech. Mater. **794**, 470–477 (2015)
12. Morel, L., Dupont, L., Boudarel, M.-R.: Innovation spaces: new places for collective intelligence?. ISTE Ltd. (2018)
13. García-Ruiz, M.-E., Lena-Acebo, F.-J.: FabLabs: the road to distributed and sustainable technological training through digital manufacturing. Sustainability **14**(7), 7 (2022)
14. Hennelly, P.A., Srai, J.S., Graham, G., Meriton, R., Kumar, M.: Do makerspaces represent scalable production models of community-based redistributed manufacturing? Prod. Plan. Control **30**(7), 540–554 (2019)
15. Hausleitner, B., Hill, A., Domenech, T., Muñoz Sanz, V.: Urban manufacturing for circularity: three pathways to move from linear to circular cities. In: Amenta, L., Russo, M., van Timmeren, A. (eds.) Regenerative Territories: Dimensions of Circularity for Healthy Metabolisms, pp. 89–103. Springer, Cham (2022)

A KPI-Based Approach to Evaluating Sustainable Land Use and Adaptable Infrastructure in Industrial Production Sites

Oskay Ozen(✉), Robin Zink, Michael Frank, and Matthias Weigold

Institute for Production Management, Technology and Machine Tools (PTW), Technical University of Darmstadt, Otto-Berndt-Street 2, 64287 Darmstadt, Germany
o.ozen@ptw.tu-darmstadt.de

Abstract. To prepare for future regulations, technologies, and sustainability goals, German companies are shifting their site planning approaches towards long-term flexibility. This is achieved through means like adaptable factories, multi-use facilities, and a future-proof energy infrastructure. As part of the DELTA Energy Laboratory, transformation strategies for producing companies are being analyzed and implemented. To this end, key performance indicators (KPIs) were developed around a metalworking facility to evaluate land use and densification of localized production sites, based primarily on high-level information. The four KPIs are: site area utilization, proportional and absolute floor space utilization, and proportional base area utilization. Considered building types include factories, energy utilities, office spaces, storage facilities, data centers, and roads, supplemented by additional site area used for green spaces, bodies of water, and social infrastructure. These KPIs were applied to production sites in the metalworking and chemical industries. The results for the latter showed that, based on specified company goals, future transformations should prioritize expanding production facilities, while slightly reducing ecological spaces without affecting social spaces.

Keywords: sustainability · land use · site planning

1 Introduction

As part of the European Green Deal, the European Union (EU) has set itself the goal of lowering greenhouse gas emissions to net zero by 2050 [1]. These goals reflect a commitment to absolute sustainability [2]. A survey conducted by the Credit Institute for Reconstruction shows that sustainability is anchored in German corporate culture: over half the responses indicate that sustainability has high priority for companies, with a tendency for large enterprises to rate its importance higher than small and medium enterprises [3]. However, industrial site planning encompasses aspects of sustainability that are not easily measured

H. Kohl et al. (Eds.): GCSM 2025, LNME, pp. 22–30, 2026.
https://doi.org/10.1007/978-3-032-21154-5_3

in absolute terms, such as social aspects. This is reflected in EU-wide corporate sustainability reporting, wherein companies report on their environmental, social, and governance (ESG) impacts [4]. To this end, sustainability in this work refers to all three of its pillars: environmental, economic, and social.

Adaptability to future regulations and new technologies is also gaining importance in factory and site planning [5]. Nevertheless, adaptability in literature typically refers to factory operations, i.e., production processes, such as per Eymüller et al. [6] and Schulz et al. [7]. Adaptability in terms of physical infrastructure is less common, but can be found in works such as those by Kaucher et al. [8]. Land-based key performance indicators (KPIs) addressing adaptability and sustainability are limited to building regulations and certifications schemes. Motivated by this gap, this work introduces new land-based sustainability KPIs.

2 Research Framework

As part of the living laboratory DELTA, a city-wide research project conducted in Darmstadt, Germany, transformation strategies of companies are being investigated with the goal of enabling successful and sustainable transformations [9]. The core results of this work were developed in a sub-project of DELTA, where a team of two energy consultants, three sustainability managers, two facility managers, and three public researchers have investigated the level of information necessary for facilitating site transformations towards sustainable site usage. The information was gathered through expert interviews among the team, supplemented by an ongoing qualitative study involving 14 enterprises of varying sizes and industrial branches. From this, a set of suitable indicators was developed to assess sustainability at a given site, with the added requirement for them to be accessible to parties with limited on-site knowledge. Since these indicators are may reflect a company's corporate vision for sustainability, they were conceptualized as KPIs that allow for continuous monitoring of the production site with regards to corporate sustainability goals.

The ETA Factory, a DELTA partner, is a research facility with a metalworking production line, a highly efficient energy supply system, and integrated offices, was used for the initial KPI development [10]. The ETA Factory is located on the Lichtwiese Campus of the Technical University of Darmstadt, allowing for both individual, factory-based KPIs as well as site-wide KPIs to be considered.

A chemical industrial production site was used to evaluate the transfer of KPIs to real production sites and their applicability to other industrial sectors.

2.1 Legal Framework and Pre-existing Indicators

At the forefront of industrial site planning lies an inherent competition for land use. With profit as the overall goal of production sites (leading to economic sustainability), a maximization of production facilities, as well as supporting structures (e.g., energy infrastructure, warehouses), may seem beneficial. However, frameworks such as ESG reporting and company certifications (e.g., ISO

14001 for environmental management systems) give incentives for companies to allocate productive area for other uses, i.e., non-productive areas or areas that only indirectly affect productivity, like green or social spaces. Each site transformation, including expansions to production or increased sustainability measures, changes the land allocation at a given site. In terms of land use, indicators used to assess such changes are commonly found in federal law regulations, such as the German Building Use Regulations (BauNVO), or criteria for sustainable construction, such as the ones used by the German Sustainable Building Council (DGNB). Both the BauNVO and DGNB criteria were considered during the conceptualization of this work's KPIs.

BauNVO and DGNB. The BauNVO serves as the German legal basis for building construction by defining the type and extent of building use (BauNVO, Sects. 1 and 2), construction methods and coverable property areas (BauNVO, Sect. 3), as well as transitional and concluding provisions (BauNVO, Sect. 5) [11]. Paragraph 19 covers the Base Area Ratio, which is defined as the ratio of permitted base area of a building to the total property area [11]. This number limits the total built-up area at a site. Paragraph 20 covers the Floor Area Ratio, which is defined as the ratio of gross floor area (the sum of all the floors' base areas) to the total property area [11]. This number limits the height of buildings at a site.

The DGNB is a German non-profit organization that offers international building certifications. Other certification systems include LEED (Leadership in Energy and Environmental Design) by the United States Green Building Council and BREEAM (Building Research Establishment Environmental Assessment Method) by the United Kingdom Building Research Establishment. While LEED and BREEAM focus on environmental sustainability, like energy efficiency, DGNB weights all sustainability aspects (environmental, economic, social) equally [12]. An important DGNB KPI for sustainable land use is the Soil Sealing Factor [13]. This indicator awards certification points based on the level of soil sealing (a minimum of 0 points for over 80% sealed surface area, a maximum of 20 points for less than 50% sealed surface area), with additional points awarded for implementing compensatory measures (10 points) [13].

3 Sustainable Site KPIs

The following KPIs were developed based on consolidated knowledge from the expert interviews and pre-existing indicators. Following Dehning's peripheral order system [14], this work treats factories (f) as main process enablers. The production periphery further includes: energy utilities (e), office spaces (o), storage facilities (w), data centers (d), and roads (r). The periphery is expanded by another level for site area allocated to green spaces (g), bodies of water (b), social infrastructure (s), and unused area (n). The index i used in this section includes all aforementioned structures for production and production periphery.

Similar to facility layout planning, production sites are commonly separated into regularly shaped areas [15]. The following KPIs were initially conceptualized under the assumption that such regular areas (henceforth referred to as blocks) are typically rectangular. It was also assumed that factories and contained rooms are typically rectangular, too. To calculate the following KPIs, it is enough to have visited the site to gain a basic understanding of the production periphery.

3.1 Site Area Utilization (ζ)

ζ describes the Site Area Utilization, using a range between 0–1, to denote the percentage of block area that is being utilized for production when viewed from an aerial perspective. In order to assess sustainability, utilization considers not only built-up areas used directly or indirectly for production, but also ecological spaces (green areas, bodies of water) as per Aizizi et al. [16], social spaces. Unused land is also considered, though it negatively affects utilization. A ζ of 1 is achievable, as legal frameworks like the BauNVO limit the total built-up area, hence allowing for remaining property to be used for socio-ecological means.

When used without further criteria, ζ only shows whether parts of the property have not been yet designated for specific uses. As such, ζ was supplemented with information from the production periphery, for which ζ_{EC} (economic), ζ_{EN} (environmental), ζ_{SO} (social), and θ_{NA} (unutilized) were introduced. Production structures and periphery contribute to ζ_{EC} (e.g., $\zeta_{\mathrm{EC,w}}$ for warehouses), ecological spaces to ζ_{EN} (e.g., $\zeta_{\mathrm{EN,g}}$ for green spaces), social spaces to ζ_{SO}, and unutilized areas to θ_{NA}. The primary functionality of an area determines its ζ categorization. ζ thus allows for a closer examination of the types of facilities present. The benefit lies primarily in allowing for an analysis of ζ with respect to the corporate sustainability goals, as discussed in Sect. 4. The aim for site planning is to maximize ζ_{EC}, ζ_{EN}, and ζ_{SO}, while minimizing θ_{NA}.

$$\zeta = \zeta_{\mathrm{EC}} + \zeta_{\mathrm{EN}} + \zeta_{\mathrm{SO}} \leq 1 \;, for\ \theta_{\mathrm{NA}} \geq 0 \tag{1}$$

3.2 Floor Space Utilization (FSU)

When looking at production structures and their periphery, not every structure is dedicated to fulfilling a single purpose. Particularly on smaller blocks, multi-use structures can be found combining different functions, such as factories with integrated offices. Due to ζ only assessing sites from an aerial perspective, it is limited when it comes to such buildings.

Proportional Floor Space Utilization ($pFSU$). $pFSU$ was conceptualized to make the sub-functions of a structure visible. It is defined as

$$pFSU_{\mathrm{i}} = \frac{\text{gross floor area by type}}{\text{base area of building}} \tag{2}$$

where i refers to the production periphery. $pFSU_{\mathrm{i}}$ is typically smaller than 1 for multi-use buildings, though it may exceed it with enough verticality (i.e., multiple floors of a single structure type i). $pFSU$ needs to be smaller than or equal to regulatory parameters such as the Floor Area Ratio.

Absolute Floor Space Utilization ($aFSU$). Factories in Germany are often constructed as single-floor buildings; this is due to vertical transport being estimated to cost ten times as much as horizontal transport [17]. However, recent shortages of affordable properties have led higher densification measures on existing sites, along with warranting consideration of vertical factories during factory planning [17]. To express the verticality of block structures, $aFSU$ is defined as the sum of $pFSU$, with i referring to the production periphery:

$$aFSU = \sum_{\mathrm{i}} pFSU_{\mathrm{i}} \geq 1 \tag{3}$$

3.3 Base Area Utilization (BAU)

Since neither $pFSU$ nor $aFSU$ give information over densification or verticality of individual structure types, the BAU is defined as

$$BAU_{\mathrm{i}} = \frac{\text{gross floor area by type}}{\text{projected base area by type}} \geq 1 \tag{4}$$

where i refers to the production periphery. Projected base area by type refers to the area covered by a structure type when it is projected onto the ground floor.

4 Use Case

4.1 Metalworking Site

All KPIs were first applied to the ETA Factory. To analyze this block, a color overlay was defined with specific colors designating specific site structures (e.g., BGR [0, 0, 192] for factories). This overlay was applied to an aerial image of the site before running the image through a color detection algorithm that returned the composition of colors (i.e., structure types) on the block, as can be seen in Fig. 2 [18]. Due to a high degree of densification and socio-environmental work by the buildings department, a ζ of 1 was recorded, albeit with a lower focus on economic sustainability than would be expected of an industrial site:

$$\zeta = \zeta_{\mathrm{EC}} + \zeta_{\mathrm{EN}} + \zeta_{\mathrm{SO}} = 0.608 + 0.365 + 0.027 = 1$$

To gain a better understanding of the space utilization inside the ETA Factory, $pFSU$, $aFSU$, and BAU were calculated for the building. An aerial image was used to determine the base area of the ETA Factory, supplemented by a general understanding of the layout within the building to determine the gross floor area by type. As shown in Fig. 1, the production hall and offices have equal FSU. Due to this, the BAU is used to gauge verticality, highlighting the lower usage of building base area for offices compared to the production hall, with

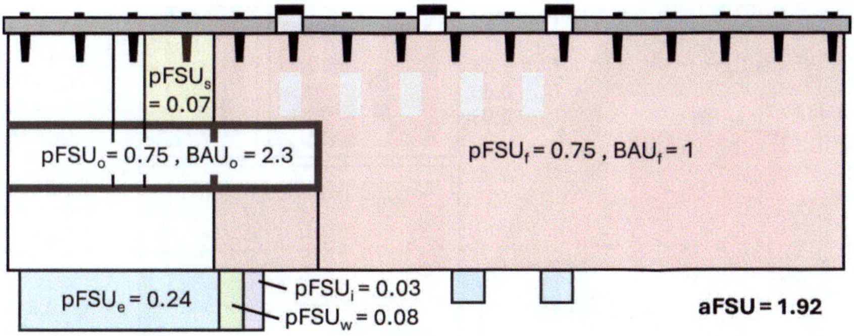

Fig. 1. FSU and BAU at the ETA Factory using a sectional view of the building.

$BAU_o = 2.3$ as opposed to $BAU_f = 1$. All other i have a BAU_i of 1, as they are spread across single floors. As can be seen in Fig. 1, the office space is increased by extending into the production facility on the second floor. Such construction measures are only feasible if they do not interfere with the intended utilization of the production hall, as is the case for the ETA Factory.

Based on a current overabundance of offices, a break room may be designated in an unused room, so long as the number of employees remains constant. Coupled with ζ, this would increase social spaces across the block. The production hall would also benefit from a higher BAU, as only the ground floor is in use.

4.2 Chemical Site

Due to confidentiality, only ζ will be discussed in this section. Three production blocks were analyzed for site sustainability using an aerial image of the site. ζ was calculated with basic knowledge of present block structure types.

While ζ is intended as a transparency measure to assist brownfield and greenfield transformations during the planning phase, it gains further importance when analyzing conformity with corporate sustainability goals. In this case, desired values for ζ_{EC}, ζ_{EN}, and ζ_{SO} for this site were defined as:

$$\zeta = \zeta_{EC} + \zeta_{EN} + \zeta_{SO}\ , with \begin{cases} \zeta_{EC} & 0 \leq \zeta_{EC} \leq 0.8 \\ \zeta_{EN} & 0 \leq \zeta_{EN} \leq 0.12 \\ \zeta_{SO} & 0 \leq \zeta_{SO} \leq 0.08 \end{cases} \tag{5}$$

It is important to note that desired values for ζ need to be defined by enterprises for specific production sites (e.g., as per [13]) while accounting for regulatory boundaries such as the the Base Area Ratio and green spaces defined in the enterprise's development plan. This effectively limits ζ_{EC} (and in part ζ_{SO}), thus leaving room for investments to further ζ_{EN}.

Despite using a limited data basis, ζ was successfully calculated for the blocks. As can be seen in Fig. 2, individual blocks do not fulfill the requirements of equation (5). However, the total ζ was 0.863 across these blocks, with ζ_{EC} =

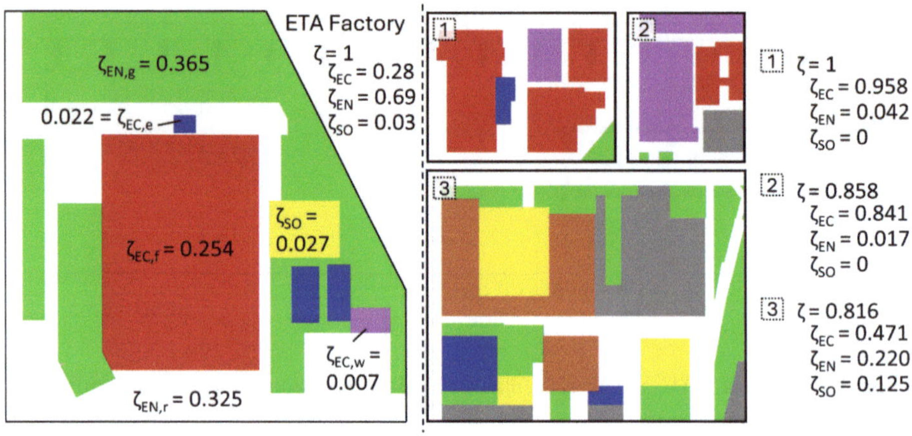

Fig. 2. Left: ζ for irregular ETA Factory block. Right: ζ for three chemical site blocks. Color scheme as per [18] (ζ_{EC}: red production, white roads, purple storage, blue energy supply, brown office. ζ_{EN}: green ecological. ζ_{SO}: yellow social. Gray unused space).

0.631, $\zeta_{\mathrm{EN}} = 0.152$, and $\zeta_{\mathrm{SO}} = 0.079$. Based on equation (5), further block transformations should thus prioritize the expansion of production facilities while slightly reducing ecological spaces without affecting the size of social spaces.

5 Discussion and Outlook

Site Area Utilization (ζ), proportional and absolute Floor Space Utilization (pFSU and aFSU), and Base Area Utilization (BAU) are introduced to address the literature gap in high-level, land-based sustainability KPIs for production sites. They act as transparency measures, to be used during transformation projects to align sites with corporate sustainability goals. They are functional with limited knowledge of the site, though they gain accuracy by using land use plans, floor plans, or building information models. In turn, the KPIs can be used during planning phases to adapt such documents to match desired site sustainability. From this work, it became apparent that a $\zeta = 1$ is not always ideal to strive for, as companies often reserve land for future expansion projects. In such cases, a $\zeta \leq 1$ may be desired. Similarly, $\zeta = 1$ could also be achieved by sealing all surfaces, which does not reflect true sustainability, leading to the necessity of sub-requirements as per Sect. 4.2. Furthermore, as seen in Sect. 4.2, not every block needs to fulfill all site sustainability goals, so long as site-wide sustainability is achieved, which is in line with the DELTA vision of cross-sectoral optimization of systems. Limitations for all KPIs lie in the quality and validity of the data basis with which the site is analyzed. Advantages lie in the KPIs being applicable regardless of block shape, as was displayed in Fig. 2.

For the future, an initial structure recognition on aerial images could be implemented via machine learning algorithms, leaving only a manual classification of structure types by appropriately coloring structure segments. Additionally, a weighting system could be introduced for different types of ecological spaces based on their environmental contribution. A similar weighting system could be introduced for production peripheries based on their contribution to production, allowing for better allocation of space to different structure types for FSU.

References

1. European Union: Document 52021DC0550. 'Fit for 55': delivering the EU's 2030 Climate Target on the way to climate neutrality (2021)
2. Hauschild, M.Z., Kara, S., Røpke, I.: Absolute sustainability: challenges to life cycle engineering. CIRP Ann. (2020). https://doi.org/10.1016/j.cirp.2020.05.004
3. Gerstenberger, J., Bauer, A.: Unternehmensbefragung (2024). https://www.kfw.de/PDF/Download-Center/Konzernthemen/Research/PDF-Dokumente-Unternehmensbefragung/Unternehmensbefragung-2024.pdf
4. European Union: Document 32022L2464. Directive (EU) 2022/2464 of the European Parliament and of the Council of 14 December 2022 amending Regulation (EU) No 537/2014, Directive 2004/109/EC, Directive 2006/43/EC and Directive 2013/34/EU, as regards corporate sustainability reporting (2022)
5. Kiemel, S., et al.: Current state and best practices on the way to zero emission in the manufacturing industry: an empirical survey in the Germany-Austria-Switzerland region. Procedia CIRP (2023). https://doi.org/10.1016/j.procir.2023.02.073
6. Eymüller, C., Hanke, J., Hoffman, A., Poeppel, A., Wanninger, C., Reif, W.: Towards a real-time capable plug & produce environment for adaptable factories. IEEE-ETFA (2021). https://doi.org/10.1109/ETFA45728.2021.9613729
7. Schulz, D., Simora, L., Lithoxopoulos, C.: Mass Individualization with Lot-Size One Order-controlled production in adaptable factories. ATP Maga. (2021)
8. Kaucher, C., Erlach, K., Bauernhansl, T.: Conception of future-proof factory buildings via software-based scenario creation and evaluation. CPSL (2024). https://doi.org/10.15488/17705
9. Kohne, T., Beck, M., Bishara, N., Seyfried, S., Schneider, J., Weigold, M.: The Darmstadt energy laboratory for technologies in application: overview paper – living lab DELTA. TUprints (2023). https://doi.org/10.26083/tuprints-00026384
10. Abele, E., Schneider, J., Maier, A.: ETA - die Modell-Fabrik: Energieeffizienz weiter gedacht. ETA, Darmstadt (2018)
11. Bundesministerium der Justiz: Verordnung über die bauliche Nutzung der Grundstücke (Baunutzungsverordnung - BauNVO). Accessed 3 July 2023
12. Ferreira, A., Pinheiro, M.D., de Brito, J., Mateus, R.: A critical analysis of LEED, BREEAM and DGNB as sustainability assessment methods for retail buildings. J. Build. Eng. (2023). https://doi.org/10.1016/j.jobe.2023.105825
13. German Sustainable Building Council: DGNB System: Criteria Set - New Construction Building. Version 2023. DGNB e.V. (2023)
14. Dehning, P.: Steigerung der Energieeffizienz von Fabriken der Automobilproduktion. Springer Wiesbaden (2017). https://doi.org/10.1007/978-3-658-19098-9
15. Pérez-Gosende, P., Mula, J., Díaz-Madroñero, M.: Facility layout planning. an extended literature review. Int. J. Prod. Res. (2021). https://doi.org/10.1080/00207543.2021.1897176

16. Aizizi, Y., Kasimu, A., Liang, H., Zhang, H., Zhao, Y., Wei, B.: Evaluation of ecological space and ecological quality changes in urban agglomeration on the northern slope of the Tianshan Mountains. Ecol. Ind. (2023). https://doi.org/10.1016/j.ecolind.2023.109896
17. Friedrich, S., Weitz, S., Luchsinger, Y.: 01006 Planung der vertikalen Fabrik. TÜVRheinland Praxis Produktionsmanagement. Accessed 03 Apr 2025. https://pxp.tuev-media.de/xhtml/document.jsf?docId=docs/pxp_0000001006
18. Ozen, O., Zink, R.: Code for Site Area Utilization v2.0. TUdatalib (2025). https://tudatalib.ulb.tu-darmstadt.de/handle/tudatalib/4527

Impact Assessment of Digitalization: Cost and Environmental Aspects Using the Example of Digital Factory Initiatives

Linda Maria Antonia Becker[1], Timo Fleschutz-Balarezo[2], and Semih Severengiz[1](✉)

[1] Bochum University of Applied Sciences, Am Hochschulcampus 1, 44801 Bochum, Germany
semih.severengiz@hs-bochum.de
[2] HTW Berlin University of Applied Sciences, Treskowallee 8, 10318 Berlin, Germany
timo.fleschutz-balarezo@HTW-Berlin.de

Abstract. With IT contributing nearly 4% to global CO_2 emissions, it presents both risks and opportunities for achieving sustainability goals. While digital technologies can enhance resource efficiency, they also increase energy consumption and environmental costs. This paper presents a framework for evaluating the environmental impact of digital factory initiatives in their early planning stages. The approach combines methods from Green IT and Life Cycle Assessment (LCA) to enable a structured comparison of digitalization alternatives. Since ecological assessments alone may overlook broader organizational concerns, the framework integrates sustainability criteria into existing economic evaluation systems, supporting decisions that are both environmentally and financially sound. The result is a framework for stakeholders in product management, manufacturing, and sustainability, helping them select digital solutions that balance performance, cost, and ecological impact.

Keywords: Digital factory initiatives · Green IT · Green IoT · TEI · BSC · Costs

1 Introduction

Industrial sectors - including production and energy - are responsible for over 30% of global greenhouse gas emissions, a figure that rises to more than 40% when transportation is included [1]. In response to this significant environmental impact, digital factory initiatives have emerged as a promising approach to make production processes more sustainable, efficient, and adaptable. These initiatives are diverse in both their technological design and implementation strategies, but they share a common goal: increasing resource efficiency in manufacturing. Researchers emphasize that circular economy principles, advanced data analytics, and AI-based optimization are essential for reducing waste and improving efficiency [2–4]. Opazo-Basáez et al. show that combining digital and green servitization in the automotive industry enhances productivity and sustainability—especially when approached in an integrated manner [5].

H. Kohl et al. (Eds.): GCSM 2025, LNME, pp. 31–39, 2026.
https://doi.org/10.1007/978-3-032-21154-5_4

Digital factory initiatives are enabled by a broad spectrum of technologies, which include cyber-physical systems (CPS), the internet of things (IoT), cloud computing, big data analytics, virtual and augmented reality (VR/AR), intelligent robotics, artificial intelligence (AI), and additive manufacturing [6]. These technologies form the infrastructural basis for data-driven, adaptive production systems. Overall, it is evident that digital factory initiatives do not represent a single unified concept, but rather a spectrum of options for integrating digital technologies according to specific business and sustainability goals. At the same time, the deployment of information and communication technology (ICT) systems brings its own environmental impact: The global ICT sector's greenhouse gas footprint currently lies between 1.5% and 4% and is expected to increase [7]. The increasing use of AI is reinforcing this trend. The training of ChatGPT3 alone emitted around 552 tons of CO_2-eq [8]. Moreover, as Chen et al. point out, the mere availability of data through IoT and CPS does not automatically lead to reductions in energy or resource consumption. Technological advances such as more powerful and energy efficient IoT devices or AI may further drive productivity, but they can also lead to increased overall energy demand - a phenomenon known as the rebound effect. From the economic point of view historical research on IT investments revealed the "productivity paradox," where organizations struggled to link IT spending to tangible productivity gains due to high costs and mismatched technologies [9]. Over time, this paradox was mitigated as studies highlighted the positive correlation between IT capabilities and organizational performance [10].

Nonetheless given these complexities, there is a clear need for an assessment methodology that can evaluate digitalization initiatives in the early planning and conceptual stages. Existing assessment methods often lack the comprehensive perspective required for the implementation of digital factories, as they tend to focus on isolated aspects rather than integrating ecological, economic, and strategic objectives. The aim of this work is therefore to develop an integrated framework that enables the structured comparison of digitalization alternatives and supports decision-makers in selecting solutions that are both environmentally and economically sound.

2 Methodology

This study employs a combined methodological approach (argumentative-deductive analysis) [11] to develop a comprehensive framework for evaluating the environmental impacts of digital solutions in the early phases of digital factory initiatives. Beginning with a conceptual foundation, we integrate ecological assessments into existing mainly economical IT evaluation systems, drawing from green IT practices and life cycle assessment (LCA) literature. Our analysis is based on a review of related literature, which provides insights into existing frameworks and methodologies for evaluating digital transformation and its environmental impacts. By synthesizing the findings from this literature review, we derive an integrated evaluation framework that focuses specifically on the initial phases of digital factory initiatives, enabling a more targeted and effective assessment of sustainability in digital transformation.

3 Related Works

3.1 Environmental Impact Assessment

A systematic literature review by Chen et al. found that most of the research focuses on the ecological benefits within the immediate production process—the "visible" outcomes of digitalization. In contrast, the ecological costs associated with the use of digital technologies themselves are often neglected—particularly those arising from the production, operation, and disposal of hardware, as well as the energy consumption of systems like IoT, AI or robotics [6].

In response, the fields of Green IT and Green IoT have developed various tools and frameworks to assess the environmental impact of digital systems. Notable among these is the Green Software Measurement Model (GSMM) by Guldner et al., which combines different methods and tools into a unified approach for evaluating software-related environmental effects [12]. This is complemented by established certification systems such as the German Blue Angel (Blauer Engel) eco-label for resource- and energy-efficient software products (DE-UZ 215), which defines systematic sustainability criteria for software [13]. Tools like the Green Algorithms Calculator further support the estimation of energy use and emissions [14]. Another well-established approach is Life Cycle Assessment (LCA), which enables a comprehensive evaluation of environmental impacts across all stages of a product's life [15], guided by international standards such as ISO 14040 [16], ISO 14044 [17] and the ICT-specific ETSI ES 203 199 V1.4.1 [18].

In the context of digital factories, environmental assessments must consider not only efficiency gains in production - such as optimized material flows and reduced defect rates - but also the significant impacts associated with additional hardware. The production, transportation, and disposal of digital components contribute to environmental degradation, exacerbated by low recycling rates and the presence of hazardous materials like lead and cadmium [19]. Furthermore, digitalization increases electricity demand both locally and through external data centers, especially for energy-intensive technologies such as AI. Projections indicate that ICT systems could account for up to 20% of global electricity demand by 2030 [19, p.14].

In the literature, environmental impact of ICT is most commonly assessed using Global Warming Potential (GWP), as recommended by the ETSI LCA standard [18], but other categories such as Abiotic Depletion Potential (ADP) and Cumulative Energy Demand (CED) are also considered to capture resource use and/or enable more comparable assessments across different energy grid mixes[8, 21]. Going beyond that the Green Cloud Computing (GCC) methodology developed by the German Environment Agency evaluates data center operations across ADP, GWP, CED, and water consumption [22].

3.2 Economical Evaluation

Common methodologies for assessing investments under uncertainty include dynamic approaches such as Net Present Value (NPV), sensitivity analysis, simulations, and decision tree analyses [23, 24]. These techniques provide organizations with structured mechanisms to evaluate the financial feasibility and sustainability of their initiatives. Life Cycle Costing (LCC) and Total Cost of Ownership (TCO) are particularly relevant around information technology (IT) investments. LCC provides a comprehensive view of all costs incurred over the lifecycle of an asset. Similarly, TCO, often viewed as a more user-centric cost management approach, summarized the costs associated with the ownership of an IT asset, e.g. software application including infrastructure. However, these methods focus mainly on financial impact and require extensive and detailed information, making them less suitable for the early phases of digital factory initiatives when data availability is often limited and/or based on estimates.

To evaluate the economic impact of IT investments, frameworks including non-financial criteria such as the Balanced Scorecard (BSC) and Total Economic Impact™ (TEI) have gained prominence. The BSC, introduced by Kaplan and Norton, provides a comprehensive framework that enables organizations to evaluate initiatives incorporating financial and non-financial performance metrics by considering perspectives such as financial outcomes, customer satisfaction, internal processes, and organizational learning and growth [25]. The TEI methodology, developed by Forrester Research, offers a more focused approach to assessing the implications of IT investments. TEI evaluates costs, benefits, flexibility, and risks associated with IT initiatives, providing a structured framework for decision-making. It also quantifies business benefits, such as improved productivity and customer satisfaction, while addressing future flexibility and risks through strategic evaluation tools [26].

3.3 Integration and Decision-Making Models

Decision models are a cornerstone of decision research, defined by essential elements such as goal systems, alternatives, environmental states, and outcome functions. Within this context, multi-criteria methods are often employed, with two principal approaches emerging in the literature: multi-attribute decision-making (MADM) and multi-objective decision-making (MODM). MADM involves ranking or categorizing known alternatives across multiple attributes, while MODM seeks optimal solutions within a set of constraints [27].

A practical example of this integration can be seen in the work of Yadla and Kulkarni, who employed a multi-attribute method, specifically the Analytic Hierarchy Process (AHP), to compare investment alternatives in the context of digital manufacturing. The evaluated framework considered criteria such as costs, ROI, reliability, safety, efficiency, flexibility, complexity, quality, communication, and space utilization [28]. This approach demonstrates the utility and adaptability of decision models in addressing multifaceted and competing objectives.

4 Concept for an Evaluation Framework

To ensure sustainability-oriented decision-making for digital factory initiatives, this framework applies a multi-attributive decision-making approach, as it allows for a structured comparison of digitalization options based on diverse evaluation criteria, which is particularly suitable in early planning phases where alternatives are known but complex to assess.

As a preparatory step, we utilize and expand a combination of the BSC and TEI methodologies. Originally designed for assessing the return on investment for technology projects, TEI is used as base model extended here to cover four key dimensions: financial, organizational, strategic, and ecological. Especially the organizational is inspired by the broader perspective of BSC. Similarly the risk assessment of TEI is complemented by an integrated evaluation of not only risk, but also flexibility and learning potential across all categories. The risk evaluation identifies uncertainties, implementation hurdles, and unintended consequences—whether financial, operational, or ecological. Conversely, flexibility and learning potential evaluation highlights opportunities for long-term innovation, sustainability contribution, and strategic advantage to capture ecological impacts in a structured and comparable way, we rely on the ICT-specific Life Cycle Assessment (LCA) standard by ETSI (ETSI ES 203 199 V1.4.1). Within this standard digital factory initiatives can be viewed as IT services. This framing allows for a dual-perspective analysis: on the one hand, it captures the environmental impact arising from additional hardware and electricity consumption; on the other, it considers the potential ecological benefits stemming from optimized production processes.

Within this structure, Global Warming Potential (GWP) and Cumulative Energy Demand (CED) are applied as core impact categories due to their methodological robustness and relevance to digital systems. Abiotic Depletion Potential (ADP) and Water Consumption are also recognized as important categories but are considered optional due to limited data availability. To enhance data collection and analysis at a more granular level, complementary tools such as the Green Software Measurement Model, the Green Algorithms Calculator, and eco-label criteria (e.g., the Blue Angel) may be integrated into the evaluation process. These instruments serve as auxiliary modules that support modular adaptation of the framework to different planning and conceptualization phases.

All together these integrative assessments build a synthesized understanding of each initiative's strengths and limitations. Table 1 summarizes the evaluation criteria for digital factory initiatives and their origin within different methods.

The accompanying illustration (Fig. 1) visualizes the interplay of the different evaluation criteria, color-coded according to their respective evaluation aspects. Within the ecological dimension, mandatory impact categories are marked in dark green, while optional ones appear in a lighter shade to reflect their conditional application based on data availability. Together with the integrated risk, flexibility and learning assessment, the financial, organizational, and ecological indicators are aggregated to support a comprehensive multi-attributive decision-making process. This approach allows for a transparent comparison and prioritization of digital factory alternatives, ultimately enabling well-founded, sustainability-aligned decisions early in the project lifecycle and contributing to the responsible digital transformation of industrial production.

Table 1. Evaluation Criteria for Digitalization Initiatives

Evaluation dimensions	Evaluation criteria	Example for digital factory initiatives	Referring method
Financial	Cost Analysis	Total lifecycle costs of implementing new technologies	TEI [26]
Financial	Business Benefits	Financial returns from improved production efficiency	TEI [26]
Organizational	Internal Business Processes	Optimization through data analysis or simulations	BSC [25]
Organizational	Customer Perspective	Improvements in product adaptation or quality	BSC [25]
Ecological	Global Warming Potential (GWP)	Impact from hardware and energy consumption	ETSI [18]
Ecological	Cumulative Energy Demand (CED)	Total energy use across lifecycle phases	Berthelot et al. [21], Dokic et al. [8]
Ecological	Abiotic Depletion Potential (ADP)*	Resource scarcity linked to digital components	Berthelot et al. [21], Dokic et al. [8]
Ecological	Water Consumption*	Water use in manufacturing and operation phases	GCC [22]
Strategic	Risk Evaluation	Assessment of uncertainty of expected outcome	TEI [26]
Strategic	Flexibility and Learning	Adaptability of management and employees	TEI [26], BSC [25]

* Optional: subject to data availability.

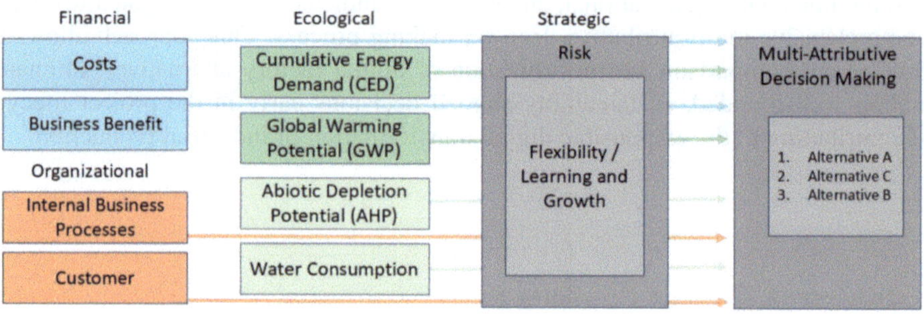

Fig. 1. Concept visualization of evaluation framework

5 Conclusion

This paper presents a tailored evaluation framework for digital factory initiatives that builds upon existing ecological and economic assessment approaches. By integrating established methods such as TEI and the ETSI LCA standard into a streamlined, application-oriented structure, the framework offers a practical and holistic tool for early-stage decision-making in industrial digitalization.

Beyond ecological and economic factors, future work will need to broaden the scope to include social dimensions—such as workforce implications, data security, and digital ethics—that are increasingly critical in digital transformation processes. Additionally, the framework's effectiveness depends on how well it accommodates the perspectives and needs of diverse stakeholders, including operators, customers, and regulators.

It's true value, however, will be demonstrated through real-world applications. A promising next step lies in testing the framework across different industrial sectors to assess its robustness and adaptability under varying operational conditions. A dedicated research project at Bochum University of Applied Sciences will focus on validating the ecological assessment components of the framework over the next two years.

References

1. Net-Zero Industry Tracker 2023 Edition. World Economic Forum, Geneva (2023)
2. Rahmani, R., Jesus, C., Lopes, S.I.: Implementations of digital transformation and digital twins: exploring the factory of the Future. Processes **12**, 787 (2024). https://doi.org/10.3390/pr12040787
3. Chen, S., Yang, Y., Wu, T.: Research on the impact of digital transformation on green development of manufacturing enterprises. Front. Energy Res. **10**, 1045328 (2023). https://doi.org/10.3389/fenrg.2022.1045328
4. Fraga-Lamas, P., Lopes, S.I., Fernández-Caramés, T.M.: Green IoT and Edge AI as key technological enablers for a sustainable digital transition towards a smart circular economy: an industry 5.0 use case. Sensors. **21**, 5745 (2021). https://doi.org/10.3390/s21175745
5. Opazo-Basáez, M., Vendrell-Herrero, F., Bustinza, O.F.: Uncovering productivity gains of digital and green servitization: implications from the automotive industry. Sustainability **10**, 1524 (2018). https://doi.org/10.3390/su10051524
6. Chen, X., Despeisse, M., Johansson, B.: Environmental sustainability of digitalization in manufacturing: a review. Sustainability **12**, 10298 (2020). https://doi.org/10.3390/su12241 0298
7. Bieser, J.C.T., Hintemann, R., Hilty, L.M., Beucker, S.: A review of assessments of the greenhouse gas footprint and abatement potential of information and communication technology. Environ. Impact Assess. Rev. **99**, 107033 (2023). https://doi.org/10.1016/j.eiar.2022.107033
8. Dokic, D., Maass, W., Groen in't Woud, F.: Towards sustainability of AI: a systematic review of existing life cycle assessment approaches and key environmental impact parameters of artificial intelligence. In: Proceedings of the 57th Hawaii International Conference on System Sciences, Honolulu, pp. 4301–4310 (2024). https://doi.org/10.24251/HICSS.2024.518
9. Buer, S.-V., Semini, M., Strandhagen, J.O., Sgarbossa, F.: The complementary effect of lean manufacturing and digitalisation on operational performance. Int. J. Prod. Res. **59**, 1976–1992 (2021). https://doi.org/10.1080/00207543.2020.1790684

10. Moeuf, A., Pellerin, R., Lamouri, S., Tamayo-Giraldo, S., Barbaray, R.: The industrial management of SMEs in the era of Industry 4.0. Int. J. Prod. Res. **56**, 1118–1136 (2018). https://doi.org/10.1080/00207543.2017.1372647
11. Wilde, T., Hess, T.: Methodenspektrum der Wirtschaftsinformatik, Arbeitsbericht Nr. 2/2006, Institut für Wirtschaftsinformatik und Neue Medien, LMU München, https://epub.ub.uni-muenchen.de/14146/1/hess_14146.pdf. Accessed 01 July 2025
12. Guldner, A., et al.: Development and evaluation of a reference measurement model for assessing the resource and energy efficiency of software products and components—Green Software Measurement Model (GSMM). Futur. Gener. Comput. Syst. **155**, 402–418 (2024). https://doi.org/10.1016/j.future.2024.01.033
13. Blauer Engel: Ressourcen- und energieeffiziente Softwareprodukte (DE-UZ 215). https://www.blauer-engel.de/de/produktwelt/software. Accessed 01 July 2025
14. Green Algorithms project: Green Algorithms calculator. https://calculator.green-algorithms.org/. Accessed 01 July 2025
15. Hauschild, M.Z., Rosenbaum, R.K., Olsen, S.I. (eds.): Life Cycle Assessment: Theory and Practice. Springer, Cham (2018). https://doi.org/10.1007/978-3-319-56475-3
16. ISO: Environmental management – life cycle assessment – principals and framework. International Standard ISO 14040 (2006)
17. International Organization for Standardization: Environmental management – life cycle assessment – requirements and guidelines. International Standard ISO 14044, Geneva (2006)
18. ETSI: Environmental Engineering (EE); Methodology for environmental Life Cycle Assessment (LCA) of Information and Communication Technology (ICT) goods, networks and services (2025)
19. Bieser, J.C.T., Coroamă, V.C.: Direkte und indirekte Umwelteffekte der Informations- und Kommunikationstechnologie. Nachhaltigkeits Manag. Forum **29**, 1–11 (2021). https://doi.org/10.1007/s00550-020-00502-4
20. Hedberg, A., Šipka, S., Bjerkem, J.: Creating a digital roadmap for a circular economy. European Policy Centre, Brussels, Belgium (2019)
21. Berthelot, A., Caron, E., Jay, M., Lefèvre, L.: Estimating the environmental impact of Generative-AI services using an LCA-based methodology. Procedia CIRP. **122**, 707–712 (2024). https://doi.org/10.1016/j.procir.2024.01.098
22. Gröger, J., Liu, R., Stobbe, L., Druschke, J., Richter, N.: Green Cloud Computing: Lebenszyklusbasierte Datenerhebung zu Umweltwirkungen des Cloud Computing. Umweltbundesamt, Dessau-Roßlau (2021)
23. Wöhe, G., Döring, U.: Einführung in die Allgemeine Betriebswirtschaftslehre. Presented at the (2008)
24. Franke, G., Hax, H.: Finanzwirtschaft des Unternehmens und Kapitalmarkt. Springer, Heidelberg (2004). https://doi.org/10.1007/978-3-662-07225-7
25. Kaplan, R., Norton, D.P.: Using the balanced scorecard as a strategic management system. Harvard Bus. Rev. **74** (1996)
26. Gliedman, C.: The Foundation of Sound Technology Investment: The Total Economic Impact^{TM} Methodology (2003). https://faculty.uml.edu/dstephenson/technology_class/forrester_reports/Forrester_TEI_Method.pdf. Accessed 01 July 2025
27. Krstić, A., Figueira, J., Greco, S., Ehrgott, M.: Multicriteria decision analysis: state of the art surveys. Ekon horizonti. **20**, 189–191 (2018). https://doi.org/10.5937/ekonhor1802189K
28. Yadla, V., Kulkarni, A.: Digital manufacturing framework for enhanced efficiencies. In: 2023 IEEE Engineering Informatics, pp. 1–11 (2023). https://doi.org/10.1109/IEEECONF58110.2023.10520527

Energy Flexibility as a Factory Design Parameter

Stefanie Samtleben[1(✉)], Thomas Sobottka[2], and Kerim Torolsan[3]

[1] Fraunhofer IFF, Sandtorstraße 22, 39112 Magdeburg, Germany
Stefanie.Samtleben@iff.fraunhofer.de
[2] Fraunhofer Austria Research GmbH, Wien, Austria
[3] Fraunhofer IPA, Stuttgart, Germany
http://iff.fraunhofer.de/

Abstract. The increasing share of renewable energy sources makes energy supply more weather-dependent, challenging supply quality and system stability. To address these challenges without extensive infrastructure expansion, this paper presents a method for estimating and optimizing the energy flexibility potential of factory designs by correlating factory energy demand with the residual load—the portion of demand not met by renewables and a key indicator of local grid constraints. Using the *relocation coefficient* as a metric, we identify factory layouts that maximize flexibility and demonstrate the approach in an aluminum foundry, achieving reduced system stability costs without compromising production performance. Building on this, we introduce an operational optimization framework that integrates market and grid congestion signals into production planning via a digital twin, enabling real-time, multi-criteria optimization. Together, these methods support manufacturing enterprises in adapting to volatile energy systems and leveraging flexibility markets.

Keywords: energy flexibility · factory planning · optimization · production planning and control

1 Energy Flexibility: A Problem Description

Addressing the challenge of leveraging energy flexibility in energy-intensive production enterprises requires consideration from two perspectives: the production system and the energy supply system. Energy flexibility is defined as the ability of a system to rapidly and cost-effectively adjust its energy demand in response to external factors. The two main external drivers are: (a) the availability of renewable energy in the grid often indicated by current carbon intensity and reflected in short-term market price signals such as those from the day-ahead market and (b) technical constraints, notably grid congestion, which are typically localized and not represented in broader market signals. A production system is considered energy flexible if it can absorb fluctuations in the energy supply

H. Kohl et al. (Eds.): GCSM 2025, LNME, pp. 40–48, 2026.
https://doi.org/10.1007/978-3-032-21154-5_5

without requiring changes to its structure, technology, or capacitys [2,3]. Consequently, the inherent structure, technology, and capacities of a factory determine its potential for energy flexibility. It is therefore essential to account for, or anticipate, a factory's energy-flexible behaviour during the planning and design of production systems and facilities. Energy demand has long been acknowledged as a factor in production systems, with early efforts to integrate energy as a production variable evident in holistic modelling approaches [4]. However, energy has historically played a secondary role in factory planning and simulation models, where priority has been given to optimizing traditional control variables such as inventory reduction, setup times, and throughput [5]. These models typically assume unlimited energy availability, neglecting its constraints or variability. While recent work by Stock [6] and Pfeilsticker [7] explores energy efficiency trade-offs, they lack actionable frameworks for incorporating energy dynamics into factory planning. Similarly, Graßl's energy flexibility indicator [1], though detailed, demands exhaustive machine control data and is impractical for early-stage planning. Analytical approaches like Kabelitz's production functions [8] remain theoretically sound but face implementation barriers due to data requirements. Few studies directly address energy market integration [9] or provide scalable methodologies [10], highlighting a critical gap in aligning factory planning with evolving energy systems.

Studies examining loads from the perspective of the electrical energy system indicate that both industrial and residential sectors can economically market their energy flexibility. This is most commonly achieved through participation in energy exchanges or global flexibility markets [11,12]. Aggregator models often focus on households, as demand-side management is generally less complex in this sector. However, grid-friendly incentives are rarely incorporated into these models. Recent regulatory developments, such as the Act on the Adaptation of Energy Industry Law to EU Requirements and the Amendment of Other Energy Law Provisions (effective 22 December 2023), have granted the Federal Network Agency broader authority to structure the grid fee system (Sect. 21 (3) sentence 4 no. 3 EnWG). In this context, new approaches-such as the introduction of localized dynamic grid charges-are being evaluated to incentivize energy-flexible behaviour. [13] Austria is piloting an innovative flexibility market mechanism aimed at mitigating grid congestion. This system functions as a localized day-ahead market, matching energy adjustments across grid regions based on day-ahead market results and grid simulations that predict congestion events. Through this mechanism, grid operators can procure "positive" flexibility (increased generation or reduced demand) or "negative" flexibility (reduced generation or increased demand), thus balancing congestion while maintaining market equilibrium.

2 Planning Energy Flexibility in Energy-Intensive Production Systems

Schneeweiß [14] argues that flexibility is not an inherent feature of production systems, but rather an opportunity to respond to external factors. The first step is to define the external signal-such as energy price fluctuations or grid constraints-and the threshold for system response. The second step is to determine how the production system can react to this signal, and what structures, technologies, or capacities support such flexibility. To address this, Schneeweiß introduces hierarchical planning, a method that breaks down complex decision-making into smaller, more manageable problems. In this section, we adapt Schnee-weiß's approach to energy flexibility and present a method to quantify a factory's energy flexibility potential.

2.1 The Energy Flexibility Potential

For planning energy flexibility, Schneeweiß employs a hierarchical approach that distinguishes between two levels: the top level and the base level. At the top level, the production system is considered strategically; factory planners design the layout-structure, technology, and capacities-based on the production program, influencing investment decisions. Evaluating these design choices requires anticipating factory operations at a high level. The resulting factory layout then serves as the starting point for base-level planning, which provides detailed feedback on flexibility, either through monitoring real operations or via simulation and optimization. This feedback informs iterative improvements at the strategic level, creating a continuous optimization loop.
In our approach to energy flexibility planning, we use the residual load-the difference between total system load and renewable energy supply-as the external signal (for more details see [15]). Residual load, managed by grid operators to balance local grid input and output, serves as an effective signal for factory planning, especially in the absence of local energy markets, as it reflects the local grid's capacity constraints.

2.2 The Value of Energy Flexibility

In 2024, the primary cost drivers for maintaining system stability in Germany are grid congestion management, the provision of control power, and energy losses, all of which have increased in recent years [16]. As shown in Fig. 1, these costs are influenced by the mix of renewable and conventional energy sources in the supply system. To mitigate these rising expenses, measures such as grid expansion and enhanced energy flexibility can be considered. Energy flexibility, in particular, can reduce costs by balancing local supply and demand. This potential can be quantified using Blarke's Relocation Coefficient R_c (see Eq. 1), which measures the correlation between the residual load of the energy supply system (e) and the load of the energy-intensive production system (d), based on their mean values (e_m, d_m) [17].

$$R_c = \frac{\sum(e-e_m)(d-d_m)}{\sqrt{\sum(e-e_m)^2\sum(d-d_m)^2}} \tag{1}$$

Blarke originally applied this coefficient to size energy storage and supply capacities for production. In this study, we adapt the coefficient to optimize the sizing of buffers and production equipment in factory layouts. An optimization program is developed with the Relocation Coefficient as the objective function, aiming to maximize the alignment between production demand and residual load. Additional key performance indicators are included as required by the use case, enabling a multi-criteria comparison of various layouts and energy supply scenarios to identify the most energy-flexible production system. Different layouts, involving varying amounts of production equipment, result in different investment levels, which must be justified by the benefits of increased energy flexibility. The Relocation Coefficient ranges from -1 to 1, where 1 indicates perfect alignment between production demand and residual load-within the maximum flexibility the system can offer. The value of energy flexibility is ultimately estimated by multiplying the Relocation Coefficient by the production system's energy demand over a defined time horizon and the system stability costs per kilowatt-hour, which ranged from €0.007/kWh in 2022 to €0.013/kWh in 2023 (own calculations based on the data of [16]).

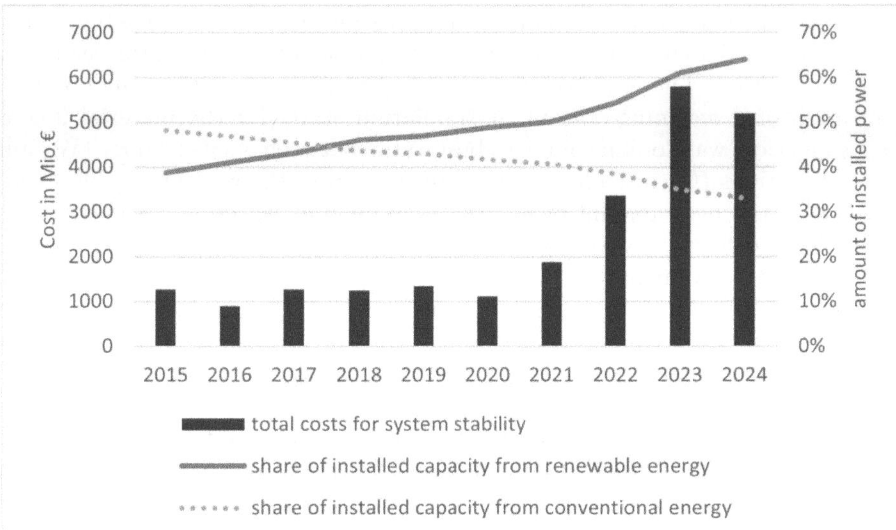

Fig. 1. System stability costs and installed renewable capacity in Germany [16].

3 Case Study: Aluminium Foundry

Metal melting is a highly energy-intensive process, particularly for aluminium, which is essential for lightweight applications in the automotive and construction industries.

To demonstrate the proposed method, it was applied to the melt preparation process in an aluminium foundry operating across three shifts. This process comprises three main steps: melting, transport, and holding/casting. A technological innovation increased the flexibility of the transport and holding stages by using insulated pots of limited capacity. These pots can maintain the aluminium at casting temperature for about one hour without additional energy input and can be transported by crane or forklift. Factory configurations are characterized by the number of pots (W), melting stations (S), and warming stations (G) (see Fig. 2). Three product types were considered, and system behaviour was analysed over a six-month period to capture seasonal variations in residual load. Due to the limited capacity of the pots, the amount of energy that can be shifted in time is also constrained. For computational efficiency, the time interval could be subdivided, though the full six months were modelled.

The residual loads studied differ in renewable energy share: 61% for Germany as a whole [16], 57% for the rural district in Thuringia [18,19], and 100 percent for the site area (based solely on photovoltaic generation, which is insufficient to meet total demand) [18].

The results show that the relocation coefficient improves as more operating resources are available, regardless of the residual load scenario. A notable increase occurs when an additional warming station (G) is added, due to its independent energy demand. An additional melting pot (W) increases the relocation coefficient as well. The improvements are very small compared to the potential of an additional warming station. A simulation study for the material flow of this production was looking for the best ratio between melting pots (W) and warming stations (G) to look for a high utilisation of the ressources. The study showed that the best amount of melting pots for four warming stations (4G) is nine (9W) and for five warming stations (5G) it is eleven melting pots (11W) [20]. So the improvements arise out of a better utility of the melting stations (S). The highest relocation coefficient is observed for the rural district scenario, likely due to its lower share of renewables and the smoothing effect of a major pump-storage power plant. Considering additional key performance indicators, such as lead time and average inventory, the optimal configuration comprises two melting stations (2S), five warming stations (5G), and seven pots (7W). Since the relocation coefficient is normalized (takes only values between minus one and one) and a value at one means a perfect fit of demand and supply, so no costs for system stability occur, the total avoided costs can be determined by multiplying the energy demand with the avoided costs per kWh (see Sect. 2.2) and the relocation coefficient. For the analysed period, the estimated avoided system stability costs are approximately €20,000 per month, providing a strong basis for negotiations and energy trading with suppliers.

4 Discussion: Energy-Flexibility-Based Capacity Planning

In this study, we demonstrate that the relocation coefficient serves as a robust metric for assessing energy flexibility and for the optimal sizing of production

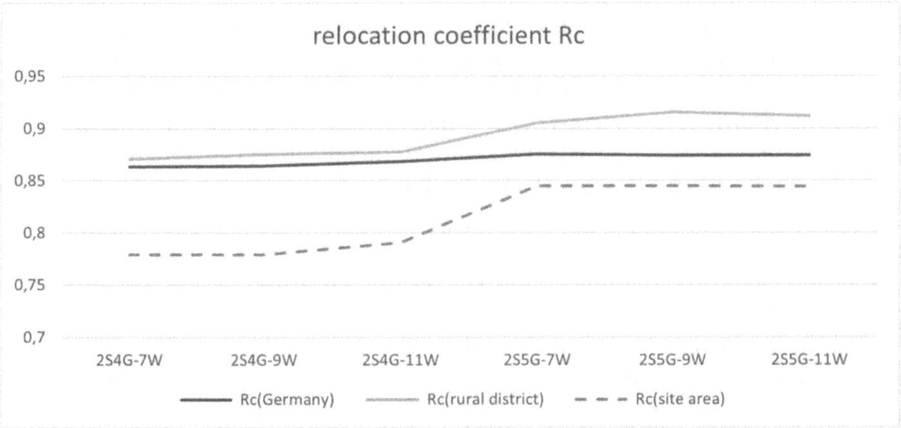

Fig. 2. Factory layouts and their relocation coefficients under varying residual loads.

capacities within manufacturing facilities. Typically, capacity planning in factories is predominantly driven by cost-effectiveness, guided by forecasts of product demand and utilising operational performance indicators such as equipment utilisation and potential bottlenecks. This planning can be executed either statically or dynamically through simulation techniques [23]. However, the absence of a local flexibility market precludes the feasibility of a purely cost-effective approach. A notable advantage of the proposed optimisation programme is its focus on a singular objective value, thereby mitigating the potential for results to be skewed by arbitrary weight adjustments across multiple objectives. Nevertheless, it remains feasible to initiate a decision-making process based on the decision variables, which can subsequently be employed to evaluate various performance indicators.

Our objective is to ascertain the minimum revenue required for a factory to effectively integrate energy flexibility into its operations. We have successfully quantified the value of energy flexibility, allowing for the consideration of enhanced production capacities that foster a more grid-friendly operational profile in investment decision-making. However, our application of this method is constrained by the reliance on historical data, particularly regarding the residual load of the energy supply system and the costs associated with system stability. That's why we like to give an outlook on the connection to the factory operation.

4.1 Outlook: Connection Between Factory Planning and Operation

While the relocation coefficient is a valuable metric for designing factories with enhanced energy flexibility and for forecasting flexibility potential, it is less suitable for real-time operational optimization. To address this, we propose an adjusted optimization framework tailored to factory operations. During this phase, the primary signals available for energy flexibility are forecasts from energy markets (mainly the day-ahead and intra-day markets, which are not localized) and localized grid congestion forecasts. Energy price forecasts are already well-established, leveraging historical market data and publicly available weather information. In contrast, grid congestion forecasting is an emerging area, with several EU-wide initiatives underway to incentivize grid-friendly behaviour - with the dynamic grid charges (Germany) and localized flexibility market (Austria) (see Sect. 1.2) as two major examples. These congestion-related signals are expected to become increasingly predictable, utilizing known grid bottlenecks, weather forecasts, and day-ahead market results, allowing market participants to prepare offers for both positive and negative flexibility contributions. This concept has been previously demonstrated by the authors for energy market-based optimization [24] and is currently being extended to incorporate grid congestion signals (flexibility markets or dynamic localized grid charges).

4.2 Conclusion

The introduced method is the connection between factory planning and factory operation as well as factory planning and flexibility incentive generation of the energy supply system. Both is needed for a successful change in the energy supply system. The enterprises need to understand their energy flexibilty as a value for the energy supply system and how to manage energy flexibility in the operation of their facility. In conclusion, the combination of a relocation-coefficient-based energy flexibility capability optimization of factories, and digital-twin-enabled operational flexibility optimization based on market forecasts, contributes a comprehensive methodology for manufacturing enterprises to adapt to an increasingly volatile energy system. This approach significantly enhances system-wide flexibility, reducing the need for grid expansion, centralized energy storage, and grid stabilization measures. However, successful implementation requires high data availability and quality, as well as advanced digitalization and controlled processes within the enterprise. Furthermore, the currently quickly evolving regulatory landscape for energy flexibility markets must be closely monitored, as clear incentives for decentralized flexibility are critical for widespread adoption.

References

1. Reinhart, G., Reinhardt, S., Graßl, M.: Energieflexible Produktionssysteme, wt Werkstattstechnik online, 622–628 (2012)
2. Unterberger, E. and Glasschröder, J., Reinhart, G.: A systems engineering based method to increase energy flexibility. Procedia CIRP 254–259 (2017). https://doi.org/10.1016/j.procir.2017.03.357
3. Samtleben, S., Schleich, C., Schenk, M.: Identifying energy flexible manufacturing layouts in a light metal foundry. Procedia CIRP **104**, 1589–1594 (2021). https://doi.org/10.1016/j.procir.2021.11.268
4. Wirth, S.: Flexible Fertigungssysteme: Gestaltung und Anwendung in der Teilefertigung. Verl. Technik, Berlin (1989)

5. Nyhuis, P., Wiendahl, H.: Logistische Kennlinien Technical report. Springer, Heidelberg (2012). https://doi.org/10.1007/978-3-540-92839-3
6. Stock, T.: Multi-objective shop floor scheduling using monitored energy data. Procedia CIRP **26**, 510–515 (2015). https://doi.org/10.1016/j.procir.2014.07.178
7. Pfeilsticker, L., Colangelo, E., Sauer, A.: Energy flexibility - a new target dimension in manufacturing system design and operation. Procedia Manuf. **33**, 28–51 (2019). https://doi.org/10.1016/j.promfg.2019.04.008
8. Kabelitz, S., Streckfuß, U.: Energieflexibilität in der Produktionstheorie. ZWF Zeitschrift für wirtschaftlichen Fabrikbetrieb **109**, 43–45 (2014). https://doi.org/10.3139/104.111082
9. Schoonenberg, W.C.H., Farid, A.M.: A dynamic model for the energy management of microgrid-enabled production systems. J. Clean. Prod. **164**, 816–830 (2017). https://doi.org/10.1016/j.jclepro.2017.06.119
10. Dinkelbach, W., Rosenberg, O.: Erfolgs- und umweltorientierte Produktionstheorie. Springer, Heidelberg (2002). https://doi.org/10.1007/978-3-642-97908-8
11. Arlt, M., Neumann, D., Rajagopal, R.: Curtailment contract design for HVAC systems. In: IEEE American Control Conference (2019)
12. Feuerriegel, S., Neumann, D.: Integration scenarios of demand response into electricity markets: load shifting, financial savings and policy implications. Energy Policy **96**, 231–240 (2016). https://doi.org/10.1016/j.enpol.2016.05.050
13. Bertsch, J., et al.: Ausgangsbedingungen für die Vermarktung von Nachfrageflexibilität: Status-Quo-Analyse und Metastudie: Kopernikus-Projekt Synergie (2019)
14. Schneeweiß, C.: Planung 1: Systemanalytische und entscheidungstheoretische Grundlagen. Springer, Heidelberg (1991). https://doi.org/10.1007/978-3-642-58237-0
15. Samtleben, S.: Bewertung von Energieflexibilitätsstrategien für energieintensive Produktionssysteme. Fraunhofer Verlag (2024). https://doi.org/10.24406/publica-4152
16. Bundesnetzagentur: SMARD- Strommarktdaten (2024). smard.de/home/download-center/downloadmarktdaten/:
17. Blarke, M.B.: Towards an intermittency-friendly energy system: comparing electric boilers and heat pumps in distributed cogeneration. Appl. Energy **91**(1), 349–365 (2012). https://doi.org/10.1016/j.apenergy.2011.09.038
18. ThEGA: Energieatlas Thüringen (2023). https://karte.energieatlas-thueringen.de/. Accessed 20 Jan 2023
19. Fraunhofer-Gesellschaft (Hrsg.): Energy Charts (2023). https://energy-charts.info/charts/power/_simulated/chart.html
20. Projektträger Jülich (Hrsg.) ETAL: Entwicklung neuartiger Technologien, Anlagenkomponenten und Logistik in Leichtmetall-Gießereien: Abschlussbericht (2022)
21. Fausto, P., Gunkel, A., Skytte, K., Bergaentzl'e, C. McKenna, R.: Designing taxes and tariffs for electricity systems with complex flexible actors. In: International Conference on the European Energy Market. IEEE Xplore (2019)

22. O'Reilly, R., Cohen, J., Reichl, J.: Achievable load shifting potentials for the European residential sector from 2022–2050. Renew. Sustain. Energy Rev. **189** (2024). https://doi.org/10.1016/j.rser.2023.113959
23. Schenk, M., Wirth, S., Müller, E.: Factory Planning Manual, Situation-Driven Production Facility Planning. Springer, Heidelberg (2014)
24. Sobottka, T., Kamhuber, F., Heinzl, B.: Simulation-based multi-criteria optimization of parallel heat treatment furnaces at a casting manufacturer. JMMP **4**(3), 94 (2020). https://doi.org/10.3390/jmmp4030094

Towards a More Sustainable Manufacturing Environment: A First Proposal on SDG-Based Design Guidelines for Planning an Assembly Line

Jonte Pietsch(✉), Arne Wendt, and Thorsten Schüppstuhl

Institute of Aircraft Production Technology, Hamburg University of Technology TUHH, Denickestr. 17, 21073 Hamburg, Germany

{jonte.pietsch,arne.wendt,schueppstuhl}@tuhh.de

Abstract. Manufacturing systems impact sustainable development among others through resource consumption and emissions, but also through working conditions and economic value creation. While planning and designing of production systems traditionally prioritizes economic efficiency, a more holistic approach incorporating environmental and social sustainability is needed. This research develops comprehensive design principles for more sustainable assembly lines based on their potential contribution to key Sustainable Development Goals. Sustainability criteria are systematically transferred into a framework that supports practical implementation in a general planning process. The resulting framework provides specific guidance for both high-level planning and detailed planning phases of assembly systems. The developed principles are a first attempt to enable companies to balance economic requirements with environmental resource conservation and social considerations.

Keywords: SDGs · sustainability · sustainable manufacturing · circular economy · sustainability criteria · assembly line · manufacturing process

1 The Importance of Sustainable Planning

The transformation towards a sustainable society and economy is a necessity. The vision of a sustainable global community is currently reflected in the UN's 'Agenda 2030' Sustainable Development Goals (SDGs). They establish a universally acknowledged framework of sustainability goals, endorsed by all UN member states [1]. The SDGs are comprised of 17 overarching goals, which are further subdivided into a total of 169 targets. Ensuring their relevance across all domains, the aforementioned objectives encompass a wide range of social, economic and ecological dimensions. Moreover, a total of 231 indicators have been delineated, providing a standardized metric for comparison between countries, regions and sectors. The majority of the SDGs' indicators are of a quantitative nature, with few being qualitative in nature (i.e. gender equality) [2, 3]. The Sustainable Development Goals, as a universally recognized framework with

H. Kohl et al. (Eds.): GCSM 2025, LNME, pp. 49–56, 2026.
https://doi.org/10.1007/978-3-032-21154-5_6

global reach, should also function as a foundation for the consideration of sustainability issues. In particular, as the Sustainable Development Goals address the issue of industrial production directly through the twelfth goal 'Sustainable consumption and production', it is vital that engineering activities be given due consideration within this context [3].

To contribute to these goals, we aim to incorporate the consideration of sustainability criteria into the process of assembly line planning. In this work a planning process of a generic assembly line, and its' methodology according to REFA is considered. Into this framework, additional a priori design principles and evaluation criteria are introduced, as a tool for proactive sustainable design; reducing later iterative efforts in optimization by application of purely retroactive methods. The result of this planning process should yield as sustainable as possible, in accordance with the SDG framework.

2 A Priori Concepts in Sustainable Manufacturing

In order to enhance the sustainability of an assembly line, the following research areas have been identified: 'Sustainable Production' and 'Sustainable Manufacturing'. The following section will present the existing SDG-based approaches in these areas and then proceed to classify their relevance to the planning process.

Stark [4] defines the fields of action in sustainable manufacturing as 'Manufacturing Technologies', 'Product Lifecycles', 'Value Creation Networks' and 'Global Manufacturing Impact'. 'Manufacturing Technologies' thereby directly addresses assembly lines [4]. Considering the planning process as a product development process, approaches from the 'Product Life-cycle' field of action are also relevant. The other two fields of action are of no further relevance and not considered further.

There are two remaining approaches to incorporating sustainability aspects into production: sustainability analysis and sustainability by synthesis. The analysis evaluates the sustainability of existing concepts a posteriori. The synthesis thus enables a priori thinking. In consideration of the objective of incorporating sustainability as early as possible in the planning process, the synthesis approaches based on SDGs are of interest and will be examined below.

Life Cycle Engineering (LCE) is a key approach in the field of sustainable product development. It encompasses a range of engineering methods that can be applied in relation to the Sustainable Development Goals (SDGs). In accordance with the principles of LCE, all engineering activities are required to be subject to the life cycle principle [5]. Consequently, cycles should be incorporated into the conceptualization phase of any developmental endeavour. In light of these concepts and the overarching goals of the SDGs, the EU is pursuing a transition towards a circular economy. This suggests a transition from a linear economic model, in which a linear flow of raw materials culminates directly in the disposal of products. It is imperative that the economic system, and consequently also production as part of this system, is transformed in such a manner that raw material cycles are established [6].

In the field of 'Manufacturing Technologies', the contribution potential of prevailing production processes and methodologies to the Sustainable Development Goals is, in certain instances, pointed out. For instance, the potential for resource savings in the utilization of additive manufacturing is accentuated, and the synergies between the

'Industry 5.0' concept and Sustainable Development Goal 9 are emphasized [7, 8]. In such cases, existing processes and concepts are only applicable to specific parts of the SDGs. However, methodologies are currently under development with the aim of integrating sustainability on a holistic basis. In the domain of factory planning, studies have been conducted to identify the interrelationships between factories and the SDGs [9]. Most notably, the 'Positivfabrik' is a comprehensive concept that was developed on the basis of all SDGs [10]. The study derives properties from the SDGs that characterize a sustainable factory. Consequently, the 'Positivfabrik' is a zero-emission facility, but only in respect to the building itself, its surrounding infrastructure and the organization - no immediately influential factors on the planning of production and assembly systems are considered [10].

A comprehensive review of the literature reveals a lack of adequate approaches. The prevailing focus of LCE-based concepts, such as the circular economy, on resource consumption is inadequate for achieving a balanced cooperation of social, economic and ecological factors [9]. The inadequacy of a priori concepts in manufacturing for addressing the entire SDGs is also evident in other areas. Moreover, there is an absence of assembly line specific concepts and no integration into a planning methodology. Nevertheless, the implementation of design principles would facilitate the process of sustainable transformation.

3 Sustainable Design Principles

In accordance with the condition that the SDGs be evaluated in a holistic manner, a set of criteria was formulated for an assembly line in a preceding work [11]. By focusing on key goals and analysing the relationships between influenceable attributes of an abstract assembly line and SDG-derived influence factors, a set of sustainability criteria was developed. These criteria cover the areas 'economic efficiency', 'Resources', 'Workplaces' and 'Providing sustainability relevant data'. They can be used in two ways: First, they can be used as a starting point for sustainability analysis. Second, they can be used for sustainable development through synthesis [11]. In this present work, the previously developed criteria are to be incorporated into a general planning process to be able to integrate sustainability a priori.

3.1 A Planning Process According to REFA

The planning methodology according to REFA divides the process into six stages: analysing the initial situation, detailing the planning task, high-level planning of the production system, detailed planning of the production system, implementing the system and operating the system [12]. In the first stage, the initial situation is analysed. The second step involves the application of analysis to the development of objectives and criteria, which are subsequently categorized and weighted. In stage three, the system is roughly designed and the first basic solution concepts are developed. From these variants, the favoured one is selected and detailed in stage four. Stage four, with detailed planning, specifies the individual subsystems from a technical and ergonomic point of view so that key figures such as personnel requirements can be derived. A schedule for introducing the system is also drawn up. Stage five, system implementation, includes procurement, installation, and other measures such as employee training to realize the planned system. Stage six, system operation, comes at the end of the planning process [12]. This process is illustrated schematically in Fig. 1.

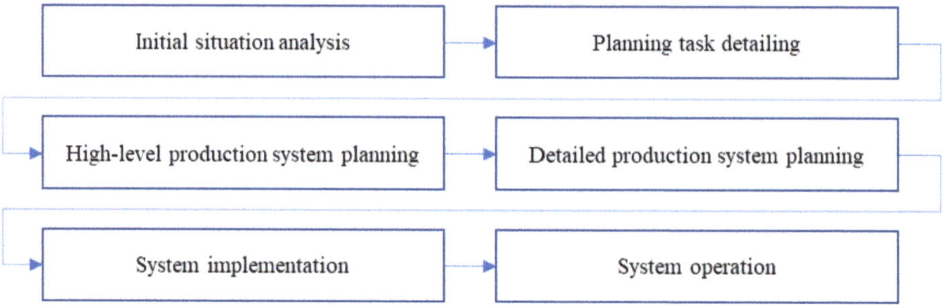

Fig. 1. Schematic planning process according to REFA [12]

In the context of the objective to develop the end product 'assembly line' sustainably, the question arises as to which part of the assembly line is influenced in which of the steps. In the initial stage, the existing situation is analysed without the formulation of decisions pertaining to the assembly line, thereby precluding the possibility of influencing sustainability. Conversely, in stage 2, the line is influenced by the objectives of the project that have been defined. At this point, the criteria that have been developed previously can be utilized. However, the actual design of the assembly line is undertaken in stages 3 and 4. The integration of sustainability considerations is imperative in this context, and as such, design principles are developed for these phases. It is evident that no substantial alterations can or should be made to the line following Phase 4. Consequently, these phases are deemed unimportant in terms of the sustainability of the end product.

3.2 Design Principles for High-Level Planning

In the REFA planning scheme, the high-level planning phase involves the generation and analysis of concepts through the integration of potential designs. This encompasses the identification of work processes, the preliminary design of the work system with respect to performance indicators, the estimation of stresses and hazards, the evaluation of working conditions, and the clarification of capacity requirements.

The configuration of work processes and the work system can be tailored during the phase of assembly planning, thereby exerting a degree of influence on the processes, the process chaining, and the requisite machinery and work equipment. However, the technical detailing is only incorporated into the detailed planning of the assembly system. At this time, the primary focus is on addressing the higher-level criteria. Moreover, it is only possible to make a recommendation for the majority of criteria if one is in possession of precise information regarding the work task and the product. Consequently, this section will solely address the design principles for a selection of the previous identified sustainability criteria and their subject areas. The issue of economic efficiency is already the focus of extant planning methodologies and is therefore included, but not further detailed. The provision of sustainability relevant information can only be addressed in the technical detailing stage. The emphasis is placed on a selection of processes and machine principles that conserve resources, in addition to the sustainable conceptualization of workplaces. The relevant criteria from the subject areas are checked for addressability, supplemented with background information from concepts like circular economy and converted into recommendations for action. Finally, the derived recommendations are summarized and aggregated. This results in the following list of principals for an assembly line for the high-level planning phase (Table 1):

Table 1. Design principles for high-level planning

Subject area	Criteria
Economic efficiency	1. Consideration of the existing design principles
Resources	2. Low raw material quantity of the work systems 3. Process selection according to 3.1 Minimized raw material consumption 3.2 Minimum energy requirement 3.3 Minimum amount of waste 3.4 Minimal waste heat 3.5 Reversibility 4. Select machine concepts according to, 4.1 Minimized raw material consumption 4.2 Minimum energy requirement 4.3 Minimum amount of waste 4.4 Minimal waste heat
Workplaces	5. Appropriate proportion of manual labour 6. Automation preferably from 6.1 Monotonous processes 6.2 Complex tasks 6.3 Physically demanding tasks 7. Process and machine concept selection according to minimum hazard potential

3.3 Design Principles for Detailed Planning

In accordance with the REFA planning scheme, the favoured solution variant is detailed after the high-level planning. This includes the final definition of the work tasks, the specification of the work systems and the layout, as well as the design of the working environment in terms of lighting, air conditioning and noise protection. At this point, the selected processes and process chains are finalized, and the basic concepts of the work systems have also been selected.

Table 2. Design principles for detailed planning

Subject area	Criteria
Economic efficiency	1. Consideration of the existing design principles
Resources	2. Precise dimensioning of the equipment 3. Check for use of second-hand equipment 4. Examination of the rationalization potential of facilities 5. Minimizing the amount of 5.1 Operating resources 5.2 Auxiliaries means 5.3 Waste 5.4 Hazardous substances 6. Select substances according to their environmental impact 7. Provide recuperation of energy and water 8. Use of electric actuators 9. Energy-efficient lighting design through 9.1 Energy-efficient technology 9.2 Minimum lighting equipment 9.3 Adaptive lighting as required 10. Consider greenhouse gas emissions 10.1 in operation 10.2 of the production of the equipment
Workplaces	11. Provide assistance systems 12. Ergonomic design according to current recommendations 13. Increase accessibility through 13.1 Increasing physical accessibility 13.2 Easy language and sign language 14. Minimization of hazards through constructive means
Providing sustainability relevant data	15. Select data to be recorded, at least 15.1 Energy demand 15.2 Resource consumption 15.3 Emissions 16. Provide suitable sensors and observe the digital data infrastructure 17. Provide information system with production-related sustainability data

High-level planning can influence some criteria such as the consumption of raw materials. However, the detailed planning enables a considerable influence on the final key figures through the technical design of the work systems. By deciding on the work systems, criteria such as energy requirements or raw material consumption can be influenced once again. It is also possible to address criteria relating to the provision of sustainability information or accessibility of workplaces during detailed planning. Influences on economic efficiency are not considered in more detail here, as current planning projects are already based on financial considerations at this stage. Similar to the design principles for high-level planning, the criteria from the previous work are also taken as a basis here. They are again checked for addressability, enriched with background information and converted into recommendations for action, which in turn are summarized and aggregated to result in the following list of design principles (Table 2):

4 Discussion of Results

The design principles for the planning process that have been presented are based on the SDGs, and thus on a framework that has been recognized on a global scale. The integration of the SDGs into engineering tasks is supported, with these tasks being directly addressed in the target definition of the SDGs. Nevertheless, it should be emphasized that relatively generic guidelines have been drawn up due to the very generic initial situation and the consideration of an abstract assembly system. Depending on the application, additional factors must be taken into account and included accordingly in the sustainability considerations. In addition, a validation would have to be carried out in order to be able to assess the extent to which a more sustainable planning result emerges from the planning process.

5 Conclusion

The present study aims to address the discrepancy between the theoretical objectives of sustainability and their practical implementation within manufacturing systems. The developed principles represent a preliminary endeavour to empower companies to make informed decisions that balance economic requirements with environmental resource conservation and social workplace considerations. As an a priori approach, it is possible to consider the necessary sustainability aspects from the outset. However, planning the assembly line without taking into account other circumstances such as company policy is only a small part of the necessary transformation. In order to achieve the required sustainable production environment in the future, a broader scope of consideration is required in further engineering activities.

Acknowledgments.

Supported by: Federal Ministry for Economic Affairs and Climate Action on the basis of a decision by the German Bundestag

This work was supported by the LuFo VI-3 project iPREFER – innovative Processes Enabling Fuel cEll Ramp-up (20M2219D) funded by the Federal Ministry for Economic Affairs and Climate Action (BMWK) based on the decision by the German Bundestag.

References

1. UN - United Nations: Transforming our world: The 2030 Agenda for Sustainable Development (2015)
2. Die Bundesregierung, Die 17 globalen Nachhaltigkeitsziele verständlich erklärt. https://www.bundesregierung.de/breg-de/schwerpunkte-derbundesregierung/nachhaltigkeitspolitik/nachhaltigkeitsziele-erklaert-232174. Accessed 24 Jan 2025
3. Statistisches Bundesamt, Indicators of the German Sustainable Development Strategy, https://dns-indikatoren.de/en/. Accessed 24 Jan 2025
4. Stark, R., Seliger, G., Bonvoisin, J.: Sustainable Manufacturing - Challenges, Solutions and Implementation Perspectives. Springer, Cham (2017)
5. Laurent, A., et al.: The role of life cycle engineering (LCE) in meeting the sustainable development goals – report from a consultation of LCE experts. J. Clean. Prod. **230**, 378–382 (2019)
6. Umweltbundesamt: Leitsätze einer Kreislaufwirtschaft, Dessau-Roßler (2020)
7. Muth, J., Klunker, A., Völlmecke, C.: Putting 3D printing to good use—additive manufacturing and the sustainable development goals. Front. Sustainabil. **4**, 1196228 (2023)
8. Costa, E.: Industry 5.0 and SDG 9: a symbiotic dance towards sustainable transformation. Sustain. Earth Rev. **7**(1), 4 (2024)
9. Juraschek, M., et al.: Urban factories and their potential contribution to the sustainable development of cities. Procedia CIRP **69**, 72–77 (2018)
10. Gebler, M.: Planung und Gestaltung einer Positivfabrik als Beitrag für eine absolute Nachhaltigkeit. Springer Fachmedien Wiesbaden, Wiesbaden (2022)
11. Pietsch, J., Wendt, A., Schüppstuhl, T.: Deriving sustainability criteria for an assembly line. In: Annals of Scientific Society for Assembly, Handling and Industrial Robotics, Berlin (2025)
12. Lotter, B., Wiendahl, H.-P.: Montage in der industriellen Produktion. Springer, Heidelberg (2012)

Production Planning and Control

Opportunity Costs in Production Full Cost Accounting: A Systematic Literature Review

Dennis Hertell[1(✉)], Günther Schuh[1,2], Seth Schmitz[1], and Calvin Kuhn[1]

[1] Laboratory for Machine Tools and Production Engineering (WZL) at RWTH Aachen University, Campus-Boulevard 30, 52074 Aachen, Germany
d.hertell@wzl.rwth-aachen.de

[2] Fraunhofer Institute for Production Technology IPT at RWTH Aachen University, Steinbachstraße 17, 52074 Aachen, Germany

Abstract. Increased product variety driven by shorter lifecycles, individualization, and mass customization causes inefficiencies in production, such as frequent changeovers and higher stock levels. Production full cost accounting supports the identification of inefficiencies and enables the economic sustainability of production by providing transparency across all production related costs. However, conventional production full cost accounting does not properly account for inefficiencies. As product variety increases, inefficiencies rise exponentially, contributing to opportunity costs. Opportunity costs represent revenue losses resulting from nonproductive utilization of production resources. To accurately assess production costs and optimize production resource utilization, opportunity costs due to product variety must be included in production full cost accounting. Neglecting opportunity costs distorts cost assessments, leads to uneconomic portfolio decisions, and reduces production efficiency. For this purpose, this paper conducts a systematic literature review to identify approaches for opportunity costs in production. The findings categorize relevant cost types and outline how opportunity costs are addressed in production full cost accounting. Based on the categorization, novel cost types are proposed to quantify opportunity costs in production full cost accounting, and research gaps are identified. Extending the scope of production full cost accounting contributes to more accurate cost assessments and long-term economic sustainability.

Keywords: Opportunity Costs · Full Cost Accounting · Production Variance

1 Introduction

In recent years shorter product lifecycles and mass customization have led to increasing product variety, causing increasing complexity in production processes [1]. While product variety raises market competitiveness, it also introduces process variety, leading to inefficiencies in production such as frequent changeovers, higher inventory levels, and increased process instability [2]. These inefficiencies generate opportunity costs [3], defined as the forgone value of the next best alternative when a decision is made [5]. This means that the economic impact of any decision is not merely the monetary outlay but also the foregone benefits that could have been realized in comparison

H. Kohl et al. (Eds.): GCSM 2025, LNME, pp. 59–67, 2026.
https://doi.org/10.1007/978-3-032-21154-5_7

to the next best alternative [6]. Within the context of production, where resources are limited, inefficiencies incur opportunity costs of otherwise productive usage of these limited resources. This implies that when resources are idle, underutilized, or directed towards less profitable activities, the opportunity cost is the potential revenue that could have been generated if those same resources had been employed in their most valuable alternative use. Opportunity costs rise exponentially with increasing product variety, as each additional variant disproportionately increases production complexity, setup times, and coordination effort [7]. Opportunity costs are not captured by traditional Full Cost Accounting (FCA) methods [8]. FCA captures and allocates all company-wide costs across relevant functional areas [9]. FCA restricted to the production context is referred to as Production Full Cost Accounting (PFCA). PFCA is a critical tool for achieving economic sustainability, as it provides a clear methodology for evaluating resource allocation efficiency and long-term profitability in production [10]. However, traditional approaches to determine PFCA fail to incorporate opportunity costs.

To address this, this paper conducts a systematic literature review to identify and classify the types of opportunity costs discussed in the production context and to propose their integration into PFCA. For this purpose, Sect. 2 outlines the theoretical background of PFCA. Section 3 presents the methodology of the systematic literature review. Section 4 categorizes the findings and introduces novel cost structure for PFCA. Finally, Sect. 5 provides a summary and outlines directions for future research.

2 Production Full Cost Accounting

FCA is a comprehensive accounting methodology that aims to allocate all direct and indirect costs incurred by a company across all functional areas, like production, administration, research and development, sales, and marketing [9]. Cost categories in FCA include material costs, personnel costs, capital costs, risk-related costs such as unplanned downtimes, service costs such as maintenance, costs for third-party rights such as licensing fees for production technologies, and statutory levies such as environmental charges. These cost categories enable systematic cost allocation and may vary depending on the organizational context [8].

Table 1. PFCA cost categories

Cost categories		
• Material costs • Personal costs	• Service costs • Costs for third-party rights • Statutory levies	• Capital costs • Risk-related costs

PFCA is introduced in this paper as the focus of FCA on the functional area of production that exclusively considers production costs. PFCA is therefore based on the same cost categories shown in Table 1 as FCA. This focus enables the evaluation of economic efficiency of production.

In addition, PFCA cost categories can be differentiated into explicit or implicit costs. Explicit costs involve actual payments, such as wages or rent, and are recorded in financial accounting. Implicit costs as part of opportunity costs reflect the financial value foregone by inefficient usage of internal resources like company-owned production resources, which remain invisible in conventional accounting. Ignoring implicit cost can obscure inefficiencies in resource usage. Including implicit costs in the analysis allows for a more accurate assessment of production efficiency, thus supporting economically sustainable production through improved decision-making. [11].

3 Methodology of the Systematic Literature Review

The systematic literature review is based on the methodology outlined by MOHER et al. [12]. For the systematic literature review the databases Science Direct, Scopus, Web of Science, IEEE and EBSCO are used. These databases are used because they are amongst the largest scientific, English language databases in the field of engineering. To find the relevant literature in the databases, the search string shown in Fig. 1, composed of two main domain subsets, is used. The search criteria are based on this paper's scope and focus on opportunity cost in industrial production and resulting synonyms used in literature. German and English language literature are considered.

Domain subset 1	AND	Domain subset 2
"opportunity cost" OR "implicit cost" OR "shadow cost" OR "complexity cost" OR "indirect cost"		"production" OR "manufacturing" OR "assembly"

Fig. 1. Search string configuration for the systematic literature search

The systematic literature review process, illustrated in Fig. 2, was applied to the 2,990 initially identified articles. In the first step, duplicates were removed. Subsequently, a filter of blacklisted words was applied to titles and abstracts to exclude articles that do not address opportunity costs in the context of industrial production such as articles focusing on agriculture, healthcare, or the energy sector.

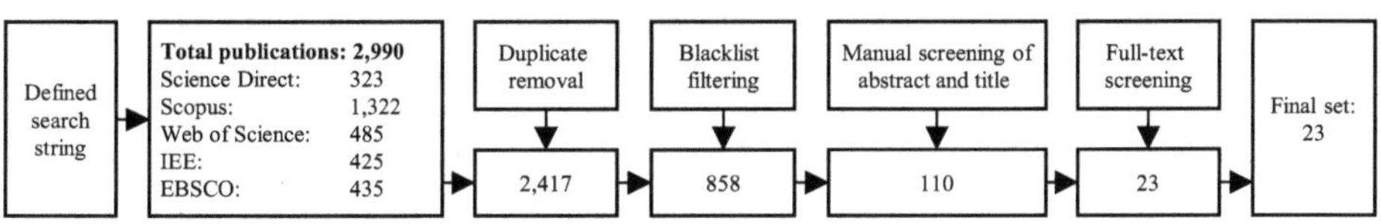

Fig. 2. Number of articles identified in the systematic literature review

The remaining articles underwent a manual title and abstract screening, resulting in 110 potentially relevant publications. A subsequent full-text screening further narrowed the selection, identifying 23 articles describing opportunity costs in production.

4 Results and Discussion

This section presents the findings of the systematic literature review on opportunity costs in production, as summarized in Table 2. A total of 23 relevant articles were identified and analyzed, from which multiple opportunity cost types were extracted and categorized into four overarching groups: capital-related, capacity and resource utilization-related, quality-related, and production logistics-related opportunity costs. Each cost type was evaluated for its current consideration within conventional PFCA as it has been introduced in Sect. 2.

The review reveals that capital-related opportunity costs such as the opportunity costs of capital tied up in inventory or facilities are generally acknowledged in PFCA through the application of imputed interest. These costs are well-documented in the literature and conceptually straightforward to integrate.

In contrast, capacity and resource utilization-related opportunity costs are only partially considered in existing PFCA methodology. While idle capacity and unplanned production downtime are often captured to some extent through overhead or risk related costs, these do not fully reflect the economic losses arising from the non-productive use of scarce resources. Setup and changeover times, likewise, are commonly recorded as explicit setup costs, but the indirect opportunity costs associated with lost throughput are not captured. These costs reflect implicit losses in decision-making contexts and thus require distinct consideration. Quality-related opportunity costs, such as those arising from scrap or rework, are partially included in risk-related costs. However, these costs are rarely treated as opportunity costs and are typically limited to explicit failure costs, neglecting the potential lost value of foregone production or revenue due to reduced product yield. Production logistics-related opportunity costs are the most underrepresented category in PFCA. Although costs such as losses from delayed delivery, production backlogs, or unfulfilled orders can significantly affect customer satisfaction and revenue, they are typically excluded from conventional cost accounting. Their omission represents a critical gap in capturing true economic impact of inefficiencies in production. To address the limitations identified in PFCA the cost structure described in Fig. 3 is proposed. The new cost structure systematically integrates opportunity costs into PFCA by distinguishing between costs for financial application and costs for decision application as two overarching categories.

The separation into financial and decision application reflects the practical distinction in their measurability and their relevance for different stakeholders. Explicit costs form the basis for calculating financial performance and reporting results, ensuring compatibility with accounting standards [9]. Although capital costs and risk-related costs are conceptually derived from opportunity costs, they remain relevant within accounting standards and are thus included in the financial application.

The additional implicit costs under decision application represent the economic value of foregone alternatives due to suboptimal resource utilization within the production system. These include opportunity costs associated with inefficient capacity and resource utilization, quality-related opportunity costs, and production logistics related opportunity costs. These implicit cost categories are based on the previously identified opportunity cost categories from the systematic literature review. Although these implicit cost categories do not appear in conventional accounting records, they are essential for capturing

Table 2. Identified opportunity cost types in production

Category	Opportunity cost type	Considered in PFCA
Capital related opportunity costs	Opportunity costs of capital tied up in inventory [3, 13–19]	Considered via imputed interest
	Opportunity costs of holding excess cash [20]	Considered via imputed interest
	Opportunity cost of capital tied up in buildings and storage facilities [18]	Considered via imputed interest
Capacity and Resource Utilization related opportunity costs	Opportunity costs of idle resource capacity [21–24]	Partially considered as idle costs
	Opportunity costs of production unplanned downtime [22, 25]	Partially considered as downtime cost allocation
	Opportunity costs of time of setup/changeovers [3, 26, 27]	Partially considered as setup costs
Quality related opportunity costs	Opportunity costs due to scrap [3, 28]	Partially considered as quality costs
	Opportunity costs related quality ensuring costs [19, 28, 29]	Partially considered as quality management costs
Production logistics related opportunity costs	Opportunity costs related to delivery performance [15, 28]	Not considered
	Opportunity cost due to reduced production speed [27]	Not considered
	Opportunity costs of products overproduced [16]	Not considered
	Opportunity costs of orders not produced [6, 13, 21, 22, 30–33]	Not considered

the true economic impact of production decisions and play a critical role in managerial decision-making [11].

Financial application		Decision application
Explicit Cost categories	Implicit Cost categories	Implicit Cost categories
• Material costs • Personal costs • Service costs • Costs for third-party rights • Statutory levies	• Capital costs • Risk-related costs	• Capacity and resource utilization related opportunity costs • Quality related opportunity costs • Production logistics related opportunity costs

Fig. 3. Proposed new cost structure for PFCA

The advantage of this structure lies in its ability to accommodate the differing informational needs of financial accounting and management. From a financial perspective, the cost structure and methodology remain unchanged. From a managerial perspective, the inclusion of implicit costs under decision application within the PFCA framework uncovers hidden inefficiencies and enhances transparency regarding the economic consequences of production-related decisions. This division allows the same cost structure to serve both financial accounting and value-oriented management decision-making. By systematically incorporating opportunity costs, the proposed cost structure enables a comprehensive PFCA that supports more informed production planning, improved product portfolio decisions, and contributes to long-term economic sustainability.

5 Summary and Outlook

This paper examines the integration of opportunity costs into Production Full Cost Accounting (PFCA). The conducted systematic literature review reveals that conventional approaches for PFCA fails to account for certain types of opportunity costs. While explicit costs such as material costs, labor costs, or overhead costs are well captured, implicit costs caused by idle capacity, changeovers, or delays are not considered in PFCA. Therefore, a new PFCA cost structure that incorporates opportunity costs is proposed. The proposed structure differentiates between financial and decision application, ensuring its utilization in both financial accounting and managerial decision making. The classification clarifies implicit cost categories and links them to decision application.

In future research, detailing the implicit cost categories for decision related application will be further explored. The calculation and detailing of the implicit cost categories will rely on simulation methods to provide deeper insights. In addition, the practical relevance needs to be validated through empirical studies. Future research will aim to develop a practicable model by linking the estimation of opportunity costs with existing fundamental work on production cost accounting. Such advancements are crucial for establishing a PFCA system that reflects not only the financial but the true economic impact on production, thereby enabling economically sustainable production strategies.

Acknowledgement. Funded by the German Federal Ministry for Economic Affairs and Climate Action within the research project "TrendAuto2030plus".

References

1. Modrak, V., Soltysova, Z.: Assessment of product variety complexity. Entropy (Basel, Switzerland) (2023). https://doi.org/10.3390/e25010119
2. Benjaafar, S., Kim, J.S., Vishwanadham, N.: On the effect of product variety in production–inventory systems. Ann. Oper. Res. **126**(1), 71–101 (2004). https://doi.org/10.1023/B:ANOR.0000012276.87248.c7
3. Myrodia, A., Hvam, L., Sandrin, E., Forza, C., Haug, A.: Identifying variety-induced complexity cost factors in manufacturing companies and their impact on product profitability. J. Manuf. Syst. **60**, 373–391 (2021). https://doi.org/10.1016/j.jmsy.2021.04.017
4. Schuh, G.: Produktkomplexität managen. Strategien - Methoden - Tools, 2nd edn. Hanser, München, Wien (2005)
5. Eatwell, J., Milgate, M., Newman, P. (eds.): The World of Economics. Palgrave Macmillan, London (1991)
6. Guo, X.: A study of production planning based on the linear programming method. In: Garengo, P. (ed.) 2020 9th International Conference on Industrial Technology and Management. ICITM 2020, Oxford, United Kingdom, 11–13 February 2020. IEEE, Piscataway (2020)
7. Salvador, F., Forza, C., Rungtusanatham, M.: Modularity, product variety, production volume, and component sourcing: theorizing beyond generic prescriptions. J. Ops. Manag. **20**, 549–575 (2002). https://doi.org/10.1016/S0272-6963(02)00027-X
8. Coenenberg, A.G., Fischer, T.M., Günther, T.: Kostenrechnung und Kostenanalyse, 9th edn. Schäffer-Poeschel Verlag, Stuttgart (2016)
9. Ernst, C., Schenk, G., Schuster, P.: Kostenrechnung klipp & klar. Springer, Heidelberg (2017)
10. Domanski, C.: Cost engineering. A practical method for sustainable profit generation in manufacturing. CRC Press, Boca Raton (2020)
11. Drury, C.: Management and cost accounting. Cengage, Australia, Brazil, Canada, Mexico, Singapore, United Kingdom, United States, Andover, Hampshire (2021)
12. Moher, D., Liberati, A., Tetzlaff, J., Altman, D.G.: Preferred reporting items for systematic reviews and meta-analyses: the PRISMA statement. PLoS Med. (2009). https://doi.org/10.1371/journal.pmed.1000097
13. Akcali, E., Bayindir, Z.P.: Analyzing the effects of inventory cost setting rules in a disassembly and recovery environment. Int. J. Prod. Res. **46**(1), 267–288 (2008). https://doi.org/10.1080/00207540600710962
14. Raturi, A.S., Singhal, V.R.: Estimating the opportunity cost of capital for inventory investments. Omega **18**(4), 407–413 (1990). https://doi.org/10.1016/0305-0483(90)90031-4
15. Corbey, M.: On the availability of "relevant costs" information in operations management. Int. J. Prod. Econ. **35**, 12–129 (1994). https://doi.org/10.1016/0925-5273(94)90072-8
16. Corbey, M., Jansen, R.: The economic lot size and relevant costs. Int. J. Prod. Econ. **30**, 519–530 (1993). https://doi.org/10.1016/0925-5273(93)90117-4
17. Singhal, V.R., Raturi, A.S.: The effect of inventory decisions and parameters on the opportunity cost of capital. J. Oper. Manag. **9**(3), 406–420 (1990). https://doi.org/10.1016/0272-6963(90)90163-8
18. Thonemann, U.W., Brandeau, M.L.: Optimal commonality in component design. Oper. Res. **48**(1), 1–19 (2000). https://doi.org/10.1287/opre.48.1.1.12445
19. Lovejoy, W.S., Sethuraman, K.: Congestion and complexity costs in a plant with fixed resources that strives to make schedule. Manuf. Serv. Oper. Manag. **2**(3), 221–239 (2000). https://doi.org/10.1287/msom.2.3.221.12348
20. Kajjoune, O., Aouam, T., Zouadi, T., Dairi, M.: Dynamic lot-sizing with short-term financing and external deposits for a capital-constrained manufacturer. Int. J. Prod. Econ. **242**, 10828 (2021). https://doi.org/10.1016/j.ijpe.2021.108281

21. Fry, T.D., Philipoom, P.R., Leitch, R.A.: Accounting-based heuristics for order processing and review in a stochastic environment. Int. J. Prod. Res. **46**(6), 1667–1684 (2008). https://doi.org/10.1080/00207540600928531
22. Rad, R.H., Razmi, J., Sangari, M.S., Ebrahimi, Z.F.: A note on the joint economic lot size model with stochastic demand and variable lead time. IJLSM **23**, 76–97 (2016). https://doi.org/10.1504/IJLSM.2016.073298
23. Kohl, H., Seliger, G., Dietrich, F. (eds.): Manufacturing Driving Circular Economy. Lecture Notes in Mechanical Engineering. Springer, Cham (2023)
24. McLaughlin, R., Taggart, R.A.: The opportunity cost of using excess capacity. Finan. Manag. (1992). https://doi.org/10.2307/3665660
25. Park, C.S., Kim, G.T.: An economic evaluation model for advanced manufacturing systems using activity-based costing. J. Manuf. Syst. **14**(6), 439–451 (1995). https://doi.org/10.1016/0278-6125(95)99916-2
26. Selen, W.J., Heuts, R.M.J.: Operational production planning in a chemical manufacturing environment. Eur. J. Oper. Res. **45**(1), 38–46 (1990). https://doi.org/10.1016/0377-2217(90)90154-4
27. Oh, K., Hong, Y.S.: Optimizing product variety for balancing market share and complexity cost in product family design. IEEE Access **12**, 108810–108824 (2024). https://doi.org/10.1109/ACCESS.2024.3438273
28. Soares, J.C., Tereso, A.P., Sousa, S.D.: A decision-making model for the rework of defective products. IJQRM **38**, 68–97 (2021). https://doi.org/10.1108/IJQRM-06-2019-0185
29. Gu, Y., Guo, L., Sun, Q.: The research of quality cost efficiency with considering the implicit cost based on non-parametric analysis. In: 2009 Second International Symposium on Electronic Commerce and Security. 2009 Second International Symposium on Electronic Commerce and Security, Nanchang City, China, 22–24 May 2009, pp. 303–306. IEEE (2009). https://doi.org/10.1109/ISECS.2009.244
30. Mahanty, A.K.: Theory of costs. In: Intermediate Microeconomics with Applications, pp. 211–239. Elsevier (1980)
31. Jensen, O.W.: Opportunity costs: their place in the theory and practice of production. Manag. Decis. Econ. **3**, 48–51 (1982). https://doi.org/10.1002/mde.4090030110
32. Srebranig, S.F., Krishnamurthi, M.: A microeconomic modelling and analysis methodology for production quantity determination. Int. J. Prod. Res. **34**, 1493–1514 (1998). https://doi.org/10.1080/002075498193138
33. Ioannou, G., Sullivan, W.G.: Use of activity-based costing and economic value analysis for the justification of capital investments in automated material handling systems. Int. J. Prod. Res. **37**, 2109–2134 (1999). https://doi.org/10.1080/002075499190932

An Analytical Framework for Literature Reviews on Manufacturing Efficiency Methods Application for Environmental Impact Reduction in Fashion and Textile Industries

Côme Citroën[1(✉)], Andrée-Anne Lemieux[1], and Steve Evans[2]

[1] Institut Français de La Mode, 34 Quai Austerlitz, 75013 Paris, France
ccitroen@ifmparis.fr
[2] Institute for Manufacturing, University of Cambridge, Cambridge CB3 0FS, UK

Abstract. Over the past decades, ever increasing production volumes have made the fashion industry a major contributor to anthropogenic climate change and numerous other types of pollution. As a result, the garment industry has been urged by scientists, lawmakers and consumers to reduce its environmental impact. With the majority of impacts being caused at the production stage, ambitious and rapid environmental gains can be aimed for through the improvement of manufacturing practices. While a large number of manufacturing methods - such as Lean Manufacturing, Six Sigma or Agile Method - have been thoroughly applied and studied in sectors such as automotive and electronics, academic literature on their use in the fashion industry remains relatively lacking, let alone for environmental impact. As a result, a cross-analysis framework was developed, aiming to qualify industrial performance methods in terms of potential environmental impact improvement and their applicability to the fashion industry. Subsequently, the literature review was conducted and showed major gaps in the literature, highlighting opportunities for future intervention research on the implementation of sustainable manufacturing practices in fashion supply chains, including through full implementation of Industry 5.0 in the fashion industry, and the role of worker engagement in sustainability performance.

Keywords: Sustainable development · Manufacturing · Energy efficiency

1 Introduction

In the past decades, manufacturing efficiency methods have radically transformed the way factories operate in certain industries such as automotive or electronics, allowing for massive efficiency gains, partially compensating their considerable usage growth.

In the meantime, the fashion industry has grown at a fast pace, with global fibre consumption quadrupling in absolute and doubling per capita in 50 years [1], but little research on the implementation or potential of manufacturing methods for environmental impact reduction has been conducted.

H. Kohl et al. (Eds.): GCSM 2025, LNME, pp. 68–76, 2026.
https://doi.org/10.1007/978-3-032-21154-5_8

This paper aims at establishing a broad view of extant academic literature combining manufacturing and resource efficiency methods, demonstrated sustainability results, and the fashion, textile/garment industry. This is achieved through an identification and comparison process between methods, industries and types of environmental impacts.

2 Methodology

This paper proposes a framework for an analytical literature review based on a set of keywords and Boolean connectors, as recommended by Tranfield et al. [2].

The research methodology of this paper consists in identifying academic papers at the intersection of a set of selected manufacturing methods, industry terms and types of environmental gains defined hereafter.

2.1 Manufacturing Concepts and Methods

While no ultimate classification of the major manufacturing efficiency practices and methods exists, it can be challenging to determine a discrete list of methods to study. This section presents a number of major industrial paradigms from the past and current decades that were selected for study as empirically part of the most prominent.

Lean manufacturing, **also** known as the Toyota Production System, is recognised as the best known and most widely used manufacturing efficiency method [3]. Lean green (or lean-green) is an alternative term when lean is used for environmental impact.

Six Sigma (SS) is a quality management methodology which uses statistical tools to reduce the defect rate in the production. It is typically based on a 5-step Define, Measure, Analyse, Improve, Control (DMAIC) process, and has variants such as Define, Measure, Analyse, Design, Verify (DMADV) [4].

Lean Six Sigma (LSS) is a hybrid method developed in the 2000s combining the approach and tools from lean and the advanced data analysis from six sigma [5].

Agile Manufacturing is a set of manufacturing techniques aimed at robustness, resilience, responsiveness, flexibility and innovation rather than efficiency [6]. It also has a hybrid version with lean, usually called "leagile", which was used as a search term.

Industry 4.0 (I4.0) is a concept denoting the 4th industrial revolution, originating in the 2000s and aiming at radical transformation of industry through smart factories and technologies, connecting machinery, robots and other key digital technologies. [7].

Industry 5.0 (I5.0) separated from I4.0 in 2018, notably adding elements of human-machine interaction, societal-driven industrial transformation and sustainability. The three pillars of I5.0 are Sustainability, Resilience and Human-Centricity. [8].

2.2 Definition of Key Industry Terms

While the terms fashion, textile & garment are often used interchangeably in literature, they have very different meanings and implications when defined strictly.

A strict definition of the fashion industry integrates items that can be worn (including accessories such as handbags), involving different types of materials, textile or not, but it also encompasses a sociological phenomenon beyond practical use through constant renewal, as highlighted by Serdari [9]. More operationally, this implies high rates of change, frequent changes in the products manufactured and with short development times, all of which being potential obstacles to continuous improvement, manufacturing efficiency and more generally, sustainability and operational efficiency [6].

Garments on the other hand, are any item of clothing and therefore are coming from a more practical point of view, and may be composed of basic items of clothing which are produced continuously. They are unlikely to include accessories.

The textile industry finally is both broader and more restricted, as it only includes materials which are fabrics or knits, as opposed to leather for example, but also encompasses many non-clothing items, such as bedsheets, tablecloths, or even baby wipes.

The specificities of fashion compared to other industry denominations, make the fashion industry relevant to study separately from the others, which could have also included the clothing or apparel industry, but the paper was restricted to the fashion industry on one side, and the textile and garment industries on the other.

2.3 Types and Sources of Environmental Impact

As this paper focuses on different types and sources of environmental impact, it is essential to define them and explain why they were selected.

Emissions cover all emissions into the air, including Greenhouse Gases (GHG), particulates and other substances. Climate change and atmospheric pollution are major issues globally and industry directly generates 24% of global GHG emissions [10].

Energy, although not a direct environmental impact, is a source of impact and a frequent focus in terms of manufacturing efficiency. Energy, (mostly electricity) represents a major source of indirect emissions for factories. For reference, energy was responsible for 34% of global GHG emissions in 2019 [10].

Materials, as well as waste, circularity and scrap, are a focus as they are all related to material efficiency, and are important metrics for operational efficiency due to cost and for sustainability as materials, parts and subcomponents often constitute one of the largest sources of environmental impact of a manufacturing business.

Defects, rework and quality were also selected as keywords, since the environmental impact of quality issues is observed as rework which requires additional machine time, and purchasing replacement materials/items are an example of wasted resources.

Water was selected as industry is a major water user with factories coming under increasing regulatory pressure to reduce their water consumption. This tendency is likely to intensify, as climate change will lead to increased water scarcity [10].

2.4 Search Details

The search was performed by conducting several series of Google Scholar and Science Direct searches, using different structures: the first approach used the following format: "[manufacturing method]" AND"[industry name] industry" AND Sustainability (voluntarily left as a thematic, not strict term). This allowed to have a first set of papers, among which those with relevant titles were selected. If this search resulted in hundreds of results, with most of these not deemed relevant, the search was narrowed down to only papers with the name of the manufacturing method in the title of the paper. Keywords related to each of the selected types of impact were then searched within each of the papers, and retained if relevant findings were identified.

With the goal being to identify the current state of research and gaps, instead of being an exhaustive list of all papers related to the topic, additional searches were conducted when no relevant papers could be identified for a combination of method, industry and type of impact, in which case, the following search format was used: "[manufacturing method]" AND "[industry name] industry" AND "[type of environmental gain]".

Finally, snowballing was used to enlarge the base of relevant papers.

3 Results

The results from the search are summarised in Table 1. Since case studies and application cases necessarily come after some form of theory, be it at least a framework, conjecture or application model, all areas in the table which are marked as having case studies also imply some form of theory.

Table 1. A comprehensive overview of identified literature and gaps from the framework – Blue for areas with case studies, red for only theoretical papers and white for absence of related papers

F: Fashion, G: Garment/Textile

	Emissions		Energy		Materials		Defects		Water	
	F	G	F	G	F	G	F	G	F	G
Lean										
SS										
LSS										
Agile										
I4.0										
I5.0										

For lean manufacturing, both theoretical and practical demonstrations of the potential for impact in the fashion industry were identified by Bilancia et al. [11] in their literature review focusing on 30 papers showing sustainability impacts of a range of lean practices in the luxury fashion industry. Although the types of impacts were not categorised in the paper, the present review identified all of the selected types to be covered. This is also demonstrated by Bilancia et al. [12] through an interview-based case study on two factories. Another case study conducted in Peru shows the different types of impact observed when using lean green in the textile industry, showing impacts across all categories and opportunities for replication. [13].

For six sigma, while numerous papers were found to show improvement on the defects side in the garment industry, including numerous case studies [14–17], no application was found strictly for the fashion industry. However, Melosi et al. [18] introduced a derivative of DMADV which was applied to leather cutting for fashion and achieved some waste reduction. Kurnia et al. [15] also show reduced scrap rate in a socks factory with six sigma, therefore demonstrating this potential for the garment industry. Finally, a Master's thesis from the National University of Science and Technology of Pakistan, presented a set of positive results on emissions, energy, waste and water, through the implementation of six sigma in a textile factory [19].

Similarly, papers on the environmental impact of Lean Six Sigma (LSS) in the fashion industry are absent, while those relevant in textile are concentrated around waste [20–22] and quality issues [20–23], mostly in case studies or literature reviews which are not sustainability-oriented.

Only Lv and Cho [24] were found to link Agile Manufacturing, with textile industry and specific sustainability impacts (quality and carbon emissions), through a case study. This paper also highlights an inherent tension between agility and sustainability. No paper was found to link Agile Manufacturing directly with specific environmental impacts in the fashion industry.

For Industry 4.0, Glogar et al. [25] made a recent comprehensive literature review intersecting the most common practices in Industry 4.0, with sustainability in the textile and fashion industries (used interchangeably). Showing applicability for all selected impact types, this paper also highlights the fact that research on the topic is mostly theoretical through frameworks and requires more full (as opposed to partial or incremental) implementation cases. Additionally, Gazzola et al. [26] conducted a bibliometric study on "digitainability" in the fashion industry, which it classifies both as I4.0 and I5.0, and underlines the possible impact of additive manufacturing technologies on Materials Waste and Emissions reduction. On top of that, Caballero-Morales et al. [7] developed an implementation model for I4.0 deployment in SMEs in the textile sector, which notably shows potential for a reduction in water consumption and waste production in the dyeing process.

Finally, on Industry 5.0, several theoretical papers, with different frameworks and application models, but without any meaningful case study showed potential for sustainable impact in textile and fashion, e.g. Glogar et al. [25] and Gazzola et al. [26]. On the textile side, Al Amin et al. [27] developed an applicative framework for I5.0 in synergy with Green Supply Chain Management for the garment industry in Bangladesh, and demonstrating potential for impact on Emissions, Energy, Waste and Quality.

4 Discussion, Implications for Future Research, Limitations

This non-exhaustive search led to the identification of 22 fully relevant academic papers, covering 20 method/industry/impact intersections with practical and theoretical background, 20 more with solely theoretical findings, and leaving the final 20 unaddressed (Table 1).

This literature review shows the literature combining manufacturing, sustainability and fashion industry is less exhaustive than that related to the textile or garment industry. This may have many different causes, but might include: the fact that the fashion industry is less mature on manufacturing topics than the garment and textile industries, that fashion might not be as much academically studied, or might have a lower potential for manufacturing efficiency due to its high rate of novelty [6]. Another option, which was verified in some papers, is that the terms clothing, garment, textile and fashion are often used interchangeably, therefore rendering impossible this kind of nuance.

Similarly, as shown by the number of impacts/manufacturing methods combinations leading to environmental impact reduction identified in this paper, there is a proven potential for manufacturing efficiency methods to be used in the textile industry for sustainability, while the demonstrated sustainability potential of lean for fashion seems to indicate that a similar pattern applies to the fashion industry.

This shows the need for more research on the use of manufacturing methods for sustainability in the fashion industry, preferably within a strict definition.

A further observation is that some types of environmental impact have been more studied than others. For instance, while quality and waste improvements have been studied for textile or fashion for almost every methodology tested, their potential for impact on water consumption or quality is very rarely researched. As freshwater scarcity is a concern in certain parts of the world, and will become increasingly important in all regions, this topic would gain greatly in being studied more in detail.

Furthermore, the absence of reliable case studies in fashion for Industry 4.0 and 5.0 despite extensive theoretical background indicates a need for field study, notably under the form of transformative research through intervention-research, co-creating knowledge as these new industrial paradigms unroll into factories.

Additionally, the lack of papers involving Agile Manufacturing and Sustainability, as well as some indications from the existing papers, seems to indicate that agility and sustainability are seen either as unrelated or opposed to each other. Since Industry 5.0 has both resilience and sustainability as pillars, and resilience is a key aspect of agility, it shows an opportunity to reconcile this tension and combine both aspects, which is another motivation to perform intervention-research in Industry 5.0. Lean 5.0, as introduced by Fani et al. [28] and first tested in the fashion industry, is worth of study.

Finally, certain papers highlighted the interaction between social sustainability and manufacturing methods, both in terms of the potential for these methods to generate social impact, and the need for worker engagement and competences in order to implement them. Notably, Glogar et al. [25] highlight the critical role of workers skills and knowledge in I4.0 deployment, which implies that implementing I4.0 would require training and therefore would lead to worker empowerment. Similarly, Melosi et al. [18] demonstrates that competence mapping and education can lead to positive results in terms of manufacturing sustainability. While this connects to the work of Karaosman et al. [29], it would be interesting to investigate the interaction between implementation of manufacturing methods, human factors and social sustainability.

The main limitations to this paper are that the search was conducted only on papers in English on two search engines, that many manufacturing methods were not included in the study, that the framework focused on the method level and did not go down to practices, and finally but most importantly, that many practices that happen on the field in the fashion industry do not end up in academic literature for a variety of reasons.

5 Conclusion

A framework for analysing the state of academic knowledge on 6 manufacturing methods (Lean, Six Sigma, Lean Six Sigma, Agile Manufacturing, Industry 4.0 and Industry 5.0), applied to the fashion, textile and garment industries, and 5 types of environmental impacts (emissions, energy, materials, quality and water) was introduced and applied.

The need for research, including intervention-research and case studies for Industry 4.0 and Industry 5.0 in the fashion industry, was established, as well as the need for studying the potential of manufacturing efficiency methods for water consumption reduction in fashion and textile. Relationships between worker engagement, change management and manufacturing method implementation are also worthy of study. Field work and pilot projects are in progress and will be the subject of future papers.

References

1. Textile Exchange, "Materials Market Report 2024," 2025
2. Tranfield, D., Denyer, D., Smart, P.: Towards a methodology for developing evidence-informed management knowledge by means of systematic review. Br. J. Manag. **14**(3), 207–222 (2003)
3. Ohno, T., Bodek, N.: Toyota Production System: Beyond Large-Scale Production, Productivity Press, 2019
4. Montgomery, D., Woodall, W.: An overview of six sigma. Int. Stat. Rev. **76**(3), 329–246 (2008)
5. Jiju, A., Snee, R., Hoerl, R.: Lean six sigma: yesterday, today and tomorrow. Int. J. Qual. Reliab. Manag. **34**(7), 1073–1093 (2017)
6. Lemieux, A.-A., Pellerin, R., Lamouri, S., Carbone, V.: A new analysis framework for agility in the fashion industry. Int. J. Agile Syst. Manag. **5**(2), 175–197 (2012)
7. Caballero-Morales, S.-O., Cuautle-Gutierrez, L., Cordero-Guridi, J.-D.-J., Alvarez-Tamayo, R.-I.: Six-sigma reference model for industry 4.0 implementations in textile SMEs. Sustainability **15**(16) (2023)
8. Van Erp, T., Pereira Carvalho, N., Gerolamo, M., Gonçalves, R., Rytter, N., Gladysz, B.: Industry 5.0: a new strategy framework for sustainability management and beyond. J. Clean. Prod. **461** (2024)
9. Serdari, T.: Defining Fashion, Luxury, and Luxury Fashion. In: Rethinking Luxury Fashion, Palgrave Advances in Luxury. Palgrave Macmillan, Cham, pp. 3–7, 2020
10. IPCC, "Climate Change 2023: Synthesis Report. Contribution of Working Groups I, II and III to the Sixth Assessment Report of the Intergovernmental Panel on Climate Change [Core Writing Team, H. Lee and J. Romero (eds.)]," IPCC, Geneva, Switzerland, 2023
11. Bilancia, A., Costa, F., Staudacher, A.P.: Achieving sustainability and circular economy in the luxury fashion industry through lean practices: a systematic literature review. Comput. Ind. Eng. (2025)
12. Bilancia, A., Costa, F., Staudacher, A.: Sustainable lean practices in the luxury fashion industry: a case study. In: Thürer, M., Riedel, R., von Cieminski, G., Romero, D. (eds.) Advances in Production Management Systems. Production Management Systems for Volatile, Uncertain, Complex, and Ambiguous Environments. APMS 2024. IFIP Advances in Information and Communication Te, 2024
13. De-la-Flor, A., Vigil, M., Ruiz-Ruiz, M.F.: Improvement of the sustainable performance in a textile company using the lean-green methodology. Int. J. Prod. Manag. Eng. **12**(1), 105–116 (2024)

14. Kankariya, P., Valase, K.: Performance improvement in Garment industries by reducing defects using six sigma methodologies. Int. J. Sci. Res. Eng. Technol. (IJSRET) **6**(3), 228–236 (2017)
15. Kurnia, H., Jaqin, C., Purba, H., Setiawan, I.: Implementation of six sigma in the DMAIC approach for quality improvement in the knitting socks industry. Tekstil ve Mühendis **28**(124), 269–278 (2021)
16. Hussain, T., Jamshaid, H., Sohail, A.: Reducing defects in textile weaving by applying Six Sigma methodology: a case study. Int. J. Six Sigma Compet. Adv. **8**(2), 95–104 (2014)
17. Mridha, J., Hasan, S., Shahjalal, M., Ahmed, F.: Implementation of six sigma to minimize defects in sewing section of apparel industry in Bangladesh. Glob. J. Res. Eng. J. **19**(3) (2019)
18. Melosi, F., Campana, G., Cimatti, B.: Competences mapping as a tool to increase sustainability of manufacturing enterprises. Procedia Manuf. **21**, 806–813 (2018)
19. Shah, A.: Improvement in condensate recovery for energy conservation using six sigma methodology in a textile industry, Master's thesis at National University of Sciences and Technology, Islamabad, 2024
20. Kurnia, H., Tumanggor, O., Jaqin, C.: Lean six sigma: literature review and implementation for textile and textile product (TTP) industries. In: 3rd Mercu Buana Conference on Industrial Engineering - MBCIE 2021, Universitas Mercu Buana, 2021
21. Karthi, S., Devadasan, S., Selvaraju, K., Sivaram, N., Sreenisava, C.: Implementation of lean six sigma through ISO 9001:2008 based QMS: a case study in a textile mill. J. Text. Inst. **104**(10), 1089–1100 (2013)
22. Alie, S., Sarjono, H.: Implementation of lean six sigma in textile industry: a systematic literature review. In: Proceedings of the 3rd South American International Industrial Engineering and Operations Management Conference, Asuncion, Paraguay, July 19–21, 2022, 2022
23. Fibriani, L., Garside, A., Amallynda, I.: Lean six sigma approach to improve the production process in the garment company: a case study. J. Eng. Manag. Ind. Syst. **11**(2), 153–168 (2023)
24. Lv, D., Cho, E.: Carbon-zero agility: enabling carbon-zero organizations through agile management and ambiguous feedback algorithms. Environ. Model. Softw. **177** (2024)
25. Glogar, M., Petrak, S., Naglic, M.: Digital technologies in the sustainable design and development of textiles and clothing—a literature review. Sustainability **17**(4) (2025)
26. Gazzola, P., Grechi, D., Iliashenko, I., Pezzetti, R.: The evolution of digitainability in the fashion industry. Kybernetes **53**(13), 101–126 (2024)
27. Al Amin, M., Chakraborty, A., Baldacci, R.: Industry 5.0 and green supply chain management synergy for sustainable development in Bangladeshi RMG industries. Clean. Logist. Supply Chain **14** (2025)
28. Fani, V., Bucci, I., Rossi, M., Bandinelli, R.: Lean and industry 4.0 principles toward industry 5.0: a conceptual framework and empirical insights from fashion industry. J. Manuf. Technol. Manag. **35**(9), 122–141 (2024)
29. Karaosman, H., Brun, A., Morales Alonso, G.: Environmental and social impact-free stance and consequences on the operational performance. J. Text. Eng. Fashion Technol. **1**(4), 153–163 (2017)

Intralogistic Value Creation – Potential of Mobile Assembly Processes

Lukas Christ[1](✉), Lennart Lamers[2], Kai Gündisch[1], and Bernd Kuhlenkötter[1]

[1] Chair of Production Systems, Ruhr-University Bochum, Industriestraße 38C, 44894 Bochum, Germany
christ@lps.rub.de

[2] Production Automation, RIF Institute for Research and Transfer e. V., , Joseph-Von-Fraunhofer-Straße 20, 44227 Dortmund, Germany

Abstract. Current challenges in the industry, like the demand for sustainable production and the shortage of adequately skilled labor, necessitate more resource-conserving and therefore more efficient and sustainable processes and innovations. One promising approach to enhance production efficiency is intralogistic value creation, which involves parallelizing traditionally stationary value-creating production processes and currently non-value-creating intralogistic transport activities. By leveraging mobile robotic systems, such as mobile robot manipulators, it becomes possible to combine transport tasks with value-creating production operations. However, implementing such processes introduces technical challenges, including oscillation behavior, constrained workspaces, limited payload capacities, and compatibility with applicable joining processes.

For this reason, this paper presents an approach for evaluating the potential of assembly processes for mobile applications with mobile robots to implement intralogistic value creation. This potential analysis uses a multi-stage evaluation based on a standardized process description of the applications under the assessment, considering technological and economic aspects. This approach aims to establish a standardized evaluation method that enables a quantifiable and, thus, objectively comparable assessment. This allows only processes with high success potential to be considered for more detailed and resource-intensive feasibility studies, whether in simulations or practical experiments.

Keywords: Production Efficiency · Resource Utilization · Mobile Assembly · Intralogistic Value Creation · Mobile Robot Manipulator

1 Introduction: Intralogistic Value Creation

The persistent and evolving challenges in industrial shortage of skilled labor and cost dynamics necessitate innovative approaches to enhance process efficiency. This is crucial for conserving resources in the context of sustainability, maintaining competitiveness amid cost developments, and ensuring the execution of all essential processes despite the skilled labor shortage. In the industrial sector, this also applies to intralogistic processes.

H. Kohl et al. (Eds.): GCSM 2025, LNME, pp. 77–85, 2026.
https://doi.org/10.1007/978-3-032-21154-5_9

An innovative approach in this regard is intralogistic value creation, which means parallelizing stationary value-creating production processes and non-value-creating intralogistic transport activities [1]. [2] and [3] also discuss the concept of integrating transport, specifically logistics, with value creation. [2] emphasizes its economic attractiveness, while [3] focuses on efficiency improvement through waste reduction, particularly by saving time. The approach of intralogistic value creation can be considered either a component or an extension of Smart Logistics. Smart Logistics integrates logistics activities with the application of cyber-physical systems, such as robotics and the Internet of Things [4].

2 State of the Art: Assembly and Process Evaluation

The transition to more automated assembly presents several challenges. The need for cost minimization for delivering raw materials, transporting unfinished goods, and removing finished products is challenging for the implementation of automated assembly [5]. Further challenges in production occur through the market change of high-volume/low-mix to high-mix/low-volume (HMVL) production [6]. Stationary robots are characterized by low flexibility [5], which are frequently utilized in automation solutions for assembly, is not optimal. Additionally, the mismatch between production and material supply induces waiting times, thus lowering productivity and therefore leading to a low suitability of stationary robots for HMLV productions [7]. Further flexibility is constrained through internal logistics and space [5]. Therefore, a transition from rigid to flexible production concepts is necessary [8]. Autonomous mobile robot manipulators (AMRM) enable a shared working environment between robots and humans for greater flexibility and more efficient usage of the available workspace [9]. Furthermore, AMRM systems could minimize the costs through improved coordination and synchronization of assembly and logistic processes [7, 10]. Based on modular solutions, this can enable flexible and agile automated production processes to produce a diversity of products with a simple setup and short cycle times [10, 11].

Methods for process evaluation are also crucial tools for assessing assembly processes, such as their suitability for automation. The automation-potential analysis aims to evaluate manual assembly processes and subsequently develop possible automation solutions. [12] It can also be used to identify opportunities for rationalization in intralogistic processes. However, the planning and evaluation of assembly concepts are used to evaluate the performance of new assembly concepts [13]. Another evaluation method is PROTEOCE, which determines the most suitable process for automation in an organization [14]. In hybrid assembly, the KOMPI Quick Check should be mentioned, as it enables the evaluation of the prospects of success for implementing human-robot collaboration assembly systems [15]. Each of these methods, along with others in the literature, has been evaluated for its suitability in optimizing or expanding processes in various contexts. However, none of them assess the applicability for mobile assembly applications and thus intralogistic value creation. For this reason, developing a method for evaluating the potential of mobile assembly processes and intralogistic value creation is required, resulting in an innovative approach compared to existing methods.

3 Evaluating the Potential of Assembly for Mobile Applications

The implementation of intralogistic value creation has a high economic potential if this approach is applied to suitable processes. When implementing this, however, both technical challenges and economic aspects must be considered. [16] describes an approach to implement intralogistic value creation through the implementation of mobile production processes using the example of an assembly process. This approach is based on four basic modules: *1. Process Analysis*, *2. Oscillation Analysis*, *3. Oscillation-Adaptive Process Planning* and *4. Practical Commissioning*, illustrated in Fig. 1. Within the process analysis, already automated processes are analyzed concerning their properties, parameters, and suitability for mobile applications in *1.2 Analysis of Considered Process*, which is elaborated upon in the following sections. Afterwards, if the process is suitable, it is broken down into sub-processes in *1.3 Classification in Sub-Processes.*

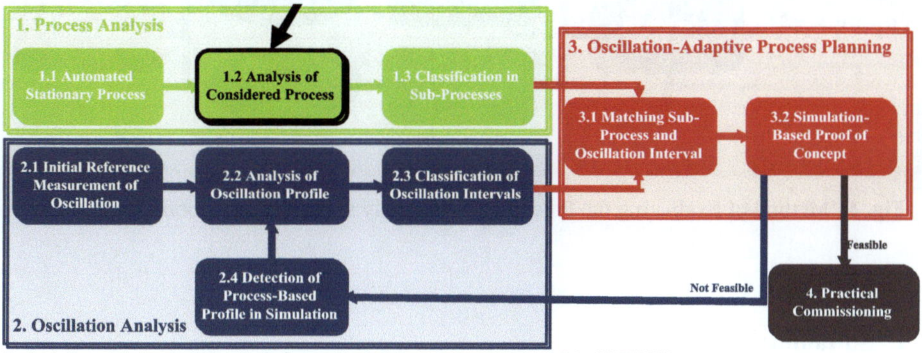

Fig. 1. Oscillation-Adaptive Sequence Planning in Production Processes cf. [16].

The *Method of Evaluating the Potential of Assembly Processes for Mobile Applications* (EPAMA) addresses module *1.2 Analysis of Considered Process*, as highlighted in Fig. 1, focusing on assembly processes. It is intended to serve as the basis for assessing the suitability of processes for selection. The method is a pre-filter for a more detailed feasibility study in module 3.2 *Simulation-Based Proof of Concept*. This method, and consequently the pre-filter, serve to structure and automate the workflow associated with the process analysis. Furthermore, the effort required for detailed feasibility studies—both simulation-based and practical—in subsequent modules is minimized by restricting feasibility assessments to particularly suitable processes through a score-based categorization. The procedure of EPAMA is illustrated in Fig. 2. A standardized process description functions as the foundation for process evaluation. In an initial implementation, this standardized process description is implemented as a template that must be filled out manually, but at the same time generates a CSV file automatically, thus guaranteeing machine readability. This ensures automated data processing for step 2. In this step, the *Suitability Quick Check*, all data relevant to the quick check is automatically extracted from this description. The most suitable robot is selected by an automated parameter and requirement matching based on a database of mobile robot manipulators, which includes systems available on the market. Requirements and parameters that are not met or only partially fulfilled are explicitly highlighted. If the analyzed process passes the *Suitability Quick Check*, it undergoes a more detailed evaluation in the next stage. This evaluation is divided into two distinct assessments: *3.1 Technical Evaluation* and *3.2 Economic Evaluation*. As a result, an evaluation generates a suitability score for implementing the process as a mobile application. Based on this score, an automated decision is made regarding whether to conduct a detailed feasibility study in the fifth step. The following section provides a detailed examination of EPAMA. However, due to the scientific value and space constraints, the focus is placed on the technical analysis, *2. Suitability Quick Check* and *3.1 Technical Evaluation*.

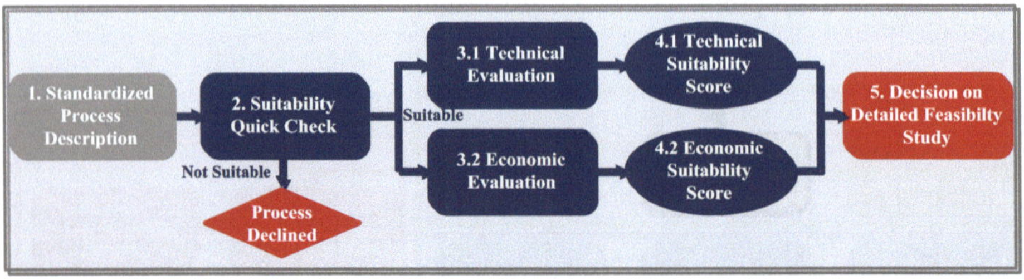

Fig. 2. Method of Evaluating the Potential of Assembly Processes for Mobile Applications

3.1 Suitability Quick Check

Within the evaluation methodology, the quick check serves to compare the mandatory requirements for the process and the mobile robotic manipulator, defined in the standardized process description, with the technical capabilities of current systems stored in a database, as well as the conditions of the application environment. During this comparison, data from the process description is automatically extracted and processed to identify the most suitable system. Various technical criteria are applied in this selection process, which are listed in Table 1.

Table 1. Quick Check Criteria for Process Evaluation.

No.	Quick Check Criteria	No.	Quick Check Criteria
1	Position accuracy manipulator	7	Workspace manipulator
2	Position repeatability manipulator	8	Loading area
3	Position accuracy mobile platform	9	Accessibility of the workstations in the plant
4	Position repeatability mobile platform	10	Joining method
5	Payload manipulator	11	Required safety guidelines
6	Payload mobile platform	-	-

If certain criteria do not meet the quantified requirements specified in the process description, this discrepancy is highlighted in the output of the most suitable system. An expert decision is required to determine the next steps in such cases. The process may either be fundamentally rejected or, if the deviations are minor, additional technical solutions with an acceptable level of effort may be implemented to fulfill the requirements. However, if a predefined quantitative quick check threshold for unmet requirements or a qualitative quick check threshold for a single unmet requirement is exceeded, the process is automatically excluded from further consideration.

3.2 Technical Evaluation of EPAMA

The technical evaluation is conducted as step 3.1, following a Quick Check that classifies a process as suitable. This evaluation involves a detailed analysis of the process to assess its suitability for transferring the stationary automated process to a mobile system. For this purpose, 35 specifically developed criteria are applied. These criteria encompass the relevant parameters for process adaptability while also considering the unique requirements of a mobile process, which arise due to mobility and its associated influencing factors. The 35 criteria are categorized into six groups: Platform, Manipulator, Material, Finished Parts, Process, and Order. Each criterion is evaluated based on either quantifiable data extracted from the standardized process description or a qualitative expert assessment. The suitability score ranges from 1 to 5, where "1" indicates non-suitability and "5" represents full suitability. To enhance the objectivity of expert evaluations, comparative examples are provided for each criterion and score during the assessment process. For instance, in the case of criterion 12, *Robustness of the end effector against errors caused by oscillation*, five different end effector examples are presented, each categorized based on its robustness against oscillation. This allows experts to assess relative, ensuring a more objective and consistent evaluation. The technical criteria used for evaluating EPAMA are presented in Table 2. Since the criteria groups and individual criteria do not all hold the same level of importance for evaluating the process transferability, a two-stage pairwise comparison was conducted by 14 robotic experts. The weighting of the criteria is derived from the average of the expert evaluations. In the first stage, the criteria groups were compared against each other, followed by comparing individual criteria within each group. The final weighting of each criterion is determined by multiplying the weighting of its respective criterion group by the weighting of the individual criterion, presented in Table 3.

A normalized suitability score is calculated based on evaluating the criteria and their respective weightings. This score is categorized into predefined ranges: unsuitable, conditionally suitable, and suitable. If a process is rated as not suitable in the technical evaluation, it is excluded from further consideration, and the result of the economic evaluation no longer influences the decision in Step 5 regarding whether a detailed feasibility study will be conducted. However, if the process is classified as conditionally suitable or suitable, the technical suitability score contributes to the final decision on the feasibility study. If the economic evaluation determines that the process is not suitable, the same principle applies as in the case of a technical non-suitability, leading to the exclusion of the process from further assessment. The categorization thresholds, unsuitable, conditionally suitable, and suitable, and the quick check thresholds are iteratively updated based on the results of feasibility studies. This includes processes assessed as conditionally suitable and suitable, as well as samples of processes classified as unsuitable. This enables an information feedback loop, ensuring continuous optimization of the evaluation method.

Table 2. Technical Criteria for Process Evaluation.

Group	No.	Technical Criteria
Platform	**1**	Implementability of the required area for the required amount of finished parts
	2	Implementability of the required area for the required amount of material
	3	Simplicity of the safety concept for the components to be transported
	4	Robustness against oscillation excitation from the ground
Manipulator	**5**	Adaptability of the complete manipulation concept (movement sequences)
	6	Adaptability of parts of the manipulation concept (movement sequences)
		Simplicity of - **7** accessibility of the poses, **8** handling the material, **9** handling the finished parts
	10	Adaptability of the complete end effector development
	11	Adaptability of parts of the end effector development
		Robustness of the end effector against – **12** oscillation, **13** acceleration
	14	Optionality of an end effector change
	15	Simplicity of an end effector change
	16	Autonomy of an end effector change
Material	**17**	Homogeneity of the materials to be assembled
	18	Simplicity of material supply
	19	Adaptability of the material supply concepts
		Robustness of the material against – **20** oscillation, **21** acceleration
Finish parts	**22**	Simplicity of finished parts storage
	23	Adaptability of the finished parts storage concept
		Robustness of the finished part against – **24** oscillation, **25** acceleration
Process		Adaptability of – **26** process monitoring concept, **27** communication concept, **28** technical availability, **29** organizational availability, **30** quality performance, **31** set-up time, **32** calibration frequency, **33** adjustment frequency
Order	**34**	Optionality of the variety of variants
	35	Implementability of the variety of variants

Table 3. Weighting of the criteria determined by the pairwise comparison of robotic experts.

No	Weight	No	Weight	No	Weight	No	Weight	No	Weight	No	Weight
1	5,34%	7	2,76%	13	3,14%	19	1,93%	25	5,04%	31	1,88%
2	**6,15%**	8	2,43%	14	1,73%	20	4,14%	26	2,15%	32	1,33%
3	3,89%	9	2,36%	15	1,35%	21	3,55%	27	1,55%	33	1,22%
4	**6,05%**	10	1,72%	16	1,81%	22	2,58%	28	2,75%	34	2,26%
5	2,00%	11	1,71%	17	2,08%	23	1,37%	29	2,13%	35	4,64%
6	2,08%	12	3,35%	18	2,58%	**24**	**5,77%**	30	3,18%	Σ	100%

4 Conclusion and Outlook

To address current industry challenges, innovative approaches are needed that not only enhance the efficiency and economic viability of production systems but also advance sustainability and resource conservation. Therefore, this paper presents an approach that serves as a key component for efficiently implementing intralogistic value creation and thus addressing the sustainable development goals nine (*industry, innovation and infrastructure*). The evaluation method EPAMA assesses automated assembly processes in terms of their suitability for mobile applications, enabling a preselection before implementing significantly more resource-intensive measures, such as simulation-based and practical feasibility studies. This ensures that only processes with sufficient suitability undergo resource-intensive examination. EPAMA itself is also a two-stage procedure designed to filter out processes that do not meet the fundamental requirements for mobile implementation before conducting a detailed technical and economic evaluation. The 35 criteria developed for the technical assessment have been weighted by experts, with criteria 2, 4, and 24 having the most significant impact on technical process suitability. Consequently, the robustness of the system and its components, and sufficient space for materials, are the most influential factors in determining technical process suitability. Due to the significant potential of intralogistic value creation, it is advisable to develop EPAMA further and enhance its predictive accuracy. This requires a comprehensive analysis of various suitable, conditionally suitable, and unsuitable processes to establish a broad data foundation for optimizing categorization thresholds and evaluation criteria. Furthermore, expanding and validating the method for industrial processes in collaboration with companies is recommended to refine both the technical and, particularly, the economic assessment.

Acknowledgement. This work was supported by the Federal Ministry of Research, Technology and Space (BMFTR) under the grant number 03HY114B within the research project H_2Giga – HyPLANT100 and under the grant number 03HY113B within the research project H_2Giga – FertiRob.

References

1. Christ, L., Schachtsiek, J., Bier, A., Kuhlenkötter, B., Glogowski, P.: Manipulator vibration of mobile robots in the context of intralogistic value creation. Procedia CIRP **130**, 942–951 (2024). https://doi.org/10.1016/j.procir.2024.10.189
2. Unger, H., Markert, T., Müller, E.: Evaluation of use cases of autonomous mobile robots in factory environments. Procedia Manuf. **17**, 254–261 (2018). https://doi.org/10.1016/j.promfg.2018.10.044
3. Siegert, J., Schlegel, T., Bauernhansl, T.: Matrix fusion factory. Procedia Manuf. **23**, 177–182 (2018). https://doi.org/10.1016/j.promfg.2018.04.013
4. Verma, P., Dixit, V., Kushwaha, J.: Risk and resilience analysis for industry 4.0 in achieving the goals of smart logistics: an overview. In: Proceedings of the International Conference on Industrial Engineering and Operations Management, pp. 151–157 (2020)
5. Michalos, G., Makris, S., Papakostas, N., Mourtzis, D., Chryssolouris, G.: Automotive assembly technologies review: challenges and outlook for a flexible and adaptive approach. CIRP J. Manuf. Sci. Technol. **2**(2), 81–91 (2010). https://doi.org/10.1016/j.cirpj.2009.12.001
6. Gan, Z.L., Musa, S.N., Yap, H.J.: A review of the high-mix, low-volume manufacturing industry. Appl. Sci. **13**(3) (2023). https://doi.org/10.3390/app13031687
7. Li, M., Huang, G.Q.: Production-intralogistics synchronization of industry 4.0 flexible assembly lines under graduation intelligent manufacturing system. Int. J. Prod. Econ. **241**, 108272 (2021). https://doi.org/10.1016/j.ijpe.2021.108272
8. Costa, R., Silva, F., Campilho, R.: A novel concept of agile assembly machine for sets applied in the automotive industry. Int. J. Adv. Manuf. Technol. **91**(9–12), 4043–4054 (2017). https://doi.org/10.1007/s00170-017-0109-4
9. Madsen, O., et al.: Integration of mobile manipulators in an industrial production. Ind. Robot: Int. J. **42**(1), 11–18 (2015). https://doi.org/10.1108/IR-09-2014-0390
10. Ghodsian, N., Benfriha, K., Olabi, A, Gopinath, V., Arnou, A.: Mobile manipulators in industry 4.0: a review of developments for industrial applications. Sensors (Basel) **23**(19) (2023). https://doi.org/10.3390/s23198026
11. Hvilshøj, M., Bøgh, S., Skov Nielsen, O., Madsen, O.: Autonomous industrial mobile manipulation (AIMM): past, present and future. Ind. Robot: Int. J. **39**(2), 120–135 (2012). https://doi.org/10.1108/01439911211201582
12. Neb, A.: Automatisierungspotenzialen auf der Spur. Fraunhofer IPA Interaktiv (2021/3):33–34 (2022)
13. Weidner, R.: Wissensbasierte Planung und Bearbeitung von Montagesystemen in der Luftfahrtindustrie. Dissertation, Universität der Bundeswehr Hamburg (2014)
14. Nguyen, T., et al.: Automation? Yes … but where to begin? In: 2021 IEEE International Conference on Industrial Engineering and Engineering Management (IEEM), pp. 435–441 (2021)
15. Hees, F., Müller-Abdelrazeq, S.L., Langer, T., Voss, M., Schmitt, R., Hüttemann, G. (Hrsg): Projektatlas Kompetenz Montage. Kollaborativ und wandlungsfähig. RWTH Aachen University Lehrstuhl für Informationsmanagement im Maschinenbau, Aachen (2019)
16. Christ, L., Krieger, S., Kuhlenkötter, B.: Oscillation-adaptive sequence planning in production processes for mobile robots. In: 2025 11th International Conference on Control, Automation and Robotics (ICCAR). IEEE (2025)

Impact of Path Planning on Energy Use and Emissions in Four Way Shuttle Storage Systems

Marco Ricci(✉), Riccardo Accorsi, Ilaria Battarra, Giacomo Lupi, Riccardo Manzini, and Gabriele Sirri

Department of Industrial Engineering, Alma Mater Studiorum Università di Bologna Viale Risorgimento, 2, 40136, Bologna, Italy
marco.ricci42@unibo.it

Abstract. Automated shuttle-based storage and retrieval systems are widely used in warehouses and distribution centers to improve efficiency, accuracy, and space utilization. A special class is represented by the four-way shuttle-based storage and retrieval system, where handling robots can move in four directions (forward, backward, and sideways) within the racking system.

This study introduces a methodology to assess energy consumption and the emissions of multiple handling robots in a shuttle-based automated storage and retrieval system. The application of a path planning simulation model assigns the actual routing to shuttles and allows for the computation of the corresponding impacts, according to a set of equations. Moreover, the paper presents the application to a realistic industrial environment using a multi-scenario comparative and competitive analysis. This analysis identifies the most critical energy-efficiency leverages in the path planning of multiple handling robots.

Keywords: Environmental sustainability · throughput · deep-lane storage system · logistics · warehousing systems · energy-based · life cycle assessment LCA · CO_2eq

1 Introduction

One of the most significant challenges our society faces today is the growing threat of global warming and climate change. Across all its stages, the industrial sector is one of the most significant contributors to total global emissions [1]. The growing recourse to industrial automation enhances performance and reliability and demands energy-consuming technologies that generate emissions [2].

Internal logistics is crucial throughout industrial and manufacturing processes [3]. Material handling and warehousing activities involve raw materials, components, and final products. Several new solutions have been presented for storage systems in the last few years, including shuttle-based automated storage and retrieval systems (SBS/RS), as described by Battarra et al. [4].

H. Kohl et al. (Eds.): GCSM 2025, LNME, pp. 86–94, 2026.
https://doi.org/10.1007/978-3-032-21154-5_10

SBS/RS comprises a warehouse structure with independent levels, connected by an elevator system that provides vertical movement. Each level has one or more storage and retrieval (S/R) vehicles, called shuttles, that move back and forth along the main double-sided cross aisle.

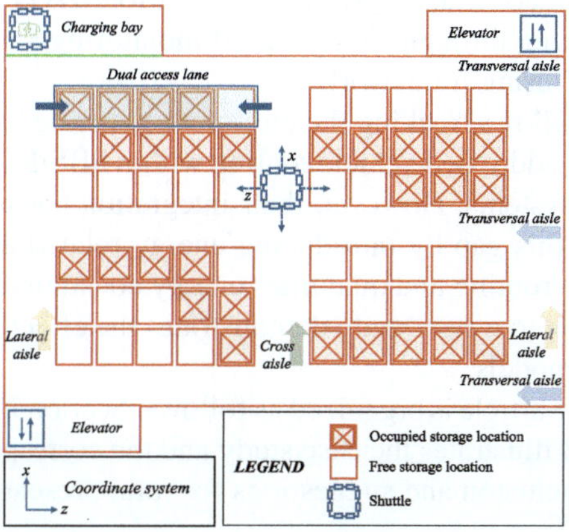

Fig. 1. Four-way shuttle-based AS/RS layout

Four-way shuttle-based storage and retrieval systems (FSS/RS) rely on four-way shuttles, vehicles that can move in four directions (forward, backward, and sideways). These systems represent an innovative and high-efficiency storage solution, enabling high-density storage. Nowadays, they are widely used in intensive storage demand sectors such as food and beverage. Each FSS/RS tier has a main cross-aisle that divides the storage lanes into two sides. This is the main pathway for shuttle movement across the tier, as shown in Fig. 1. Each lane is a channel where one (single depth) or more (multi-depths) unit loads (ULs) can be stored (See Dual access lanes in Fig. 1). Lateral aisles allow dual access to storage lanes. Transversal aisles provide additional passages between cross aisles and lateral aisles. Figure 1 shows that shuttles can move along the lateral or cross-aisles (x-direction), inside the lanes, and along the transversal aisles (z-direction). Additional elements, such as elevators and charging bays, are distributed in the perimeter of the layout. Lateral and transversal aisles facilitate shuttle movements, enabling flexible multidirectional routing and enhancing operational flexibility in FSS/RS [5]. In such a warehouse, S/R tasks can be dynamically assigned and executed across multiple paths, making route planning a pivotal component of system performance.

A common framework for addressing this challenge is the lifelong Multi-Agent Path Finding (MAPF) problem, which coordinates the movements of multiple shuttles, each performing multiple tasks, avoiding conflicts and maximizing throughput. The MAPF approach involves assigning tasks to agents and planning collision-free paths to reach the goals [6]. In this context, Mao et al. [7] apply genetic and A*-based methods to coordinate four-way shuttle systems without conflict.

The MAPF solution enables the movement of shuttles within the tier and permits the completion of all the assigned tasks. Since the shuttles are electrically powered, analyzing individual routes allows for an accurate assessment of energy consumption at both the agent and overall system levels. Previous studies have proposed methodologies to estimate energy-related performance, including analytical models that predict average travel times and energy consumption based on system design parameters [8], and analyses of shuttle velocity and acceleration profiles to identify energy-efficient configurations under throughput constraints [9].In addition, Mourtzis et al.[10] estimate energy consumption in machining through sensor-based monitoring and case-based reasoning within a cloud manufacturing framework.

Energy calculation allows tracking the emissions generated by storage operations.

Recent research has addressed vehicle routing within SBS/RS and the energy-related assessment of storage systems. However, their integration has not yet been addressed. This study aims to fill this gap by introducing energy-related equations into a MAPF algorithm that performs routing and quantifies energy consumption and emissions in an FSS/RS. The application to a case study demonstrates the effectiveness of the proposed approach and adopted models.

The remainder of the article is organized as follows. Section 2 examines the proposed methodology. Section 3 illustrates the case study and the corresponding results. Finally, Sect. 4 presents the conclusion and suggestions for future research.

2 Methodology

This study assesses energy consumption and environmental emissions in an FSS/RS. The proposed methodology is based on a 3-step approach, as illustrated in Fig. 2.

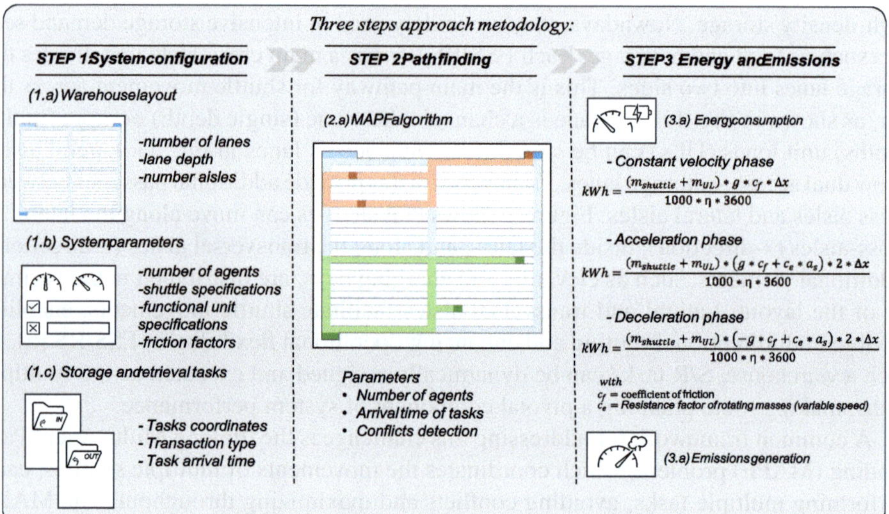

Fig. 2. Three-step approach methodology

The first step is the system configuration and involves defining the warehouse layout and setting key parameters (e.g., warehouse specifications, system settings, and S/R tasks). The second step introduces the MAPF algorithm. The third step evaluates energy consumption and emissions across shuttle motion phases.

Additionally, the proposed approach supports comparative and competitive analyses by tuning key parameters, enabling the exploration of how different system features affect performance.

Each step is described in detail in the following subsections.

2.1 System Configuration

The first step is the system configuration (Step 1 in Fig. 2). It relies on a defined set of features, including the warehouse layout (e.g., number of lanes, lane depth, and number of internal aisles), system parameters (e.g., number of agents, shuttle specifications including speed, acceleration and weight, unit load characteristics and equipment friction factors), and the task set, defined by goal coordinates, task type (e.g., storage or retrieval), and arrival time.

2.2 Multi-Agent Path Finding

The second step of the proposed methodology focuses on computing and assigning a path to each shuttle (Step 2 in Fig. 2). In this context, the MAPF problem is addressed in a grid-based environment. The storage layout is an undirected graph, where nodes correspond to discrete, traversable positions, and edges represent valid movements between adjacent cells. Shuttles navigate the grid by moving to one of the neighboring nodes or waiting in place at discrete time steps. Path planning is handled using the A* algorithm, which balances optimality and computational efficiency. For each shuttle, A* computes the lowest-cost path from start to goal by exploring nodes with the lowest estimated total cost. This cost combines the travel distance from the starting point with a heuristic estimate of the remaining distance to the goal (computed using Manhattan distance). To ensure conflict-free coordination, a structured three-phase collision management strategy is implemented, consisting of: (i) collision detection, identifying potential path intersections via predicted trajectories; (ii) priority assignment, applying rule-based criteria to agents (e.g., task urgency, proximity to destination); and (iii) collision avoidance, dynamically replanning affected paths with A*. This layered approach supports real-time, scalable coordination, enhancing throughput while ensuring system safety. The leverages that allow for generating different and comparative scenarios are:

Number of agents: modification of the operational capacity of the system.
Task arrival time: push option (all tasks available at the beginning of the algorithm computation), pull option (tasks available according to an exponential distribution arriving time)
Conflict resolution: when enabled ensures collision-free, deployable paths. When disabled, it allows potential conflicts, supporting evaluating the system's optimal but not realistic response, which can serve as a benchmark solution.

2.3 Energy and Emission Calculation

The final step consists of a set of equations to assess energy consumption and emissions generated by the shuttles in an FSS/RS (Step 3 in Fig. 2).

The following notation is introduced to support the subsequent analysis.

W	Energy consumption	[kWh]
TH/kWh	Throughput per kilowatt-hour	[task/kWh]
E	Emissions generated	[$kgCO_2eq$]
m_{shu}	Shuttle Weight	[kg]
m_{UL}	UL Weight	[kg]
g	Gravitational acceleration	[m/s^2]
a_s	Acceleration of the shuttle	[m/s^2]
c_f	Coefficient of friction	
c_e	Resistance factor of rotating masses (variable speed)	
	Availability of the shuttle	
	Distance traveled by the shuttle	[m]

The equations, defined and validated by Ekren et al. [8], adopted to quantify energy consumption (kWh) are the following:

Energy consumption for the acceleration motion of the shuttle:

$$W = \frac{(m_{shu} + m_{UL}) \cdot (g \cdot c_f + a_s \cdot c_e) \cdot 2 \cdot \Delta x}{1000 \cdot \eta \cdot 3600} \tag{a}$$

Energy consumption for the shuttle's deceleration phase:

$$W = \frac{(m_{shu} + m_{UL}) \cdot (-g \cdot c_f + a_s \cdot c_e) \cdot 2 \cdot \Delta x}{1000 \cdot \eta \cdot 3600} \tag{b}$$

Energy consumption for the shuttle's constant-velocity phase:

$$W = \frac{(m_{shu} + m_{UL}) \cdot g \cdot c_f \cdot \Delta x}{1000 \cdot \eta \cdot 3600} \tag{c}$$

The energy consumption equations applied in the analysis address the fundamental aspects of motion for a physical object: starts, stops, constant velocity motion, accelerated, and decelerated motion. Since the system consists of shuttles moving within a rail structure, the analysis involves weight, inertial forces, and static and dynamic friction. These physical phenomena constitute the key assumptions for the valid application of the energy consumption formulas used in the study. Each shuttle operates within the tier layout, following its assigned path. The shuttle comes to a stop whenever its direction of movement changes. Based on stops, the model identifies linear and continuous segments traveled. Depending on the length of the segment, the equation only includes acceleration (a) and deceleration (b) components if the shuttle does not reach its maximum velocity. Otherwise, the energy calculation incorporates the constant-velocity motion component (c). Figure 3 shows the two possible velocity profiles and corresponding components in the equation.

The total energy consumption of the shuttle is the sum of all the single linear and continuous segment components. The system's total energy consumption is the sum of all the shuttles of the tier. An additional energy efficiency metric is also assessed,

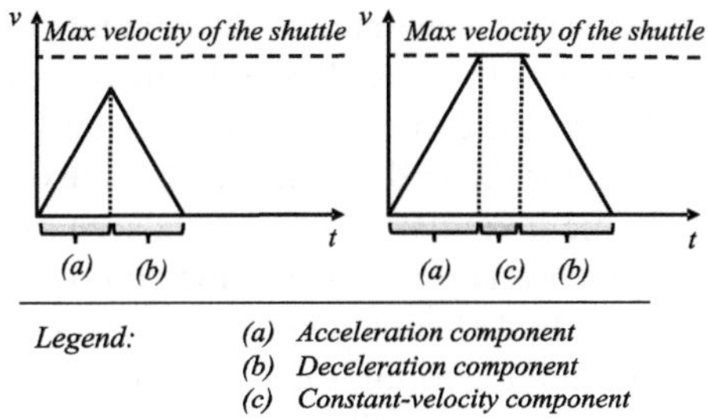

Fig. 3. Linear continuous segment velocity profile

measured using throughput per kWh (TH/kWh) and expressed as the number of tasks completed per kilowatt-hour.

Based on energy consumption, the model computes the resulting emissions through the SimaPro 7.3 software platform. This study adopts the ReCiPe Midpoint method, which enables a detailed evaluation of environmental impact by converting various emissions (e.g., CO_2, CH_4) into CO_2 equivalents. This approach was selected due to its widespread adoption across multiple applications for assessing the environmental impact of complex systems, as shown in Battarra et al. [11].

3 Case Study and Results

This section presents a case study application where the proposed methodology presented in Sect. 2 is applied to an FSS/RS in the food and beverage industry. Moreover, the case study introduces a multi-scenario analysis to compare the results of possible different system setups.

The main characteristics of the selected system are the following:

- Warehouse layout (FSS/RS):
  - Aisles: two lateral aisles, one central cross aisle, eight transversal aisles
  - Lanes: ninety lanes, crossed by the central aisle
  - Lane-depth: 11 m (eleven storage locations) per side of the cross aisle
  - Tiers: ten
- Input parameters:
  - Shuttle specifications: velocity 1 m/s, acceleration and deceleration 1 m/s^2, availability η: 1, weight m_{shu}: 300 kg
  - UL weight m_{UL}: 800 kg
  - Coefficient of friction c_f:0.01
  - Resistance factor of rotating masses (variable speed) c_e: 1.15
- Task list: 2800 S/R tasks (average daily tasks with a 24-h-a-day operation).

The study discloses 12 scenarios, generated by the combination of 3 features in STEP 2 of the methodology: the number of agents (4, 5, 6 shuttles), the arrival time of the tasks (pull option or push option), and the conflict detection (enabled and disabled).

Table 1 reports the results for the pull-based scenarios (exponential arrival time of tasks). The total execution time is similar across the six scenarios, as task arrival times lead to comparable throughput levels, expressed as the number of served tasks per hour (TH in Table 1). The first line of Table 1 contains scenarios A and D, characterized by four shuttles. In scenario A, the MAPF algorithm solves potential path collisions. According to the computed routing, the system's overall energy consumption is 132.1 kWh. Energy efficiency (TH/kWh) is 21.19 tasks per kilowatt-hour. The resulting emissions are estimated at 90.12 kgCO2eq. A comparison with scenario D, which shares the same system setup but neglects conflict detection, shows a slight performance improvement. In this instance, energy consumption is 128.16 kWh (–2.98%), emissions are 87.42 $kgCO_2eq$ (–2.99%), and TH/kWh increases to 21.84 (+3.06%).

Table 1. Energy consumption W[kWh] and emissions E[$kgCO_2eq$]. Exponential arrival time of tasks (Pull option).

	Conflict detection = Yes					Conflict detection = No				
# Sh	***Scen***	TH	W	TH/kWh	E	***Scen***	TH	W	TH/kWh	E
4	A	117.1	132.1	21.19	90.12	D	117.2	128.16	21.84	87.42
5	B	117.1	136.4	20.53	93.04	E	117.1	132.66	21.10	90.50
6	C	116.6	140.1	19.98	95.54	F	117.1	134.01	20.89	91.42

This trend is confirmed in scenarios with 5 (B, E) and 6 (C, F) shuttles, showing similar variations of –2.74% and –4.33% in energy consumption, –2.73% and –4.31% in emissions, and + 2.78% and + 4.55% in TH/kWh, respectively. These results indicate that conflict resolution introduces more complex paths and additional stops, degrading system performance. The comparison between scenarios A (4 shuttles), B (5 shuttles), and C (6 shuttles) shows that increasing the number of shuttles leads to higher energy consumption and emissions (132.1, 136.4, and 140.1 kWh; 90.12, 93.04, and 95.54 $kgCO_2eq$) and lower energy efficiency with TH/kWh values of 21.19, 20.53, and 19.98, respectively. The same pattern holds for scenarios without conflict resolution (D, E, F). As the calculation accounts for operational dynamics, this increase is primarily due to layout congestion. Required path deviations cause additional stops, ultimately reducing performance.

Table 2 presents scenarios where all tasks are available from the start of the MAPF algorithm (push option). As a result, shuttles are continuously engaged from the beginning of the simulation, and increasing their number reduces total execution time, as reflected by TH values. The trends observed in the pull configuration are confirmed, except for a slight drop in results from 4 (G, L) to 5 (H, M) shuttles.

Table 2. Energy consumption W[kWh] and emissions E[$kgCO_2eq$]. Tasks available at time = 0.

	Conflict detection = Yes					Conflict detection = No				
# Sh	*Scen*	TH	W	TH/kWh	E	*Scen*	TH	W	TH/kWh	E
4	G	138.8	129.5	21.61	88.38	L	144.5	123.23	22.72	84.06
5	H	169.5	126.6	22.11	86.35	M	179.8	121.83	22.83	83.11
6	I	193.9	131.8	21.24	89.93	N	207.7	123.40	22.69	84.18

A comparison between pull (Table 1) and push (Table 2) configurations shows improved performance under the latter, with all other parameters held constant. Scenario G reports lower energy consumption (−1.92%), lower emissions (−1.93%), and a higher TH/kWh (+1.98%) compared to scenario A. These trends hold when increasing the number of shuttles or disabling conflict detection. This improvement is primarily driven by the early availability of all tasks, which enables more efficient goal assignment to shuttles. As a result, path deviations and idle times are reduced, leading to lower energy consumption and emissions, and an increase in energy efficiency.

4 Conclusions

The industrial sector heavily contributes to global emissions, and advanced automation may increase its impact, underscoring the need for factory-level emission assessment. This study proposes a three-step methodology to assess energy consumption and emissions in FSS/RS, define system configuration, assign shuttle paths to fulfill tasks, and evaluate storage operations' energetic and environmental impact. A case study applies the methodology. The multi-scenario approach allows for the investigation of the impacts of several system features. Specifically, increasing the number of shuttles or introducing a time-distributed arrival of tasks leads to lower environmental performance.

Future developments may enable statistical analyses across multiple case studies, leading to more evidence-based trends. In addition, this methodology allows for assessing environmental externalities, based on an efficiency-driven approach to the operation strategy. Future research may integrate environmental criteria, such as energy or emission-based cost functions, directly into path planning to support an alternative and more sustainable decision-making. Moreover, this study addresses only the travel component, as it accounts for more than 50% of the handling activity. A cradle-to-grave emission assessment of the system may enhance and complete the analysis.

References

1. Sixth Assessment Report, Chapter 6: Energy systems. Accessed 07 May 2025
2. Ferreira, B., Reis, J.: A systematic literature review on the application of automation in logistics. Logistics **7**(4), 80 (2023)

3. Ferraro, S., Cantini A., Leoni L., De Carlo F.: Sustainable logistics 4.0: a study on selecting the best technology for internal material handling. Sustainability **15**(9), 7067 (2023)
4. Battarra, I., Accorsi R., Lupi B., Manzini R.: A design framework for shuttle-based automated storage systems. In: 18th IFAC Symposium on Information Control Problems in Manufacturing INCOM 2024, pp. 1216–1221, Elsevier, Vienna, Austria (2024)
5. Mao, J., Cheng J., Li X., Zhao H., Lin C.: Modelling analysis of a four-way shuttle-based storage and retrieval system on the basis of operation strategy. Appl. Sci. **13**(5) (2023)
6. Stern, R., et al.: Multi-agent path finding – an overview. Artif. Intel. (2019)
7. Mao, J., Cheng, J., Li, X., Cao, B.: Research on scheduling optimization of four-way shuttle-based storage and retrieval systems. Sci. Rep. **13**, 3999 (2023)
8. Ekren, B., Akpunar, A., Sari, Z., Lerher, T.: A tool for time, variance and energy related performance estimations in a shuttle-based storage and retrieval system. Appl. Math. Model. **63**, 109–127 (2019)
9. Liu, Z., Wang, Y., Jin, M., Wu, H., Dong, W.: Energy consumption model for shuttle-based storage and retrieval systems. J. Clean. Prod. **282**, 124480 (2021)
10. Mourtzis, D., Vlachou, E., Milas, N., Dimitrakopoulos, G.: Energy consumption estimation for machining processes based on real-time shop floor monitoring via wireless sensor networks. In: 49th CIRP Conference on Manufacturing Systems, pp. 637–642, Stuttgart, Germany (2016)
11. Battarra, I., Accorsi, R., Lupi, G., Manzini, R.: Life cycle assessment of an automated multi-cellular case-picking system. J. Manuf. Syst. **79**, 419–434 (2025)

Enhancing Production Planning and Control by Integrating the Circular Strategy 'Repair'

Jonah Schulz[1(✉)], Jost Runte[1], Roman Krämer[1], Alexander Rokoss[1], and Matthias Schmidt[2]

[1] Leuphana University Lueneburg, Universitätsallee 1, 21337 Lueneburg, Germany
Jonah.Schulz@leuphana.de
[2] Leibniz University Hannover, An Der Universität 2, 30823 Garbsen, Germany

Abstract. Production planning and control (PPC) is a central task for manufacturing companies to align the logistical performance of production to customer requirements. Alongside well-established customer requirements such as delivery time and quality, sustainability-related criteria are gaining importance for maintaining competitiveness. In response, many manufacturing companies are increasingly adopting circular economy principles, also known as circular strategies. A promising approach to integrate circular strategies into production processes is the 'repair' strategy, in which end-of-life products are restored to full functionality to extend their lifespan. However, due to fluctuations in the availability and quality of end-of-life products, the resulting interactions with existing processes lead to new challenges for PPC. Despite repair's increasing role in circular production systems, its specific impact on PPC and related interdependencies remains unexplored. For this reason, the present paper deals with the systematic identification and deductive derivation of resulting interdependencies within PPC that emerge from integrating the 'repair' strategy and embeds these findings within an existing PPC framework model. Ultimately, this paper contributes to a deeper understanding of how repair as a circular strategy affects PPC processes and provides a foundation for future framework development in this research area.

Keywords: Circular Economy · Production Planning and Control · Circular Strategies · Repair

1 Introduction

Production planning and control (PPC) is a central task in industrial value creation, significantly contributing to the efficiency, flexibility and competitiveness of manufacturing companies by ensuring optimal resource utilisation and targeted process coordination [1]. Considering global climate change and growing resource scarcity, the transition to sustainable economic systems is gaining momentum, with the circular economy (CE) offering a viable alternative to the linear "take-make-waste" model through the application of circular strategies [2]. This transition impacts existing PPC processes by increasing planning complexity due to volatile reverse material flows [3]. Therefore, this

H. Kohl et al. (Eds.): GCSM 2025, LNME, pp. 95–103, 2026.
https://doi.org/10.1007/978-3-032-21154-5_11

paper extends an existing PPC framework model based on the circular strategy 'repair', to show interdependencies within the internal supply chain (ISC). The paper is structured as follows. Section 2 first presents the basics of PPC and selects a suitable PPC framework model, followed by a detailed presentation of the circular strategy 'repair'. Section 3 examines and expands the ISC as a foundation for modelling interdependencies. Section 4 presents the deductively derived interdependencies. In Section 5, the results are summarised, the limitations are outlined, and an outlook on further research is provided.

2 Theoretical Background

2.1 Production Planning and Control

PPC encompasses the planning and coordination of the entire production process with all relevant sub-areas, such as work preparation, procurement, and manufacturing [4] and aims to ensure the achievement of defined production goals despite unavoidable disruptions [5]. Within PPC, these production targets translate to logistical objectives, such as utilisation, throughput time and schedule compliance. The objectives exhibit inherent trade-offs, resulting in conflicting objectives. Therefore, companies must position themselves in this area of conflict and derive an appropriate PPC configuration. Various PPC framework models support this positioning, including the PPC model according to HACKSTEIN [4], the Aachen PPC Model [6], the Model of Manufacturing Control [7] and the Hanoverian Supply Chain Model (HaSupMo) [8]. This article uses the HaSupMo as a basis for the model extension, as it represents the latest evolutionary state of the PPC framework development.

The HaSupMo divides the ISC into the core processes 'Procurement', 'Production', and 'Dispatch'. The core process 'Production' can also be segmented into a 'Preliminary' and 'End Production Stage' through the inclusion of an 'Interim Storage'. The material flow through the ISC is planned and controlled by eleven PPC main tasks. These include, for example, *Manage Order*, *Plan Secondary Requirements*, *Plan Production*, and *Control Production*. Each of these PPC main tasks comprises individual operational steps, referred to as PPC tasks, which must be carried out at the operational level. For this purpose, the HaSupMo provides a set of procedures and methods to execute these PPC tasks systematically [8].

PPC main tasks and PPC tasks affect the material flow within the ISC, thereby influencing logistical objectives. The interdependencies between the core processes and the PPC main tasks are illustrated in the HaSupMo based on the Model of Manufacturing Control [7]. By addressing different planning levels, the HaSupMo constitutes a holistic PPC framework model. However, in its original form, it is not designed to represent circular material flows [9, 10].

2.2 Circular Economy and the Circular Strategy 'Repair'

The core of the CE is to replace the 'end-of-life' concept with the targeted reduction, reuse and recovery of materials [11]. This is implemented through circular strategies, whose definitions vary across models, ranging from three to ten strategies depending on the specific modelling framework [12, 13]. This article adopts the model and definitions by POTTING ET AL. [13], as its granular distinction between strategies enables the consideration of specific material cycles in the context of PPC.

In this framework, the circular strategy 'repair' is defined as "repair and maintenance of defective product so it can be used with its original function" [13]. This definition is distinguishable from other circular strategies that extend product lifespans, which often lack clear differentiation in other models. Firstly, it differs from 'reuse' where a used product is not reconditioned. Secondly, it differs from 'refurbishment' in which the used product is upgraded. Thirdly, it differs from 'remanufacturing' as the product identity remains unchanged, resulting in comparatively lower disassembly effort.

2.3 Description of Research Demand

For companies, implementing 'repair' poses significant challenges, primarily due to planning uncertainties arising from fluctuating inflows of used products in terms of quantity and quality [3]. Although 'repair' is the most frequently applied circular strategy [14], there is currently little scientific research. Notably, the development of a holistic framework for PPC, as previously called for by GUIDE ET AL. [15], remains an open research gap. A review of the relevant literature on the different planning levels of PPC, as discussed in Section 2.1, further highlights this gap (Table 1):

Table 1. Literature review regarding the extension of PPC towards a holistic framework

Source	Considering all Core Processes of the ISC according to HaSupMo	Holistic View on PPC Main Tasks	Linkage of PPC Tasks and Possible Methods	Modelling the Interdependencies with logistical objectives	Considering the circular strategy 'repair´
[16]	○	○	○	○	●
[15]	◐	◐	●	○	○
[17]	◐	○	●	○	○
[10]	●	◐	○	◐	○
[18]	●	●	●	●	○
[9]	●	●	●	●	○

It becomes evident that there is extensive preliminary work on the topic of expanding the HaSupMo regarding circular strategies, although the circular strategy 'repair´ has not yet been addressed. This research gap is addressed in the following chapters by extending the ISC and deriving the interdependencies within PPC for 'repair'.

3 Extension of the Internal Supply Chain

Within the model by POTTING ET AL. [13], 'repair' represents a moderate level of intervention in the product lifecycle. It offers a higher degree of circularity than 'remanufacturing', but less than 'reuse', regarding resource preservation and product integrity. In contrast to 'reuse', which requires minimal intervention and enables early closure of the material loop [19], 'repair' involves targeted restoration of functionality through inspection and selective replacement of components [20]. Compared to 'remanufacturing', 'repair' demands significantly less effort, as it involves reduced disassembly and excludes reprocessing. The definitional positioning and existing model extensions for 'reuse' and 'remanufacturing' serve as a basis for deriving the ISC. This approach ensures compatibility with the HaSupMo and its extensions while representing the associated material cycles. The core processes required for integrating 'repair' into the ISC can consequently be determined as shown in Fig. 1:

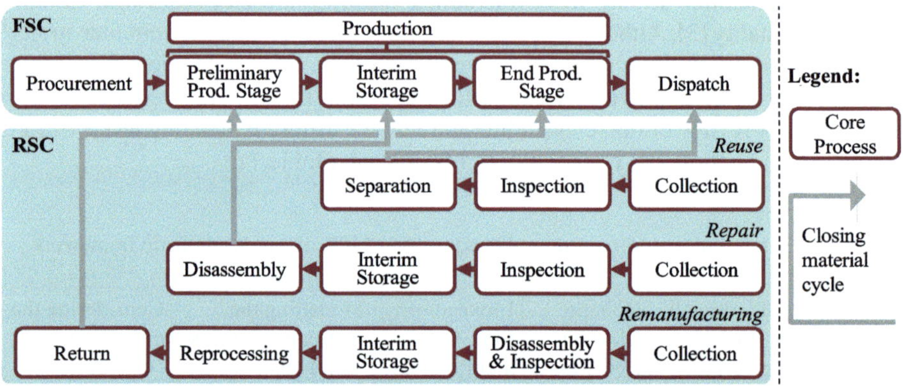

Fig. 1. Extension of the ISC regarding 'repair'

As shown in Fig. 1, the core processes of the forward supply chain (FSC) remain unchanged, whereas those of the reverse supply chain (RSC) differ depending on the circular strategy. Although 'Collection' consistently marks the first step, differences already emerge in the subsequent core process. For 'reuse' and 'repair', an inspection of the entire product is conducted to assess its condition and determine its suitability for the respective strategy [18]. In contrast, remanufacturing begins with disassembly to extract parts and components, followed by an inspection to evaluate their condition [9]. The third core process across all strategies is a storing core process, which optionally decouples upstream and downstream core processes in terms of quantity and timing. However, in the case of 'reuse', a separation step occurs, where usable and non-usable products are distinguished [18]. In the fourth core process, 'repair' focuses on disassembling defective parts and components from the product to restore it to a usable semi-finished state. By contrast, 'remanufacturing' includes an optional reprocessing step if the quality is insufficient to restore the previously disassembled parts and components to the required quality [9]. While the semi-finished products resulting from 'repair' enter the interim storage stage in the core process 'Production' in the FSC before being completed in the final production step, the reprocessed parts and components in remanufacturing are fed back into the core process 'Production' as secondary requirements, forming an additional storage stage parallel to procurement and interim storage [9].

The derived structure and naming of the core processes in the FSC and the RSC also align with the model by SCHÄFERS AND WALTHER [10], the publication by FLEISCHMANN ET AL. [21], and the extended SCOR model from VEGTER ET AL. [22].

4 Interdependencies Within the Defined Core Processes

As indicated by prior research [23], no new PPC main tasks are emerging when implementing 'remanufacturing'. Since 'repair' requires less effort and intervention, it is plausible to assume that no new PPC main tasks arise in this context either. This assumption is supported by CASTIGLIONE ET AL., who emphasise that it is the increased complexity of existing tasks, rather than the emergence of new ones, that presents the main challenge [16]. Additionally, JABBOUR ET AL. [24] and BAG ET AL. [25] highlight the need to adapt planning capacities and structures when integrating repair processes without identifying additional PPC main tasks.

Analogous to the extension of the ISC for 'repair', the resulting interdependencies show parallels to the HaSupMo extensions for 'reuse' and notable similarities with the extension for 'remanufacturing', as the underlying core processes closely resemble each other. The following paragraphs analyse how the interdependencies between core processes and the PPC main tasks differ across the FSC and RSC.

Within the core process 'Collection', the interdependencies largely mirror those observed in the 'remanufacturing´ extension of HaSupMo. However, a difference lies in the possibility of different storage strategies [26]. Since 'repair' does not necessarily involve a change of ownership, make-to-order and assemble-to-order strategies can be applied. Therefore, the planned output is additionally influenced by the PPC main task *Manage Order* compared to 'remanufacturing'.

The interdependencies within the core processes 'Inspection' and 'Disassembly' in the RSC for 'repair', shown in Fig. 2, resemble those in 'Disassembly & Inspection' and 'Reprocessing' in the RSC for 'remanufacturing' [9]. These core processes are grounded in the same logistical building block of production defined in the HaSupMo, as both involve the usage of resources to restore the value of recovered components for continued use in production [8]. Despite the process-specific differences in the two circular strategies, the associated planning stages are similar.

Additionally, as with the implementation of 'remanufacturing', 'repair' introduces additional material into the FSC in the form of disassembled products. However, while 'remanufacturing' generates secondary requirements through individual parts and components, 'repair' generates semi-finished products that can be directly re-integrated into upstream production stages. This new material source modifies, and in some areas reduces, the requirements within the FSC, as fewer semi-finished products need to be initially produced and planned.

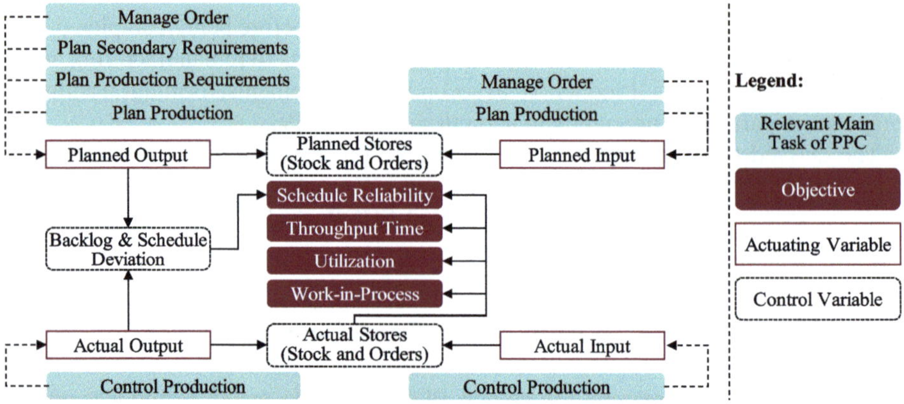

Fig. 2. Interdependencies within the core processes 'Inspection´ and 'Disassembly´ [8, 9]

5 Conclusion and Limitations

This article analyses the circular strategy 'repair' and conceptualises a model extension based on the HaSupMo. The deductively derived interdependencies are illustrated and compared to other model extensions. The core result is the extended model that embeds the identified interdependencies within an established PPC framework to ensure structural coherence. As the extension is based on the HaSupMo, its PPC methods remain applicable, enabling their application within the context of 'repair'. However, their effects and suitability in the RSC may differ from those in the FSC, indicating the need for further investigation. The comparative analysis of existing HaSupMo extensions and the derivation of a new repair-focussed extension show that only minor adaptations were needed. This indicates the potential to merge all extensions into a unified, holistic model showing interdependencies for all circular strategies. The presented extension thus provides a foundation for further research and iterative refinement.

Nevertheless, the results are subject to limitations, as the derivation of interdependencies is primarily based on existing model extensions due to the limited availability of specific literature on PPC in the context of 'repair'. Moreover, the validation of the interdependencies has not yet been carried out. Given these limitations, future research should focus on validating the derived model extension, including the proposed interdependencies. Although the literature distinguishes between empirical validation using real production data and simulation-based validation under controlled conditions [27], neither approach has been applied to this model extension nor to the existing extensions for 'remanufacturing' and 'reuse'. Given the current lack of detailed and comprehensive real-world data, especially in the context of circular strategies such as 'repair' [28], simulation offers a suitable and readily available alternative. It enables the targeted variation of input parameters directly linked to individual PPC tasks [29], allowing specific interdependencies to be examined under controlled, reproducible conditions. Future research should therefore prioritise developing a dedicated simulation environment tailored to circular production systems. However, when real-world data is available in sufficient depth, it should be used to complement simulation results to further strengthen the validity of the model and assess its practical applicability in real production contexts [30].

Such an approach would support the ongoing development of PPC research at the process level and enable researchers to embed both existing and newly developed PPC methods in this proposed model extension to assess the effect on logistical objectives throughout the ISC.

Acknowledgements. The project 'TrICo - Transformation through Innovation and Cooperation in Communities' was funded by the Federal Ministry of Education and Research under grant number 03IHS284A. The author is responsible for the content of this publication.

References

1. Hopp, W.J., Spearman, M.L.: Factory Physics: Foundations of Manufacturing Management, 2nd edn. McGraw-Hill international edition, Boston (2000)
2. Kronenberg, C.: Circular Economy, in *Nachhaltigkeit und Digitalisierung – (k)ein unternehmerisches Dilemma*, p. 47. Springer, Berlin Heidelberg, Berlin, Heidelberg (2023)
3. Ritzén, S., Sandström, G.Ö.: Barriers to the Circular Economy – Integration of Perspectives and Domains vol. 64, p. 7 (2017)
4. Hackstein, R.: Produktionsplanung und -steuerung (PPS): Ein Handbuch für die Betriebspraxis. VDI-Verlag, Düsseldorf (1984)
5. Lödding, H.: Verfahren der Fertigungssteuerung. Springer, Berlin Heidelberg, Berlin, Heidelberg (2016)
6. Schuh, G., Brosze, T., Brandenburg, U.: Aachener PPS-Modell, in Produktionsplanung und -steuerung 1, p. 11. Springer, Berlin, Heidelberg (2012)
7. Lödding, H.: Handbook of Manufacturing Control: Fundamentals, Description, Configuration. Springer, Berlin, Heidelberg (2013)
8. Schmidt, M., Nyhuis, P.: Produktionsplanung und -steuerung im Hannoveraner Lieferkettenmodell: Innerbetrieblicher Abgleich logistischer Zielgrößen. Springer Vieweg, Berlin, Heidelberg (2021)
9. Schulz, J., Hofmann, H., Rokoss, A., Schmidt, M.: Understanding resulting interdependencies within production planning and control in hybrid manufacturing/remanufacturing systems. In: Proceedings of the Conference on Production Systems and Logistics: CPSL 2025, Hannover publishing (2025)
10. Schäfers, P., Walther, A.: Modelling Circular Material Flow and the Consequences for SCM and PPC vol. 11, p. 91 (2017)
11. Kirchherr, J., Yang, N.-H.N., Schulze-Spüntrup, F., Heerink, M.J., et al.: Conceptualizing the Circular Economy (Revisited): An Analysis of 221 Definitions 194 (2023)
12. Jawahir, I.S., Dillon Jr., O.W., Rouch, K.E., Joshi, K.J., et al.: Total life-cycle considerations in product design for sustainability: a framework for comprehensive evaluation. In: 10th International Research/Expert Conference "Trends in the Development of Machinery and Associated Technology", p. 1 (2006)
13. Potting, J., Hekkert, M., Worrell, E., Hanemaaijer, A.: Circular Economy: Measuring Innovation in the Product Chain. PBL Netherlands Environmental Assessment Agency, The Hague (2017)
14. Schuh, G.: PPS-Report 2023: Studienergebnisse. RWTH Aachen University (2023)
15. Guide, V.R.: Production planning and control for remanufacturing: industry practice and research needs **18**, 467 (2000)

16. Castiglione, C.: Circular economy strategies at the manufacturing system scheduling level: the impacts on Makespan. In: Italian Manufacturing Association Conference, Materials Research Forum LLC, p. 250 (2023)
17. Kenné, J.-P., Dejax, P., Gharbi, A.: Production planning of a hybrid manufacturing–remanufacturing system under uncertainty within a closed-loop supply chain **135** *(*2012)
18. Schulz, J., Rokoss, A., Schmidt, M.: The Influence of Product Reuse on Production Planning and Control **130**, 1543 (2024)
19. Jayaraman, V.: Production planning for closed-loop supply chains with product recovery and reuse: an analytical approach **44**, 981 (2006)
20. Kravchenko, M., Pigosso, D.C., McAloone, T.C.: Towards the ex-ante sustainability screening of circular economy initiatives in manufacturing companies: consolidation of leading sustainability-related performance indicators **241** (2019)
21. Fleischmann, M., Krikke, H.R., Dekker, R., Flapper, S.D.P.: A characterisation of logistics networks for product recovery **28**, 653 (2000)
22. Vegter, D., van Hillegersberg, J., Olthaar, M.: Supply chains in circular business models: processes and performance objectives **162** 2020.
23. Schulz, J., Hillnhagen, S., Schmidt, M.: Investigation on the influence of remanufacturing on production planning and control – a systematic literature review. In: Proceedings of the Conference on Production Systems and Logistics: CPSL 2024, Hannover publish-Ing (2024)
24. Lopes de Sousa Jabbour, A.B., Jabbour, C.J.C., Godinho Filho, M., Roubaud, D.: Industry 4.0 and the circular economy: a proposed research agenda and original roadmap for sustainable operations **270**, 273 (2018)
25. Bag, S., Gupta, S., Kumar, S.: Industry 4.0 adoption and 10R advance manufacturing capabilities for sustainable development **231** (2021)
26. Nyhuis, P., Wiendahl, H.-P.: Fundamentals of production logistics: Theory, tools and applications; with 6 tables. Springer, Berlin, Heidelberg (2009)
27. Woschank, M., Dallasega, P., König, A., Hoffelner, M.: Potentials of using real-time data to increase the update frequency of production planning and control strategies in MTO: a discrete event simulation study **36**, 760 (2024)
28. Acerbi, F., Sassanelli, C., Terzi, S., Taisch, M.: A systematic literature review on data and information required for circular manufacturing strategies adoption **13** (2021)
29. Mourtzis, D.: Simulation in the design and operation of manufacturing systems: state of the art and new trends **58**, 1927 (2020)
30. Muehlbauer, K., Rissmann, L., Meissner, S.: Machine learning decision support for production planning and control based on simulation-generated data. In: Knowledge Discovery, Knowledge Engineering and Knowledge Management, p. 257. Springer Nature Switzerland, Cham (2023)

Towards Simulation-Driven Framework for Optimising Rail Depot Operations and Maintenance Scheduling

Cansu Kandemir[1](✉), Hassna Louadah[2], Christopher Haynes[1], Gareth Tucker[3], Adam Bevan[3], and Marco Franchino[1]

[1] Advanced Manufacturing Research Centre (AMRC), The University of Sheffield, Sheffield, UK
c.kandemir@amrc.co.uk

[2] School of Engineering and Sustainable Development, De Montfort University, Leicester, UK
hassna.louadah2@dmu.ac.uk

[3] Institute of Research, University of Huddersfield, Huddersfield, UK

Abstract. Efficient and sustainable Train Fleet Maintenance (TFM) is essential for realising the full environmental benefits of rail transport. This paper presents an integrated framework that combines short-term maintenance scheduling optimisation with Discrete Event Simulation (DES) to enhance the adaptability, reliability, and sustainability of depot operations, including maintenance activities. The scheduling tool, underpinned by Mixed Integer Programming and a formal ontological model, generates cost-effective maintenance plans that consider operational constraints. These plans are validated using a detailed DES model built in AnyLogic Software, which replicates depot layouts, rules, and capacities. Simulation enables early conflict detection, proactive adjustments, and reductions in unnecessary movements, energy use, and emissions. A realistic case study demonstrates how the integrated approach supports continuous improvement and improves asset utilisation. The results highlight the potential of combining optimisation and simulation to support greener and more resilient maintenance planning. Future work will focus on formal validation and integration with broader decision-support systems to accelerate the transition to low-carbon rail networks.

Keywords: Scheduling · Rolling Stock Maintenance · Rail Depot Management

1 Introduction

Rail transport is widely regarded as one of the most sustainable modes of travel, offering a significantly lower environmental footprint compared to road and air transport. However, realising these benefits depends heavily on the efficiency and reliability of train fleet maintenance (TFM), which remains a complex and resource-intensive operation. Rail depots not only carry out maintenance activities but also handle inspections, cleaning, refuelling, minor repairs, and stabling each adding to the operational demands of an already spatially constrained environment. Inefficient depot operations can substantially

H. Kohl et al. (Eds.): GCSM 2025, LNME, pp. 104–112, 2026.
https://doi.org/10.1007/978-3-032-21154-5_12

undermine the sustainability of rail services. For example, delayed maintenance may lead to unnecessary train idling, rerouting, and fuel use. According to the International Energy Agency, well-maintained rail systems consume up to 30% less energy per passenger-kilometre than poorly managed ones [1]. Moreover, unreliable rail services risk a modal shift toward more carbon-intensive modes, such as road and air transport.

In the UK, inefficiencies in depot operations are increasingly evident. Train idling alone is estimated to account for over 100,000 tonnes of CO_2 emissions annually [2], while maintenance-related delays rose by 20.5% in 2024 [3]. These figures highlight the urgent need for smarter, more adaptive TFM strategies that not only improve operational efficiency but also contribute to national decarbonisation targets.

To address this, we propose an integrated framework that combines optimisation and simulation to support more responsive, reliable, and sustainable depot operations. Specifically, the framework integrates a short-term maintenance scheduling tool with a Discrete Event Simulation (DES) model of depot activities [4]. The scheduling component is based on the ontology-driven Virtual Depot tool [4, 11], which enables the generation of cost-efficient, operationally feasible maintenance plans. These are then validated using a detailed DES model that captures the dynamic, constrained nature of real-world depot operations.

The proposed framework aims to reduce resource conflicts, cut emissions, and improve asset utilisation supporting the transition to greener, more resilient rail services and contributing to the UK's net-zero objectives.

2 Background

To fully realise the environmental benefits of rail transport, it is essential to optimise the efficiency of key supporting infrastructure, particularly depot and maintenance operations. This section outlines key operational constraints within rail depots and reviews relevant optimisation and simulation techniques for TFM, laying the foundation for the integrated approach proposed in this paper.

2.1 Depot Constraints and the Need for Adaptability

Depot operations are inherently complex, shaped by factors such as limited space, access restrictions, variable train arrivals, and layout bottlenecks. Trains must adhere to strict sequencing rules to enter or exit tracks, and ongoing maintenance activities can further restrict access. These constraints necessitate maintenance schedules that are flexible and capable of adapting to real-time changes, such as train arrival times, without compromising reliability or efficiency. At the same time, maintenance activities themselves must be planned in a way that does not disrupt other depot operations, ensuring that service-ready trains are not delayed, and overall train service reliability is maintained.

2.2 Optimisation and Simulation in Train Fleet Maintenance (TFM)

To improve the efficiency of maintenance operations, various optimisation approaches have been explored. Techniques such as Mixed Integer Programming (MIP) and heuristic-based scheduling can effectively generate cost-efficient plans while adhering to resource constraints. However, traditional optimisation approaches often fall short in accounting for the real-time variability and complexity of depot environments. Static models may not capture disruptions caused by late arrivals, unexpected asset unavailability, or physical constraints within the depot layout [5].

To address this, simulation-based techniques, particularly DES, have gained popularity. DES provides a dynamic, system-level view of depot operations, enabling planners to test the feasibility of proposed maintenance schedules under realistic conditions. For instance, Chiu et al. demonstrated that simulation-based scheduling could significantly reduce energy consumption while enhancing overall system performance in urban transport [6]. Additionally, a Siemens-led project employed Plant Simulation to evaluate depot schedules dynamically [7], though it did not incorporate adaptive optimisation mechanisms.

2.3 Towards Integrated Solutions

Combining optimisation and simulation offers a promising pathway to overcoming the limitations of standalone methods. Optimisation ensures efficient task allocation and resource usage, while simulation validates these plans against the operational realities of the depot. This integration enables proactive conflict detection, scenario testing, and continuous refinement which are key capabilities for resilient and sustainable depot management.

Crucially, improved maintenance planning also supports broader sustainability objectives. Just-in-time scheduling minimises idle time, reduces wear and emissions, and ensures rolling stock longevity, all aligning with circular economy principles and national decarbonisation targets [8, 9]. The framework proposed in this paper builds on this integrated vision, using a combination of MIP-based scheduling and DES to deliver operationally feasible, environmentally responsible maintenance strategies.

3 Methodology

The proposed framework addresses the challenge of optimising short-term TFM while ensuring sustainable and resilient depot operations. It integrates two core components: a short-term maintenance scheduling optimisation tool and a DES model of depot activities. Together, these components provide a dynamic approach for generating, validating, and adjusting operational plans, with a focus on efficiency, reliability, and minimising resource use.

The short-term maintenance scheduling tool is based on the methodology proposed by Louadah et al. [10, 11]. It incorporates a formal ontological model that captures key aspects of depot infrastructure, rolling stock characteristics, and maintenance tasks. Each maintenance task is defined by its type, duration, required resources (e.g., tracks), and associated lateness or tardiness penalties. Using an MIP approach, the tool aims to minimise overall maintenance costs while respecting resource availability, operational windows, and regulatory constraints. Short-term maintenance needs are captured through unit-specific due dates, ensuring tasks are scheduled within their correct operational window to avoid unnecessary downtime. The scheduling tool maintains a register of assets (train units), each uniquely identified and linked to its maintenance history, characteristics, and associated operational constraints. This register forms the basis for associating tasks with specific rolling stock and enables accurate scheduling tailored to each unit's needs.

A key feature of the tool is its ability to automatically generate optimised schedules once the planning horizon is set. Users define a start date and scheduling window, and the tool outputs a ready-to-implement schedule. This enhances planning flexibility and supports management of depot resources, contributing to more sustainable maintenance operations by reducing idle time and improving asset utilisation. An example of a short-term schedule generated by the tool is shown in Fig. 1, which combines both the tool's input parameters and an example of the generated schedule.

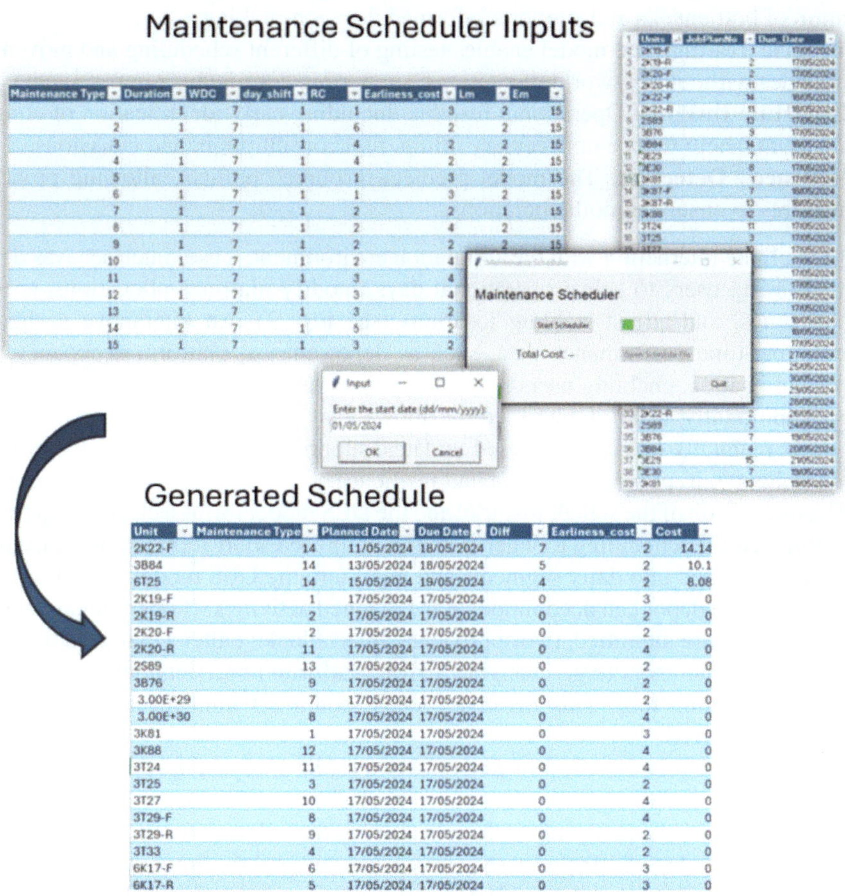

Fig. 1. Short-Term Scheduling Tool: Inputs and Example of Generated Schedule.

To validate the feasibility of the generated schedules, a detailed DES model replicates depot activities and fleet movements. Developed in AnyLogic Simulation Software using its rail library, the model accurately reflects the depot layout, operating rules, and resource capacities. Key inputs to the model include track layouts, maintenance shed capacities, train timetables, staff availability, operational rules (e.g., speed limits), and depot-specific constraints. The simulation model also incorporates train timetables, which are linked via Excel files maintained and updated by depot operators. These timetables include expected arrival and departure times for each train unit, identified by asset numbers, enabling the simulation to reflect real-time depot occupancy and future arrivals or departures. This integration ensures that users maintain visibility over which trains are present in the depot, which are expected, and when units are due to depart, a critical feature for both proactive decision-making and rapid conflict resolution.

The DES model offers several sustainability-related advantages:

- **Visualisation**: 2D and 3D visualisation of depot operations allows intuitive identification of bottlenecks and supports informed decision-making.
- **Scenario Testing**: The model enables testing of different scheduling and movement strategies without real-world disruptions, supporting continuous improvement.
- **Real-Time Insights**: Operational insights, including early identification of congestion points, help reduce unnecessary idling, fuel consumption, and emissions.
- **Bottleneck Detection**: The model predicts resource conflicts, allowing proactive measures to ensure smooth operations.

To facilitate interaction with the simulation environment, a user interface was developed, allowing users to select operational days, modify platform allocations, specify transfer times, and adjust stabling locations (see Fig. 2). An interactive dashboard presents real-time performance data, such as depot entries, platform occupancy, and fuel line utilisation, enabling users to evaluate operational performance over 12-h windows. Additionally, a rail network editor allows easy modifications to the depot layout, either manually or via CSV-based updates, supporting rapid testing of alternative configurations to improve throughput and resilience.

The integration of the scheduling tool and the DES model forms a closed-loop framework. Initially, the scheduling tool generates an optimised short-term maintenance plan. This plan is divided into daily segments and input into the DES model. The simulation tests whether the depot can accommodate the planned activities. If operational conflicts or infeasibilities are identified, two approaches are available: either a manual adjustment of simulation parameters (e.g., changing arrival platforms) or a complete rescheduling via the optimisation tool. This iterative process continues until a feasible, validated schedule is produced.

To support this process, the tool generates two levels of reporting. First, intermediate reports highlight specific issues such as platform clashes or unmet resource requirements, allowing planners to manually adjust the schedule before full re-simulation. These reports typically include a list of affected assets, time windows, and the nature of the conflict. If issues persist, a comprehensive conflict report is produced, detailing all infeasible slots and their causes. This enables users to either modify operational assumptions or trigger a rescheduling, streamlining the iteration process between planning and execution.

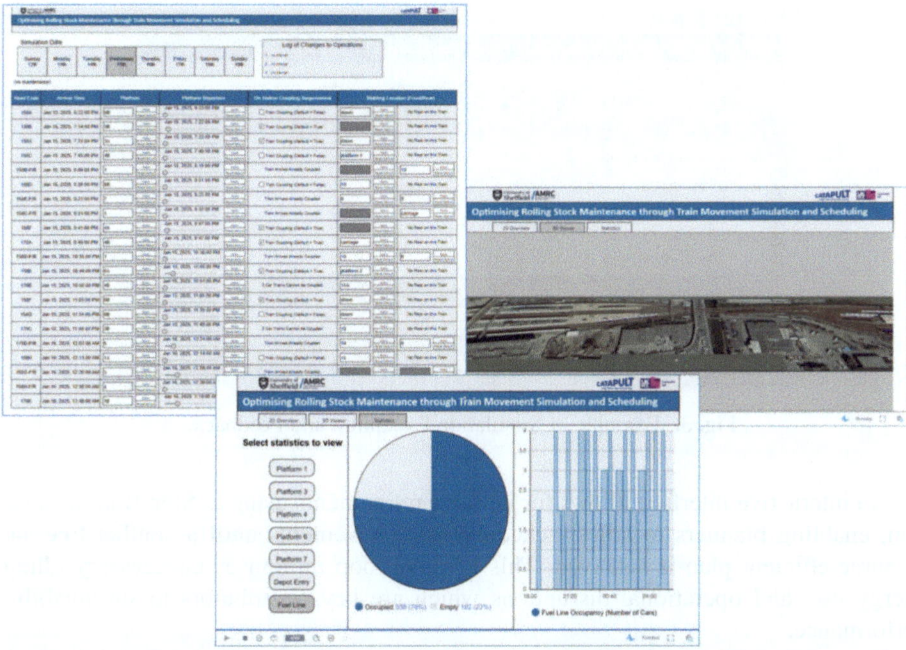

Fig. 2. Model User Interface, 3D visualisation and Dashboard.

This integrated approach enables not only cost and time optimisation but also more sustainable depot management. By reducing idle times, preventing operational conflicts, and enabling proactive adjustments, the framework promotes more efficient use of depot infrastructure and rolling stock assets. Ultimately, it contributes to broader sustainability goals by improving reliability, supporting a modal shift from less sustainable transport options, and minimising resource consumption within maintenance operations.

4 Application

Preliminary trials were conducted using a realistic case study to evaluate the effectiveness of the proposed framework in delivering more sustainable and efficient TFM and depot operations. By simulating the real-time execution of optimised short-term maintenance plans, the integrated approach enables leaner operations through reduced train movements, fuel consumption, and scheduling conflicts ultimately supporting more environmentally responsible depot management.

The DES model was used to proactively identify operational conflicts such as scheduling overlaps and platform clashes that could arise during the execution of the maintenance schedule generated by the optimisation tool. Upon detecting a conflict, the system issues an immediate error message with detailed diagnostic information to assist users in understanding and resolving the issue (see Fig. 3). Users can then review platform availability and make targeted adjustments.

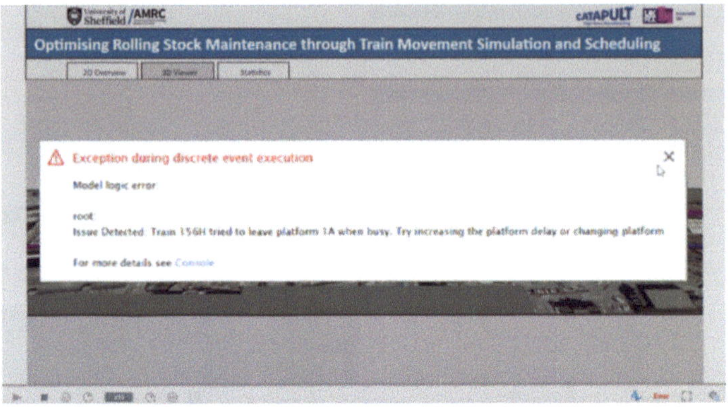

Fig. 3. Example of Simulation Exception and Feedback.

An interactive interface allows for iterative refinements using a 'Stop Button' function, enabling planners to adjust schedules and movements until a conflict-free and resource-efficient plan is achieved. This iterative loop minimises unnecessary idling, energy use, and operational disruptions which are key contributors to sustainability performance.

To systematically assess the framework's impact, all schedule modifications made during the trials were recorded and summarised in Table 1. These adjustments illustrate how the simulation component supports responsive and sustainable decision-making by flagging inefficiencies early and enabling data-driven corrections. Platform changes were the most frequent, indicating that initial maintenance plans often conflicted with platform availability. The spike in changes on Thursday suggests either high utilisation or tightly packed scheduling windows that required manual redistribution. Although limited, increased platform delays in transferring train units into the depot were recorded on Thursday and Friday, likely due to congestion or delayed movements during resource reallocation, issues that would have remained undetected without simulation-based foresight. Meanwhile, changes in maintenance locations involved reassigning trains to alternative sheds or tracks, typically because originally assigned areas were occupied or blocked due to overlapping activities.

Table 1. Number of changes made through the DES model to optimise the depot operations.

	Monday	Tuesday	Wednesday	Thursday	Friday
Platform Changes	1	3	1	5	2
Increase Platform Delays	0	0	0	2	2
Maintenance Location Changes	0	0	2	1	1

These findings highlight the value of integrating simulation with optimisation, demonstrating the system's ability to identify operational conflicts early and enable planners to resolve them before implementation. This ensures maintenance plans are not only cost-effective but also operationally feasible, resilient, and aligned with sustainability objectives.

5 Conclusions and Future Work

This research proposes an integrated framework that bridges the gap between short-term maintenance scheduling and operational simulation for TFM, supporting the drive towards more sustainable and efficient rail systems. By facilitating evidence-based decision-making, the framework enhances train service reliability, optimises depot resource utilisation, aligning closely with broader environmental and sustainability objectives.

The DES model not only validates the feasibility of maintenance schedules but also identifies operational constraints early, reducing unnecessary train movements, energy/fuel consumption, and emissions. Its flexibility ensures applicability across various depot layouts, providing a robust and adaptable tool for stakeholders aiming to optimise maintenance operations while advancing sustainability targets.

Future research directions include the development of a comprehensive validation framework for short-term schedules, wherein all possible depot operational scenarios are systematically evaluated to generate pass/fail outcomes. Moreover, future work will focus on integrating the proposed framework with wider control systems to enable real-time coordination between rail operations, maintenance scheduling, and resource management. This would ensure that maintenance activities are not only operationally feasible but also strategically aligned with overall service and network demands.

Acknowledgements. This work was carried out as part of the project, "*Improving the reliability and resilience of train fleet maintenance – virtual maintenance depot and operations optimisation*", led by Hassna Louadah, and supported through Tranche 1 of the Innovation Launchpad Network+ Researcher in Residence scheme [4]. It is a collaboration between DMU - De Montfort University, IRR - Institute of Railway Research, and Advanced Manufacturing Research Centre (AMRC), University of Sheffield part of High Value Manufacturing Catapult (HVMC), funded by the Engineering and Physical Sciences Research Council's Innovation Launchpad Network+ Researcher in Residence scheme [grant numbers EP/W037009/1, EP/X528493/1].

References

1. International Energy Agency (IEA): The Future of Rail – Opportunities for energy and the environment. https://www.iea.org/reports/the-future-of-rail. Accessed 28 June 2025
2. Rail Safety and Standards Board (RSSB): Carbon Reduction through Improved Depot Operations. https://www.rssb.co.uk/research-catalogue/C1120. Accessed 28 June 2025
3. Network Rail: Delay Attribution Data. https://www.networkrail.co.uk/who-we-are/transparency-and-ethics/transparency/open-data-feeds. Accessed 05 May 2024
4. RiR Tranche 2: Improving the Reliability and Resilience of Train Fleet Maintenance – Virtual Maintenance Depot and Operations Optimisation (2024)
5. Huang, Y., et al.: An integrated approach for the energy-efficient driving strategy optimization of multiple trains by considering regenerative braking. Comput. Ind. Eng. **126**, 399–409 (2018). https://doi.org/10.1016/j.cie.2018.09.041

6. Chiu, C.-C., Huang, H., Chen, C.-F.: A simulation-based optimization approach for the recharging scheduling problem of electric buses. Transp. Res. E Logist. Transp. Rev. **192**, 103835 (2024). https://doi.org/10.1016/j.tre.2024.103835
7. Donhauser, T., Natterer, N., Genc, E.: Modularer Simulationsbaukasten zur Modellierung und Optimierung der Instandhaltungsabläufe in Zugdepots. In: Simulation in Produktion und Logistik 2021, pp. 15–17. Erlangen (2021)
8. Geissdoerfer, M., Savaget, P., Bocken, N.M.P., Hultink, E.J.: The circular economy – a new sustainability paradigm? J. Clean. Prod. **143**, 757–768 (2017). https://doi.org/10.1016/j.jclepro.2016.12.048
9. Department for Transport: Decarbonising Transport: "A Better, Greener Britain" (2021)
10. Louadah, H., McCluskey, L., Tucker, G., Papadakis, E.: Towards rolling stock preventive maintenance scheduling – short-term scheduling optimisation. In: CEUR Workshop Proceedings (2023)
11. Louadah, H., Papadakis, E., McCluskey, T.L., Tucker, G.: Supporting the management of rolling stock maintenance with an ontology-based virtual depot. Appl. Sci. (Switzerland) **14** (2024). https://doi.org/10.3390/app14031220

Direct and Indirect Electrification Concepts Based on Induction Heating for the Substitution of Gas-Intensive Production Processes

Alexander Fröhlich, Alexander Hackert, Anja Rautenstrauch(✉), Martin Kroll, and Till Clausmeyer

Professorship Forming Technology, Chemnitz University of Technology, Chemnitz, Germany
anja.rautenstrauch@mb.tu-chemnitz.de

Abstract. The substitution of gas-intensive production processes with electrothermal technologies represents a crucial step toward sustainable manufacturing. Although, this technological transformation still faces challenges and limitations. Next to the carbon footprint, relative price and the available grid infrastructure, electric energy often has to facilitate high-temperature, homogeneous or selective and rapid heating processes in different atmospheres. This paper presents two concepts for direct and indirect electric heating, based on electromagnetic induction, that address these fundamental challenges in industrial manufacturing processes and initiate climate-friendly and sustainable production. It showcases a solution for highly efficient hot gas generation with variable atmospheres that allow homogeneous heating and a solution for selective rapid sheet metal heating.

Keywords: gas-intensive production · electrothermal technology · heating system

1 Introduction

The transition to sustainable energy use, particularly the substitution of fossil fuels with renewable energies, is a pivotal step in achieving climate targets and fostering an environmentally sustainable industry [1]. Natural gas is one of the main heating sources in industrial applications, where it contributes CO_2-emissions of around 200 g/kWh [2]. A sustainable replacement of natural gas therefore requires a lower emission factor. Electric energy can provide this alternative, if produced by a lower carbon footprint (Fig. 1). In the generation of electric energy, the trend towards a reduction of CO_2-emissions is global. On average, this is already the case e.g., in Spain, France, Scandinavia, Canada and Brazil [3]. Apart from the CO_2-footprint and additional benefits such as logistics and required infrastructure that come along with strategic dependencies, the energy transition must ensure the competitiveness of industry.

Due to the reduction of prices for renewable energies, esp. Electrical energy, and increased costs for fossil fuels, esp. Gas in Europe, the gap between the costs for natural gas and renewable energies is narrowing again (Fig. 2). In some countries, the costs in

H. Kohl et al. (Eds.): GCSM 2025, LNME, pp. 113–121, 2026.
https://doi.org/10.1007/978-3-032-21154-5_13

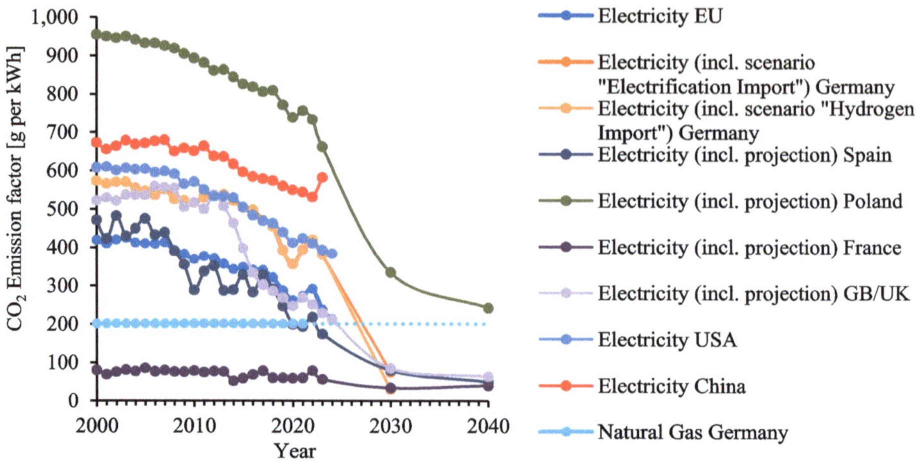

Fig. 1. Carbon intensity of electricity and natural gas for various states over time [2–8]

2023 have almost been equal. In this year, the ration between industrial gas and electricity prices was 0.99 in Finland, 0.95 in Moldova and 0.86 in Sweden [8, 9]. In other regions, electric energy be must be utilized particularly efficiently.

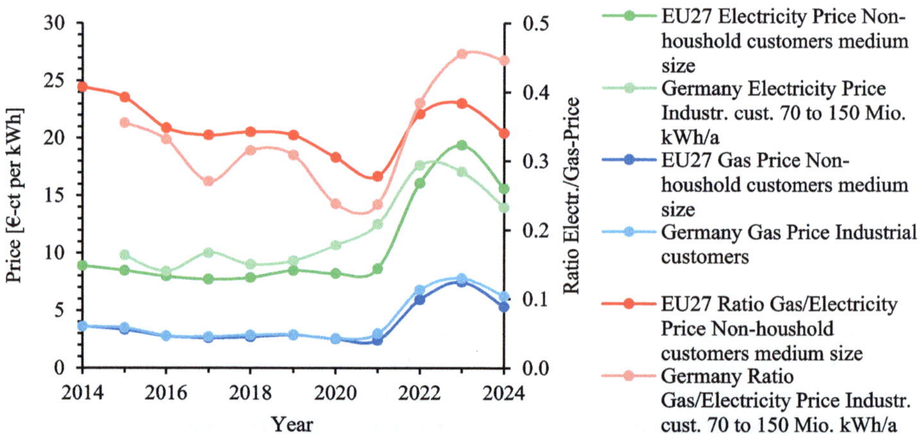

Fig. 2. Prices of electricity (EU27 incl. Electricity taxes, Germany incl. Non-refundable taxes, duties and levies, excl. Refundable taxes) and gas in Eur-ct per kWh and ratio of gas and electricity price for industrial customers in Germany over time [2, 3, 10]

Ratio of gas and electricity prices for industrial customers in various European countries showed major differences in 2023. Especially in countries such as Finland or Sweden with relatively low electricity costs, the implementation of industrial solutions using electric heat pays off particularly quickly.

The automotive industry exemplifies the implementation of the abovementioned transition, as evidenced by its ongoing efforts to reduce its carbon footprint. In numerous

European production facilities, electricity is sourced entirely from renewable sources, and balance sheet CO_2 neutrality is either being demonstrated or will be implemented in the near future [11]. This transition is further evidenced by the strategic initiatives of prominent automotive manufacturers, such as Volkswagen's "Zero Impact Factory" and Mercedes Benz's "Climate Transition Action Plan" [12] which underscore the centrality of decarbonization in their respective corporate strategies. However, despite these strides, numerous production processes continue to rely heavily on fossil fuels.

The decentralized generation of hot air with gas burners and the centralized provision of heat with hot water generators and decentralized hot air heat exchangers is the state of the art worldwide and requires enormous amounts of fossil fuels (gas). In addition, the complexity of such a system results in enormous energy losses, leading to very high operating costs.

Conventional gas-based heat generation is widespread worldwide, but it is inefficient and heavily dependent on fossil fuels. Its complexity causes high energy losses and operating costs. Switching to electrical systems with renewable energies, closed air circuits, heat recovery, and electric heating can replace fossil fuels, reduce energy consumption, and significantly cut CO_2 emissions.

2 Fundamentals of Induction Heating

Induction heating is a contactless method of heating electrically conductive materials. A generator produces a high-frequency alternating current that flows through a coil (inductor). This current flow generates an alternating electromagnetic field around the coil [13]. Induced eddy currents generate heat in electrically conductive materials due to ohmic resistance. In ferromagnetic materials (e.g., steel up to the Curie point), the remagnetization in the alternating magnetic field also contributes to heating (hysteresis losses) (Fig. 3) .

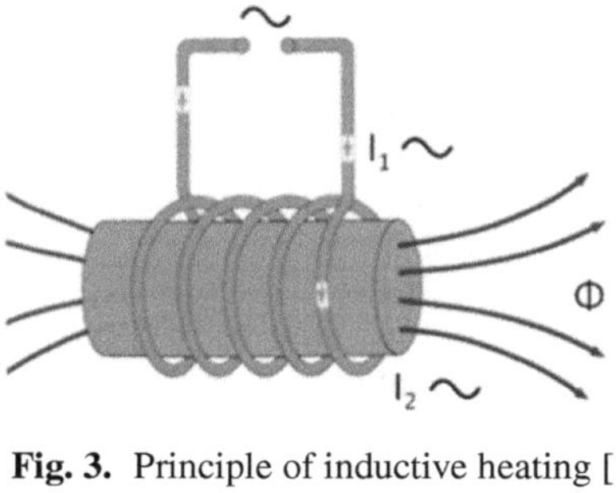

Fig. 3. Principle of inductive heating [13]

Above the Curie point, ferromagnetic materials lose their magnetic properties. With increasing frequency, the current density generated decreases radially inward from the edge of the workpiece. This decrease results from overlapping eddy currents and the resulting high resistance inside the workpiece. Penetration depth δ is the depth at which the current density J has fallen to 1 /e (=0.368). The higher the frequency, the thinner the surface layer through which the current flows [14].

3 Electrification Concepts for Heat Treatment in Production Processes

3.1 Indirect Electrification

Concept Presentation – Particularly in the area of process heat, this technology offers the potential to replace fossil fuels with electrical energy and thus significantly reduce CO_2 emissions. The basic principle is to heat a material (susceptor) that is permeable to the medium and homogeneously by induction heating. As the medium passes through the susceptor, it absorbs thermal energy and transports the heat outwards to the desired location. Particularly in the area of process heat, this technology offers the potential to replace fossil fuels with electrical energy and thus significantly reduce CO_2 emissions. There have already been studies in this context, whereby aluminum foam as a porous medium offers particular advantages for heat transfer in air, as it can act both as a susceptor and as an effective transfer medium. Investigations into the effects of special insulating materials and flow field improvement on the performance of induction air heaters have already achieved efficiency increases of 87.1% [15, 17]. Another novel concept in the form of a honeycomb induction air heater developed for high-temperature process heating applications above 900 °C was also presented [23]. The feasibility of an efficient and durable induction gas heater was demonstrated for the first time by investigating heating elements made of silicon carbide (SiC) ceramics.

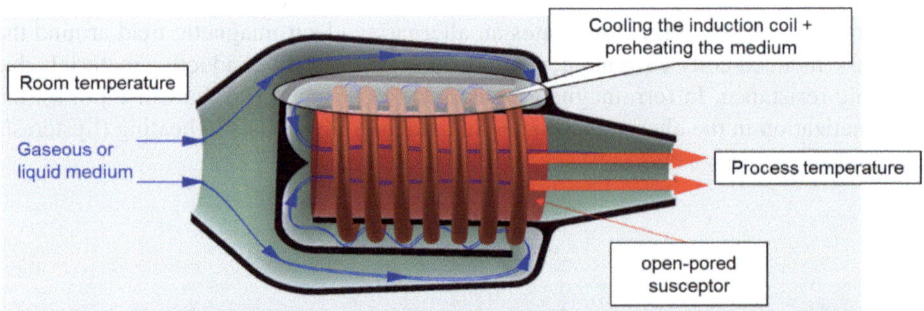

Fig. 4. Technical concept - inductive generation of hot gas

The concept shown in Fig. 4 features an innovative operating principle. Almost all of the electrical power is converted into usable heat energy. At the same time, the external water cooling of the inductors required in conventional systems is replaced by integrated cooling with the medium to be heated [18]. This means that the waste heat generated in the system from the induction coil is fed directly back into the process, resulting in high energy efficiency. A material is required that can be used with long-term stability even at higher temperatures and also offers efficient inductive heating. Impressive results were achieved using the composite material C/C-SiC. It is a silicon carbide reinforced with carbon fibres, whereby the filling level must be sufficiently high and the fiber orientation to achieve percolation within the fibre-structure. At Chemnitz University of Technology, Hackert et al. were able to heat C/C-SiC samples to temperatures above

1800 °C using induction heating [16]. Furthermore, it is crucial to equip the susceptor with specially designed channels so that the thermal energy is transferred efficiently. Thanks to innovative inductor cooling concepts, electrical energy can be converted into heat with virtually no loss, opening up new possibilities for the electrification of industrial processes. In the field of process heat in particular, this technology offers the potential to replace fossil fuels with electrical energy, thereby significantly reducing CO_2 emissions.

In the next step, the authors want to transfer this new approach to a functional demonstrator and thereby achieve significantly higher process temperatures of up to 1,200 °C, thus opening up industrial applications in the high-temperature range. With the new flow concept shown in Fig. 4, a thermal efficiency of up to 96% is aimed for.

Figure 4 shows the innovative concept whereby the medium introduced is guided past the induction coils in a targeted manner. The effect of this is that the coils are cooled by the medium, which is still relatively cold in this section, and the medium can be preheated in return. The water cooling normally required by the induction coil can therefore be omitted and significantly increases the efficiency of the overall system. This indirect heating of a susceptor offers the advantage over conductive heating systems that the geometry, surface properties and contact length of the susceptor can be specially designed for optimum heat transfer [15, 17].

Technical Risks - The implementation of inductive direct heating is associated with certain technical challenges. In particular, there is a risk of oxidation of the coil and susceptor in oxygen-containing media at temperatures above 800 °C, which can significantly affect the material durability and lifespan of the components. Additionally, inadequate cooling of the coils—especially when using gaseous media—can lead to overheating and, in the worst case, destruction of the coils.

Expected added value – In the European Union (EU) in 2022, the industry sector accounted for 25.1% of the final energy consumption. Thereof, natural gas accounted for 31.2% of final energy consumption in the EU's industry sector which amounts to 822 TWh. This consumption also emitted 165 Mio. t CO2 [19]. If only 10.2% of the natural gas consumption in the EU industrial sector could be replaced by electric heating, which represents the share of the iron and steel industry on the total final energy consumption of the industry sector, 10.1 Mio. t CO2 could be saved with an electric energy CO2 emission factor of 80 g/kWh. This factor is in the area of 2030 projections for Germany (75.0 g/kWh) [20], Spain (79.5 g/kWh) [8] and the United Kingdom (85.1 g/kWh) [9]. Further savings could be achieved by additional replacements of natural gas consumption in the chemical and petrochemical (166.0 TWh), the non-metallic minerals (131,9 TWh) and the paper, pulp and printing (66.1 TWh) sectors or if the electric energy CO2 emission factor could be further reduced, e.g. to the level of France (2023 projection: 33.9 g/kWh) [5].

Overall, inductive direct heating of fluids offers a promising opportunity to make industrial processes more energy-efficient and climate-friendly. The combination of high efficiency, direct energy input, and design flexibility not only reduces operating costs but also makes a valuable contribution to achieving climate targets.

3.2 Direct Electrification

Concept Presentation - This technological approach focuses on increasing the efficiency of existing manufacturing processes. Through the targeted integration of inductive heating technology, the cold shear cutting process used to date is to be replaced by warm cutting, which is combined directly with heat treatment within a progressive die, see in Fig. 5.

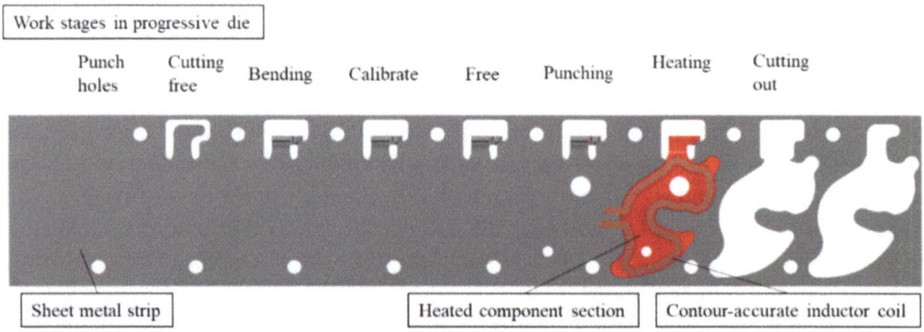

Fig. 5. Technical concept - Full surface – sequence of work steps progressive die

This allows small components to be manufactured under optimized thermal conditions. Integrating heat treatment directly into the progressive die eliminates the need for separate, downstream heat treatment. This significantly shortens the process chain and enables more efficient use of resources. The new process route also offers further advantages: longer tool life, improved component quality, and a reduction in primary energy consumption and energy-related CO_2 emissions. Instead of the previous conventional process chain—consisting of progressive die operations, followed by heat treatment and cleaning—an integrated process combination is implemented. This combines progressive die production and induction-assisted press hardening in a single step, followed only by cleaning.

Technical Risks - The implementation of the induction-based process chain involves various technical risks, particularly in the areas of integration into the forming process, electrical supply, as well as process monitoring and quality assurance. For example, the installation of the inductor and cooling system requires additional space, and the heating time must be precisely synchronized with the press cycle. Electrically, the high energy demand and potential electromagnetic interference caused by induction fields present significant challenges. Furthermore, there are risks related to process stability, such as uneven heating of components, overheating at the edges, and difficult temperature control, which necessitates real-time monitoring of temperature profiles.

Expected Added Value – The integration of inductive heating technology into the process chain offers considerable potential for increasing energy efficiency and reducing the environmental impact of industrial heat treatment processes. Currently, heat treatment of manufactured components is carried out in external hardening shops, which have an average energy requirement of around 4.835 billion watt hours per year in Germany.

This corresponds to a specific consumption of around 508 kWh per ton of steel, assuming a total consumption of 9.4 million tons per year. In comparison, induction hardening requires only 187 kWh per ton—a reduction in energy consumption of around 63% [21]. Extrapolated, this results in a theoretical savings potential of around 1,786 GWh per year if induction hardening were to be implemented across the board [22]. This energy saving not only represents an economic advantage, but also has significant ecological implications. For example, an estimate by an OEM showed that the manufacture of 100 small components using inductive heating could save around 4,176 g of CO_2 equivalents. With an annual production volume of 150,000 such parts, this results in a total saving of 6,264 kg of CO_2eq per year. A technology-integrating company predicts that the gradual expansion of the process to two additional sites will result in a reduction of 18,792 kg CO_2eq in the first year and 37,584 kg CO_2eq in the following year. These figures illustrate the sustainable added value of the new process, not only in ecological terms but also in economic terms. The replacement of conventional processes with energy-efficient induction hardening thus makes an important contribution to the decarbonization of industrial manufacturing processes while supporting the achievement of ambitious climate targets.

Despite technical challenges in integration, energy supply, and process monitoring, the induction-based process chain offers great potential for increasing efficiency and reducing CO_2 emissions. By directly combining forming and heat treatment in a single tool, process steps are eliminated, energy consumption is reduced by up to 63%, and a significant contribution is made to sustainable industrial production.

Introduction

The introduction of the induction-based process chain involves technical challenges such as limited installation space, high energy demand, electromagnetic interference, and complex temperature control. However, initial feasibility studies and successful preliminary work have demonstrated that integration of the technology is fundamentally possible and yields promising results in terms of energy efficiency and component quality. Based on this, the technology can be further optimized through targeted research to prepare it for industrial application, significantly reduce CO_2 emissions, and sustainably improve process chains.

References

1. Cleanenergywire Homepage. https://www.cleanenergywire.org/factsheets/germanys-greenhouse-gas-emissions-and-climate-targets. Accessed 21 Apr 2025
2. Umweltbundesamt, "Kohlendioxid-Emissionsfaktoren für die deutsche Berichterstattung atmosphärischer Emissionen," Umweltbundesamt, Dessau (2022)
3. Ember, "Energy Institute - Statistical Review of World Energy (2024) – with major processing by Our World in Data," Ember and Energy Institute. https://ourworldindata.org/grapher/carbon-intensity-electricity. Accessed 2 May 2025
4. Sandbag Climate Campaign CIC, EMBER. https://ember-energy.org/data/electricity-data-explorer. Accessed 2 May 2025
5. Aurora Energy Research: Carbon intensity outlook of the power sector in France from 2020 to 2050 (in grams of CO2 per kilowatt-hour) (2020). https://www.statista.com/statistics/1190067/carbon-intensity-outlook-of-france/. Accessed 2 May 2025

6. Luderer, G., Kost, C., Sörgel, D.: Deutschland auf dem Weg zur Klimaneutralität 2045 - Szenarien und Pfade im Modellvergleich, Kopernikus-Projekt Ariadne, Potsdam (2021)
7. Aurora Energy Research: Carbon intensity outlook of the power sector in Poland from 2020 to 2040 (in grams of CO2 per kilowatt-hour). https://www.statista.com/statistics/1190077/carbon-intensity-outlook-of-poland/. Accessed 2 May 2025
8. Aurora Energy Research: Carbon intensity outlook of the power sector in Spain from 2020 to 2050. https://www.statista.com/statistics/1190075/carbon-intensity-outlook-of-spain/. Accessed 2 May 2025
9. Aurora Energy Research: Carbon intensity outlook of the power sector in Great Britain from 2020 to 2050 (in grams of CO2 per kilowatt-hour). https://www.statista.com/statistics/1189677/carbon-intensity-outlook-of-great-britain/. Accessed 2 May 2025
10. European Union: Final energy consumption in industry - detailed statistics, eurostat, 04 2024. https://ec.europa.eu/eurostat/statistics-explained/index.php?title=Final_energy_consumption_in_industry_-_detailed_statistics. Accessed 02 May 2025
11. https://www.volkswagen-group.com/de/zero-impact-factory-17307. Accessed 21 Apr 2025
12. https://group.mercedes-benz.com/dokumente/investoren/berichte/geschaeftsberichte/mercedes-benz/mercedes-benz-climate-transition-action-plan-2024-de.pdf. Accessed 21 Apr 2025
13. https://inductoheat.eu/was-ist-induktion/. Accessed 21 Apr 2025
14. Wohlfahrt, D., Jürgens, R.: Induction heating of thin slabs and sheets in the rolling line, https://www.db-thueringen.de/servlets/MCRFileNodeServlet/dbt_derivate_00018417/IWK_2006_7_0_29.pdf. Accessed 21 Apr 2025
15. Unver, U., et al.: Analysis of a novel high performance induction air heater. Therm. Sci. **22**(Suppl. 3), S843–S853 (2018)
16. Hackert, A., et al.: Inductive heating of ceramic matrix composites (CMC) for high-temperature applications. Materials, **17**, 2175 (2024). https://doi.org/10.3390/ma17102175
17. Demonstration of the First-of-its-kind Ceramic Induction Air Heater, International UIE-Congress, 14 (2024). https://doi.org/10.1016/j.csite.2019.100440
18. Baig, W., Ali, H.M.: An experimental investigation of performance of a double pass solar air heater with foam aluminum thermal storage medium. Case Stud. Thermal Eng. **14** (2019). https://doi.org/10.1016/j.csite.2019.100440
19. Bartsch, C.-O.: EINE VORRICHTUNG ZUR HEISSGASERZEUGUNG UND VERFAHREN ZU DEREN BETRIEB. Europe Patent EP 4 456 668 A1 (2023)
20. Final energy consumption in industry – detailed statistics: https://ec.europa.eu/eurostat/statistics-explained/index.php?title=Final_energy_consumption_in_industry_-_detailed_statistics. Accessed 21 Apr 2025
21. Luderer, Kost, Sörgel (2021), Ariadne Report, PIK
22. Brill, S., Schibisch, D.M.: Induction hardening versus case hardening—a comparison. Elektrowaerme Int. **3**, 67–75 (2014)
23. Belik, S., Khater, O., Zunft, S.: Induction heating of a fluidized pebble bed: numerical and experimental analysis. Appl. Sci. **13**, 2311 (2023). https://doi.org/10.3390/app13042311

Deep Reinforcement Learning for Control Strategy Optimization of an Industrial Cooling Supply System in the Chemical and Pharmaceutical Industry

Tobias Lademann(✉), Arthur Stobert, Heiko Ranzau, Jonas Klingelhöfer, Manuel Scharfe, and Matthias Weigold

Institute for Production Management, Technology and Machine Tools (PTW), Technical, University of Darmstadt, Otto-Berndt-Str. 2, 64287 Darmstadt, Germany
t.lademann@ptw.tu-darmstadt.de

Abstract. Control optimization of industrial energy supply systems is essential for the global energy transition and the reduction of energy costs for industry. Deep Reinforcement Learning (DRL) can optimize the control strategy of complex energy supply systems considering volatile external influences such as electricity prices. This work implements and evaluates the DRL application for an industrial cooling supply system from the chemical and pharmaceutical industry. The simulation model of the system for training and performance evaluation is based on datasheets and historical data. Parameter identification is used to reduce the relative model error from 19% to 10% with respect to cumulative electric energy consumption over a five-month period. The DRL based operation is benchmarked against the conventional rule-based control strategy in the simulation model. The results show that the application of DRL is suitable for identifying effective and energy efficient operating strategies. However, not all operating cost factors can be reduced equally and the trade-off between different optimization targets is challenging. Based on the implementation of this work, the performance of the DRL strategy should be improved in future research.

Keywords: Energy Efficiency · Machine Learning · Parameter Identification

1 Introduction

Achieving a carbon-neutral European economy by 2050 requires a significant reduction in industrial energy consumption and a rapid shift towards renewable energy sources. With the industrial sector accounting for approximately 20% of Europe's CO_2 emissions [1], industrial energy supply systems (IESS), which convert external energy sources such as electricity into useful energy for production processes, such as cooling power, are a particularly relevant target [2]. Cooling systems alone account for 19% of the electricity consumption in the EU-28 [3] and 14% in Germany [4]. Despite increasingly volatile energy prices and competing objectives of complex IESS, significant cost and energy savings can be achieved by optimized operational strategies [5].

H. Kohl et al. (Eds.): GCSM 2025, LNME, pp. 122–130, 2026.
https://doi.org/10.1007/978-3-032-21154-5_14

Deep Reinforcement Learning (DRL) describes a machine learning paradigm based on deep learning, where agents learn to take actions A_t based on a system state S_t in an environment to maximize a cumulative reward R_t [6, 7] and offers a promising alternative to conventional control approaches [5, 8–10]. Therefore, this work implements and evaluates the DRL application for an industrial cooling supply system from the chemical and pharmaceutical industry. Section 2 describes related work while Sect. 3 presents the methodology. Section 4 details the use case, simulation model, and DRL implementation. Section 5 shows the results and Sect. 6 concludes with an outlook.

2 Related Work

Kohne et al. compared the performance of a DRL-based controller with alternative control approaches on two IESS simulation models of increasing complexity, highlighting difficulties to learn an advantageous strategy in complex systems [8]. However, the simulation models were not exact representations of real systems and were therefore not parameterized with real data [8]. Weigold et al. developed a method for the control optimization of IESS with DRL [9]. The method was validated through the application to an industrial cooling supply system where the simulation-based evaluation shows a reduction in energy costs and CO_2 emissions [9]. Ranzau et al. applied this method of Weigold et al. [9] to the IESS of a research factory including a real-world application where robust and cost-effective control was observed [10]. However, the error of the simulation model was not considered [10]. He et al. applied a recurrent network architecture to control a cooling supply system using thermal load predictions [11]. The control strategy is benchmarked based on a simulation considering historical weather data, although the simulation model is parameterized using simulated load profiles instead of actual load profiles [11]. Luo et al. optimized the operational control of a multi energy grid, consisting of different gas and electricity based supply systems as well as an electrical and thermal storage [12]. The performance was evaluated in a simulation with a synthetically generated dataset [12]. Zhang et al. implemented a DRL control for a heating, ventilation and air-conditioning (HVAC) system, whereby the Sim2Real-Gap of the simulation model of the office building was minimized by identifying selected model parameters using Bayesian calibration [13]. Li et al. conducted a systematic review of DRL applications in smart manufacturing for various applications and highlighted the simulation to reality (Sim2Real) problem as a key research challenge [7].

In summary, we identify a lack of literature addressing real IESS and quantifying as well as minimizing the Sim2Real-Gap, which limits the transferability of results from simulation environments to the actual IESS and neglects challenges related to the high complexity of these systems.

3 Methodology

The goal of this work is to develop an optimized control strategy of an industrial cooling supply system, for which we adapt the approach of Weigold et al. [9]. To transfer the operating strategy from the simulation model to the real system in the last step, the validation and Sim2Real-Gap minimization of the simulation model is of particular importance [7, 10]. We therefore extend the approach to include parameter identification (PI) methods to identify uncertain simulation model parameters. This increases the validity of the results in terms of robustness and cost savings in the real system and thus increases user acceptance. The transfer of the optimized strategy to the real system requires an extensive expert review and further methods and is therefore out of scope for this work [9]. The chosen methodology based on Weigold et al. [9] is: **Step 1:** System boundary, data accumulation and framework, **Step 2:** system modelling, parameter identification and model validation, **Step 3:** Implementation of DRL algorithm, **Step 4:** Comparative study and analysis.

4 Use Case

In this work, a cooling supply system from the chemical and pharmaceutical industry with an average electrical energy consumption of 340 MWh per month is optimized. This chapter describes steps 1–3 of the methodology.

4.1 System Overview

The thermal energy supply system consists of a cooling water and cold water circuit, operating between 16.0–26.5 °C and 5.0–7.5 °C, respectively. The IESS is shown in a simplified hydraulic scheme in Fig. 1. The cooling water circuit consists of four parallel cooling towers (CT), a bypass valve and one buffer storage. The cold water circuit consists of three compression chillers (CC), one absorption chiller (AC) and a buffer storage. An additional CC is used to operate an ice storage system. A pump group consisting of four parallel pumps supplies the CC from the cooling water buffer and another similar pump group supplies the production processes from the cold water buffer. Each CC is equipped with a separate pump in the return flow on the cold water side, which is not visualized in the scheme. Depending on the required flow temperature, manufacturing processes are supplied by either the cold or cooling water buffer, both represented by thermal loads.

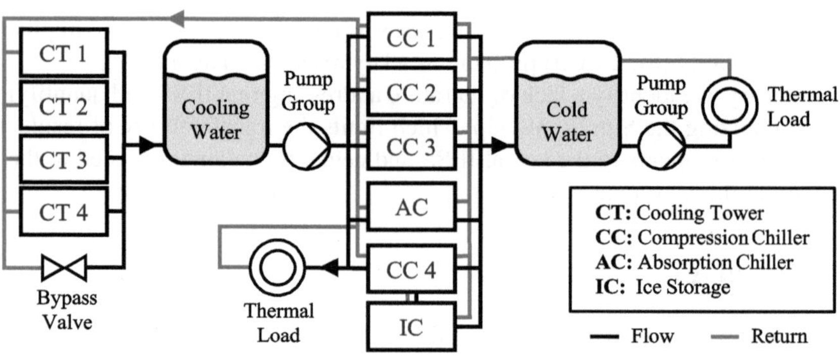

Fig. 1. Simplified scheme of the hydraulic circuit: the system boundary includes the depicted components

The conventional control is realized via PI-Controllers and cascaded hysteresis control depending on temperature setpoints in the storages. The system pressure is controlled via a PI controller for the pumps in combination with cascaded control of the cooling towers. The bypass valve is used to ensure that the cooling water temperature does not fall below the required minimum. However, the bypass valve is also operated manually, whereby no generally applicable strategy is known. The ice storage system is operated manually as well, when peak loads occur.

In step 1 of the methodology, *eta_utility* is chosen as a framework [14]. Data is aggregated from 1 August–09 December 2024 with a 15-min resolution, which is used for modelling and validation of the system. Since the actual electricity price is unknown, the Day-Ahead Price and CO_2 intensity is aggregated from *Agorameter* [15].

4.2 Simulation Model and Parameter Identification

This section describes step 2 of the methodology. The IESS is modeled using *Modelica* [16]. The relevant parameters such as nominal thermal and electrical power of the components (CC, CT, AC, pumps, buffer and ice storage) are extracted from data sheets. The model receives weather data (ambient temperature, relative humidity and air pressure) and thermal load as inputs. Simulation outputs contain the electrical power and temperatures for cold and cooling water. The conventional control strategy is implemented in *Modelica* to avoid small step sizes and thus increased simulation times.

To evaluate the accuracy of the model, it is simulated using the available real input data. The thermal load in the cooling water is assumed to be 440 kW based on temporary measurements, since no load profiles are available. Two data gaps from 9–12 October and 27 November–1 December are filled by carrying forward the last available value. The initial relative error of the simulation model for cumulative electricity energy over the entire period is 18.52% (see Table 1 and Fig. 2).

To reduce this gap, a systematic parameter identification (PI) approach is conducted. The PI uses the work of Ranzau [17], which offers an implementation of the method *curve_fit* with the Trust Region Reflective-Algorithm from the python library *Scipy* [18]. This function performs non-linear least squares reduction to fit a function to data by varying given parameters in a defined value range. Cumulative energy consumption is chosen as the function to be fitted because reducing energy consumption is also the main objective in control optimization and should therefore be modelled accurately to avoid biasing the DRL agent's prediction during training.

The parameters chosen for PI are the nominal electrical power of the 4 CC's as well as the nominal thermal power of CC 3 and CC 4, since these six model parameters are not available in data sheets but have a significant influence on cumulative energy consumption. The parameter range is set to $\pm$ 25% around the baseline values. No identified value falls on the boundary limit, which indicates a sufficient parameter range. To reduce computing time, three 2-week periods are selected for the PI, resulting in three identified parameter sets. Each parameter set is used to simulate both the respective 2-week period and the entire period. Additionally, a simulation over the whole period is conducted with the average of the three parameter sets.

Table 1 shows the relative errors of electric energy for each parameter set. The parameter set Training 2 scores the lowest error for the whole dataset and is therefore

Table 1. Relative model errors for cumulative electric energy

Parameter set	Initial Error	Error training dataset	Error whole dataset
Initial	18.52%	-	18.52%
Training 1 (Aug. 01–15)	15.72%	0.71%	9.94%
Training 2 (Sep. 01–15)	7.54%	3.50%	**9.82%**
Training 3 (Oct. 05–19)	23.33%	20.71%	21.11%
Average (Training 1–3)	-	-	13.83%

determined as the final parameter set. The high error in Training 3 is likely due to minimal system runtime or the existing data gap. Overall, the relative model error in cumulative electric energy consumption can be reduced from 18.52% to 9.82%.

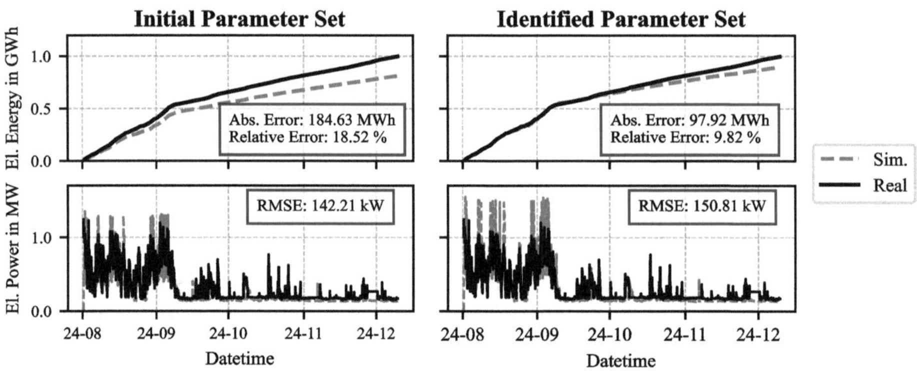

Fig. 2. Simulated electrical energy and power for the initial and identified parameter set

The comparison of the initial and final parameter sets is depicted in Fig. 2. Note that the root mean square error (RMSE) of the electric power is higher with 150.81 kW compared to 142.21 kW with the identified parameter set. This is attributed to the higher load peaks and differing switching times. Uncertainties in the simulation model remain regarding the assumption of a constant load in the cooling water and the operating strategy of the bypass valve and IC as described in the previous section.

4.3 Deep Reinforcement Learning Implementation

In step 4 the observation and action spaces are defined and the DRL algorithm and hyperparameters are selected. The action space is multi-discrete and is derived from the actual building automation implementation. The following systems are operated with a relative set point: CT, bypass-valve, pumps of CC 1–3 and the pump of the AC. The CC and AC action sets the flow temperature between 4–6 °C and 5–6 °C, respectively. The IC operates in three modes: *off*, *charge* and *discharge*. An additional action switches CC 4 between direct supply and charging the IC. The observation space consists of the flow temperatures in cold and cooling water circuits, flow temperature of CT, AC and CC, volume flow rate of the AC and CC as well as flow temperature and state of charge of the IC. The observation space also includes the previous system state, electricity price and weather data. A total of 15 actions and 38 observations are defined. The action space allows over 2×10^{10} possible combinations.

Proximal Policy Optimization (PPO) is selected as the DRL algorithm [19] and hyperparameters are adapted from Ranzau et al. [10]. A multilayer perceptron (MLP) with a size of 500, 400 and 300 is used for each the actor and critic network. The episode length is 3 days with a step size of 180 s. A linear schedule for the learning rate from 4×10^{-4} to 2×10^{-5} with 32 parallel environments is used.

The consideration of the relevant terms and their weights in the reward signal $\mathrm{R_t}$ is essential to obtain a strategy that meets the requirements [9]. For this work, the function is adapted from Ranzau et al. [10] due to the similarity in system complexity and shown in the following equation.

$$R_{\mathrm{t}} = R_{\mathrm{t}}^{\mathrm{Energy}} + R_{\mathrm{t}}^{\mathrm{Temperature}} + R_{\mathrm{t}}^{\mathrm{Switching}} + R_{\mathrm{t}}^{\mathrm{Fallback}}$$

The reward represents the overall operating costs and includes costs for electricity (Day-Ahead Price), costs for temperature violations in the cooling and cold water supply exceeding the requirements described in Sect. 4.1 (1 € per timestep), costs for switching (1–20 €) and costs for the activation of the fallback strategy when temperature far exceed the limits (50 €). The selected costs are adapted from Ranzau et al. [10].

5 Results and Discussion

The DRL agent is trained on a split of the available dataset, with the first and last week of the dataset dedicated to evaluation (step 4) considering different weather and load conditions. With 400 training episodes of 6 days each, training takes 31.3 h on a server computer. Figure 3 shows the DRL actions as well as the feed temperature in the cold and cooling water for both the DRL and the conventional strategy for the first evaluation week. The DRL strategy reduces energy consumption by 2% from 82.2 MWh to 80.5 MWh and CO_2 emissions by 0.6%. However, there is an increase of 0.2% in energy costs. The temperature limits are exceeded less frequently in cooling water than by the conventional strategy, but more frequently in cold water, which leads to higher overall operating costs of 15 623 €, which is an increase of 16.9%.

For the second evaluation week, the overall operating costs are reduced by 3.9%. However, the DRL agent shows an increase of 83.7% in energy costs, 84% in energy consumption and 83.6% in CO_2 emissions compared to the conventional control. This is due to the operation of both the AC and CC1, while the conventional controller only activates the AC. This strategy by the DRL agent can reduce costs for temperature violation by 76.1%. In absolute terms, the temperature violation costs of 4 063 € in conventional operation accounts for a significant share of the total costs of 7 576 €.

In both scenarios no effective use of the IC can be observed, which could have been an advantage of the DRL due to manual conventional operation. The results indicate that the complexity of the IESS and the reward function weighting cause the DRL agent to focus on operational stability, instead of effective strategies to reduce energy costs.

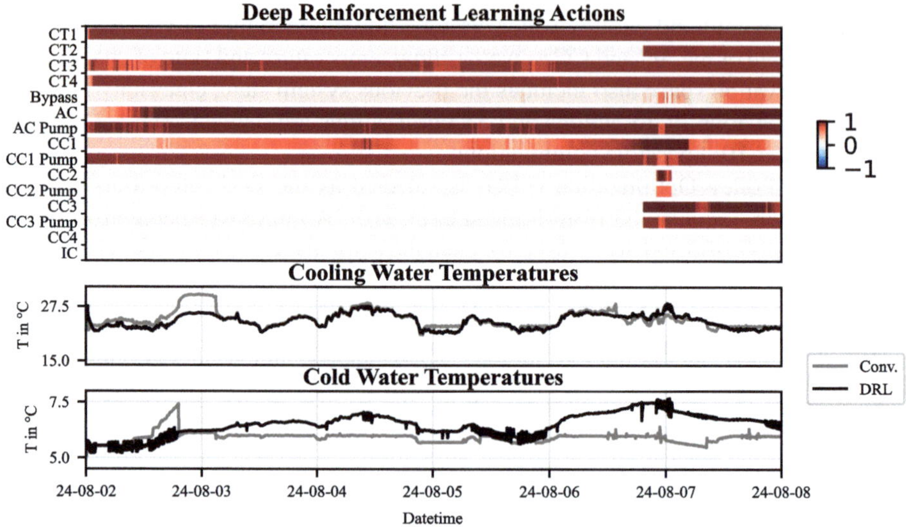

Fig. 3. Simulation of DRL agent for evaluation week 1 with DRL actions *on* (1), *off* (0) and *discharge* (–1, only IC) and the temperatures compared to the conventional (conv.) simulation

6 Outlook

The results show that the application of DRL with the applied method is suitable for identifying effective and energy efficient operating strategies in complex IESS. However, not all cost terms can be reduced and the trade-off between different optimization targets is challenging. Therefore, the performance of the DRL agent should be optimized in the next step. To balance the different optimization targets and cost weights, expert knowledge and review is required. We expect that the simple reduction of temperature costs will lead to instabilities in the training process, which must be addressed accordingly [5, 7]. To prevent inefficient or undesirable strategies, expert knowledge can be integrated, e.g. by expert demonstrations, which can support finding suitable and accepted strategies [7]. To handle the complexity of the IESS, multi-agent reinforcement learning (MARL) can be implemented [5, 7]. To optimize the strategy for the IC, thermal load forecasts should be included [11]. Finally, an optimized strategy should be transferred to the real system, whereby the remaining model error should be evaluated and further reduced if needed [9].

Acknowledgements. The authors gratefully acknowledge the financial support of the project EISKIG (Grand Number 03EN6012A) and SynErgie (Grant Number 03SFK3A0–3) and thank Projektträger Jülich (PtJ) for the project supervision.

References

1. European Environment Agency. EEA greenhouse gases — data viewer. https://www.eea.europa.eu/ds_resolveuid/5e8d1c3335a342b1822a39b5eb964715. Accessed 15 Apr 2025

2. Buoro, D., Casisi, M., De Nardi, A., Pinamonti, P., Reini, M.: Multicriteria optimization of a distributed energy supply system for an industrial area. Energy **58**, 128–137 (2013)
3. Reitze, F., Toro, F.A., Fleiter, T., Rehfeldt, M., Schön, M.: Prozesskältebedarf in Industrie und Gewerbe - wo liegen die Bedarfsschwerpunkte in Europa?: Themenblock 5: Industrie 2017
4. VDMA, 2020. Energiebedarf für Kältetechnik in Deutschland: Eine Abschätzung des Energiebedarfs von Kältetechnik in Deutschland nach Einsatzgebieten (2017)
5. Panten, N.: Deep Reinforcement Learning zur Betriebsoptimierung hybrider industrieller Energienetze, Darmstadt (2019)
6. Sutton, R.S., Barto, A.G.: Reinforcement Learning: An Introduction, 2nd edn. MIT Press, Cambridge (2018)
7. Li, C., Zheng, P., Yin, Y., Wang, B., Wang, L.: Deep reinforcement learning in smart manufacturing: a review and prospects. CIRP J. Manuf. Sci. Technol. **40**, 75 (2023)
8. Kohne, T., Ranzau, H., Panten, N., Weigold, M.: Comparative study of algorithms for optimized control of industrial energy supply systems **3** (2020)
9. Weigold, M., Ranzau, H., Schaumann, S., Kohne, T., Panten, N., Abele, E.: Method for the application of deep reinforcement learning for optimised control of industrial energy supply systems by the example of a central cooling system. CIRP Ann. **70**(1), 17–20 (2021)
10. Ranzau, H., Lademann, T., Borst, F., Ozen, O., et al.: Application of Deep Reinforcement Learning for the Control of a Complex Industrial Energy Supply System, in Decarbonizing Value Chains, p. 351. Springer, Cham (2025)
11. He, K., et al.: Predictive control optimization of chiller plants based on deep reinforcement learning. J. Build. Eng. **76**, 107158 (2023)
12. Luo, Y., Liu, C., Lai, Q.: Optimal scheduling for a multi-energy microgrid by a soft actor-critic deep reinforcement learning. In: 2022 IEEE Power & Energy Society General Meeting (PESGM), p. 1. IEEE, Piscataway (2022)
13. Zhang, Z., Chong, A., Pan, Y., Zhang, C., Lam, K.P.: Whole building energy model for HVAC optimal control: a practical framework based on deep reinforcement learning. Energy Build. **199**, 472 (2019)
14. Grosch, B., Ranzau, H., Dietrich, B., Kohne, T., et al.: A framework for researching energy optimization of factory operations **5** (2022)
15. Agora Energiewende, 2023. Agorameter: Modellversion 3.0, Berlin
16. Modelica Association. Modelica Standard Library. https://github.com/modelica/ModelicaStandardLibrary. Accessed 24 April 2025
17. Ranzau, H.: Anwendung von DRL zur Betriebsoptimierung industrieller Energieversorgungssysteme - Programmcode und ausgewählte Ergebnisse. TU Darmstadt (2024)
18. Virtanen, P., Gommers, R., Oliphant, T.E., Haberland, M. et al.: SciPy 1.0: fundamental algorithms for scientific computing in Python. Nat. Methods, **17**, 261 (2020)
19. Schulman, J., Wolski, F., Dhariwal, P., Radford, A., Klimov, O.: Proximal policy optimization algorithms (2017)

Designing Energy-Efficient Industrial Machines Based on Parametrised Life Cycle Assessment

Anastasiia Timofeeva[1(✉)], Giampaolo Campana[2], Gregory Peters[3], and Maurizio Fiorini[1]

[1] Department of Civil, Chemical, Environmental and Materials Engineering, University of Bologna, Via U. Terracini 28, 40131 Bologna, Italy
anastasiia.timofeev2@unibo.it

[2] Department of Industrial Engineering, University of Bologna, Viale del Risorgimento 2, 40136 Bologna, Italy

[3] Division of Environmental Systems Analysis, Department of Technology Management and Economics, Chalmers University of Technology, Chalmersplatsen 4, 41296 Gothenburg, Sweden

Abstract. This paper addresses the lack of data on the total energy demand of industrial washing processes. It presents an approach to designing energy-efficient industrial automated machines based on a parametrised Life Cycle Assessment (LCA) as a key analytical tool. In particular, the environmental performance of an industrial washing machine designed for the pharmaceutical sector is evaluated throughout its LCA performed in a cradle-to-grave boundary using the ReCiPe 2016 midpoint calculation method. Parametrisation in LCA enables modelling of different energy consumption scenarios based on various factors of the machine's washing cycle and usage mode. This method allows manufacturers to compare energy-saving designs, operational improvements and sustainable energy strategies without rebuilding the entire LCA model. Furthermore, a Total Estimated Energy Consumption (TEEC) framework, including operational states, has been applied to achieve a more comprehensive energy demand. Indeed, determining energy parameters for different operating states allows for a more detailed environmental impact evaluation.

This work supports the development of sustainable industrial machines, aligning with circular economy principles and eco-design strategy. The findings revealed that the use phase contributes the most to environmental impact, mainly due to energy consumption. The analyses allowed for the identification of critical components for their improvement. Besides, optimisation scenarios are proposed to enhance the machine energy efficiency.

Keywords: Sustainable Manufacturing · Life Cycle Assessment · Industrial Washing Machines · Energy Efficiency · Eco-Design

1 Introduction

Life Cycle Assessment (LCA) is today a fundamental approach for evaluating the environmental impacts caused by materials, products and processes over their whole life cycles. This necessitates the estimation of material and energy flows, which may mean

H. Kohl et al. (Eds.): GCSM 2025, LNME, pp. 131–138, 2026.
https://doi.org/10.1007/978-3-032-21154-5_15

consulting databases, such as the Ecoinvent, Sphera and Agri-footprint databases. Additionally, parametrised LCA is an approach that estimates variable parameters in the foreground systems to assess environmental impacts under varying conditions.

Although LCA is commonly used to assess environmental performance, its implementation for specific industrial machines remains limited, especially in the context of operational data from automated systems, which constrains the potential for detailed sustainability assessment at a professional level. Indeed, industrial automated machines are complex systems designed to perform specific manufacturing or processing tasks with minimal human intervention by integrating mechanical, electrical, and digital components. These machines are commonly used in various sectors such as pharmaceuticals, automotive, electronics, and food production, where they execute functions like assembly, material handling, packaging, or cleaning with high precision and consistency. The primary advantages of industrial automation include increased productivity, improved product quality, reduced operational costs, and enhanced workplace safety. In the pharmaceutical sector, automated washing machines ensure sterilisation and product safety, preventing contamination and meeting regulatory standards [3]. However, they have significant environmental impacts due to high energy consumption during the use stage.

Many researchers have focused on developing energy consumption models by associating them with different process parameters. Zeulner et al. presented an Extended Energy Modeling Approach, that has been developed for the machining process of vertical pick-up turning machines. This approach allows for the use of a single measurement campaign to model the energy demand of process variations on the machine tool, thereby reducing the need for further measurements [4]. Kibira et al. developed a systematic approach for identifying and applying Key Performance Indicators (KPIs) aimed at evaluating, monitoring, and enhancing the environmental performance of manufacturing processes [5, 6]. The Consumption Performance Sustainability Index (CPSI), a dimensionless index developed originally for additive manufacturing, has been adapted for CNC (Computerised Numerical Control) machining to serve as a complementary tool to LCA, incorporating process-specific parameters often excluded from conventional assessments [7].

The ISO 14955 standards provide a specialised methodological framework for assessing and enhancing the energy efficiency of machine tools by considering their operational behaviour, design parameters, and energy flows. It not only enables the calculation of total energy consumption but also allows for a detailed breakdown of energy use across different machine tool functions and subsystems. This functional decomposition makes it possible to identify specific components or operational states responsible for the highest energy demand [8]. Since ISO 14955 is specifically designed for machine tools that operate across multiple functions and distinct operational states, its methodological approach is particularly suited for analysing energy flows at the scale of industrial systems. Therefore, it is reasonable to consider extending the model to industrial automated equipment, which similarly performs diverse operations and exhibits variable energy consumption patterns depending on the operating state.

This study aims to quantify the environmental impacts of automatic industrial washing machines associated with life cycle stages, thereby identifying critical areas for improvement. For this purpose, a design efficiency approach, mainly based on a parametrised LCA, has been implemented on a washing machine for the pharmaceutical sector with the aim of identifying critical areas that can be the object of improvements.

2 Methodology

The implemented LCA adopts cradle-to-grave system boundaries, which include the extraction and production of raw materials, machine assembly, transportation, usage, and end-of-life stages. Primary data was collected from the manufacturer, accessible datasheets, and the Ecoinvent v3.10 database, which provides the background data. The system was modelled with SimaPro v9.6 using the ReCiPe 2016 Midpoint (H) impact assessment method. The Functional Unit (FU) was defined as the washing process of mechanical parts.

The designing for energy-efficient machines has been implemented according to ISO 14955, adapted and applied to an industrial automatic washing machine [8]. The average performance of the machine tool represents the estimated amount of energy supplied according to the time shares of defined operating states of the machine. These shares are determined based on a single production year.

The assessment of the time share of operating states is based on the individual distribution of machine states, which reflects the machine's actual usage scenario. It includes both active and inactive periods of machine and divided into following states that will be sub-indexed as: "Off", "Standby", "Setup", "Ready", "Processing" and "Emergency Stop". For instance, the "Emergency Stop" state means that the operator has stopped the machine for safety reason. In the following paragraphs it will be sub-indexed as "Estop". If individual time shares are unavailable, the default state distribution provided in ISO 14955 shall be applied [8].

To define the operating state shares, the manufacturer provided data describing a full-year reference working scenario. It includes the durations of both active and inactive periods and measured values of power and time for the "Off" and "Processing" operating states. Data for the remaining operating states were determined using default ISO values, supplemented by empirical modelling of the power consumption, based on values from the scientific literature review (Table 1) [9]. Based on the measured power values and the real-time shares, the estimated annual energy supply of the machine is calculated according to Eq. 1 [8]:

$$\begin{aligned} \mathrm{E}_{\mathrm{year}} = (&\mathrm{S}_{\mathrm{off}} \times \mathrm{P}_{\mathrm{off}} + \mathrm{S}_{\mathrm{standby}} \times \mathrm{P}_{\mathrm{standby}} + \mathrm{S}_{\mathrm{setup}} \times \mathrm{P}_{\mathrm{setup}} + \mathrm{S}_{\mathrm{ready}} \times \mathrm{P}_{\mathrm{ready}} \\ &+ \mathrm{S}_{\mathrm{processing}} \times \mathrm{P}_{\mathrm{processing}} + \mathrm{S}_{\mathrm{Estop}} \times \mathrm{P}_{\mathrm{Estop}}) \times (365 \times 24) \end{aligned} \tag{1}$$

where "S" is for the share value of machine operating states per year, and "P" is the average power during machine operating states. The calculated annual energy demand E_{year} for each operating state identifies which are energetically significant. This allows for the Functional-Oriented Analysis (FOA) to be conducted in accordance with [8].

Furthermore, the obtained annual energy demand of the machine was scaled to the lifetime of the equipment using Eq. (2–4), see Sect. 3.2, and subsequently integrated into the use stage of the LCA model. This enables a quantitative assessment of the energy-related environmental burden for a specific production case, which is particularly relevant when comparing technological alternatives. This approach ensures the integration of real operational performance into the LCA, rather than relying on generic or average data approximations.

To assess energy consumption during the industrial washing process, a parametrised LCA was conducted based on two energy scenarios. The first, baseline scenario considered only the energy consumed by the equipment during active operational phases, such as washing, rinsing, and drying. In contrast, the second, extended scenario included energy consumption encompassing all operational states. This approach allowed for a more accurate estimate of the Total Estimated Energy Consumption (TEEC) during use stage and, consequently, a more comprehensive evaluation of the environmental impact.

LCA was further parametrised using the TEEC data obtained from the extended operational analysis. To evaluate the influence of the electricity source on environmental performance, three distinct electricity mixtures were incorporated into the model. Each mix reflects a different energy profile based on regional supply characteristics: Mix_1 combines electricity from natural gas and alpine reservoir hydropower; Mix_2 is dominated by hydropower; and Mix_3 primarily relies on natural gas. This parametrisation allows for a more detailed assessment of the environmental impacts associated with electricity consumption, highlighting the sensitivity of LCA results to the energy mix composition, particularly in terms of greenhouse gas emissions.

3 Results and Discussion

3.1 Identification of Share Values for Each Operation State

To accurately estimate the operational energy behaviour of a machine within its annual production context, a detailed time allocation analysis was performed based on a representative industrial scenario. The modelled production schedule assumes 250 active production days per year with 16 h of daily operation, of which 11.16 h are dedicated to the processing state. The remaining 8 h of each working day are categorised as non-operative – "Off" state, and an additional 115 non-working days per year are considered entirely inactive, as 24 h/day in "Off" state. Based on this structure, the total yearly duration, equal to 8760 h, was decomposed into the mentioned operating states (Table 1). Time shares for non-processing states were distributed proportionally according to ISO 14955 default values, enabling the derivation of empirical annual hour distributions per state.

The resulting share of each operating state over one year is shown in Table 1. It reveals the dominance of the "Off" state, which accounts for approximately 56.1% of total time, due to both scheduled downtimes and production breaks. The "Processing" state contributes 31.9%, reflecting the core value-generating activity. Other states represent minor shares: "Standby" (5.2%), "Setup" (3.5%), "Ready" (1.7%), and "Emergency Stop" (1.7%). These time-based distributions are critical for calculating annual energy consumption per operational mode and serve as the basis for FOA of energy performance.

The average power consumption was determined across operating states using a hybrid data acquisition strategy, combining measured data with literature-derived estimations (when measurements are unavailable). Specifically, the average power values for the "Off" and "Processing" states were obtained directly from the washing machine

Table 1. Parameters for the calculation of supplied energy for the share of operating states

	Off	Standby	Setup	Ready	Processing	Emergency stop
Share per year (%)	56.1	5.2	3.5	1.7	31.9	1.7
Share value per year, *S*	0.561	0.052	0.035	0.017	0.319	0.017
Average power *P*, kW	0.02	3.945	2.63	1.35	15.78	1.35

manufacturer, based on standardised operating conditions. These values provide a high-confidence baseline for the most energetically dominant states within the annual cycle. In contrast, for states such as "Standby", "Setup", "Ready", and "Emergency Stop", no direct measurements were available. Therefore, average power values for these states were modelled based on empirical data from published studies [9, 10], including prior implementations of ISO 14955 and related energy performance analyses of comparable machines. This approach ensures methodological consistency while acknowledging data availability limitations, thereby enabling a complete estimation of the machine's annual energy profile.

3.2 Total Estimated Energy Consumption Over the Lifetime of the Machine

The estimated energy supplied by a machine for one production year under the given circumstances is defined as the sum of the power consumed during a given time and can be expressed mathematically based on Eq. 1 and Table 1, the result of which is presented in Eq. 2.

$$E_{year} = 4.72 \times 10^4 kWh \quad (2)$$

When multiplied by its expected lifetime, a machine's annual total energy demand yields a cumulative value representing TEEC over the entire use stage (Eq. 3–4):

$$E_{total} = E_{year}(kWh) \times \mathrm{t}_{\mathrm{lifetime}}(year) \quad (3)$$

$$E_{total} = 4.72 \times 10^4 * 20 = 9.44 \times 10^5 kWh \quad (4)$$

E_{total} can be directly integrated into an LCA model as a quantitative input for the use stage. Incorporating lifetime energy demand allows a more accurate estimation of environmental impacts, particularly in categories such as global warming potential (GWP) and resource scarcity.

3.3 Parametrised Life Cycle Assessment

The results of LCA demonstrate that the use phase is the dominant contributor across many environmental impact categories. The results indicate that impact categories such as freshwater eutrophication (99.9%) and stratospheric ozone depletion (98.7%) are probably driven by using detergent during the operational phase. Similarly, water consumption (98.4%) can be attributed to the intensive use of water during the washing

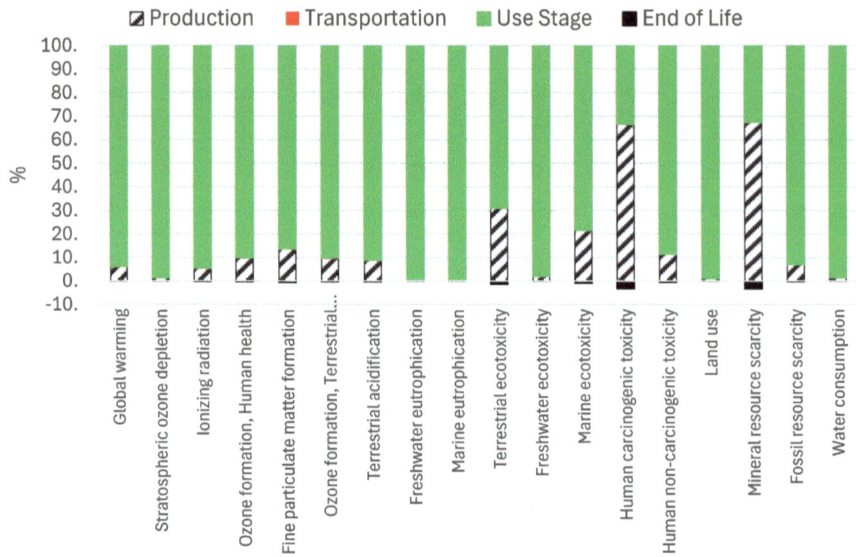

Fig. 1. LCA of an automated machine within cradle-to-grave system boundaries.

process. Furthermore, the high contribution to GWP (93.9%) is primarily associated with elevated electricity consumption during equipment operation (Fig. 1).

Conversely, impact categories such as human carcinogenic toxicity and mineral resource scarcity show a relatively minor contribution at the use stage, indicating that materials extraction and production processes also represent significant environmental hotspots. The transportation stage makes the most minor contribution, and its impact is negligible. On the other hand, the disposal stage recovers the environmental impact through a high rate of recycling of the metal materials, which are the dominant components in the production stage. These findings underscore the critical importance of operational efficiency and energy optimisation during the use stage to reduce the overall environmental footprint of the machine.

The comparative analysis of LCA at the use stage, Fig. 2, reveals that modelling with estimated annual energy supply results in notably higher environmental impacts across all damage categories that are most relevant for the impact assessment of energy, compared to the baseline scenario.

These differences emphasise the crucial significance of including total estimated energy consumption of machine during the use stage in the LCA. Neglecting this parameter leads to an underestimation of impact and reduces the accuracy of sustainability assessments, especially for energy-intensive systems operating over a long lifetime.

The comparative evaluation of the electricity mixtures reveals significant differences in their environmental performance. Mix_3 exhibits the highest impact across nearly all categories, particularly for GWP, ozone formation, toxicity, and resource depletion, indicating a high reliance on carbon-intensive or polluting energy sources (Fig. 3).

In contrast, Mix_2 demonstrates the lowest overall impact, suggesting a more significant share of renewable energy sources in its composition. Mix_1 shows intermediate

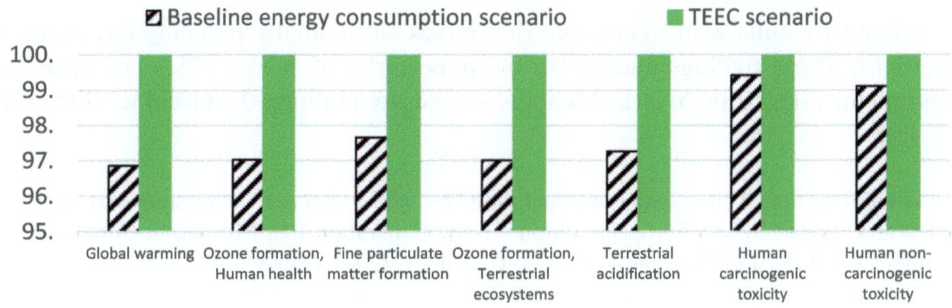

Fig. 2. Use stage comparison of annual energy demand.

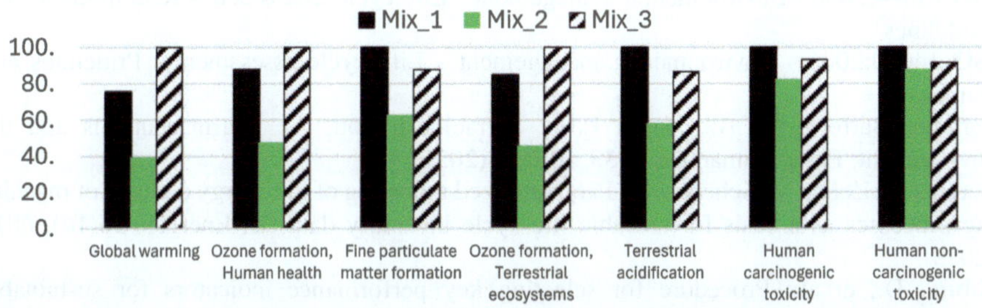

Fig. 3. LCA comparison based on energy sources.

results, with notable impacts in particulate matter formation and toxicity. These findings underline the crucial necessity of LCA parameterisation, especially for energy-intensive systems, and emphasise that integrating renewable energy sources can substantially mitigate life cycle environmental damage.

4 Conclusion

The analysis demonstrates that the use stage is the primary driver of environmental impact in the life cycle of the industrial automatic machine under investigation, primarily due to its high energy consumption. This result confirms what is generally expected for production machines.

Accurate modelling of operational time shares and energy demand is essential for reliable LCA outcomes. Precise measurements are necessary to acquire the needed data for this analysis. Alternatively, when some data cannot be directly measured, models can be considered in order to evaluate the missing data. This analysis is a first step to identify operational stages that need attention and can be considered to reduce energy consumption.

To identify possible scenarios related to relevant aspects, for example, the energy mix, a parametrised LCA has been implemented. The choice of electricity mix significantly affects impact results, with renewable-rich mixes substantially reducing environmental burdens. These findings underscore the importance of energy efficiency measurements and the integration of cleaner energy sources to minimise the machine's life cycle footprint.

Acknowledgements. This research was supported by the Ministry of Universities and Research (MUR), and funded by the European Union - Next Generation EU, Mission 4, Component 2, Investment 3.3 (Ministerial Decree 117/2023) CUP J33C22001470009.

References

1. ISO 14044:2006 - Environmental management - Life cycle assessment - Requirements and guidelines
2. ISO 14040:2006 - Environmental management - Life cycle assessment - Principles and framework
3. Sammut Bartolo, N., Azzopardi, L.M., Serracino-Inglott, A.: Pharmaceuticals and the environment. Early Human Dev. **155**, 105218 (2021)
4. Zeulner, J., Zeller, V., Schebek, L.: Parameterized modeling of the energy demand of machining processes as a basis for reusable life cycle inventory datasets. Energies **16**(16), 6011 (2023)
5. Kibira, D., et al.: Procedure for selecting key performance indicators for sustainable manufacturing. Proc. ASME MSEC **2017**, 1–8 (2017)
6. Kibira, D., et al.: Procedure for selecting key performance indicators for sustainable manufacturing. J. Manuf. Sci. Eng. **140**(011005), 1–7 (2018)
7. De Bernardez, L., et al.: A proposal for an extension of the consumption perfor-mance sustainability index to machining processes. In: LNME, pp. 721–729, Springer, Cham (2025)
8. ISO 14955 Machine tools — Environmental evaluation of machine tools
9. Zhou, L., et al.: Energy consumption model and energy efficiency of machine tools: a comprehensive literature review. J. Clean. Prod. **112**, 3721–3734 (2016)
10. Loffredo, A., Frigerio, N., Matta, A.: Energy efficiency improvement of industrial parts washers using state control. J. Manuf. Lett. **35**, 1303–1311 (2023)
11. Dybinski, O., et al.: Modeling industrial scale washing machine process using artificial neural networks. J. Environ. Integr, **9**(4), 1933–1946 (2024)

Enhancing Paint Utilization in Railcar Manufacturing Using a Sustainable Deep Deterministic Policy Gradient Approach

Olugbenga Adegbemisola Aderoba[1,2(✉)], Tshifhiwa Nenzhelele[1], Jan Adriaan Swanepoel[1], and Genevieve Bakam[1]

[1] Department of Industrial Engineering, Tshwane University of Technology, Pretoria, South Africa
olugbenga.aderoba@gmail.com

[2] Department of Mechanical Engineering, Landmark University, Omu-Aran, Nigeria

Abstract. Efficient application of paints is crucial in railcar manufacturing to minimize waste and environmental impact while ensuring high-quality coatings. This study presents a sustainable approach to optimizing paint utilization in rail car manufacturing by applying the Deep Deterministic Policy Gradient (DDPG) algorithm. This was achieved by dynamically adjusting painting parameters such as the spray pattern, spray flow rate, nozzle angle, and distance to minimize the consumption of paints and solvents while maintaining uniform coverage and quality. Real-time sensor data from automated painting systems and historical painting records were utilized to learn optimal spraying policies that balance resource conservation with production requirements. Through simulations and experiments in a virtual manufacturing environment, the approach's effectiveness in improving paint application efficiency and reducing waste generation was demonstrated. The scalability and adaptability of the DDPG-based optimization framework are evaluated, highlighting its potential for practical implementation in real-world railcar manufacturing facilities. This research contributes to advancing sustainable practices in railcar manufacturing by providing a data-driven approach to optimize paint utilization, enhancing efficiency, and reducing the environmental footprint in painting processes.

Keywords: Railcar manufacturing · Deep Deterministic Policy Gradient · Paint utilization · Sustainable Manufacturing · Optimal spraying policies

1 Introduction

The railcar manufacturing industry is constantly seeking ways to optimize its processes for enhanced efficiency, cost reduction, and minimal environmental impact. An important process in railcar manufacturing is the painting process, which helps enhance the aesthetic appeal of railcar geometry and its parts together while protecting against environmental factors such as corrosion, weathering, and wear [1, 2]. This is key to the production of railcars as the process of coating surfaces with protective and aesthetic

H. Kohl et al. (Eds.): GCSM 2025, LNME, pp. 139–147, 2026.
https://doi.org/10.1007/978-3-032-21154-5_16

layers of paint increases their durability [3]. The painting process faces several challenges regarding its utilization, including over-spraying, which is excessive paint application leading to material waste and environmental pollution [4], inefficient coating, which leads to inconsistent coverage and thickness, compromising coating quality and durability, and high paint consumption, which contributes to resource depletion and emissions. Inefficient paint usage not only escalates production costs but also generates substantial waste, posing environmental challenges. Addressing these challenges requires a holistic approach that integrates advanced technologies and sustainable practices into the manufacturing process. Existing conventional methods often rely on manual adjustments and fixed parameters, leading to suboptimal paint usage and environmental waste.

Emerging technologies such as automatic spray-painting gantry systems and sophisticated robots have improved performance and productivity, thereby reducing manpower costs. These systems understand spraying dynamics and railcar geometries for optimum robot movement [2]. Aerial robots have also been developed to address the wastage challenges of paint application over openings in railcars [5]. However, the incorporation of data-driven technologies using artificial intelligence is still needed to enhance the utilization of paint and improve coating precision and accuracy. Deep reinforcement learning, which is an artificial intelligence technique, has emerged as a powerful tool for optimizing complex industrial processes [6]. Deep Deterministic Policy Gradient (DDPG), a deep reinforcement learning algorithm, is capable of learning optimal control policies in continuous action spaces, making it well-suited for enhancing the utilization of paint. This study hereby presents an approach that uses DDPG to enhance paint utilization in rail car manufacturing, thereby reducing costs and environmental impact. The main contribution of this study involves integrating sustainability considerations to reduce paint waste, minimize environmental impact, leading to cost savings, and align with the United Nations' sustainable development goals 9 and 12 of resource efficiency and sustainable manufacturing.

2 Related Work

Before the advent of artificial intelligence-based approaches, traditional methods, which rely more on manual intervention and empirical knowledge rather than automated decision-making algorithms, have been used to adjust spray parameters in painting applications. Modern methods now use automated guided vehicles integrated with painting robots to aid the coating process, which improves paint accuracy [7]. However, intelligent spraying and coating methods need to be incorporated into existing methods to ensure that the entire process is sustainable in the manufacturing environment. Previous studies on optimizing paint utilization in manufacturing processes have been conducted to determine paint strategies, parameters, and paths. The Taguchi method and fuzzy logic have been employed to optimize the painting process by analyzing the surface finish and determining the best robot painting paths [8]. This method adjusts the paint pattern, robot speed, and nozzle distance to yield the optimum surface quality, however, pressure adjustment was not considered. The pressure adjustment of the spray gun is considered one of the most important components in the painting process, as it determines the force with which paint is dispensed from the nozzle [9]. Higher pressure typically results in finer atomization of paint particles, allowing for better coverage and penetration into surface irregularities. Overshooting the optimum pressure can lead to overspraying and waste. Hence, fine-tuning the pressure and flow rate to the optimal level for sprayed paint and surfaces achieves better control and efficiency, and reduces waste in paint applications. Furthermore, optimal trajectory planning for flying spray-painting robots has also been developed to combat intricate railcar body geometries [5]. Multiple nozzles are optimized using NSGA-III multi-objective evolutionary optimization algorithms to address paint distribution and trajectory issues of complex objects [10]. These methods have shown tremendous success in improving paint efficiency but often lack adaptability to dynamic manufacturing environments and may struggle to account for complex interactions between process variables.

Previous research has explored various sustainability initiatives in manufacturing, including green supply chain management, renewable energy integration, and waste reduction strategies [11]. Optimizing paint utilization in the form of improving spraying efficiency in rail car manufacturing remains an effective way to attain sustainability goals by achieving energy savings, reducing costs, and protecting the environment [4]. With increasing regulatory pressures and the global push towards sustainability, manufacturers still seek innovative solutions to optimize paint usage. Advanced methods, such as Proximal Policy Optimization (PPO) and multi-objective evolutionary algorithms, have shown promise in similar robotic-spraying tasks [10], but they often struggle with non-stationary reward functions. In contrast, Deep Deterministic Policy Gradient (DDPG) directly handles continuous action outputs through its actor critic networks that adapts more stably to the high-dimensional, feedback-driven requirements of real-time paint application [6]. Hence, the DDPG algorithm becomes well-suited for systems to learn optimal control policies through interaction within a dynamic and uncertain manufacturing environment. However, its application to paint utilization in rail car manufacturing remains relatively unexplored.

3 Methodology

3.1 Problem Formulation

The two main objectives are to maximize paint coverage on rail car surfaces while minimizing paint waste and reducing the amount of Volatile Organic Compounds (VOC) and other pollutants. VOC constitutes one of the greatest sources of emissions in the painting process during manufacturing, as it forms ozone and smog through photochemical reactions when exposed to intense sunlight [12].

3.2 Environment Design

The railcar surface geometry, including other paint parameters, and application equipment such as a simulated robotic arm with spray guns, is modeled as a Markov Decision Process (MDP).

State Representation. The state space, S, encompasses state variables, $s_t \in S$, such as 3D scanning of the railcar body, 3DS, to generate its surface position and representation

[3], measurement of paint thickness, PT in (μm), surface preparation quality, SPQ, painting robot speed, PRS,(m/s), spraying angle, SA (degrees), paint flow rate, PFR [10], in (L/min), spray nozzle average distance, SND, from the surface in (mm), and spray pattern, SP, at any given time step, t. Environmental factors such as ambient temperature, T, in degrees Celsius (°C) and relative humidity, H, (%) in paint booths are considered critical as they strongly determine the paint viscosity [1]. Incorporating these factors ensures that the agent receives enough information to make informed decisions regarding paint application. The state space is represented in (1)

$$S = \{3DS, x_t, y_t, z_t, \varnothing_t, \theta_t, \delta_t, PT_t, SPQ_t, PRS_t, SA_t, PFR_t, SND_t, SP_t, T_t, H_t\} \quad (1)$$

where x_t, y_t, z_t are the coordinate positions of the painting robot, and where $\varnothing_t, \theta_t, \delta_t$ are the orientations of the painting robot. The high dimensionality of 16 state variables and their nonlinearities prompted the use of the DDPG approach to model complex, smooth policies.

Action Space Definition. The action space is defined to encompass the control parameters relevant to paint application, including but not limited to spray nozzle settings. The action space, A, consists of continuous control actions, a_t, such as adjusting spray nozzle distance from the surface in millimeters (between 150 mm and 200 mm) [5]; adjusting painting robot arm speed in (m/s); adjusting spraying angle in degrees (°); ; adjusting the spray pattern (p) in a linear, zig-zag, or circular pattern [8]; and adjusting the paint flow rates (liters/min).

Reward Function Design. A reward function incentivizes efficient paint utilization while penalizing excessive paint usage and deviations from quality standards, and it is carefully crafted to balance maximizing coverage and minimizing waste. This study designed positive rewards for achieving the desired paint thickness [10] with minimal waste and negative rewards for overuse or underuse of paint and emissions. To promote eco-friendly practices, environmental impact metrics are also included in the reward function. The overall reward function, R, uses the weighted sum of three individual reward functions namely, maximizing Paint Utilization Efficiency, R_{pue}; achieving high paint coverage quality, R_{pcq}; and minimizing environmental impact, R_{ei}, as developed. R_{pue} denotes the ratio of paint that effectively covers the railcar surface to the total paint used as represented in (2)

$$R_{pue} = k_{pue}\chi\left[-\left(\frac{total\ paint\ dispensed - effective\ paint\ used}{total\ paint\ used}\right)\right] \quad (2)$$

R_{pcq} denotes uniform and adequate application of paint as represented in (3)

$$R_{pcq} = k_{pcq}\chi\left[-\left(\frac{1}{N}\sum\nolimits_{i=1}^{N}|T_i - T_{target}|\right)\right] \quad (3)$$

where N is the number of sampled points on the railcar surface, T_i is the actual thickness of paint at point i, and T_{target} is the desired thickness. Additionally, R_{ei} represents the rewards for reducing overspray and excess paint use as represented in (4). Where k_{pue},

k_{pcq}, and k_{ei} are scaling factors used to normalize the respective reward components. The overall reward function, *R*, is represented in (5)

$$R_{ei} = k_{ei}\chi\left[-\left(\frac{overspray\ volume\ of\ paint\ used}{total\ paint\ used}\right)\right] \tag{4}$$

$$R = \alpha R_{pue} + \beta R_{pcq} + \gamma R_{ei} \tag{5}$$

where the weights α, β, and γ are tuned on the basis of their relative importance through domain expertise and empirical testing. A total negative reward denotes the need for improvement in paint utilization.

3.3 DDPG Algorithm and Training

The DDPG algorithm learns a policy that maximizes the cumulative reward with an actor-critic deep neural network using policy approximation (actor) and the value function (critic) as shown in Fig. 1. The actor-network generates the control actions given the current state, whereas the value critic network estimates the expected future rewards [13]. The actor-network, $\mu(s|\theta^{\mu})$, determines the optimal action in the form of adjusting spray parameters for a given state with a policy function, μ and parameterized by θ^{μ}. The output layer uses a tanh activation function to ensure the actions remain within a feasible range. The actor-network is regularly updated using the sampled policy gradient according to (6)

$$\nabla_{\theta^{\mu}} J \approx \frac{1}{N}\sum\nolimits_{i}[\nabla_{a} Q(s, a, \theta^{Q})|_{s=s_i,a=\mu(s)_i} \nabla_{\theta^{\mu}}\mu(s|\theta^{\mu})|_{s=s_i}] \tag{6}$$

The critic network, $Q(s_i, a_i|\theta^{Q})$, parameterized by θ^{Q} evaluates the action by estimating the Q-value (expected reward). It takes a state-action pair *(s, a)* as input and outputs the expected cumulative reward. The critic network is also updated by minimizing the loss, *L* using (7).

$$\mathrm{L} = \frac{1}{N}\sum\nolimits_{i}(y_i - Q(s_i, a_i|\theta^{Q}))^2 \tag{7}$$

where target values, y_i are computed using (8)

$$y_i = R_i + \gamma Q'(s_{i+1}|\theta^{\mu'})|\theta^{Q'}) \tag{8}$$

The experience replay buffer stores past experiences (s_t, a_t,r_t,s_{t+1}) to break the correlation between consecutive samples and stabilize learning. The target networks, μ' and Q' stabilize training by slowly tracking the learned networks and updating copies of the actor and critic networks as shown in (9) and (10).

$$\theta^{\mu'} \leftarrow \tau\theta^{\mu} + (1-\tau)\theta^{\mu'} \tag{9}$$

$$\theta^{Q'} \leftarrow \tau\theta^{Q} + (1-\tau)\theta^{Q'} \tag{10}$$

where $\tau \ll 1$ is a small update rate

Training and Testing. Initial training was conducted using a dataset of manually controlled painting operations using PPO to provide a baseline for the DDPG agent. Data were also collected from existing automated painting systems and stored in an experience replay buffer. These data were preprocessed to normalize and standardize the inputs for the neural networks. Actor-critic neural networks together with target networks were initialized with random weights and trained to maximize the cumulative reward. During training for 500 episodes, the agent iteratively adjusts its actions on the basis of the received rewards and observed states. For each time step *t,* an action $\mu(s|\theta^{\mu})$ + N_t is taken, where N_t is added for exploration. This study adopted the Ornstein-Uhlenbeck process exploration strategy to explore the action space efficiently. Each executed action, a_t in the environment, obtains a reward R_t, to yield the next state, s_{t+1}, stored in the replay buffer, *B.*

Performance Evaluation. The trained DDPG agent was evaluated on a simulated rail car manufacturing environment using ABB RobotStudio Painting PowerPac. The Painting PowerPac is add-on tool in the ABB commercial-grade software designed and validated by ABB for accurate simulation of paint application processes with validated use cases that fall within a ± 5% error margin when compared to real-world industrial settings. The performance of the agent was evaluated using paint utilization efficiency, coverage quality, and adherence to environmental and quality constraints. This study defines paint utilization efficiency as the ratio of the amount of paint that effectively covers the railcar surface to the total paint used. The coverage quality was defined as a measure of how uniformly and adequately the paint covers the railcar surface, whereas the environmental impact is the reduction in paint wastage and overspray, which contributes to lower environmental pollution.

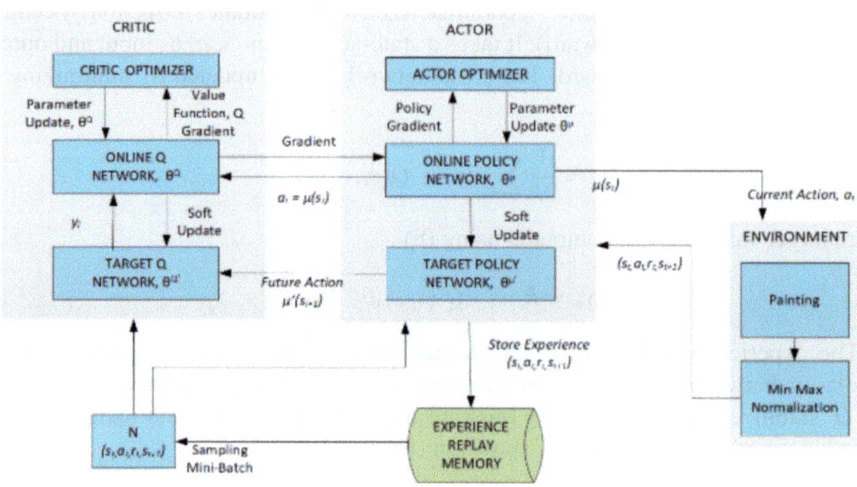

Fig. 1. DDPG System Architecture

4 Results and Discussion

Our results demonstrate significant improvements in paint utilization efficiency compared to baseline methods.

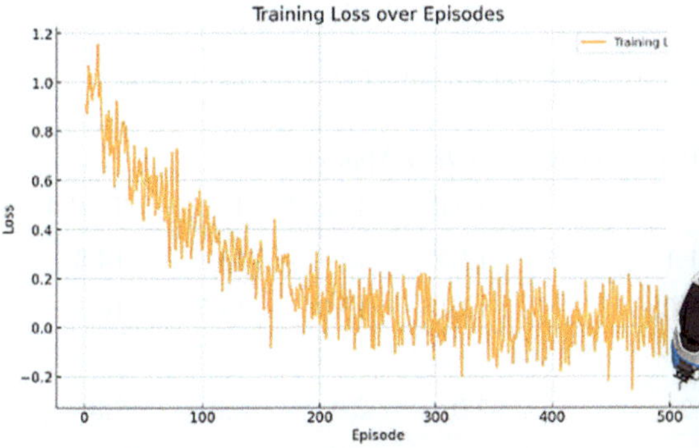

Fig. 3. Training Loss over episodes

Fig. 2. Simulated Railcar

Paint Utilization Efficiency (PUE). The simulation results, as shown in Table 1 demonstrated a 25% improvement in paint utilization efficiency, significantly exceeding the simulation's inherent error margin. This highlights the algorithm's robust ability to optimize the paint application process effectively across various railcar types. Maintaining a consistent reduction in paint waste correlates with existing studies on achieving maximal paint utilization efficiency [10].

Coverage Quality (CQ). This was assessed through detailed inspections of the painted simulated railcar surface model as shown in Fig. 2. The DDPG approach ensured a uniform paint thickness across various parts of the railcar, reducing over-sprayed and under-sprayed areas using metrics such as surface roughness and gloss levels as presented in Table 1. The standard deviation across runs was consistently below 2.1% for all key metrics, which supports the statistical validity of our results. This finding is consistent with existing studies on paint thickness measurements in which a lower surface roughness gives a smoother surface, invariably indicating a more uniform paint thickness [3]. Surface roughness detects oversprayed areas through thicker uneven coatings, and undersprayed areas through irregularities in insufficient coverage through roughness data. For the gloss level metric, the acceptable level for the railcar exterior surface body was 75 Gloss units (GU) at a 60-degree angle, enhanced from the baseline approach of 65 GU.

Environmental Impact (EI). The DDPG approach resulted in a 30% decrease in paint waste, leading to a significant reduction in the environmental footprint. This improvement aligns with sustainable manufacturing practices and regulatory compliance for environmental protection [4].

Training Loss. The training loss over 500 episodes is shown in Fig. 3. The loss generally decreases as the number of episodes increases, following an exponential decay pattern with some noise from 1.1 to –0.1. This trend confirms that the algorithm progressively learns to optimize paint application using continuous control actions from the action space, reducing the loss and improving the efficiency over time, which is typical in reinforcement learning models [13], where early stages involve exploration and rapid learning, followed by gradual refinement.

Table 1. Paint utilization efficiency, coverage quality, and environmental impact

Railcar Type		PPO Approach	DDPG Approach			
			Trial 1	Trial 2	Trial 3	Trial 4
TC1, TC2 M1, M2 M3, M4	**Paint Utilization Efficiency (%)**	11.1	12.2	12.7	13.5	14.8
		12.08	13.1	13.9	14.7	16.1
	Coverage Quality					
	Average Surface Roughness (μm)	3.2	3.0	2.9	2.8	2.8
	Gloss Level (Gloss Units)	65	70	71	73	75
	Environmental Impact					
	Average Paint waste (litres)	15.4	14.2	12.6	11.3	10.78
	Standard Deviation (litres)	2.3	2.1	1.7	1.3	1.1

5 Conclusions and Recommendations

This study presented an approach that uses DDPG to enhance paint utilization efficiency in rail car manufacturing processes. Our experimental results demonstrate the effectiveness of the proposed approach in reducing paint waste while maintaining high-quality coating standards. Our method offers significant cost savings of a 30% decrease in paint waste and environmental benefits, contributing to the advancement of sustainable practices in the manufacturing industry. With further refinement and implementation, our DDPG approach holds promise for revolutionizing paint utilization in railcar manufacturing and beyond. Future research could focus on extending the approach to real-world manufacturing environments to enhance its practical applicability and explore collaborative robotic systems for enhanced paint application efficiency.

References

1. Kebisek, M., Tanuska, P., Spendla, L., Kotianova, J., Strelec, P.: Artificial intelligence platform proposal for paint structure quality prediction within the industry 4.0 concept. IFAC-PapersOnLine **53**(2), 11168–11174 (2020). https://doi.org/10.1016/j.ifacol.2020.12.299

2. Gabbar, H.A., Idrees, M.: ARSIP: automated robotic system for industrial painting. Technologies (Basel), **12**(2) (2024). https://doi.org/10.3390/technologies12020027
3. Edelvik, F., Tiedje, O., Jonuscheit, J., Carlson, J.S.: SelfPaint-A self-programming paint booth. Procedia CIRP **72**, 474–479 (2018). https://doi.org/10.1016/j.procir.2018.03.167
4. Liu, J., Zhao, L., Wang, Q., Jiang, N., Liu, J.: PIV experimental analysis of the escape phenomenon of air spraying jet in factories. J. Build. Eng. **82**, 108384 (2024). https://doi.org/10.1016/j.jobe.2023.108384
5. Vishal, S.: Design of flying robot painting system for Indian railway passenger coaches. Mater. Today: Proc. **46**, 9630–9635 (2021). https://doi.org/10.1016/j.matpr.2020.07.147
6. Lillicrap, T.P., et al.: Continuous control with deep reinforcement learning. In: International Conference on Learning Representation (2015)
7. Qie, J., Miao, Y., Liu, H., Han, T., Shao, Z., Duan, J.: Design and process planning of non-structured surface spray equipment for ultra-large spaces in ship section manufacturing. J. Mar. Sci. Eng. **11**(9), 1723 (2023). https://doi.org/10.3390/jmse11091723
8. Santhosh, R., Sut, D.J., Uma, M., Sethuramalingam, P.: Optimizing IRB1410 industrial robot painting processes through Taguchi method and fuzzy logic integration with machine learning. Int. J. Intell. Robot. Appl. **8**(2), 380–400 (2024). https://doi.org/10.1007/s41315-024-00325-2
9. Jakkula, R.V.S.K., Sethuramalingam, P.: Enhanced coating layer resistance to corrosion and UV degradation with robotic nanospray painting. Arab. J. Sci. Eng. **48**(9), 11913–11926 (2023). https://doi.org/10.1007/s13369-023-07612-3
10. Liu, X., Feng, Y., Zeng, Q., Hu, X., Yang, Z.: Multi-objective optimization of spraying trajectory planning for large ship blocks using evolutionary computation. Procedia CIRP **104**, 530–535 (2021). https://doi.org/10.1016/j.procir.2021.11.089
11. Choi, J.K., Thangamani, D., Kissock, K.: A systematic methodology for improving resource efficiency in small and medium-sized enterprises. Resour. Conserv. Recycl. **147**, 19–27 (2019). https://doi.org/10.1016/j.resconrec.2019.04.015
12. Bysko, S., Krystek, J., Bysko, S.: Automotive paint shop 4.0. Comput. Ind. Eng. **139**, 105546 (2020). https://doi.org/10.1016/j.cie.2018.11.056
13. Zhang, H., Xu, J., Zhang, J., Liu, Q.: Network architecture for optimizing deep deterministic policy gradient algorithms. Comput. Intell. Neurosci. **2022**, 1–10 (2022). https://doi.org/10.1155/2022/1117781

Value Retention Processes

Disassembly Depth as a Determinant of Reassembly Performance in Remanufacturing Systems

Maik Nübel[1(✉)], Jonah Schulz[2], Carl-Philipp Immanuel Oltmanns[1], and Matthias Schmidt[1]

[1] Institute of Production Systems and Logistics, Leibniz University Hanover, An der Universität 2, 30823 Garbsen, Germany
nuebel@ifa.uni-hannover.de

[2] Institute for Production Technology and Systems, Leuphana University, Universitätsallee 1, 21335 Lüneburg, Germany

Abstract. Remanufacturing offers the potential to reduce dependency on raw material suppliers and represents an element in the transition toward a sustainable circular economy. A key challenge is the synchronization of disassembly and assembly processes, which is essential for the efficiency and effectiveness of remanufacturing systems. One promising approach is to increase productivity by reducing the disassembly depth through the omission of certain disassembly steps. However, such a reduction must be closely aligned with the requirements of the subsequent assembly process.

This paper investigates the impact of varying disassembly depths on the assembly process. Therefore, the process is presented, and the corresponding workload distributions are systematically analysed. The findings demonstrate that decreasing disassembly depth results in reduced throughput times and enhanced workload flexibility within the process. This flexibility arises from the dynamic use of subassemblies and contributes to workload balancing within the assembly system. A final simulation-based validation highlights these effects and demonstrates the potential role of subassemblies in generating efficient and effective remanufacturing systems.

Keyword: Remanufacturing · disassembly depth · load flexibility

1 Introduction

Trade conflicts, military conflicts, and distribution conflicts driven by climate change - all conflicts increase the vulnerability of global supply chains and highlight the urgent need to reconsider the current linear economic model. Therefore, the importance of integrating a sustainable economy is growing rapidly. In this context, the Circular Economy (CE) plays a key role by enabling the recovery and reuse of resources [1]. CE contributes to extending product life cycles and shifts the economy towards the efficient use of resources [2]. Concurrently, CE enables the decoupling from global supply chains,

H. Kohl et al. (Eds.): GCSM 2025, LNME, pp. 151–159, 2026.
https://doi.org/10.1007/978-3-032-21154-5_17

thereby ensuring the security of raw material supply and, consequently, preventing supply uncertainties [3]. Beyond addressing resource shortage, additional drivers for the implementations include potential savings in material and energy consumption, as well as the strengthening of competitive advantage [4]. In comparison with new (linear) components, the utilisation of reused products can result in a reduction of up to 90% in terms of material and energy consumption [4, 5]. Essential to the success of the CE is the implementation of highly flexible production systems, which are primarily driven by a high degree of product individuality resulting from diverse usage patterns [4]. Accordingly, flexibility is defined as the ability of a factory system to adapt to dynamic market conditions without requiring structural changes [6].

2 Theoretical Background

The circular economy, as outlined by Potting et al., is organized into ten circular strategies. Among these, remanufacturing is defined as "using parts of discarded products in a new product with the same function" [1]. This definition can be extended by the assignment of a new part identity during the process of remanufacturing [3].

The remanufacturing system is divided into the core processes of assembly and disassembly. The primary function of assembly is the manufacturing operation of joining, which establishes connections between individual parts, thereby creating greater complexity and functional integration [7]. In contrast, disassembly represents the reverse process, in which returned products are separated into subassemblies, components, and raw materials [8]. The extent to which a product is disassembled is denoted as disassembly depth and is distinguished in partial and complete disassembly. This depth offers potentials for optimisation the disassembly process by omitting certain disassembly steps [7, 9]. Through non-destructive separation of End-of-Life (EoL) products, subassemblies and components can be efficiently reintegrated into assembly processes.

Both disassembly and assembly processes are subject to significant uncertainties, which must be addressed through the implementation of flexible systems. Thereby, the flexibility of a work station is differentiated into capacity flexibility and load flexibility. According to Lödding, capacity flexibility refers to "the ability of a production system to adjust its capacity quickly, cost-effectively, and to the greatest extent possible". Load flexibility reduces fluctuations in task-specific workloads across fixed capacities [10].

Taking into account the requirements for highly flexible yet productive assembly systems, the following research question is proposed:

How can variations in disassembly depth enhance productivity and flexibility in (re)assembly within remanufacturing systems?

To address the research question, a deductive-experimental research approach is applied. This approach involves the development of a general theoretical framework, which is subsequently tested through simulation-based experimentation [11].

The research starts with an examination of remanufacturing processes under varying disassembly depths, followed by the derivation of a theoretical framework describing their qualitative impact on (re)assembly operations. This framework is then validated using a simulation model implemented in Siemens Plant Simulation. This paper concludes with a discussion of limitations and a summary of key findings. Within the scope of this study, several assumptions have been made, which are listed in the following section:

- Assembly and reassembly are performed simultaneously within the same assembly system. Regulatory requirements are not taken into account.
- The product design allows a module-specific disassembly and inspection.
- The quality of disassembled components and subassemblies is assumed to be sufficient for (re-)assembly into a new product with the same function. Therefore, the reprocessing of these components and subassemblies is excluded from the scope of this study.

3 Remanufacturing Processes by Varying Disassembly Depth

The disassembly and assembly processes are distinguished by opposing material flows, a distinction that is further reinforced by an intermediate inspection step and subsequent storage. Even within a CE, maintaining a storage of linear components remains necessary due to variations in product configurations, component quality, and market conditions. An example of the remanufacturing process by varying disassembly depth is provided in Fig. 1 and serves as a reference throughout this paper.

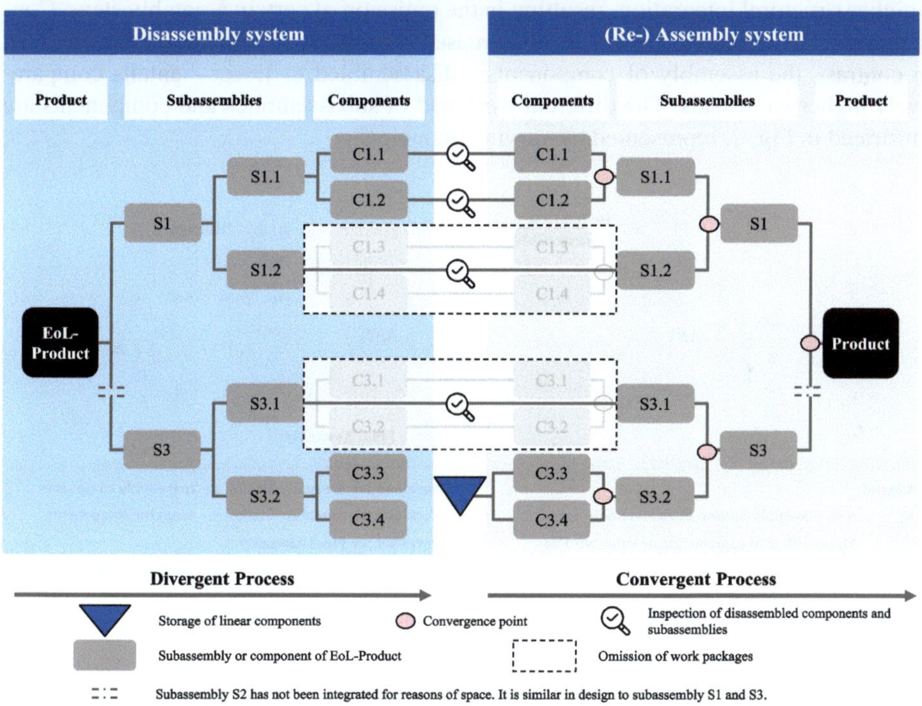

Fig. 1. Interrelation between the Disassembly system and (Re-)Assembly system by varying disassembly depth

The disassembly of EoL products is characterised by a diverging process. In this process, the EoL product is gradually disassembled, resulting in a transition to a lower-level structural configuration. The disassembly begins at the subassembly level and proceeds step by step to the individual component level [7]. In contrast, during (re-)assembly, subassemblies and components are successively joined in a converging process. So-called convergence points denote specific locations within the (re-)assembly process where all requisite subassemblies and/or components must converge to enable the formation of a more complex assembly or the final product [12].

The procedural interrelation becomes particularly evident during the transitional phases between disassembly and (re-)assembly. By decreasing the disassembly depth, specific work packages within disassembly are omitted. This directly affects subsequent assembly operations, as the corresponding work packages are also omitted. Consequently, the number of convergence points required in the (re-)assembly process is reduced. Henceforth, the terms reassembly and assembly will be used synonymously.

4 Qualitative Impact on Assembly Process

Varying disassembly depths significantly affect the assembly system configuration by creating heterogeneous workloads across assembly stations. The subassemblies exhibit a higher structural integration, resulting in the omission of certain assembly steps. Consequently, subassemblies result in a decreased workload per assembly station (AS). In contrast, the assembly of components - disassembled or linear - entails comparatively higher workloads. The different workload of subassemblies and components are illustrated in Fig. 2, represented by varying diameters.

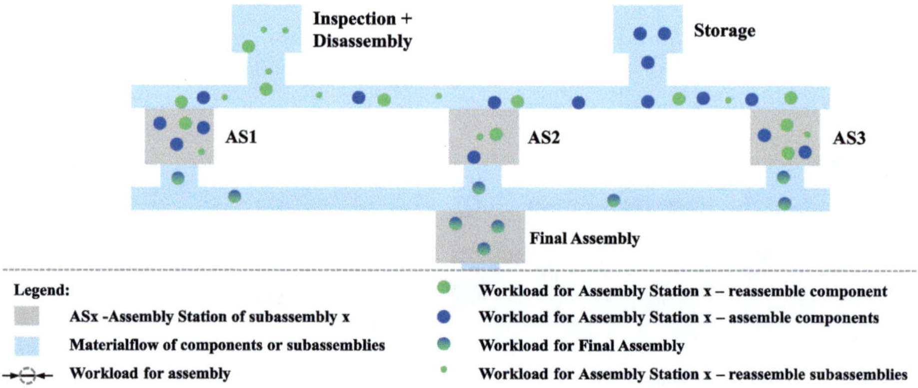

Fig. 2. Assembly process based on varying disassembly depth

The mixing of components and subassemblies at AS1 to AS3 necessitates a decision regarding whether to assemble components or subassemblies. The fluctuation in workloads gives rise to the establishment of time buffers and influence the processing time at the individual assembly stations. These effects are illustrated in Fig. 3 by comparing the processing times associated with assembly of disassembled subassemblies (Case 1 |

Case 3) and components (Case 2) with linear components. The workload is represented by varying diameters.

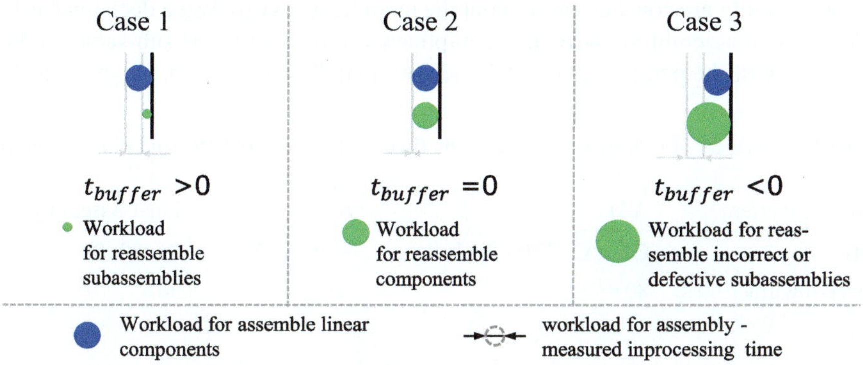

Fig. 3. Time buffer (t_{buffer}) in assembly due to varying disassembly depths

In Case 1, the workload associated with subassemblies is lower relative to linear components. This results in a time buffer (t_{buffer}), which arises due to the reduced work packages required for subassemblies and increases available capacity at the respective station.

In Case 2, the workload associated with disassembled components corresponds to the linear components. As a result, no t_{buffer} is generated, and the decision to use disassembled or linear components has no time-specific relevance.

In Case 3, an incorrect or defective subassembly is provided at the assembly station, which may be attributed to a faulty inspection process or a variant deviation. This results in additional work packages for rework or removal of the subassembly, increasing the workload and prolonging processing times. Consequently, a negative t_{buffer} emerges, tying up capacity in the assembly process.

With regard to the research question outlined in Sect. 2, the findings indicate that varying disassembly depths and the use of subassemblies lead to the emergence of t_{buffer}. These buffers can be utilized to reduce processing times at assembly stations and to absorb situational workload peaks. t_{buffer} Support workload balancing within the assembly system and enhance its ability to respond to production uncertainties -such as supply chain disruptions – through a load flexibility.

5 Validation of the Theoretical Framework

In order to validate the qualitative impact of varying disassembly depths, a simulation-based approach is applied. The remanufacturing process, derived from the assembly structure shown in Fig. 1 and Fig. 2, is modelled using the Plant Simulation environment by Siemens. Plant Simulation enables detailed analysis of assembly processes and provides insights into both procedural and logistical workflows [13]. For this purpose, three identical validation scenarios are applied, which differ solely in terms of disassembly depth and sequencing rules.

5.1 Description of the Simulation Model

In the simulation model, components and subassemblies are distributed through the respective AS1 to AS3 into the Final Assembly. The processing times per component and subassembly are constant throughout the model. At AS1 to AS3 a decision has to be made between assembling individual components or preassembled sub-subassemblies. An overview of the parameters of AS1 and the Final Assembly is shown in Table 1:

Table 1. Overview of the parts and processing times at the Assembly S1 and Final Assembly

Assembly system	AS1	AS1	Final Assembly
Parts	C1.1; C1.2; C1.3; C1.4	S1.1; S1.2	S1; S2; S3
Processing times	80 s	40 s	50 s

The parameters for AS1 to AS3 are uniform and therefore represented collectively by AS1. For the purpose of validation, the simulation is carried out by three validation scenarios. The scenarios differentiate only by the disassembly depth and the sequencing rules and are described in Table 2.

Table 2. Overview of the validation scenarios

Scenario	Disassembly Depth	Sequencing rule
V1: Complete Disassembly	Complete disassembly	Assembly of components in the order of their arrival at the work station
V2: Partial Disassembly with FIFO sequencing rule	Partial Disassembly with varying subassemblies and components	Assembly of subassemblies and components in the order of their arrival at the work station (FIFO)
V3: Partial Disassembly with load-oriented sequencing rule	Partial Disassembly with varying subassemblies and components	Assembly of components in the order of their arrival at the work station (FIFO); Assembly of subassembly in the absence of buffer in the Final Assembly

In scenario V3, the absence of a buffer at Final Assembly leads to the prioritization of sub-subassemblies with a $t_{buffer} > 0$ at upstream stations AS1 to AS3, enabling a load-oriented sequencing rule. As a result, sub-subassemblies are used to reduce processing times at these upstream stations and to mitigate part shortages at Final Assembly. To compare the qualitative effects, the three validation scenarios are based on a total of 23 equivalent orders, with subassemblies accounting for 52% of the orders in both V2 and V3. The total processing times range from 90 s for subassemblies to 130 s for individual components.

5.2 Results of the Simulation Model

The throughput time of the 23 orders and the utilization rate at the Final Assembly are considered as the key performance indicator. For the purpose of the study, the output curve (Fig. 4) and the utilization (Table 3) of the Final Assembly are visualized.

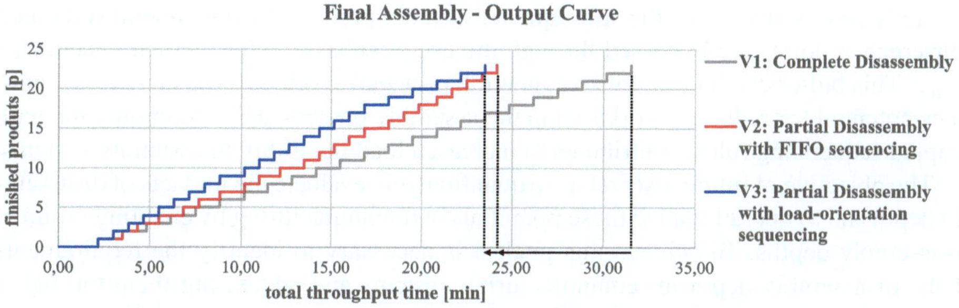

Fig. 4. Output of the Final Assembly by different validation scenarios

Table 3. Utilization rates of the Final Assembly across different validation scenarios

Scenario	V1	V2	V3
Utilization rates	60.8%	79.3%	81.6%

The findings demonstrate that the shortest total throughput time is attained in scenario V3, attributable to its highest utilization rate. This is attributed to the omission of specific assembly steps (compared to V1) and the load-oriented sequencing rule applied (compared to V2). As a result, load flexibility has the potential to balance the workload, thereby increasing overall productivity within the assembly system.

6 Limitations

This paper is subject to underlying assumptions, which may limit the applicability of the results. In particular, the integration of assembly and reassembly within a single assembly system is constrained by regulatory requirements. Additionally, product-specific requirements regarding the disassembly and inspection of subassemblies must be fulfilled. The results highlight the qualitative potential of disassembly depth for the assembly process. This potential has been demonstrated and qualitatively validated using a simulation-based approach. However, a deterministic model to quantify the impact and evaluate its feasibility in an industrial context is still pending.

7 Conclusion and Outlook

This paper investigates how varying disassembly depths can contribute to increased productivity and flexibility in remanufacturing. In this context, the interrelation between remanufacturing processes -disassembly and (re-)assembly - and disassembly depth is analysed. The analysis demonstrates that varying disassembly depths affect both the disassembly process itself and the subsequent assembly process. During assembly, certain convergence points are bypassed through the use of subassemblies, which generates a t_{buffer}. This buffer can have a positive, neutral, or negative effect. A positive t_{buffer} indicates potential for reducing workload in the assembly process and, in combination with adapted sequencing rules, contributes to increased load flexibility in assembly systems.

The objective of future research is to quantitatively evaluate the impacts of disassembly depth and to operationalize these potentials in remanufacturing by enabling minimal disassembly depths. To achieve this goal, it is necessary to identify the requirements of the disassembly depth in remanufacturing systems and addressing them through a requirement-specific design.

The findings obtained need to be incorporated into the development of new products and processes. To fully realize the potential of the circular economy, a holistic approach is essential - one that integrates product and process development with the systematic design of remanufacturing systems.

References

1. Potting, J., Hekkert, M., Worrel, E., Hanemaaijer, A.: Circular economy: measuring innovation in the product chain. Planbureau voor de Leefomgeving (2017)
2. Parker, D., et al.: Remanufacturing market study. For Horizon 2020, grant agreement No 645984, November 2015 (2015)
3. WiGeP, WGP: Update-Factory für ein industrielles Produkt-Update. Ein Beitrag zur Kreislaufwirtschaft, Aachen (2021)
4. Colledani, M., Battaïa, O.: A decision support system to manage the quality of end-of-Life products in disassembly systems. CIRP Ann. **65**(1), 41–44 (2016)
5. Steinhilper, R.: Recent trends and benefits of remanufacturing: from closed loop businesses to synergetic networks. In: Proceedings Second International Symposium, pp. 481–488. IEEE, Japan (2021)
6. Tolio, T., Valente, A.: An Approach to design an approach to design the flexibility degree in flexible manufacturing systems. In: 16th International Conference on Flexible Automation and Intelligent Manufacturing, University of Limerik, Ireland, 26–28.06.2006 (2006)
7. Lambert, A.J.D., Gupta, S.M.: Disassembly Modeling for Assembly, Maintenance, Reuse, and Recycling. CRC Press, Boca Raton (2005)
8. Vanegas, P., Peeters, J., Duflou, J.: Disassembly. In: Laperrière, L., Reinhart, G. (eds.) CIRP Encyclopedia of Production Engineering, pp. 395–399. Springer, Heidelberg (2014)
9. Langella, I.M.: Planning Demand-Driven Disassembly for Remanufacturing. DUV, Wiesbaden (2007)
10. Lödding, H.: Verfahren der Fertigungssteuerung. Springer, Berlin Heidelberg (2016)
11. Nyhuis, P., Wiendahl, H.-P.: Logistische Kennlinien. Grundlagen, Werkzeuge und Anwendungen, 3rd edn. VDI-Buch. SPRINGER VIEWEG, Berlin, Heidelberg, Dordrecht, London, New York (2012)

12. Johansson, G.: Design for Disassembly: A Framework (1997)
13. Malega, P., Gazda, V., Rudy, V.: Optimization of production system in plant simulation. SIMULATION **98**(4), 295–306 (2022)

Using Stable Diffusion Based Augmentation for Visual Wear Classification in Remanufacturing

Engjëll Ahmeti[1(✉)], Maximilian Herold[1], Jan Koller[1], and Frank Döpper[1,2]

[1] Fraunhofer Institute for Manufacturing Engineering and Automation IPA, Bayreuth, Germany
{engjell.ahmeti,maximilian.herold,jan.koller, frank.doepper}@ipa.fraunhofer.de
[2] Chair of Manufacturing and Remanufacturing Technology, University of Bayreuth, Bayreuth, Germany

Abstract. The increasing demand for raw materials and shorter product cycles have driven the advancement of the circular economy, with the European Union focusing on reducing raw material consumption and ensuring the circularity of products and components. Remanufacturing, a key element of circular economy, faces challenges in assessing the condition of products and components, a process traditionally carried out manually, requiring expert knowledge. This paper aims to address the lack of diverse datasets by generating synthetic datasets to enable AI-supported visual inspections in remanufacturing. Specifically, this paper focuses on generating realistic wear patterns, in the example of corrosion effects, with starter motors as the reference product. The research explores different diffusion models, emphasizing globally generated corrosion patterns. The outcomes include the creation of a synthetic dataset, the development of an accurate wear classification model, and a validated framework for integrating AI-based inspections in remanufacturing operations. This approach aims to improve efficiency and quality in remanufacturing processes, contributing to a circular economy.

Keywords: Remanufacturing · Machine Learning · Circular Economy

1 Introduction

Remanufacturing, a key element of circular economy, faces challenges in assessing the condition of products and components, a process traditionally carried out manually, requiring expert knowledge. During the initial sorting and inspection step of remanufacturing, products are visually evaluated for wear and reusability, with suitable parts routed for remanufacturing and unsuitable parts directed to recycling. The complexity of remanufacturing, particularly in assessing the condition of returned used products and components, poses a barrier to wider adoption of remanufacturing process in the circular economy. The reliance on manual

H. Kohl et al. (Eds.): GCSM 2025, LNME, pp. 160–168, 2026.
https://doi.org/10.1007/978-3-032-21154-5_18

visual inspection can lead to error rates of 5–30%, often due to fatigue and variability [1]. To automate, enhance, and increase the throughput of the remanufacturing process, this paper proposes a two-step approach leveraging artificial intelligence algorithms. The first step addresses the need for high-quality training datasets, specifically, image datasets that capture various types of defects commonly found in returned products destined for remanufacturing. There is a significant scarcity of data and a challenge in collecting large numbers of used parts with diverse defect types and variants, especially in the automotive industry. To address this, this study introduces a method for generating synthetic training images of global wear patterns using generative diffusion models. In the second step, we focus on identifying an effective classifier for detecting and categorizing defects. We train multiple deep learning models on real and synthetic data and evaluate their performance on real-world images of used parts. The proposed model achieves high accuracy on real test data, demonstrating its ability to reduce the time required for initial inspection, increase throughput, minimize human error associated with manual inspection, and enable possible integration in reverse logistics systems. These improvements contribute to a more efficient, scalable, and sustainable remanufacturing process.

2 Background and Related Work

2.1 Remanufacturing Process and Initial Inspection Challenges

As Europe increasingly prioritizes circular economy strategies, remanufacturing emerges as a viable and sustainable solution. Research shows that remanufacturing can reduce material consumption by up to 90% compared to the production of new components. For instance, the remanufacturing of vehicle starter motors achieves material savings of 50% and energy savings of 55% [2]. Furthermore, remanufacturing can reduce greenhouse gas emissions by 79–99% compared to manufacturing new products [1]. The economic benefits are also substantial, with studies indicating cost reductions of up to 50%, and energy savings of up to 60% [3]. Despite these advantages, the initial inspection stage remains a major challenge. The high volume, high number of variants and different conditions of the incoming used products make manual inspection time-consuming, resource-intensive, and prone to human error [4].

2.2 Visual Inspection and Synthetic Data Generation

There has been growing research into the automation of the initial inspection step in remanufacturing. Various approaches have explored the potential for digitization and automation through the use of sensors [5], 3D modeling and geometry reconstruction [6], and artificial intelligence (AI) algorithms. The latter algorithms have gained increasing attention due to their flexibility and adaptability for future enhancements. Aravapalli et al. applied object detection techniques to locate and classify defects on metallic surfaces [7]. Similarly, Ahmeti et al. used object detection to identify and classify defective pins in electrical connectors of small electrical appliances [8]. Nwankpa et al. proposed a deep convolutional neural network (DCNN) model to detect defects such as pitting, rust, cracks, and combined faults on mild steel plates [9]. Saiz et al. used a deep learning ensemble (a combination of neural network models) to detect wear on constant velocity joint cages [10]. Schlüter et al. applied machine learning to enhance the initial identification, sorting, and inspection of components using convolutional neural networks [11]. Previous research relied on self-created or open-source datasets, which often lack component and defect variety. Gathering large image volumes remains difficult due to limited data availability and the high cost of acquiring real parts. This paper proposes an approach to generate synthetic image data that simulates global wear patterns on used parts. Since capturing real-world defect variations at scale is impractical, synthetic data offers a viable alternative. Two main strategies exist: using 3D models [12] or applying generative AI methods such as Generative Adversarial Networks (GANs) for tool wear monitoring [13] and Stable Diffusion for pipeline inspection [14]. This study focuses specifically on generating synthetic wear datasets using Stable Diffusion models.

3 Methodology

3.1 Image Acquisition for Training

As no open-source datasets were available containing images of used parts with and without wear patterns, it was necessary to create a real-world dataset using old parts, in this study, starter motors were used. The goal was to develop a cost-effective imaging system for dataset creation. The setup was based on a Raspberry Pi High-Quality (RHQ) camera equipped with a 16 mm telephoto lens. For lighting, two Aputure Amaran Tri-8C lights with EZ-Box II softboxes were used. The camera settings for the image acquisition were as follows: an aperture of f/15 and an ISO value of 800. In total, 51 starter motors of the same type were analysed, of which 43 were non-corrosive and 8 corrosive. Each starter motor was captured in 20 images: 2 images with the starter flat on both sides, 1 image of the head and 1 of the tail, 8 images taken in 45-degree rotational increments of the head, and 8 of the tail. The final dataset comprised 860 images of non-corrosive and 160 images of corrosive starter motors. This dataset formed the basis for subsequent stages of the study, particularly the generation and analysis of global wear patterns. Since the native resolution of the RHQ camera is 4056×3040 pixels, preprocessing was required to prepare the images for input into Stable Diffusion models. Specifically, the images were center-cropped by trimming 508 pixels from each side of the width to produce square 3040×3040 images, which were then resized to 1024×1024 pixels.

3.2 Generation and Evaluation of Synthetic Image Using Stable Diffusion

Acquiring images of used parts suitable for remanufacturing is feasible. However, obtaining images of used parts with a variety of defects from different perspectives presents a significant challenge. To overcome this, we employed Stable

Diffusion models to generate synthetic images representing global wear patterns. Stable Diffusion is a text-to-image and image-modification diffusion model that operates by adding noise to an initial image and then learning to reverse this process to generate new visual content [14].

The goal was to synthesize wear patterns that could be globally applied across the image to produce realistic visual representations of corrosion. In this study, three different models based on Stable Diffusion were used. The first is LSAST, which introduces a novel framework for generating highly realistic, artistically stylized images while preserving the structural content of the original image using predefined prompts from the study [15]. The other two models are based on the latest Stable Diffusion XL (SDXL): SDXL Multi-ControlNet (SDXL_MC) and SDXL LoRA (SDXL_L). SDXL_MC leverages ControlNet, a neural network architecture that enables conditioning of a pretrained text-to-image diffusion model via efficient fine-tuning [16]. SDXL_L applies a Low-Rank Adaptation (LoRA) method, which allows efficient adaptation of large pretrained models to downstream tasks while significantly reducing the number of trainable parameters and memory requirements compared to full fine-tuning [17]. For each real non-corrosive starter motor image, a corresponding synthetic corrosive image was created using each of the three models, resulting in 860 corrosion-augmented images per model. The base text prompt used for SDXL_MC and SDXL_L was "rust on metal areas of the car motor starter and corrosion on the edges of the starter". An additional corrosive-based text prompt for improving quality was "blurry, deformed, low quality, over-saturated, textured-background".

Although the generated images do not have one-to-one real-image counterparts for direct comparison, quantitative evaluation metrics can still provide insight into their realism and quality. In this study, we employ two such metrics: Fréchet Distance with DINOv2 (FD-DINOv2) and Peak Signal-to-Noise Ratio (PSNR). FD-DINOv2 extends the traditional Fréchet Distance, a measure of similarity between feature distributions with DINOv2 which is a self-supervised vision model developed by Meta that enhances feature representation accuracy [18]. PSNR evaluates the ratio of signal strength to noise between original and modified images, offering a standard measure of image quality degradation [19].

3.3 Wear Classifier and Dataset Design

For wear pattern classification, several computer vision methods can be applied, including image classification, object detection, and instance segmentation. These methods differ in complexity and purpose: image classification assigns a single label to an entire image, object detection identifies and localizes multiple objects within an image, while instance segmentation performs pixel-level classification to segment and localize individual objects accurately. Since the goal of this paper is to determine whether an image of a used part shows wear, image classification is the most appropriate method. It provides sufficient performance without the added complexity of higher-level techniques. For this task, we adopt Deep Residual Neural Networks (ResNet) which is a well-established and widely used architecture for image classification. ResNet is known for its ability to train deeper networks effectively by using residual mappings and skip connections, which mitigate the vanishing gradient problem and maintain performance as depth increases [20]. To assess the impact of synthetic data and model robustness, four datasets were prepared based on the real and synthetic images generated in the previous section: D_1: 160 non-corrosive real images and 160 corrosive real images, D_2: 860 non-corrosive real images, 160 corrosive real images, and 490 synthetic corrosive images, D_3: 320 non-corrosive real images, 160 corrosive real images, and 140 synthetic corrosive images, and D_4: 860 non-corrosive real images, 160 corrosive real images, and 630 synthetic corrosive images.

4 Results and Discussion

4.1 Quantitative Results

To evaluate the performance of this study, we first assessed the quality of the generated images using the two introduced metrics: FD_DINOv2 and PSNR. Since the image generation models were fed with non-corrosive starter motor images and tasked with producing synthetic images containing corrosion, the evaluation must be interpreted in reverse. An image of corrosive used part should differ significantly from a non-corrosive one, which is reflected in the metrics, as shown in the **Data Generation** section of Table 1. High FD_DINOv2 values indicate strong dissimilarity, while PSNR values in the range of 20–30 suggest noticeable visual differences due to added corrosion. Figure 1 compares two real starter motor images (non-corrosive and corrosive) with three synthetically generated corrosive images using: i) LSAST, ii) SDXL_L, and iii) SDXL_MC.

Table 1. Descriptive statistics of synthetic generation and classifier evaluation.

Metric Type	Metric	Data	Data Generation Models			
			Real	LSAST	SDXL_L	SDXL_MC
Data Generation	PSNR	–	–	26.18	22.81	20.25
	FD_DINOv2	–	–	782.35	699.08	843.88
Classifier	F_1-score	D_1	0.48	–	–	–
		D_2	-	1.00	0.27	0.94
		D_3	-	1.00	0.00	1.00
		D_4	-	0.05	1.00	1.00

Visual inspection on the generated images (shown in Fig. 1) confirms that the generated images successfully replicate corrosion effects, aligning with the metric results and effectively augmenting the limited real dataset. Additionally, we trained a lightweight pretrained ResNet-18 model for 10 epochs with a batch size of 16 and a learning rate of 0.001 on the different datasets described in the previous section to evaluate the impact of synthetic data. As shown in the

Classifier section of Table 1, the best F_1-score was achieved using both real and synthetic data, with the SDXL_MC model producing the most effective synthetic images. Dataset D_1, which included only real data, served as a baseline and yielded an F_1-score of 0.48, highlighting the value of using synthetic images for training.

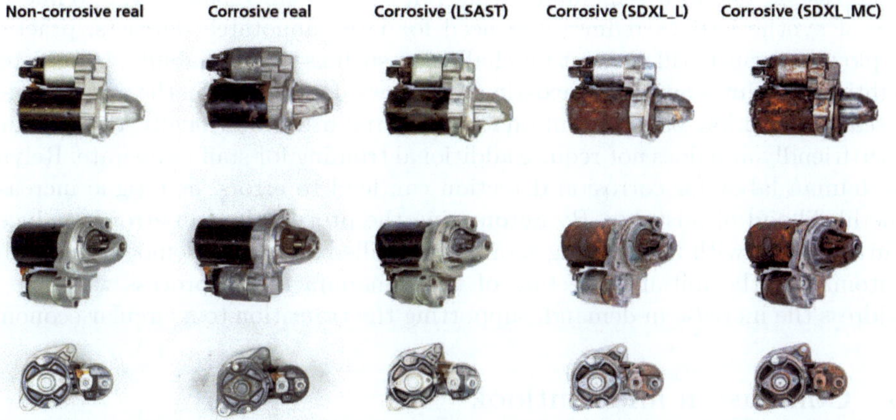

Fig. 1. Comparison of 2x3 real starter motor images (non-corrosive vs. corrosive) and 3x3 synthetic corrosive images using LSAST, SDXL_L, and SDXL_MC.

4.2 Qualitative Insights

We checked the synthetic images generated from three models and observed that the model that does not use prompts and has an integrated prompt does not add corrosion to the images. Although the metrics indicate that the real and synthetic images are dissimilar which is true, the images are dissimilar because LSAST generates cartoonish images and distorts the image (see Fig. 1). However, it does not add corrosion to the image but attempts to create a cartoonish version of the real input image. The other two models, SDXL_L and SDXL_MC, add global corrosion in a fitting manner and fully support the results of the metrics. Thus, while these two instances demonstrate that metrics are good indicators of the similarity and dissimilarity of images, there is always a need for visual inspection, as was the case with LSAST. Additionally, the dataset was created by rotating the starter motor as corrosion may appear locally or across the entire surface. If corrosion appears only on one side or section of a starter motor, and the test set contains only real images, the classifier may incorrectly classify a defective part as non-defective. Using synthetic data across training, validation, and testing introduces global corrosion on starter motors, improving accuracy. While this raises concerns for real-world use, it can be addressed by proper positioning and flagging low-confidence predictions for manual review.

4.3 Discussion

This proposed approach supports the development of an AI-based system for detecting corrosion on starter motors. It can be deployed either as a static unit or integrated with a cobot for automated inspection and sorting. Designed to be user-friendly and based on a cost-efficient Raspberry Pi camera setup, it is accessible for both remanufacturing lines and small repair shops. While the use of synthetic data reduces the need for large annotated datasets, practical deployment must still account for challenges such as part variability, inconsistent lighting, and integration into existing workflows. Despite these, the method offers a scalable and low-barrier solution for industrial use. Additionally, the system is user-friendly and does not require additional training for staff to operate. Relying on human labor for corrosion detection can lead to errors, as fatigue increases the likelihood of mistakes. By automating the process, human error is reduced. Furthermore, with the growing shortage of skilled labor and demographic shifts, automating the initial inspection of the remanufacturing process will help to address the increase in demand, supporting the transition to a circular economy.

5 Conclusion and Outlook

In this paper, an approach utilizing synthetic corrosion generation on images using Stable Diffusion models has shown promising results in the automation of corrosion detection in a remanufacturing environment. By using a small number of real images of corrosive parts (dataset D3, see Table 1), good classification results were achieved. The capability of this approach opens up possibilities for future developments. However, challenges as stated in the qualitative insights such as ensuring dataset diversity and quality should be addressed to enhance the effectiveness of the synthetic data generation process. Future research should focus on refining this process, including the exploration of additional generative models. In summary, by automating the inspection step in remanufacturing, our research helps reduce labor costs, minimize human error, and increase throughput, all of which are essential for achieving more sustainable and efficient production practices. With further research and development, synthetic data generation using Stable Diffusion models holds promising potential for more cost-effective and sustainable inspection in remanufacturing.

Acknowledgements. This work emerged from the VersAtile-KI research project and was funded by the German Federal Ministry of Education and Research (BMBF) with the funding code 01IS23020C.

References

1. Nasr, N., Russell, J.: Re-defining Value–the manufacturing revolution. Remanufacturing, Refurbishment, Repair and Direct Reuse in the Circular Economy. ISBN: 978-92-807-3720-2 (2018)
2. Köhler, D.C.F.: Regenerative Supply Chains: Regenerative Wertschöpfungsketten. Shaker Verlag, Aachen 9783844006162 (2011)
3. Salah, B., et al.: A qualitative and quantitative analysis of remanufacturing research. Process. **9**(10), 1766 (2021). https://doi.org/10.3390/pr9101766
4. Cemenska, J., Rudberg, T., Henscheid, M., Lauletta, A., Davis, B.: AFP automated inspection system performance and expectations. SAE Tech. Pap. Ser. (2017). https://doi.org/10.4271/2017-01-2150
5. Wang, Z., Xu, Y., Ma, X., Thomson, G.: Towards smart remanufacturing and maintenance of machinery - review of automated inspection. Condition Monit. Prod. Optimisation (2020). https://doi.org/10.1109/ETFA46521.2020.9212110
6. Khan, A., Mineo, C., Dobie, G., Macleod, Ch., Pierce, G.: Vision guided robotic inspection for parts in manufacturing and remanufacturing industry. J. Remanufacturing **11**(1), 49–70 (2021). https://doi.org/10.1007/s13243-020-00091-x
7. Aravapalli, S.R.M.: An automatic inspection approach for remanufacturing components using object detection. Dissertation (2022)
8. Ahmeti, E., Koller, J., Wieland, F., Bargmann, D., Döpper, F.: Automating defect detection in small electrical appliances: a machine learning approach for remanufacturing (2025). https://doi.org/10.15488/18877
9. Nwankpa, Ch., Eze, S., Ijomah, W., Gachagan, A., Marshall, S.: Achieving remanufacturing inspection using deep learning. J. Remanufacturing **11**(2), 89–105 (2021). https://doi.org/10.1007/s13243-020-00093-9
10. Saiz, F.A., Alfaro, G., Barandiaran, I.: An inspection and classification system for automotive component remanufacturing industry based on ensemble learning. Information **12**(12), 489 (2021). https://doi.org/10.3390/info12120489
11. Schlüter, M., et al.: AI-enhanced identification, inspection and sorting for reverse logistics in remanufacturing. Procedia CIRP **98**, 300–305 (2021). https://doi.org/10.1016/j.procir.2021.01.107
12. Anderson, J.W., Ziolkowski, M., Kennedy, K., Apon, A.W.: Synthetic image data for deep learning (2022). https://doi.org/10.48550/arXiv.2212.06232
13. Friedrich, M., Ahmeti, E., Dumler, J., Döpper, F.: Towards sustainable machining: synthetic data generation for efficient optical tool wear monitoring via generative adversarial networks. GCSM 2023(2025). https://doi.org/10.48550/arXiv.2212.06232
14. Ramos, L.T., Casas, E., Rivas-Echeverrïa, F.: Synthetic generated data for intelligent corrosion classification in oil and gas pipelines. Intell. Syst. Appl. **25**, 200463. ISSN 26673053 (2025). https://doi.org/10.1016/j.iswa.2024.200463
15. Zhang, Zh., et al.: Towards highly realistic artistic style transfer via stable diffusion with step-aware and layer-aware prompt (2024). arXiv
16. Zhang, L., Rao, A., Agrawala, M.: Adding conditional control to text-to-image diffusion models (2023)
17. Hu, E.J., et al.: LoRa: low-rank adaptation of large language models (2021). https://doi.org/10.48550/arXiv.2106.09685
18. Stein, G., et al.: Exposing flaws of generative model evaluation metrics and their unfair treatment of diffusion models (2023). https://doi.org/10.48550/ARXIV.2306.04675

19. Korhonen, J., You, J.: Peak signal-to-noise ratio revisited: is simple beautiful? In: International Workshop on Quality of Multimedia Experience, pp. 37–38. ISBN 978-1-4673-0726-0 (2012). https://doi.org/10.1109/QoMEX.2012.6263880
20. He, K., Zhang, X., Ren, S., Sun, J.: Deep residual learning for image recognition (2015). https://doi.org/10.48550/arXiv.1512.03385

End-of-Life Decision-Making in Circular Manufacturing: A Complex Adaptive System of Systems Perspective

Tanver Ahammad Hazari(✉) and Carla S. A. Assuad

Norwegian University of Science and Technology, Teknologivegen 22, 2815 Gjovik, Norway
tanver.a.hazari@ntnu.no

Abstract. End-of-life (EoL) decision-making in circular manufacturing is a critical challenge for sustainable industry. Traditional linear models fail to capture the dynamic and emergent behaviors of modern production systems. This paper develops a novel framework combining Complex Adaptive Systems (CAS) and System-of-Systems (SoS) perspectives to enhance EoL strategies such as reuse, remanufacturing, and recycling. Methodologically we combine CAS principle – adaptive system, self-organizing agents and nonlinear dynamics - with SoS principle which addresses interdependence among production networks, supply chains, and markets. The proposed framework offers practical decision support for optimizing resource recovery, reducing waste, and designing adaptive circular supply chains.

Keywords: End-of-life management · circular manufacturing · Complex adaptive system of systems

1 Introduction

The transition to a circular economy represents a paradigm shift, with effective End-of-Life (EoL) product management at its core. EoL decision-making in circular manufacturing is complex, requiring a holistic view of interconnected systems. This paper examines EoL decisions through the combined lenses of Complex Adaptive Systems (CAS) and Systems of Systems (SoS), offering an integrative framework to address key challenges.

EoL assessment is vital for sustainable manufacturing, guiding choices that conserve resources, reduce emissions, and support regulatory compliance, such as Extended Producer Responsibility (EPR). It also enables economic recovery from products and materials. However, manufacturers face internal barriers—like reverse logistics costs, disassembly constraints, and quality issues with recycled materials—and external ones, including conflicting stakeholder interests [1]. These challenges reflect the dynamic complexity of circular systems, characterized by feedback loops, trade-offs, nonlinearities, and self-organizing actors [2].

H. Kohl et al. (Eds.): GCSM 2025, LNME, pp. 169–176, 2026.
https://doi.org/10.1007/978-3-032-21154-5_19

A systems-based approach is essential for navigating this complexity. The diverse actors, shifting market conditions, and regulatory pressures in modern production systems challenge linear decision models [3]. Systems thinking identifies leverage points for strategic intervention but requires suitable methodologies to capture the adaptive and dynamic nature of industrial ecosystems. The SoS perspective highlights the interdependence of autonomous systems pursuing shared goals, yet it also calls for stronger focus on learning and adaptation [4]. CAS theory complements this by emphasizing local interactions and emergent behaviors among agents like manufacturers, recyclers, and consumers [5]. Together, CAS and SoS offer a layered understanding of circular manufacturing: SoS captures macro-level system interconnections, while CAS reveals micro-level adaptive behaviors and emergent dynamics arising from individual interactions.

This paper proposes a simulation model architecture for EoL decision-making from the manufacturer's viewpoint. Section 2 outlines the theoretical foundations; Sect. 3 introduces the EoL assessment framework; Sect. 4 presents the simulation model architecture; Sect. 5 provides a critical discussion; and Sect. 6 concludes with key insights and future research directions.

2 Theoretical Background

2.1 System of System and Complex Adaptive System Theory

While both CAS and SoS theories address multi-component interactions and emergent behaviors, they stem from distinct traditions and emphasize different dimensions of complexity.

SoS theory conceptualizes independent, operationally autonomous systems collaborating toward integrated goals. Maier [6] identifies five defining characteristics: management independence, operational independence, geographical distribution, evolutionary development, and emergent behavior—especially the first two. The ABCDE framework—Autonomy, Belonging, Connectivity, Diversity, and Emergence—is often used for SoS modeling [7]. However, modeling emergence remains a key challenge: it is either overlooked in mathematical models [8] or addressed only at the system level, not within constituent subsystems [9]. In System Dynamics, for instance, emergence is not explicitly represented but arises from the structure and feedback loops—posing limitations when system structure is not well defined and bottom-up modeling is required.

CAS, in contrast, involves numerous interacting, adaptive agents whose nonlinear dynamics produce emergent, often unpredictable behaviors. Agents learn and evolve through interaction with their environment, enabling self-organization over time [10]. Although CAS lacks a universal definition [11], core characteristics include emergence, adaptation, nonlinearity, learning, and co-evolution, typically captured using agent-based modeling [12].

Despite shared emphasis on interaction and emergence, CAS takes a bottom-up view, focusing on decentralized, adaptive agent behavior, whereas SoS applies a top-down lens, integrating autonomous systems for coordinated functionality. CAS embraces unpredictability and observes emergent patterns from local interactions without requiring full system comprehension. SoS assumes decomposability and external knowledge to guide system integration. Although CAS–SoS integration has been explored in domains such as healthcare [13], pandemic modelling [14], and Naval warfare [15], its application to EoL management in circular manufacturing remains, to the best of the authors' knowledge, unexplored.

2.2 EoL Management

A key gap in literature is the absence of a comprehensive EoL assessment framework that fully leverages multiple strategy alternatives for sustainable impact. This stems from two issues: (1) most EoL assessments focus on a single sustainability pillar—commonly environmental—neglecting specifically the social dimension most of the time [16]; (2) many assume a preferred EoL option (e.g., recycling) without evaluating alternatives [17, 18]. While [19] introduces a multi-indicator approach to compare options, it is limited to individual products, failing to consider emergent behaviors from aggregate EoL processing or supply–demand mismatches.

Managing returned products at EoL involves tightly linked complexities—such as disassembly [20], production planning [21], and technology selection [22] —requiring an integrated assessment framework capable of revealing emergent patterns and systemic interdependencies. Such a framework should support decision-making at both micro and macro levels, enabling more robust, adaptive EoL strategies.

3 Framework for EoL Assessment

Figure 1 illustrates a simplified material flow diagram of a product, incorporating three end-of-life (EoL) alternatives—Reuse, Remanufacturing, and Recycling—alongside key stakeholders. Following the definitions in [23] Reuse refers to extending product life through an additional use cycle with the product remaining functional and requiring minimal processing, while Remanufacturing involves restoring a product to a condition equivalent to or better than new. The core EoL assessment challenge depicted in this figure is the selection of the most suitable EoL alternative, indicated by red arrows.

A typical EoL decision problem involves selecting the optimal alternative from *n* options ((here n = 1, 2, or 3 for Reuse, Remanufacturing and Recycling, respectively) by jointly considering economic, environmental and social factors. For a given product *i*, the total impact $EoLI_{in}$ of an alternative *n* is defined as:

$$EoLI_{in} = w_1 Ec_{in} + w_2 En_{in} + w_3 Sc_{in}$$

where $Ec_{in}, En_{in}, Sc_{in}$ represent economic gain, environmental impact (as a negative influence), and social gain for the nth alternative, and w_1, w_2, w_3 are weighing factors set by the manufacturer. Although each of the $En_{in}, Ec_{in}, Sc_{in}$ may include multiple indicators, they are aggregated for simplicity.

For any product *i*, the best EoL option can be selected as follows:

$$EoLI_i = \max(EoLI_{i1}, EoLI_{i2}, EoLI_{i3})$$

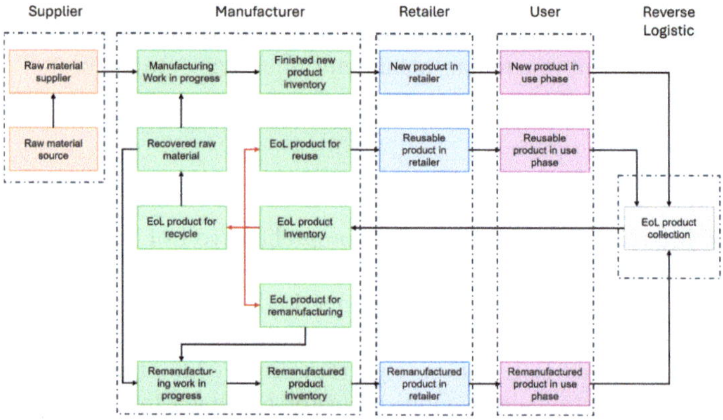

Fig. 1. Simplified material flow of a product considering three EoL alternatives, relevant stakeholders are differentiated with boundaries.

When decisions are made for s number of products at the EoL inventory at time t_1 for alternative selection, the maximum gain from these s products could be calculated as:

$$EoLI_{max} = \sum\nolimits_{i=1}^{s} EoLI_i$$

However, several constraints—both external and internal—must be accounted for in the EoL assessment process. External constraints include supply–demand alignment, regulatory compliance, third-party processing capacity, and variability in product returns. For instance, the number of products designated for reuse or remanufacturing must not exceed the corresponding market demand at time t_2, introducing dependency across stakeholders. Internal constraints involve production planning, resource availability, disassembly capacity, storage limitations, and technical compatibility. Together, these constraints underscore the complexity of EoL decision-making within circular manufacturing systems.

Although the problem may appear amenable to purely mathematical modeling, this is not the case due to uncertainties inherent in the quality of returned products. Variability in product condition can necessitate differing levels of processing, thereby affecting both cost and lead time. As a result, a simulation-based approach is more suitable. This study, therefore, employs simulation optimization as a more appropriate technique for addressing the EoL assessment problem under these dynamic and uncertain conditions.

4 Simulation Model Architecture

The proposed simulation model architecture, aligned with the framework outlined in the previous section, supports simulation-optimization techniques for EoL decision-making in circular manufacturing (Fig. 2). The system is modeled hierarchically, with the supply chain at the top level and the manufacturer—who owns the decision problem—at the next. This focus reflects both the central role of the manufacturer and the practical challenge of limited access to granular data from other supply chain actors.

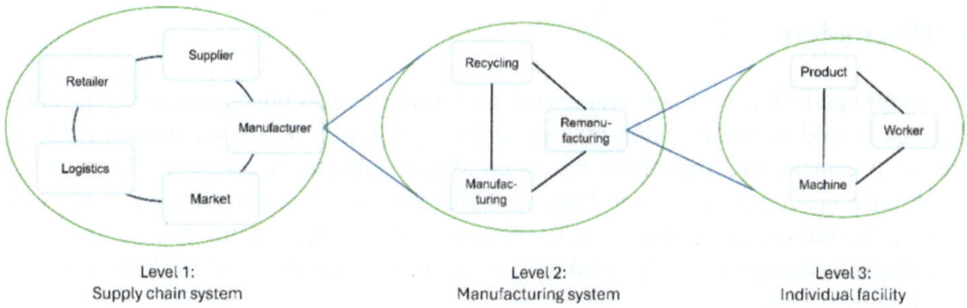

Fig. 2. Example of simulation architecture proposed for EoL assessment.

At the core are manufacturing agents responsible for decision-making under uncertainty and complexity. The manufacturing system comprises multiple semi-autonomous units (Level 2), each modeled as a complex system with its own internal structure and decision logic. These units, composed of individual components or agents (Level 3), are modeled using Agent-Based Modeling (ABM), Discrete Event Simulation (DES), or hybrid methods. This aligns with CAS theory, in which localized rules and agent interactions give rise to emergent behaviors.

At the macro level (Level 1), SoS theory provides a suitable abstraction for modeling coordination among distributed units and external stakeholders. CAS becomes less practical in this context due to limited visibility into local rules and agent-level data across organizational boundaries. SoS theory enables the representation of interoperable systems linked by shared objectives and communication protocols, without requiring detailed modeling of each actor's internal dynamics. ABM, System Dynamics (SD), or hybrid simulations can be applied depending on the system characteristics.

Overall, this architecture leverages the complementary strengths of CAS and SoS to capture multi-level complexity in circular manufacturing, offering a scalable and theoretically grounded framework for simulation-based decision support.

4.1 Hypothetical Scenario: Lithium-Ion Battery Circular Economy System

The manufacturer implements complex adaptive system (CAS) principles by conceptualizing production units (assembly lines, disassembly robots, quality control systems etc.) as autonomous agents governed by predefined decision protocols. These agents adhere to sustainability criteria, where remaining functional life determines product pathways. Adaptive behaviors emerge through, for example, optimization of disassembly sequences in response to dynamic cobalt and lithium price fluctuations. Externally, raw material suppliers, who are operationally independent, are integrated as system-of-systems (SoS) components, enabling emergent price-signal interactions. This hybrid CAS/SoS architecture facilitates bottom-up process adaptation within controlled manufacturing environments and top-down coordination across supply chain networks.

5 Discussion

The integrated EoL assessment framework and simulation architecture address the complexities of end-of-life decision-making at an aggregate level by accounting for both internal organizational dynamics and external systemic interactions. Avoiding overly deterministic assumptions about data availability, stakeholder behavior, or centralized control, the framework embraces the uncertainty and interdependence inherent in circular manufacturing systems. This enables a more context-sensitive and resilient approach to EoL assessment.

The framework's methodological foundation reflects a purposeful bifurcation between internal complexity and inter-organizational coordination, aligning with the epistemic distinctions of CAS and SoS. CAS is applied to the intra-organizational level, particularly within manufacturing firms, where decentralized interactions among adaptive agents—such as machines, departments, or decision-makers—generate emergent behaviors. These environments, characterized by feedback loops and internal heterogeneity, are well-suited to CAS modeling, which facilitates the exploration of adaptive production, resource recovery strategies, and responses to uncertain product returns.

Conversely, SoS theory is employed at the inter-organizational level to represent actors beyond the manufacturer's direct control—such as recyclers, distributors, and consumers—as autonomous yet functionally interdependent systems. Given the lack of access to micro-level behavioral rules and calibration data, CAS is less applicable in these contexts. SoS enables abstraction through shared goals, interoperability, and standardized communication, allowing for coordinated system-level modeling without requiring internal detail for each actor.

Despite their complementarity, integrating CAS and SoS presents challenges—particularly in articulating how emergent behaviors at the CAS level interface with the structured coordination assumed in SoS. CAS emphasizes localized adaptation and diversity, while SoS relies on protocol-driven alignment. Reconciling these perspectives requires careful modeling to avoid oversimplifying emergence or imposing unrealistic systemic order. Nonetheless, when rigorously applied, this integration offers a powerful lens to capture both micro-level adaptivity and macro-level coordination essential to EoL decision-making in circular manufacturing.

6 Conclusion and Future Work

In summary, integrating Complex Adaptive Systems (CAS) and Systems of Systems (SoS) offers a theoretically robust and practically applicable framework for modeling end-of-life assessment in circular manufacturing. By applying CAS to internal adaptive dynamics and SoS to inter-organizational coordination, the approach preserves the epistemic integrity of each paradigm while harnessing their complementary strengths. Despite challenges in bridging these modeling logics, the framework provides a strong foundation for advancing EoL decision-making in complex circular manufacturing systems and warrants validation through targeted case studies.

References

1. Marconi, M., Germani, M.: An end of life oriented framework to support the transition toward circular economy. In: DS 87-5 Proceedings of the 21st International Conference on Engineering Design (ICED 2017), vol. 5: Design for X, Design to X, Vancouver, Canada, 21–25 August 2017, pp. 199–208 (2017)
2. Sterman, J.D.: Business Dynamics: Systems Thinking and Modeling for a Complex World. Irwin McGraw-Hill, Boston (2000)
3. Murray, A., Skene, K., Haynes, K.: The circular economy: an interdisciplinary exploration of the concept and application in a global context. J. Bus. Ethics **140**, 369–380 (2017)
4. Samad, T., Parisini, T., Annaswamy, A.: Systems of systems. Impact Control Technol. **12**(1), 175–183 (2011)
5. Bjørnbet, M.M., Vildåsen, S.S.: Life cycle assessment to ensure sustainability of circular business models in manufacturing. Sustainability **13**(19), 11014 (2021)
6. Maier, M.W.: Architecting principles for systems-of-systems. Syst. Eng. J. Int. Council Syst. Eng. **1**(4), 267–284 (1998)
7. Jiang, T., Zhou, J., Wang, M., Li, E., Zhang, S.: SoS applications in production/manufacturing domain: a review and discussion. Int. J. Adv. Manuf. Technol. **130**(5), 2219–2239 (2024)
8. Baldwin, W.C., Felder, W.N.: Mathematical characterization of system-of-systems attributes. In: Transdisciplinary Perspectives on Complex Systems: New Findings and Approaches, pp. 1–24 (2016)
9. Hsu, J.C., Clymer, J.R., Garcia Jr, J., Gonzalez, E.: Agent-based modeling the emergent behavior of a system-of-systems. In: INCOSE International Symposium, vol. 19, no. 1, pp. 1581–1590. Wiley Online Library (2009)
10. Tate, W.L., Ellram, L.M., Bals, L.: Handbook of Theories for Purchasing, Supply Chain and Management Research. Edward Elgar Publishing (2022)
11. McDaniel, R.R., Jr., Lanham, H.J., Anderson, R.A.: Implications of complex adaptive systems theory for the design of research on health care organizations. Health Care Manage. Rev. **34**(2), 191–199 (2009)
12. Ahmad, M.A., Baryannis, G., Hill, R.: Defining complex adaptive systems: an algorithmic approach. Systems **12**(2), 45 (2024)
13. Tarpey, R.J., Mullarkey, M.T.: Engineering innovative clinical resource management by design: a guided emergent search through a complex adaptive system of systems. IEEE Trans. Eng. Manage. **70**(3), 927–941 (2021)
14. Barat, S., Kulkarni, V., Paranjape, A., Parchure, R., Darak, S., Kulkarni, V.: Fine-grained prediction and control of covid-19 pandemic in a city: application to post-initial stages. In: International Conference on Principles and Practice of Multi-Agent Systems, pp. 314–330. Springer (2022)
15. Johnson, B.: Modeling and simulation analysis of a complex adaptive systems of systems approach to naval tactical warfare. In: International Conference on Complex Systems, pp. 261–286. Springer (2020)
16. Stropnik, R., Sekavčnik, M., Férriz, A.M., Mori, M.: Reducing environmental impacts of the ups system based on PEM fuel cell with circular economy. Energy **165**, 824–835 (2018)
17. Padhamnath, P., et al.: Development of PV panel recycling process enabling complete recyclability of end-of-life silicon photovoltaic panels. Sol. Energy Mater. Sol. Cells **286**, 113571 (2025)
18. Rohit, K., Verma, A., Dhairiyasamy, R., Gabiriel, D.: A continuous supply chain management approach using SPJS-Fuzzy DEMATEL and LPWBN for automotive electric vehicles in India. Sustain. Futures **9**, 100518 (2025)

19. Alamerew, Y.A., Brissaud, D.: Circular economy assessment tool for end of life product recovery strategies. J. Remanufacturing **9**(3), 169–185 (2019)
20. Ravichandran, B., Boissie, K., Madhavan, P.: A multidimensional methodological approach to disassembly index calculation: enhancing product dismountability and sustainability. Sustain. Prod. Consum. (2025)
21. Zhang, H., Huang, Z., Wang, D., Tian, G., Wang, W.: U-shaped disassembly line balancing problem under interval Type-2 trapezoidal fuzzy set: Modeling and solution method. Eng. Appl. Artif. Intell. **145**, 110211 (2025)
22. Cao, Y., Li, G., Meng, K., Peng, X.: Managerial decisions of remanufacturing technology portfolio under a discrete choice model. Transp. Res. Part E Logist. Transp. Rev. **195**, 103959 (2025)
23. Hildenbrand, J., Dahlström, J., Shahbazi, S., Kurdve, M.: Identifying and evaluating recirculation strategies for industry in the nordic countries. Recycling **6**(4), 74 (2021)

End-of-Life Calculator: Aircraft Seat Case Study

Marco Franchino(✉), Christopher Haynes, Cansu Kandemir, and Matteo Zamarco

University of Sheffield Advanced Manufacturing Research Centre (AMRC), Sheffield, South Yorkshire S9 1ZA, UK
m.franchino@amrc.co.uk

Abstract. Manufacturing has traditionally prioritised cost, lead times, and quality over sustainability. However, increasing awareness of environmental impact demands greater accountability. The concept of sustainability often lacks clarity, and companies struggle to prioritise environmental goals.

This paper presents the development of an end-of-life (EoL) calculator based on selected key performance indicators (KPIs), divided into three areas—economic, environmental, and social—to evaluate EoL options such as disposal, recyclability, and raw material recovery, providing performance scores to support decision-making. This study includes a proof-of-concept case focused on the aviation sector, where non-operational factors contribute significantly to its carbon footprint. Aircraft cabin components, making up 10% of an airliner's empty weight and replaced multiple times during an aircraft's lifespan, create substantial waste, a significant portion of which is not recycled.

To validate the calculator, a Discrete Event Simulation (DES) model of the supply chain was developed to assess the recyclability of aircraft seats. This study highlights the challenges and opportunities of implementing circular economy practices in regulated sectors like aerospace.

Despite its limitations, the calculator establishes a practical framework for sustainable decision-making in manufacturing and supply chains.

Keywords: Circular Economy · Decision-Making · Aerospace

1 Introduction

Historically, manufacturing companies have prioritised cost, lead times, and quality over sustainability. However, increasing awareness of the environmental impact of human activities has highlighted the urgent need for industries to incorporate environmental responsibility into their definitions of success [1]. Despite this, debates continue over the concept of sustainability and how to assess it. Many companies struggle to determine the best processes for prioritising environmental responsibility and require structured methodologies to support decision-making towards a circular economy [2].

Globally, growing volumes of electrical and electronic equipment contribute to increasing end-of-life (EoL) waste [3]. Some products become obsolete due to wear

H. Kohl et al. (Eds.): GCSM 2025, LNME, pp. 177–186, 2026.
https://doi.org/10.1007/978-3-032-21154-5_20

and technological advancements, while others are deliberately designed with a limited lifespan to encourage frequent replacements [4]. This has historically contributed to a significant EoL product stream.

The traditional economic model follows a linear 'take-make-dispose' approach, where raw materials are extracted, processed, used, and ultimately discarded as waste [5]. While this system has driven economic growth, it also leads to issues such as resource depletion, pollution, and increasing landfill waste. In a linear economy, EoL typically involves discarding, sorting, and final disposal via landfill, incineration, or recycling—each with varying environmental implications [6].

In contrast, the circular economy is a sustainable model based on designing products for durability, reuse, and recyclability. Materials for new products come from old ones, minimising waste through reuse, remanufacturing, and recycling. When disposal is unavoidable, energy recovery or safe disposal methods are used as a last resort [7].

Key considerations in a circular economy include:

- **Reuse and Refurbishment:** Functional components are recovered, repaired, or resold.
- **Recycling:** When reuse is not possible, materials such as metals, plastics, and glass are separated and processed for use in new products.
- **Circularity:** A broader systemic model aiming to maintain the value of products, materials, and resources in the economy for as long as possible and minimise waste. Includes reuse, repair, remanufacturing, recycling, refurbishment, and also design-for-disassembly, reverse logistics, regulatory alignment, and business model transformation.

Disassembly is considered crucial in a circular economy, as it allows materials and components to be recovered efficiently, reducing waste and environmental impact while improving resource efficiency and sustainability. However, determining whether disassembly is worthwhile requires careful evaluation of various factors.

Currently, most of the EoL processes focus on material recycling, such as shredding, rather than reuse. While this is preferable to landfill, it still requires raw materials to be reprocessed and remanufactured. Additionally, recycling processes are often inefficient and fail to capture all recyclable materials. Many products are not designed for easy repair or reuse, leading to premature disposal.

As products reach the end of their lifecycle, they present both challenges and opportunities. Strategic decisions on disposal, recycling, and sustainable innovation are essential to reducing environmental impact and promoting circular economy principles. Proper management of EoL products can help mitigate environmental harm and drive more sustainable practices.

2 Methodology

This section outlines the structured approach taken to develop and apply a decision-support tool for evaluating end-of-life (EoL) alternatives within a sustainable manufacturing context. It introduces the rationale for selecting key performance indicators

(KPIs) and explains how these informed the creation of an EoL Calculator. The following subsections describe the process of KPI definition, the functioning of the calculator, and the results obtained from its application in a case study.

2.1 Definition of KPIs

The selection of KPIs in this study was guided by the Triple Bottom Line (TBL) framework, which emphasises the need to balance economic, environmental, and social considerations [8].

This approach is particularly appropriate for evaluating circular economy strategies because circularity inherently spans these three dimensions: it aims to preserve economic value through resource efficiency, reduce environmental burdens through waste minimisation and emissions control, and support social outcomes such as job creation, worker safety, and ethical sourcing. TBL aligns conceptually with circular economy principles by recognising that sustainability cannot be achieved through economic metrics alone. It offers a structured yet flexible foundation for identifying relevant KPIs that capture the multidimensional impact of end-of-life (EoL) decisions. While TBL itself does not prescribe a specific measurement system, it supports a holistic evaluation framework [2].

Drawing on literature, prior projects, and discussions with our industrial partners, we defined a list of criteria to evaluate products' EoL alternatives. These KPIs, grouped according to the three TBL dimensions, are presented in Table 1.

Table 1. EoL Calculator: KPIs.

TBL Areas	KPIs	Description	Source / Standard	Rationale
Economic	Value of recycled material or subcomponents	Assess the market value of recovered materials or sub-components.	[9],[10]	Promotes material circularity and supports financial viability.
	Labour Cost	Cost of labour involved in the EoL process.	[11],[12]	Key operational cost that affects economic feasibility.
	Equipment Cost	Cost of machinery or tools required.	[13]	Impacts capital expenditure and business case.
	Facilities Cost	Space or infrastructure cost for EoL processes.	[14],[15]	Influences site selection and scalability.
	Time	Time required for processing or disassembly.	[16],[17]	Disassembly time affects efficiency and economic return.
	Energy Cost	Cost of energy used in the EoL process.	[18],[19]	Reduces operational costs and environmental burden.
	Demand for recycled materials or sub-components	Market demand and customers for recovered outputs.	[20],[21]	Encourages reuse and market-driven circularity.
	Throughput	Number of products processed per hour.	[22]	Reflects system capacity and impacts profitability.
	Disposal Fees	Fees/taxes for disposal of unrecyclable waste.	[23],[24]	Encourages alternatives to landfill; affects cost.
Environmental	Safe Disposal	Ability to dispose/recycle materials without environmental harm.	[25],[26]	Ensures minimal ecological impact and compliance.
	Waste Reduction	Volume of waste diverted from landfill.	[10],[24]	Supports circular economy and sustainability targets.
	Energy Consumption	Energy used during the EoL process.	[18],[27]	Reduces carbon footprint and energy intensity.
	Process Emissions	Emissions from the EoL process (CO_2, NO_x, particulates).	[27],[28]	Addresses climate change and pollution control.
Social	Hazardous Substances	Operator exposure to hazardous materials.	[29],[30]	Protects health, meets safety regulations.
	Job Creation	Number of jobs created through EoL activities.	[31],[32]	Supports regional economic development and social value.
	Health Impact	Health impact on workers and the public.	[33],[34]	Key to ESG performance and workplace wellbeing.
	Human Rights	Risks of worker exploitation in EoL processes.	[35],[36]	Addresses global supply chain ethics.
	Infrastructure	Impact on local transport, utilities, and logistics.	[37]	Promotes sustainable local development.
	EoL Regulation	Likelihood or effect of relevant regulations.	[38],[39]	Ensures compliance, helps anticipate risks and opportunities.

2.2 EoL Calculator

We used the KPIs described in the previous section as comparison criteria to develop an EoL Calculator for assessing alternative scenarios based on a trade study analysis methodology. A trade study (or trade-off study) is a formal tool that supports decision making selection of the best or most balanced solution among the proposed alternatives [8].

Our calculator is designed to assess the viability and value of various EoL alternatives, making it easier for original equipment manufacturers (OEM) to assess the viability of alternatives to landfill. Results are presented as performance scores against KPIs, enabling users to compare and assess various scenarios effectively.

The steps to use the calculator, once a specific product and EoL alternatives are selected, are shown in Fig. 1.

Compared to tools like SimaPro, GaBi, and openLCA, our EoL Calculator offers specific advantages. It is easier to use and highly customisable—allowing users to add or modify KPIs (including qualitative ones), visualise results, and compare scenarios. It integrates logistics and cost factors (e.g. labour, energy, disposal) often absent from standard LCA tools. The calculator is transparent in the use of information, handles mixed data types (quantitative, qualitative, assumptions), and is suitable for early-stage decision-making, especially where information is scarce and detailed LCA databases are unavailable. While it does not replicate full-scale LCA capabilities, it complements them with practical features, like complete decision-making traceability.

We conducted case studies using the calculator to assess its effectiveness on different products and validate the approach. A relevant example from the aerospace sector is presented in the following section.

Fig. 1. Workflow of the EoL Calculator tool.

2.3 Case Study

Background. This case study develops and validates the EoL Calculator for aircraft seats as an example of its use. The aerospace sector is working towards carbon neutrality by 2050 in response to increasing government pressure to accelerate emissions reductions [40].

Although in-flight emissions have traditionally received the most attention, there are other opportunities within aviation to lower its carbon footprint. Non-operational factors contribute significantly, with tens of thousands of tonnes of material being consumed and discarded as waste each year. The environmental impact is closely linked to the mass of the aircraft cabin [41].

Cabin interiors account for approximately 10% of an aircraft's empty weight and are replaced four to five times over the airframe's lifespan. As a result, they represent a considerable share of an aircraft's environmental footprint [42]. However, applying circularity principles to cabin components is particularly challenging, meaning a significant proportion of aircraft materials end up in landfill or incineration [43].

Since sustainability has historically not been a priority for the industry, there is limited awareness of the environmental impact of cabin interiors beyond their effect on aircraft weight. Consequently, in modern airliners, around 50% of cabin components (by weight) are not currently recycled, with limited reuse within the aerospace sector [44]. Each cabin element presents unique technical and commercial challenges, further complicating efforts to improve sustainability.

Input Data. For this case study we focused on a single element of the aircraft cabin: the aircraft seat. The alternatives compared through the calculator are the following:

- **Landfill (Baseline)**: the product is disposed of in landfill at EoL with no processing.
- **Manual recycling**: the product is recycled by hand for its raw materials.
- **Automatic recycling**: the product is recycled automatically for its raw materials.
- **Manual disassembly for reuse**: the product is disassembled by hand by human operators to recover sub-assemblies and/or components, any remaining parts are recycled.
- **Automatic disassembly for reuse**: the product is disassembled automatically (e.g. using robots) to recover sub-assemblies and/or components, any remaining parts are recycled.

To assess and compare scenarios to support the decision-making process, we developed a Discrete Event Simulation (DES) model in AnyLogic DES software. This analysis was presented in one of our previous papers [45]. In that study, we focused on Western Europe as the geographical area of interest, considering aircraft fleets with varying seat compositions (e.g. number and types of seats) across different countries. We selected Western Europe as the geographical scope of this study due to its strategic relevance to the aerospace sector, the high density of Maintenance, Repair and Overhaul (MRO) facilities, and the availability of detailed fleet and material composition data. The model also incorporated the regular overhauls required for seat components and orders from airlines for new aircraft. Each country's aircraft fleet and corresponding seat configurations (number, type, and size with relative material composition) were incorporated into the model, with data sourced from online databases. This includes the age of the aircraft, reflecting real-life conditions at the start of the model. In the Baseline Scenario no recycled materials were considered. Seat manufacturers receive orders from the aircraft companies and supply new seat components accordingly. We calculated material usage and associated costs over a five-year period. In contrast, the alternative scenarios introduce a hypothetical Seat Recycling Centre, located in the UK. We tested several variations of these scenarios, adjusting parameters such as recycling rates, logistical efficiency, and processing capacity. Further to this, triangular distributions have been used to consider the stochasticity of processing times within the hypothetical recycling centre. After evaluating these variations, we decided to present the most balanced scenario in terms of cost-effectiveness and environmental impact. This balanced scenario offered the most practical and feasible results within the scope of our study.

We used this previous work as a source of information to validate our EoL Calculator. Product data were entered (***Step 1*** in Fig. 1). This includes the material breakdown for a typical seat and cost value of materials (Table 2). The DES model provided insights for the recycling processes and relative costs (e.g. method of manufacturing, equipment, labour, energy) (***Step 2*** in Fig. 1).

Table 2. Aircraft Seat: material breakdown and costs. All material cost data was originally in USD (as per [45]) and refers to prices as of August 2024. Since the case study focuses on the UK context, values have been converted to British Pounds Sterling (GBP) using the average exchange rate for that month (1 USD = 0.79 GBP) [46].

Material	Composition (%)	Cost
Foam	18.2	£279.87/m^{3}or £3.10/Kg
Fabric	1.3	£47.24/m^2 or £278.82/Kg
Aluminium	54.7	£3.48/Kg
Plastic	25.8	£1.98/Kg

The calculator allows users to assign relative importance to each KPI (***Step 3*** in Fig. 1). This is used to adjust the performance score assigned to an alternative, adjusting the results depending on KPIs' importance. Importance scores may vary depending on project or company requirements, an investigation with stakeholders is encouraged to properly set the scores at the beginning of any application. For this example, we chose to use a balanced policy for each KPI and consequently set all the importance scores equal.

Alternatives are scored against each KPI (***Step 4*** in Fig. 1). These scores may be of a different nature—quantitative, qualitative, or opinion-based—depending on the data availability. Quantitative and measurable scores (e.g. energy consumption in KWh, energy costs in £) are used where available data or assumptions are possible. These values are normalised on a scale from 1 (very poor performance) to 10 (excellent performance) to facilitate comparison. Scales from 1 to 10 are directly used to score qualitative and opinion-based data through workshop sessions. Justifications and references are recorded to track the decision-making process. For our example, we used open-source data and reasonable assumptions supported by workshop discussions.

Finally, results are presented on a normalised scale from 0 (worst possible performance) and 1 (best possible performance) (***Step 5*** in Fig. 1). Results are visualised through bar charts to facilitate comparison as shown in the following paragraph.

Results. The results of the case study (Fig. 2) showed that 'Manual disassembly for part reuse' scored the highest total when each KPI was weighted equally. This alternative also scored the highest for 'environmental' factors. The lowest scoring alternative was 'Landfill' which received a high score for economic factors (it is the cheapest solution) but received a low score for 'social' and 'environmental' factors.

The results of the case study alternatives were closely matched with only a 3% score difference between the manual and automated for part reuse alternatives. This highlights the influence of subjective data and KPI weighting on outcome variability, in particular, KPIs' importance scoring significantly affects the final rankings of alternatives. For instance, in a scenario where economic factors are prioritised, 'Landfill' alternative has a clear advantage. This finding underscores the importance of lowering the costs related to recycling and reuse options, for example designing products that are easy to disassemble. This also highlights the importance of governance in placing rules and regulations towards obligations towards recyclability routes.

When environmental and social considerations are given greater importance, disassembly options outperform traditional disposal methods. This indicates that companies prioritising sustainability goals may need to adjust their decision frameworks accordingly.

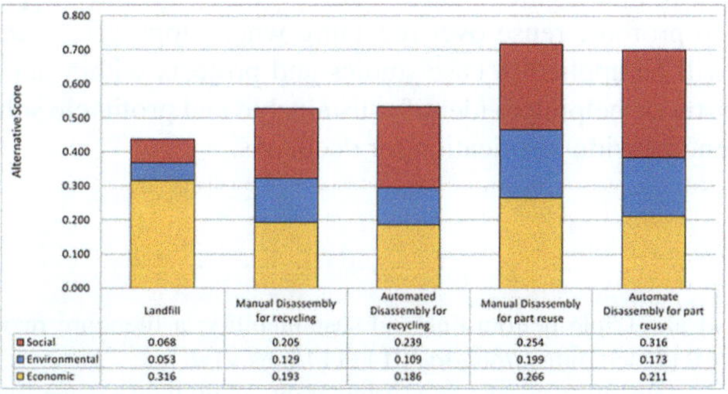

Fig. 2. Results from the EoL Calculator: performance scores across alternatives (normalised).

3 Findings

Our end-of-life (EoL) calculator has been tested across various products and scenarios. The tool has demonstrated value in supporting and informing the decision-making process.

The findings from the case study demonstrate the effectiveness of the decision-support calculator in evaluating EoL options based on economic, environmental, and social factors. The tool provides a structured assessment framework, enabling manufacturers to make data-driven decisions in line with circular economy principles.

The importance of high-quality, real-world data was evident in the analysis. While open-source datasets provided a reasonable baseline, more accurate industry-specific data would further refine the model's reliability and applicability. The case study used data provided by our research centre's aerospace partners. Where specific values were unavailable, we made reasonable assumptions using reputable academic sources, databases, and online data. While some deviations from real-world conditions may exist, the example is not intended to provide definitive answers. Instead, it serves as a proof of concept for the EoL Calculator, to facilitate discussions and offer insights to support the transition towards more sustainable industrial practices. In real-world applications, users will use their own set of data.

4 Conclusions

The calculator's structured approach allows it to be adapted across different sectors and products. Future enhancements, such as integrating real-time market data and automation trends, could improve decision accuracy and broaden its applicability across a wider range of industries. Organisations may adopt the calculator to align their EoL strategies with regulatory frameworks and sustainability targets, potentially gaining economic incentives or mitigating future compliance risks. The methodology also supports policy discussions on waste management and circular economy incentives, providing quantitative evidence to promote reuse over recycling where appropriate. In the long term, the framework will be applied to case studies and projects within our research centre and industrial partners, helping to identify sustainable and profitable solutions, quantify improvements, and provide justification for decisions.

References

1. Alhaddi, H., et al.: Triple bottom line and sustainability: a literature review. Bus. Manag. Stud. **1**(2), 6–10 (2015). https://doi.org/10.11114/bms.v1i2.752
2. Hallstedt, S.: Sustainability criteria and sustainability compliance index for decision support in product development. J. Clean. Prod. **140**, 251–266 (2017)
3. GOV.UK: WEEE in the UK. https://www.gov.uk/government/statistical-data-sets/waste-electrical-and-electronic-equipment-weee-in-the-uk. Accessed 9 July 2024
4. Mancini, M.: Design-driven obsolescence. Des. J. **22**(Sup1), 2243–2246 (2019). https://doi.org/10.1080/14606925.2019.1595019
5. Bocken, N.M.P., et al.: The sufficiency-based circular economy—an analysis of 150 companies. Front. Sustain. **4**(2), 1–15 (2022). https://doi.org/10.3389/frsus.2022.899289
6. GOV.UK: Emissions. https://www.gov.uk/government/publications/landfill-sites-impact-on-health-from-emissions/impacts-on-health-of-emissions-from-landfill-sites. Accessed 5 July 2024
7. Mamudu, U.U., et al.: Circular economy in sustainable manufacturing: challenges and strategies. Int. J. Sci. Res. Arch. **12**(2), 129–141 (2024). https://doi.org/10.30574/ijsra.2024.12.2.1217
8. Elkington, J.: Towards the sustainable corporation: win-win-win business strategies for sustainable development. Calif. Manag. Rev. **36**, 90–100 (1994)
9. Giudice, F., La Rosa, G., Risitano, A.: Product Design for the Environment: A Life Cycle Approach. CRC Press (2006)
10. ISO 14051:2011. Environmental management — Material flow cost accounting — General framework
11. ISO 15686-5:2017. Buildings and constructed assets — Service life planning — Part 5: Life-cycle costing
12. Liu, W., Zhang, M., Dong, H., Chen, X., Fujita, T.: A review of remanufacturing in China: perspectives, barriers and strategies. J. Clean. Prod. **255**, 120111 (2020)
13. EEA (European Environment Agency). Waste and material resources reports. https://www.eea.europa.eu
14. WRAP (Waste and Resources Action Programme). Resource efficiency and facility management guidance. https://www.wrap.org.uk
15. UNIDO (United Nations Industrial Development Organization). Facility Planning Guidelines (2017). https://www.unido.org

16. Bakker, C.A., Wang, F., Huisman, J., den Hollander, M.C.: Products that go round: exploring product life extension through design. J. Remanufacturing **4**(1), 1–13 (2014)
17. ISO 8887:2017. Technical product documentation — Design for manufacturing, assembling, disassembling and end-of-life processing (MADE)
18. ISO 14044:2006. Environmental management — Life cycle assessment — Requirements and guidelines
19. Das, S., Ingole, S., Ramesh, S., Rajashekar, B.: Energy and environmental impacts of recycling aluminium – a life cycle assessment perspective. Resour. Conserv. Recycl. **54**(7), 557–570 (2009)
20. Ellen MacArthur Foundation. The Circular Economy: A Transformative COVID Recovery Strategy (2020). https://ellenmacarthurfoundation.org
21. EU (European Union). Monitoring Framework for the Circular Economy (2020). https://ec.europa.eu/environment/circular-economy
22. Bahubalendruni, M.V.A.R., Varupala, V.P.: Disassembly sequence planning for safe disposal of end-of-life waste electric and electronic equipment. Natl. Acad. Sci. Lett. **44**, 243–247 (2021). https://doi.org/10.1007/s40009-020-00994-0
23. UK Government. UK Landfill Tax (2024). https://www.gov.uk/guidance/landfill-tax
24. EU. Waste Framework Directive (Directive 2008/98/EC). Official Journal of the European Union, L312, pp. 3–30 (2008)
25. Basel Convention, 1989. Basel Convention on the Control of Transboundary Movements of Hazardous Wastes and Their Disposal. https://www.basel.int
26. DEFRA (UK Department for Environment, Food & Rural Affairs). Guidance on the disposal and recycling of waste. https://www.gov.uk/guidance/waste
27. IPCC (Intergovernmental Panel on Climate Change), 2006. 2006 IPCC Guidelines for National Greenhouse Gas Inventories. https://www.ipcc-nggip.iges.or.jp/public/2006gl
28. Ecoinvent Association. Ecoinvent Database v3.8 (2021). https://www.ecoinvent.org
29. REACH Regulation (EC) No 1907/2006. Registration, Evaluation, Authorisation and Restriction of Chemicals. European Chemicals Agency
30. OSHA (Occupational Safety and Health Administration). Hazard Communication and Worker Safety. https://www.osha.gov
31. ILO (International Labour Organization). Employment and Decent Work for a Green Economy. https://www.ilo.org
32. Bocken, N.M.P., de Pauw, I., Bakker, C., van der Grinten, B.: Product design and business model strategies for a circular economy. J. Ind. Prod. Eng. **33**(5), 308–320 (2016)
33. WHO (World Health Organization). Occupational health: Health impacts of hazardous waste (2021). https://www.who.int
34. EU-OSHA (European Agency for Safety and Health at Work). Occupational Safety and Health Reports. https://osha.europa.eu
35. UN. UN Guiding Principles on Business and Human Rights. United Nations Human Rights Council (2011). https://www.ohchr.org
36. OECD. OECD Due Diligence Guidance for Responsible Business Conduct (2018). https://www.oecd.org
37. UNEP (United Nations Environment Programme). Circularity in Cities: Infrastructure Planning Guidelines. https://www.unep.org
38. UK Government. UK Environment Act 2021 (2021). https://www.legislation.gov.uk
39. European Commission, 2020. European Green Deal. https://ec.europa.eu/info/strategy/priorities-2019-2024/european-green-deal
40. Parnell, G.S.: Trade-off Analytics: Creating and Exploring the System Tradespace. Wiley, Hoboken (2016)

41. Dixon, J., Bell, K., Brush, S.: Which way to net zero? A comparative analysis of seven UK 2050 decarbonisation pathways. Renew. Sustain. Energy Transit. **2**, 100016 (2022). https://doi.org/10.1016/j.rset.2021.100016
42. www.sustainableaviation.co.uk/news/uk-aviation-industry-strengthens-commitment-to-achieving-net-zero-and-launches-first-interim-decarbonisation-targets. accessed 10 Jan 2025
43. IATA: BIPAD 1st Edition. IATA, Montreal-Geneva (2018)
44. Aerospace Technology Institute (ATI): Sustainable Cabin Design. ATI (2022)
45. Franchino, M., Haynes, C., Kandemir, C.: Supply chain development for circularity in aerospace: aircraft seat case study. In: Proceedings of the Operational Research Society Simulation Workshop 2025 (SW25), pp. 1–13. Springer, Heidelberg (2025). https://doi.org/10.36819/SW25.047
46. Bank of England: Exchange rates. https://www.bankofengland.co.uk/boeapps/database/Bank-Rate.asp. Accessed 9 Aug 2024

Sustainability Assessment of End-of-Life Strategies for PEM Fuel Cells

Sophie Victoria Bernard, Erik Alexander Recklies, and Semih Severengiz(✉)

Bochum University of Applied Sciences, Sustainable Technologies Laboratory, Am Hochschulcampus 1, 44801 Bochum, Germany
semih.severengiz@hs-bochum.de

Abstract. Green hydrogen technologies play a significant role in the transition towards clean energy systems and the achievement of global climate goals. In particular, PEM fuel cells offer a variety of potential applications in domains such as transport and energy supply. However, PEM fuel cells contain many valuable resources and materials with non-negligible ecological impacts (e.g., platinum group metals, Nafion). There is a lack of a systematic analysis for evaluating and selecting End-of-Life strategies for PEM fuel cells considering sustainability criteria. This paper aims to contribute to filling this gap and is funded by the Federal Ministry for the Environment, Nature Conservation, Nuclear Safety and Consumer Protection as part of the project "GH2GH – Green Hydrogen for Decentralised Energy Systems in Sub-Saharan Africa". The analysis is based on an extensive literature review and interviews with experts from research, politics, and industry. Different End-of-Life strategies for PEM fuel cells are presented focusing on applicability and feasibility. In this context, several R-strategies (Repair, Reuse, Recycle, and Recover) are considered. Additionally, possible criteria for a sustainability assessment of these End-of-Life strategies are elaborated and discussed.

Keywords: PEM fuel cell · Hydrogen technology · Circular economy · R-strategies

1 Introduction

Facing global climate goals, the transition towards a climate-neutral energy supply appears as one of the greatest challenges of the 21st century. With the Green Deal, the European Union is pursuing the goal of becoming climate-neutral by 2050. Green hydrogen technologies play a key role in this context by enabling sustainable energy storage [1]. There is no doubt that the number of hydrogen systems installed will rise sharply in the coming years. According to the International Energy Agency, the production of low-emission hydrogen could reach 49 million tons per year by 2030 based on planned projects. This significant growth is primarily due to electrolysis projects, for which a total capacity of around 520 GW has been announced [2].

Alongside this growth of hydrogen technologies and the increasing demand for the materials used (e.g., platinum group metals (PGMs), rare earths, cobalt, and nickel), the

H. Kohl et al. (Eds.): GCSM 2025, LNME, pp. 187–195, 2026.
https://doi.org/10.1007/978-3-032-21154-5_21

issue of viable End-of-Life (EoL) strategies for fuel cells arises. These EoL strategies can contribute to material conservation, reduce CO_2 emissions, and enable both economic and business cost savings in the context of a circular economy [3]. The topic of EoL strategies for fuel cells is therefore relevant for various target groups (e.g., manufacturers, recyclers, researchers, and political stakeholders).

Our paper engages with current discourses on EoL strategies for PEMFCs and provides insights into the perspectives of relevant stakeholder groups. A PEMFC (Proton Exchange Membrane Fuel Cell) is an electrochemical device that converts hydrogen and oxygen into electricity, water, and heat using a proton-conducting membrane (refer to Sect. 2.2 for a more detailed depiction). We address the following central research question: *What are the perspectives of different stakeholders on the EoL of PEMFCs with regard to (a) promising EoL strategies, (b) challenges associated with their implementation, and (c) their evaluation in terms of sustainability criteria?* Adopting an explicit sustainability perspective, our study contributes to the scientific discourse on EoL strategies for PEMFCs. The insights gained may support the design and facilitation of more effective and future-oriented transformation processes in this field.

2 Theoretical Background

To support the reader's understanding of our study, this section introduces and defines key terminology, presents a basic description of a PEMFC, and offers a concise overview of the current state of research.

2.1 Key Definitions EoL Strategies

A product's EoL is reached when it is no longer needed or used by its original user and for its original purpose [4]. A variety of EoL strategies aim at the prevention of waste, after discarding a product. In the context of a circular economy, R-strategies are getting mounting attention. In Table 1 the R-strategies that are relevant for our considerations are defined according to [5]. In general, these and other R-strategies focus on minimizing the use of materials and preserving the value of products for as long as possible.

Table 1. Definition and classification of selected R-strategies according to [5].

R-strategy	Definition	Classification	
Reuse	Reuse by another consumer of discarded product which is still in good condition and fulfils its original function.	Extension of the lifetime of the product or its parts	↑ Increasing circularity
Repair	Repair and maintenance of defective product so it can be used with its original function.		
Recycle	Process materials to obtain the same (high grade) or lower (low grade) quality.	Useful application of materials and resources.	
Recover	Incineration of materials with energy recovery.		

2.2 PEMFCs

A PEMFC uses hydrogen and oxygen to generate electricity through an electrochemical reaction. Based on [6], its structure can be described as follows: The stack is the core component of a PEMFC. It includes the membrane electrode assembly (MEA) and is supplemented by the balance of plant (BoP). The MEA consists of a membrane in the center, coated with a catalytic layer on both sides, microporous layers (MPL), and gas diffusion layers (GDL). MPL and GDL are often regarded as one unit, protected and supported by a sealing. Multiple MEAs can be combined in a single stack to achieve a desired performance. Additionally, the stack includes bipolar plates (BPPs), current collectors, end plates, and tie rods or strap connections. The BoP, which encompasses all supporting systems such as control electronics, the cooling system, cabling, and housing, is also essential for the operation of the PEMFC.

2.3 State of Research

Based on a literature review it can be stated that existing studies (especially Life Cycle Assessments (LCAs)) show possible positive effects of EoL strategies for hydrogen systems and their components in terms of environmental impacts. However, a review of extant LCAs on the EoL of PEMFCs reveals a tendency to prioritize individual components or only a single R-strategy (e.g., recycling). The integration of several R-strategies, such as reuse, refurbishment, or remanufacturing, has scarcely been systematically contemplated to date. Existing studies focus particularly on the recycling of the MEA [6, 7]. Further articles consider the entire stack or even the BoP but still explicitly focus on the recycling of PGMs (e.g., [3, 8, 9]). The examination of other R-strategies is limited to a few articles (Recover: [9]; Reuse: [10]).

3 Methods

The goal of our research approach is to deliver in depth insights into the subjective perspective of different stakeholders on EoL strategies of PEMFCs. Based on the literature review eight semi-structured expert interviews [11] were conducted. To gain a variety of perspectives on the topic under investigation various stakeholders were considered and contacted. As this work was conducted as part of the research project "GH2GH – Green Hydrogen for Decentralised Energy Systems in Sub-Saharan Africa" which is funded by the Federal Ministry for the Environment, Nature Conservation, Nuclear Safety and Consumer Protection and involves the collaboration with regional partners from Ghana, the stakeholders are located in these two countries. The final sample of experts included both a representative of a German and a Ghanaian recycling company, a representative of a German fuel cell manufacturer, three researchers with expertise in the context of hydrogen technologies, and two representatives from the field of Ghanaian politics. Apart from Expert A who only agreed to answer the questions in written form, all interviews were conducted via Zoom. The interviews were recorded, transcribed, and analyzed based on principles of qualitative content analysis [12]. During the analysis the inductive approach was supplemented by deductively generated categories that were derived from the literature review. Table 2 provides an overview of the interviewed experts.

Table 2. Interviewed experts by stakeholder group, country, and date of interview. * = in written form.

	Stakeholder Group	Country	Date of interview
Expert A	Research/Politics	Ghana	17.12.2024*
Expert B	Research/Politics	Ghana	16.12.2024
Expert C	Recycler	Ghana	13.12.2024
Expert D	Recycler	Germany	10.12.2024
Expert E	Research	Germany	05.12.2024
Expert F	Research	Germany	19.12.2024
Expert G	Fuel cell producer	Germany	18.12.2024
Expert H	Research	Germany	13.12.2024

4 Results of the Expert Interviews

According to the addressed research questions this section will focus on the EoL strategies for PEMFCs (3.1), challenges in the application of these EoL strategies (3.2), and on sustainability criteria for these EoL strategies (3.2). Even though the transcribed interviews cannot be attached to this conference paper, we provide references to the respective interviews by the pseudonyms given in Table 2.

4.1 EoL Strategies for PEMFCs

Four conceivable EoL strategies for each component considering various criteria were identified through the expert interviews: Repair, Reuse, Recycle, and Recover (refer to Table 1 for respective definitions).

Repair: The end of the first life of a PEMFC is reached either when an excessive loss of efficiency has occurred or the fuel cell fails due to an internal defect or a defect in one or more other BoP components. In this case, a repair is considered first (Expert G): In the case of larger systems and complete solutions, the cause of the defect must be investigated on-site. In some cases, spare parts can be replaced directly on-site. However, the fuel cell often must be sent back to a manufacturer's site to check its repairability and carry out a repair.

Reuse: If a repair of the PEMFC is ruled out, the components are dismantled and examined. Individual components are checked for their reusability. While some experts proclaim that the BPP can be cleaned and reused (Expert F, Expert G), other experts doubt this and consider recycling the BPP at material level to be more efficient (Expert D). However, in the case of design-based and technological developments in particular, components may become obsolete and therefore cannot be reused even if they are fundamentally reusable (Expert H). In general, housing parts, cables, silicone hoses, metal tubes, screws and end plates can be reused without any problems, provided they are not deformed, corroded or damaged (Expert G).

Recycle: Several components cannot be reused. This applies particularly to the MEA (Expert F) which must be sent to qualified recycling companies. The recovery of PGMs in particular has been the focus of recycling to date. Most common recycling approaches aim to burn the MEA and are only concerned with the recycling of platinum (Expert H). However, there are also recycling procedures that can be used to recover the ionomer of the membrane as well (Expert D, Expert F, Expert G, Expert H). Other components of the PEMFC (e.g., BPP, metal parts of the BoP) can also be recycled, if they are available in a pure, single-material form. The recycling of rare earths is discussed as future possibility (Expert D).

Recover: Components that cannot be recycled are incinerated. Energy is recovered from the incineration process, which can be credited to the manufacturing process. Only a few components cannot be incinerated and must therefore be sent to landfill.

Figure 1 gives a general overview of the depicted R-strategies in the context of a circular economy. Figure 2 provides a morphological box as a potential decision tool of possible R-strategies for the respective components; two exemplary pathways are illustrated for the MEA and the stack in pink and blue.

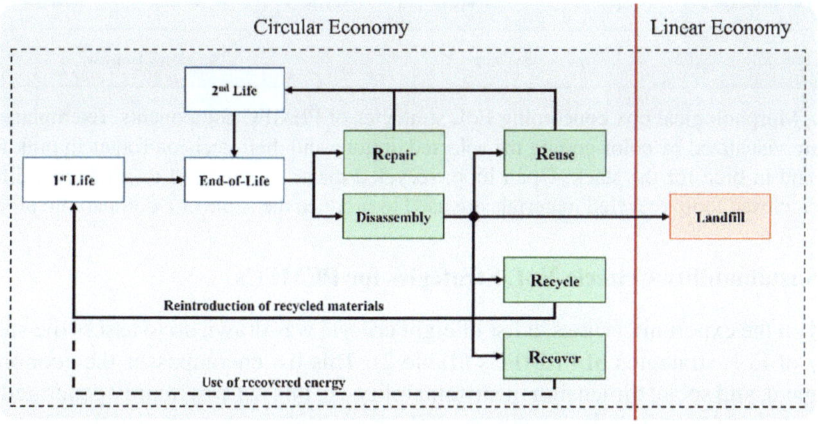

Fig. 1. Overview of R-strategies for PEMFCs at the end of their life and illustration of relationships and interdependencies between different strategies in the context of a circular economy.

4.2 Challenges in Application of EoL Strategies for PEMFCs

In addition to the potential benefits of various EoL strategies for PEMFCs, the interviewees also highlighted several challenges associated with their implementation. Of all the mentioned challenges three key challenges were identified: economic viability, complex material composition, and the ongoing technological development.

Economic viability emerged as one of the most frequently mentioned concerns. At present, there are relatively few PEMFCs reaching their EoL, which limits the economic attractiveness of applying R-strategies for recycling companies (Expert A, Expert D). More broadly, interviewees emphasized that no EoL strategy is likely to be adopted on a large scale unless it proves to be financially viable (Expert G, Expert H).

The applicability of different EoL strategies largely depends on the materials used in PEMFCs (Expert D). However, there is still significant variation in material composition across different systems. Analyzing components in terms of their material properties remains challenging due to the high labor intensity and required time investment.

Fuel cell technology is still undergoing rapid development, and it is expected that further advances in both technology and product design will occur in the coming years (Expert F). As a result, there is a risk that components may quickly become obsolete, making repair or reuse more difficult.

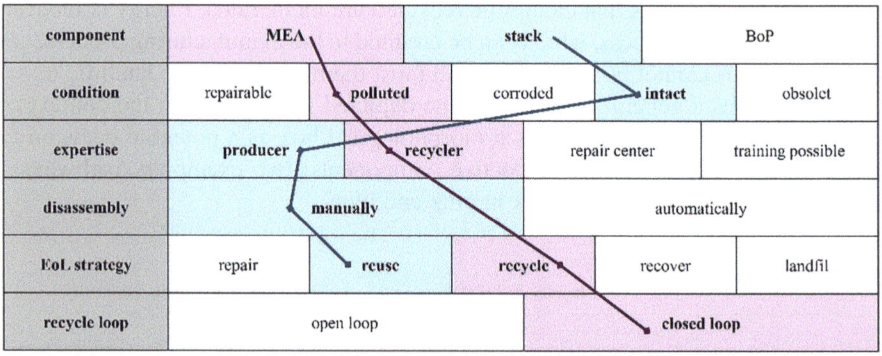

Fig. 2. Morphological box concerning EoL strategies of PEMFC components. Exemplary pathways are visualized by color-coding the selected options and their decision routes in pink for the MEA and in blue for the stack. Open loop: recycled materials are used to produce a different product, closed loop: recycled materials are used to produce the same or a comparable product.

4.3 Sustainability Criteria EoL Strategies for PEMFCs

Based on the expert interviews, a list of eight criteria was drawn up to assess the sustainability of EoL strategies of PEMFCs (Table 2). This list encompasses the economical, ecological, and social dimension of sustainability. As this list was mainly generated from the opinions of the interviewed experts, it does not claim to be exhaustive.

1. *Stakeholder involvement:* Several experts mention the importance of involving stakeholders into the entire product life chain, including EoL (Expert D, Expert E, Expert F). This participation appears to be central for determining materials, conditions, and possible EoL strategies of the respective fuel cells and for communicating benefits and challenges of these processes.
2. *Component-specific consideration of EoL strategies:* Also repeatedly mentioned is the need for component-specific consideration of EoL strategies (Expert E, Expert F, Expert G, Expert H). Depending on the respective component and its condition, different R-strategies are conceivable (Table 3).
3. *Extensive on-site recycling:* On-site recycling can help to minimize costs and emission due to reduced transport distances (Expert D, Expert G, Expert H).
4. *On-site staff training:* To guarantee successfully implemented on-site EoL treatment and to achieve technology-knowledge-transfer, it is important to train on-site staff (Expert D).

5. *Avoidance of health risks:* According to Experts D and F a central criterion in the social dimension is the avoidance of health risks: For every EoL strategy the safety of all those involved must be the top priority. A health risk (e.g., due to toxic gases) must always be ruled out.
6. *Economic viability:* As a particularly economic criterion the economic viability is discussed (Expert A, Expert F). As mentioned above, no EoL strategy appears to be promising for the future if it is not profitable.
7. *Recycling company certifications:* To guarantee that the recycling processes carried out are safe and properly monitored the recycling company should be certified (e.g., in accordance to DIN EN ISO 14001:2015) (Expert D).
8. *MEA-recycling:* The process of MEA-recycling is of particular importance as the MEA contains the materials with both the highest costs and the greatest ecological impact (e.g., PMGs, Nafion). The chosen recycling procedures for the respective component should be as efficient and as environmentally friendly as possible (Expert E).

5 Conclusion and Implications

In light of the growing significance of hydrogen technologies, an increasing number of fuel cells are expected to reach EoL in the future, necessitating the development of appropriate treatment strategies. Our study presents the results of expert interviews concerning EoL strategies for PEMFCs. The findings complement the current state of research. Comparable to the prominent role attributed to recycling in the scientific discourse, recycling also emerges as a key topic in the expert interviews. Nonetheless, the interviews encourage a more comprehensive consideration of alternative R-strategies beyond recycling. Particularly reusing components of the PEMFC emerges as a promising option as studies in other domains have already proven that reuse can lead to substantial reductions in environmental impact [13]. Based on the expert opinions, key challenges for the EoL treatment of PEMFCs could be identified: economic viability, complex material composition, and the ongoing technological development. Additionally, several criteria have been identified to assess the sustainability of EoL treatment of PEMFCs across economic, environmental, and social dimensions.

The results of this study bear implications for both practical application and future research. From an eco-design perspective, it appears necessary to design PEMFCs that can be easily disassembled for repair, refurbishment, and reuse of components. In the long-term, this can help to enhance the circularity of the PEMFC and to reduce both the product's costs and its ecological impact. Future research appears promising in terms of evaluating various EoL scenarios for PEMFCs in a comparative LCA, investigating the influencing factors on the suitability of PEMFCs for different R-strategies in more detail, and developing a more comprehensive sustainability criteria catalog for the EoL of PEMFCs.

Acknowledgment. The research work published in this paper was funded by the Federal Ministry for the Environment, Nature Conservation, Nuclear Safety and Consumer Protection under the funding program Export Initiative Environmental Protection. Funding reference: 67EXI6503A. The research work is under the project Green hydrogen for decentralized energy systems in sub-Saharan Africa-Pilot project in Tema, Ghana.

Disclosure of Interests. The authors have no competing interests to declare that are relevant to the content of this article.

References

1. European Commission: The European Green Deal. https://eur-lex.europa.eu/legal-content/EN/TXT/?uri=CELEX:52019DC0640. Accessed 21 Apr 2025
2. IEA: Global Hydrogen Review 2024. https://www.iea.org/reports/global-hydrogen-review-2024. Accessed 21 Apr 2025
3. Uekert, T., Wikoff, H.M., Badgett, A.: Electrolyzer and fuel cell recycling for a circular hydrogen economy. Adv. Sustain. Syst. **8**(4), 2300449 (2024)
4. Rose, C.M.: Design for environment: a method for formulating product end-of-life strategies. Stanford University (2001)
5. Potting, J., Hekkert, M., Worrell, E., Hanemaaijer, A.: Circular economy: measuring innovation in the product chain (2017)
6. Duclos, L., et al.: Closing the loop: life cycle assessment and optimization of a PEMFC platinum-based catalyst recycling process. Green Chem. **22**(6), 1919–1933 (2020)
7. Duclos, L., Lupsea, M., Mandil, G., Svecova, L., Thivel, P.-X., Laforest, V.: Environmental assessment of proton exchange membrane fuel cell platinum catalyst recycling. J. Clean. Prod. **142**, 2618–2628 (2017)
8. Bauer, T., Singh, K., Mandil, G., Svecova, L., Dubau, L.: Identification of the environmental hotspots of a recycling process – case study of a Pt PEMFC catalyst closed-loop recycling system evaluated via life cycle assessment methodology. Int. J. Hydrog. Energy **63**, 396–410 (2024)
9. Stropnik, R., Lotrič, A., Bernad Montenegro, A., Sekavčnik, M., Mori, M.: Critical materials in PEMFC systems and a LCA analysis for the potential reduction of environmental impacts with EoL strategies. Energy Sci. Eng. **7**(6), 2519–2539 (2019)
10. Lotrič, A., Sekavčnik, M., Kuštrin, I., Mori, M.: Life-cycle assessment of hydrogen technologies with the focus on EU critical raw materials and end-of-life strategies. Int. J. Hydrog. Energy **46**(16), 10143–10160 (2021)
11. Kaiser, R.: Qualitative Experteninterviews: Konzeptionelle Grundlagen und praktische Durchführung, 2nd edn. Springer Fachmedien, Wiesbaden (2021)
12. Mayring, P.: Qualitative Inhaltsanalyse: Grundlagen und Techniken, 13th edn. Beltz, Weinheim Basel (2022)
13. Nikolic, M., Bergmann, G., Schelte, N., Severengiz, S.: Closing the loop of small WEEE – life cycle based approach for the evaluation of end-of-life strategies on the example of coffee machines. Resour. Conserv. Recycl. Adv. **23**, 200220 (2024)

Recovery of NdFeB Magnets from the Ferrous Fraction of Shredded Waste Streams by Physical Sorting

Thorsten Ihne[1(✉)], Roman Hahn[1], Jörg Franke[2], and Florian Risch[1]

[1] Institute for Factory Automation and Production Systems (FAPS), Friedrich-Alexander-Universität Erlangen-Nürnberg, Fürther Straße 246B, 90429 Nuremberg, Germany
thorsten.ihne@faps.fau.de

[2] Institute for Factory Automation and Production Systems (FAPS), Friedrich-Alexander-Universität Erlangen-Nürnberg, Egerlandstraße 7, 91058 Erlangen, Germany

Abstract. A secure and sustainable supply of critical raw materials is crucial to the economic strategy of industrialized countries. In the context of the current transition to green technologies, this applies in particular to rare earth elements (REEs) used for neodymium-iron-boron (NdFeB) magnets, which are essential for various key applications. The European Union and the United States have classified these elements as critical because they combine significant economic relevance, poor substitutability and high supply vulnerability. Recycling can reduce dependency, but the necessary technical and organizational conditions are currently not given. Consequently, political, scientific and industrial stakeholders are intensifying their efforts to establish an effective circular economy. At least on a laboratory scale, a broad spectrum of methods for processing end-of-life (EOL) magnets exists, but magnet extraction is also a major challenge. For specific applications, dismantling or hydrogen processing are effective, but for sufficient processing capacities, the implementation of a universally applicable process based on shredding and subsequent sorting is essential. Therefore, the present paper discusses a multi-stage sorting approach for the ferrous shredder fraction using simulative and experimental investigations to fill the gap in the existing process chain. The focus is set on physical processes exploiting differences in the magnetic and mechanical properties of the relevant materials.

Keywords: Rare Earth Magnets · Critical Raw Materials · Recycling · Physical Sorting · Magnetic separation · Hard And Soft Magnetic Materials

1 Introduction and Motivation

The ongoing economic transformation in industrialized countries, driven by climate change mitigation, depends on a secure supply of specific metals and metalloids used in specialized and emerging technologies such as batteries, fuel cells, electric motors,

H. Kohl et al. (Eds.): GCSM 2025, LNME, pp. 196–204, 2026.
https://doi.org/10.1007/978-3-032-21154-5_22

photovoltaics, and semiconductors. As a result, global demand for these resources is expected to rise sharply. [1–3] Apart from their technological and economic importance, these raw materials are often characterized by a vulnerable supply chain due to low market diversification and poor substitutability [1]. Many are geopolitically dispersed, occur in low concentrations, or are mined only as by-products, affecting security of supply [4]. Recent geopolitical crises, the COVID-19 pandemic, and intensifying trade conflicts have highlighted the lack of resilience in global supply chains. Moreover, primary production is often linked to significant environmental and social implications [1]. In response, the European Union (EU-27) and the United States of America (USA) have identified Critical Raw Materials (CRMs) and developed strategies to secure sustainable and stable supply [5, 6]. A prime example of CRMs are rare earth elements (REEs) applied for high performance permanent magnets. Rare earth (RE) magnets are essential for electric vehicle (EV) drives, wind turbines, and other strategic applications with significant market growth. Therefore the global demand for RE magnets and the related RE metals is expected to rise sharply as shown in Fig. 1 [7–9].

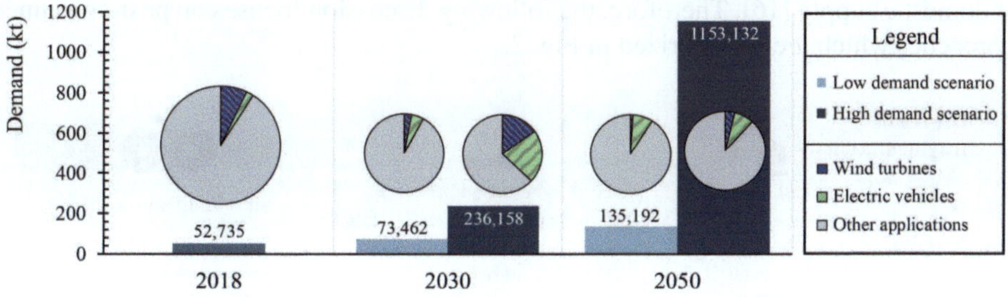

Fig. 1. Forecasted global demand for RE metals used for magnets (based on [8])

Sintered neodymium-iron-boron (NdFeB) magnets dominate due to their superior magnetic properties [9–11]. They contain about one-third REEs by weight, primarily light REEs like neodymium and praseodymium, with smaller amounts of heavy REEs such as dysprosium, and terbium, which improve specific material properties but have a significant cost impact [7, 10, 11]. The global RE magnet supply chain is poorly diversified. The People's Republic of China dominates the entire value chain with 60% market share in ore mining, 87% in ore refining, 91% in alloy production, and 94% in sintered RE magnet production [12]. In contrast the EU-27, Japan, and the USA depend heavily on imports, resulting in low supply chain resilience, price volatility, and ecological concerns due to high energy use and toxic waste in primary production [10, 11].

2 Approaches and Challenges for the Recovery of RE Magnets

To secure the supply of CRMs, various strategies exist. Short-term stockpiling is complemented by long-term measures such as demand reduction through material improvement and substitution, and diversification of supply. The latter includes the expansion of primary production, either domestically or in third countries, and the promotion of circular economy. [13] Through the "Critical Raw Materials Act" (CRMA), the EU27 is taking a comprehensive approach, setting targets for the domestic extraction, processing and recycling of CRMs by 2030, with the aim of achieving 10%, 40% and 25% respectively, in relation to annual consumption [5].

The recovery of RE magnets from relevant waste streams has high economic, geopolitical and ecological potential. However, circular economy is impeded by logistical challenges, low technology readiness and a lack of economic incentives. Due to low progress since the 2010/2011 global RE crisis, the European recycling rate has remained stagnant at approx. 1%. [9, 10, 14] Nonetheless, current market trends, policy initiatives, and technological innovations offer potential to new business models and the ramp-up of processing capacities [7, 14]. Here, the circular economy for RE magnets comprises various approaches depending on the input material, such as reuse and remanufacturing on different levels, and both short- and long-loop recycling [9, 15]. A further differentiation can be made between the processing of production scraps (pre-consumer recycling) and end-of-life (EoL) products (post-consumer recycling) [13]. Due to the low local value added in Europe and the USA, pre-consumer recycling has currently a limited relevance for domestic supply [16]. Therefore, the following discussion focuses on post-consumer approaches which are summarized in Fig. 2.

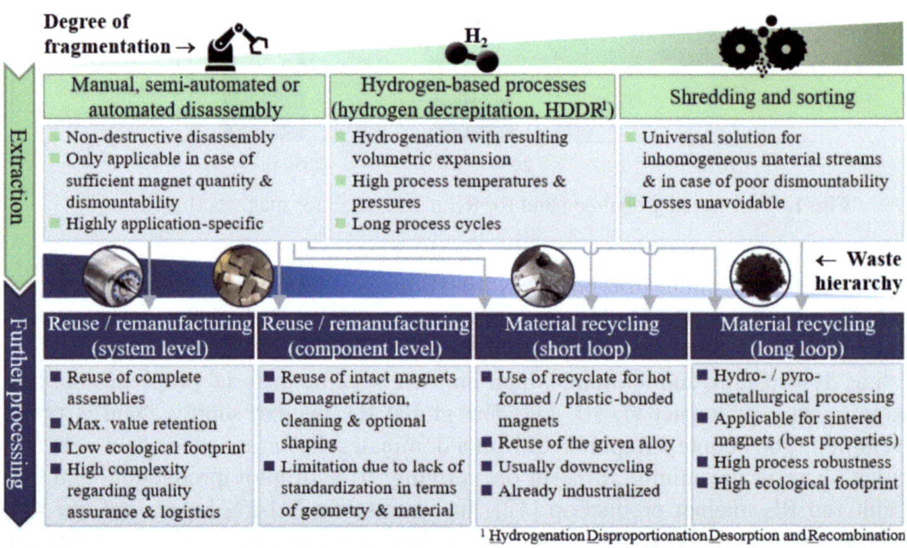

Fig. 2. Circular economy approaches for the recovery of NdFeB magnets from EoL products

Both remanufacturing and recycling require a multistage process chain, starting with the collection and pre-treatment of EoL products. The following mechanical or chemical process routes must be tailored to the specific characteristics of the individual waste stream, with efficient magnet extraction being a central challenge [9, 10].

Reuse and remanufacturing concepts are often impeded by the complexity of non-destructive dismantling, lack of standardization, small product volumes, and logistical challenges. Consequently, corresponding approaches are only applicable for certain EoL products such as EV drives, MRI scanners and wind turbines [9, 10, 17]. Short loop recycling involves the production of sintered, hot-deformed or polymer-bonded magnets based on secondary material [17]. While physical properties change, the fundamental material composition remains mainly unchanged [9]. Accordingly, it is essential that the input material is homogeneous, not oxidized and contains only a negligible amount of impurities [10, 17]. In addition to destructive dismantling, hydrogen-based processes, enabling simultaneous demagnetization and powderization, are applicable for short loop recycling of homogenous waste streams. The resulting powder can be processed by injection molding, pressing or sintering, but process sensitivity, scalability and cycle times remain challenging [10, 18, 19].

Long-loop recycling recovers REEs or RE oxides and allows greater tolerance for input material quality and high flexibility regarding the compatible magnet production routes. Pyrometallurgical methods like liquid metal extraction and slagging convert REEs into extractable phases, followed by molten salt electrolysis or metallothermic reduction. Here sufficient RE concentrations are required. In hydrometallurgical process routes, EoL magnets are chemically leached in acid, and the individual REEs are separated via solvent extraction, ion exchange, or ionic liquids. [9] Due to the ability to process inhomogeneous and contaminated waste streams, long-loop approaches enable the utilization of shredded material. Despite material losses during shredding and a higher ecological footprint, this pathway is essential for broad resource recovery (cf. Fig. 2), as alternative extraction methods are limited by poor dismantlability, product diversity and miniaturization [9, 10, 20].

3 Physical Sorting of NdFeB from Ferrous Shredder Fractions

3.1 State of the Art, Definition of Process Interfaces and Need for Research

Shredding and subsequent sorting is a standard approach for value recovery from inhomogeneous waste streams, combining flexibility, high throughput and the ability to process complex material compounds [21]. This is particularly relevant for waste electrical and electronic equipment (WEEE) originating from consumer and small industrial devices, which are often inefficiently treated, causing complete loss of CRMs [3]. During shredding, losses of RE magnets occur due to adhesion and dust formation. The NdFeB concentration in shredded WEEE granulate is typically between 100 and 500 ppm [10, 22]. The resulting, disperse material mixture undergoes several sorting stages with magnetic separation as standard method for isolation of the ferrous fraction. This material stream contains also the RE magnet fragments, which are then lost in steel production as part of the slag [9, 10].

The separation of NdFeB is currently not realized due to various factors like low concentration, magnetic agglomerations and adhesion, small particle size and the similarity to soft materials regarding many physical properties [23]. Accordingly, the research objective is to integrate a physical sorting process into the industrial process chain capable of isolating NdFeB from the ferrous fraction with sufficient purity for efficient hydrometallurgical processing. Such a process chain is currently not available. A potential approach has been investigated using comminution and screening to exploit the different fracture behavior of relevant material groups [23, 24]. The corresponding study by Peelman, et al. shows a tripling to quadrupling of the REE concentration compared to the input material. However, the study refers to hard disk drive (HDD) scrap, which is partially representative due to the low proportion of soft magnetic material.

3.2 Proposal for Separation and Enrichment Process for the NdFeB Fraction and Analytical and Experimental Pre-Analysis

A universally applicable multi-stage approach utilizing magnetic, thermal and mechanical material properties has been introduced by Ihne, et al. [20] and forms the base for further investigation. The proposed process layout is visualized in Fig. 3.

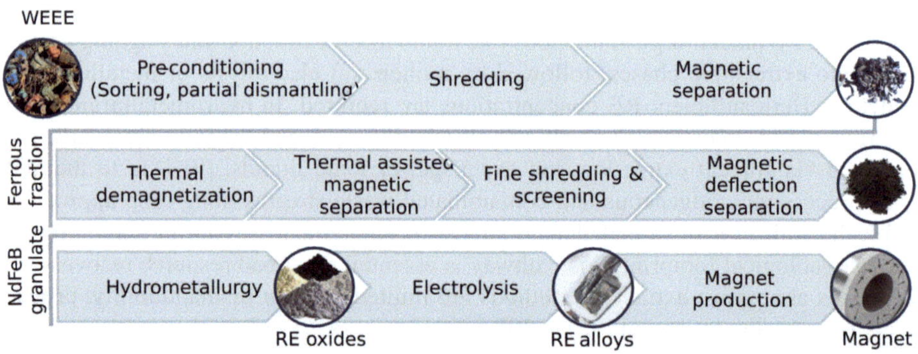

Fig. 3. Overall process chain for the recovery of NdFeB from WEE by long loop recycling

After shredding, and separation of the ferrous fraction by magnetic separation the magnetic agglomerations must be dissolved by heating to the material specific Curie temperature, causing loss of magnetization. A further effect is the change from ferromagnetic to paramagnetic behavior above the Curie temperature according to the Curie-Weiss law as shown in Fig. 4.

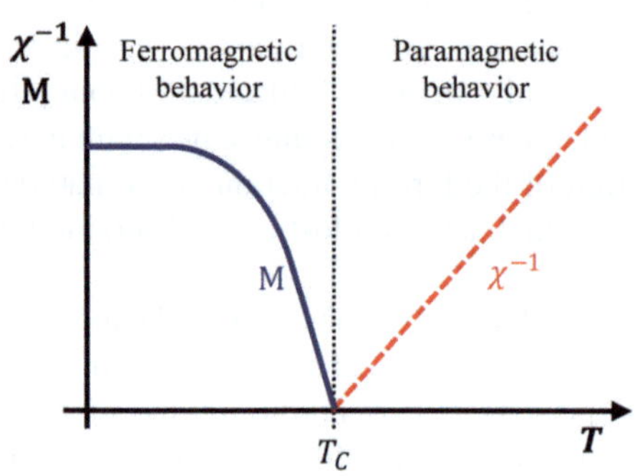

	Material	T_C (°C)	μ_r (-)
	Electrical steel	745	4000
Hard magnetic	NdFeB	310 – 380	1.04 – 1.12
	Hard ferrites	450	1.05 – 1.20
	SmCo	720 – 820	1.04 – 1.12
	AlNiCo	700 – 860	1.30 – 6.20

Fig. 4. Temperature dependence of magnetization M and susceptibility χ for ferromagnetics (l), Curie temperature (T_C) and relative permeability (μ_r) for relevant materials (r) [20]

Since NdFeB has the lowest Curie temperature among relevant materials, the effect allows selective pre-sorting. When a magnetic field is applied to the heated granulate, other ferromagnetic materials are still deposited, while the paramagnetically behaving NdFeB no longer adheres and is also not attracted at the applied field intensities. Stable processing requires a temperature range of 350–400 °C and uniform heat distribution. For an initial indication of feasibility, a reference material sample is prepared consisting of 95% steel particles and 5% shredded and already demagnetized HDD magnets. This material is the input for initial testing with the test rig shown in Fig. 5.

Fig. 5. Test rig and material sample for investigation of thermally assisted magnetic separation

The sample is heated to 350 °C and inserted into the working air gap of a coil with a hot plate preventing rapid cooling in the air. The magnetic flux density can be adjusted via the coil current and measured with a Hall sensor at a reference point. Even at low flux densities, a visible separation into soft and hard magnetic fractions occurs, with good sorting quality at 38 mT. This value is significantly below the level of commercial magnetic separators, which reach 0.15 T (weak-field) to 2.5 T (strong-field) [21].

After thermally assisted magnetic separation, further measures are recommended for enrichment and homogenization of the NdFeB containing fraction. This can be achieved by the approach of Peelman, et al. using a second comminution process combined with particle screening. These results have been validated qualitatively by processing shredded HDDs via thermal demagnetization and subsequent hammer milling. Vibration sieving of the resulting granulate with different mesh sizes separates a fine fraction (particle size <250 μm) with dominant NdFeB content as shown in Fig. 6.

Simultaneously, homogenization of the NdFeB-rich fine fraction supports reliable purification via magnetic deflection sorting, which exploits differences in susceptibility to create distinct particle trajectories in an applied magnetic field. Separation efficiency also depends on particle volume, magnetic field intensity and its local spatial gradient [21, 25]. Analytical methods demonstrate that magnetic fields generated by conventional magnetic separators create sufficiently different trajectories for soft and hard magnetic materials, as shown in Fig. 7. However, interactions between individual particles and potential remagnetization require further simulative and experimental investigation, particularly for strong-field separators, where the applied field strength may exceed the coercivity of NdFeB.

Fig. 6. Enrichment of NdFeB share in HDD scrap by crushing and screening process

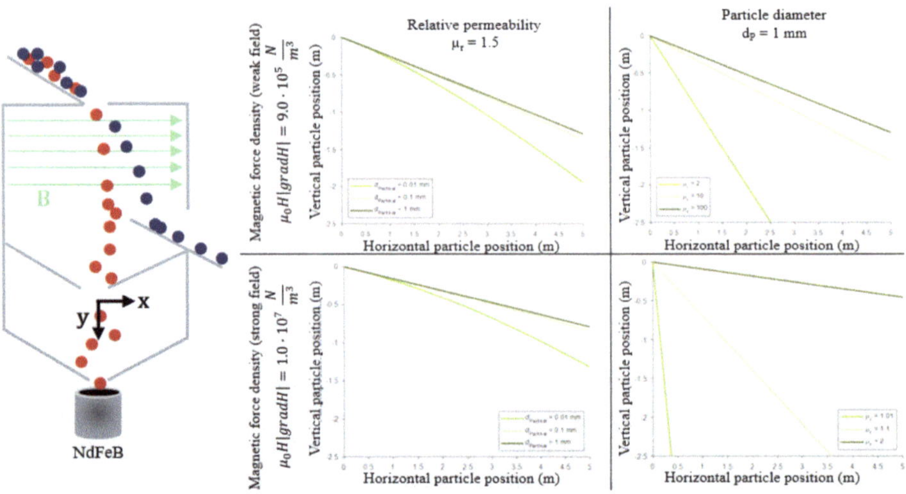

Fig. 7. Analytical investigation of separation behavior in weak and strong magnetic fields

4 Conclusion and Outlook

REEs applied in NdFeB magnets are classified as critical raw materials. Besides other measures, circular economy strategies can support a resilient and sustainable supply. Here various process routes exist that need to be selected individually for specific applications. For inhomogeneous waste streams such as WEEE, shredding combined with sorting and subsequent long-loop recycling appears effective. Currently, NdFeB magnets are separated together with the ferrous fraction, but due to lack of appropriate processes for further sorting of the disperse, soft and hard magnetic material mixture, this leads to a total loss of the RE content. In this context, this paper introduces a multistage approach for physical separating and up-concentrating of the NdFeB share. Initial feasibility has been investigated by analytical and experimental methods. These results form the basis for further simulative and experimental investigations. Another focus of future research is the assessment of the economic and environmental sustainability of the entire process chain, particularly regarding energy efficiency.

Acknowledgement. The presented results have been obtained in the context of research activities at the Institute for Factory Automation and Production Systems (FAPS) of the Friedrich-Alexander-Universität Erlangen-Nürnberg (FAU). This research is related to the project REEPRODUCE and has been funded by the European Union Grant Agreement N° 101057733.

References

1. Graedel, T.E., Harper, E.M., Nassar, N.T., Nuss, P., et al.: Criticality of metals and metalloids. Proc Natl Acad Sci U S A **112**, 4257 (2015)
2. Bobba, S., Carrara, S., Huisman, J., Mathieux, F., Pavel, C.: Critical raw materials for strategic technologies and sectors in the EU. European Union, Luxemburg (2020)
3. Ongondo, F.O., Williams, I.D., Whitlock, G.: Distinct urban mines: exploiting secondary resources in unique anthropogenic spaces. Waste Manag. **45**, 4 (2015)
4. Nassar, N.T., Graedel, T.E., Harper, E.M.: By-product metals are technologically essential but have problematic supply. Sci. Adv. **1**, e1400180 (2015)
5. European Parliament and the Council of the European Union. An EU critical raw materials act for the future of EU supply chains: 2024/1252 (2024)
6. Bauer, D., Nguyen, R., Smith, B.: Critical Materials Assessment, Washington DC (2023)
7. Gielen, D., Lyons, M.: Critical materials for the energy transition (2022)
8. Alves Dias, P., Bobba, S., Carrara, S., Plazzotta, B.: The role of rare earth elements in wind energy and electric mobility. European Union, Luxembourg (2020)
9. Schönfeldt, M., Diehl, O., Gassmann, J.: Recycling of NdFeB magnets in Germany. Federal Institute for Geosciences and Natural Resources (2024)
10. Yang, Y., Walton, A., Sheridan, R., Güth, K., et al.: REE Recovery from End-of-Life NdFeB Permanent Magnet Scrap: A Critical Review, vol. 3, p. 122 (2017)
11. Binnemans, K., Jones, P.T., Müller, T., Yurramendi, L.: Rare Earths and the Balance Problem: How to Deal with Changing Markets?, vol. 4, p. 126 (2018)
12. Noll, S.: Impact of automotive magnet and rare-earth-free e-motor trends on material shortages, Pordenone (2023)
13. Sprecher, B., Daigo, I., Spekkink, W., Vos, M., et al.: Novel indicators for the quantification of resilience in critical material supply chains, with a 2010 rare earth crisis case study. Environ. Sci. Technol. **51**, 3860 (2017)
14. Rizos, V., Righetti, E., Kassab, A.: Developing a supply chain for recycled rare earth permanent magnets in the EU: CEPS In-Depth Analysis, Brussels (2022)
15. Chowdhury, N.A., Nlebedim, I.C., Ginosar, D.M., Handwerker, C., et al.: Value Recovery Pathways for Rare Earth Elements and Nd-Fe-B Magnets from End-of-Life Products, in Rare Earth Metals and Minerals Industries, p. 423. Springer, Cham (2024)
16. Rombach, E., Friedrich, B.: Recycling of rare metals. In: Handbook of Recycling, p. 125. Elsevier (2014)
17. Schönfeldt, M., Opelt, K., Hasan, M., Gröninger, M., et al.: Functional Recycling and Reuse of Nd–Fe–B Permanent Magnets from Various Waste Streams (2025)
18. Zakotnik, M., Harris, I.R., Williams, A.J.: Multiple recycling of NdFeB-type sintered magnets, vol. 469, p. 314 (2009)
19. Habibzadeh, A., Kucuker, M.A., Gökelma, M.: Review on the parameters of recycling NdFeB magnets via a hydrogenation process. ACS Omega **8**, 17431 (2023)
20. Ihne, T., Hahn, R., Kühl, A., Franke, J.: Approaches for the Physical Sorting of Soft and Hard Magnetic Material Mixtures in the Context of Rare Earth Recycling. publish-Ing, Hannover
21. Schubert, H.: Handbuch der Mechanischen Verfahrenstechnik. Wiley (2002)

22. Bandara, H.M.D., Darcy, J.W., Apelian, D., Emmert, M.H.: Value analysis of neodymium content in shredder feed. Environ. Sci. Technol. **48**, 6553 (2014)
23. Elwert, T., Schwarz, S.: Entwicklung einer industriell umsetzbaren Recycling-Technologiekette für NdFeB-Magnete, Clausthal-Zellerfeld (2019)
24. Peelman, S., Venkatesan, P., Abrahami, S., Yang, Y.: Recovery of REEs from end-of-life permanent magnet scrap generated in WEEE recycling plants. In: Davis, B.R., et al. (eds.) Extraction 2018. TMMMS, pp. 2619–2631. Springer, Cham (2018). https://doi.org/10.1007/978-3-319-95022-8_221
25. Worrell, E., Reuter, M. (eds.) Handbook of Recycling. Elsevier (2014)

Supply Chains

Artificial Intelligence and Blockchain Used for Sustainability Reporting in Supply Chain Management – A Literature Review

Michael Fiolka[1(✉)], Tobias Jornitz[2], and Luisa Strehl[2]

[1] TU Dortmund University, Leonhard-Euler-Straße 5, 44227 Dortmund, Germany
michael.fiolka@tu-dortmund.de
[2] Fraunhofer Institute for Material Flow and Logistics, Joseph-von-Fraunhofer-Str. 2-4, 44227 Dortmund, Germany
{tobias.jornitz,luisa.marie.strehl}@iml.fraunhofer.de

Abstract. Since the emergence of the blockchain and the uprising of ChatGPT, the Distributed Ledger Technology (DLT) and Artificial Intelligence (AI) are well-discussed topics both in public and professional circles, but especially in the domain of Supply Chain Management (SCM). These subjects are tech-savvy, complicated to explain and even more complex to use. On top of that, there is a scientific discussion around synergies in combining both technologies. Together they can be useful in engaging current challenges in SCM, where transparency-related data has to be generated, processed and formed into decisions and reports. To investigate the potentials of these technologies working together in a non-financial reporting environment, we performed a systematic literature review. We also included literature focusing solely on the technological perspective. The objective is a comprehensive overview on how a combination of DLT and AI could help to solve current challenges arising from sustainability related regulations. Further, we discussed ideas around Internet of Things applications or Federated Learning approaches, that use data from different entities and can be used in sustainability reporting, exploring possibilities to enhance compliance and responsible business conduct in SCM.

Keywords: Sustainability · Blockchain · Artificial Intelligence

1 Introduction

The timeline we live in truly presents itself as an era of promising advances in digitization. Different incarnations of Artificial Intelligence (AI) and the Distributed Ledger Technology (DLT) already conquered our daily lives in form of digital currencies or interactive assistants. In academia and the industrial world, these tools are under investigation to foster the Industry 4.0 by analysing big data through AI [1, 2] or create transparency in supply chains via the Blockchain (BC) [3, 4]. There is also a scientific discussion about the combination of both technologies [5, 6]. BC represents a transparent, immutable and distributed registry, while AI is proficient in working with large

H. Kohl et al. (Eds.): GCSM 2025, LNME, pp. 207–214, 2026.
https://doi.org/10.1007/978-3-032-21154-5_23

datasets and in prediction applications [7]. While AI often operates in opacity, BC may help to make AI more explainable. This can be achieved by logging the process behind decisions or documenting the training data used, the model creation and pipeline used for testing, the flow of dependencies, participants and other metadata, which enhances trust and transparency [6, 8, 9]. Deep Learning (DL) and Machine Learning (ML) can be utilized to analyse the behaviour of BC participants and to identify majority attacks in BC networks, enhancing security [10, 11]. AI can be used to sort and classify data before it will be stored on a BC, which reduces storage efforts and streamlines documentation and retrieval of relevant data [12, 13]. These applications may prove particularly useful in handling sustainability related data in Supply Chain Management (SCM), where trustworthy datasets have to be generated at different locations by various parties (which often cannot be trusted). This is especially important for the European industries, where there are new reporting regulations on the horizon. The Corporate Sustainability Reporting Directive (CSRD) [14] pushes companies into handling sustainability related data in their sourcing processes to create compliant, non-financial reports. We are currently developing a BC and AI based software to tackle this challenge. Hence, the goal of this paper is to systematically identify already existing scientific approaches, which utilizes AI and BC in combination for sustainability reporting.

2 Related Literature

Several literature reviews and surveys were performed on AI and BC working together in different applications. *Ekramifard et al.* identified articles on integration, use cases and applications of combining BC and AI, resulting in an overview of primary study fields, with most applications in data security, healthcare, trading, Internet of Things (IoT), classification prediction and decision making [15]. *Hussain and Al-Turjman* evaluated the convergence of both technologies in characteristics, utilization and relevance, and highlighted different applications like the decentralized use of AI e.g. in the supply industry [7]. *Cai et al.* discussed in their review the mitigation capabilities of Artificial Neural Networks (ANN) and BC for environmental impacts, emphasizing the improvement of supply chain transparency, emission reduction and the promotion of circularity and the quantification of reporting data [16]. *Vărzaru* conducted a literature review on the impact of digital technologies on sustainability accounting and reporting, concluding that digital technologies like cloud computing, Big Data, IoT, and AI have a significant influence on the percentage of enterprises that are reporting [17]. A general research approach on combining both technologies is presented by *Ressi et al.*, revealing several technological methods and application areas [18]. This paper aims to investigate the existence of consisting use cases in science literature on sustainability reporting, hence our first research questions is stated as follows:

a) Which applications for sustainability reporting combining BC and AI exist in science literature?

It is possible that there are no or few current applications for sustainability reporting. To generate ideas inspired from applications outside of sustainability reporting, we also want to explore existing BC and AI applications, which can be used or transferred to the academic field in question:

b) Which applications combining BC and AI could possibly be used for sustainability reporting?

To answer these questions, we performed a systematic literature review. In the following paragraphs, we describe our methodology and documentation for the review. After presenting the results, we will discuss the results and the research questions.

3 Method

For our literature review we relied on the proceeding of *Kitchenham and Charters* [19]. First, we approached the review with an unstructured desk research using Google Scholar to find appropriate literature on the synergy of AI and BC, to get a general grasp of the topics, possible use cases and keywords that are relevant. That initial search enabled us to try different versions of search strings in the Scopus database, resulting in the final search string in table 1.

Table 1. Used search string, date and number of results

("AI and distributed ledger") OR ("AI and blockchain") OR ("deep learning and distributed ledger") OR ("deep learning and blockchain")OR ("artificial intelligence and distributed ledger") OR ("artificial intelligence and blockchain") OR ("machine learning and blockchain") OR ("machine learning and distributed ledger") AND ("sustainability" OR "circular" OR "green" OR "transparency" OR "CSR" OR "Due Diligence") AND ("Reporting" OR "Monitoring") AND "Supply chain management" - Database: Scopus, Date: 26.03.2025, Results: 554

The search resulted in 554 hits total. We evaluated all papers based on title, abstracts and keywords, while focusing on the following inclusion criteria: Description of a use-case, application or concept of a combination of AI methods and DLT approaches in the research areas related to sustainability and sustainability reporting, including ecologic, economic and social topics. We excluded papers that were not accessible to the subscriptions of TU Dortmund University. This resulted in 26 relevant papers for a full-paper analysis, with 4 papers actually reporting on this very narrow topic. Through a backwards integration of papers referred to in these 4 papers, we included 5 additional papers. To have a broader understanding of possible use cases, we included 5 more papers which are employing AI and BC applications, but had no direct implication on sustainability, resulting in a total of 14 relevant findings for this literature review.

4 Findings

We assessed the papers by the presentation of a use-case combining BC and AI, including both proven applications and conceptual use-cases, describing the roles of the respective technologies and, if available, the actual phenotype (e.g. the AI-methods used). We also focused on original papers extracted from literature reviews within our SLR-results, to embrace a comprehensive backwards integration. We assessed the use cases not directly related to sustainability reporting and integrated them into our structure.

4.1 Secure Agreements and Transactions

BC presents a strong use case for storing data from transactions to create transparent and reliable evidence. This seem to apply especially for AI-agents as shown in *Pimenidis et al.*; They described a use case instrumentalizing AI agents to negotiate on agreements which are then saved on the blockchain for monitoring and auditing quality standards of sustainable products [20]. Another application for creating evidence is presented by *Shinde et al.*, who used a BC and Natural Language Processing (NLP) based solution to store transaction data, by creating proofs for stakeholders to fight corruption and fraud in the construction sector [21]. An approach for transactions between multiple parties is presented by *Khan et al.*, who used a Hyperledger and AI-enabled framework for secure transaction execution of smart grid and related power distribution automation [22]. They opted for a real-time solution for sustainable energy industries to achieve transparency, provenance and trustworthiness. They emphasized the use of inter- and outer-chain transactions, which are protected by a proxy re-encryption to provide transaction security.

4.2 Data Integrity, Sensor Data and Data Sharing

Using reliable sensor data from IoT-systems is also a large contributor to this review, for example for segregating plastic waste types with an AI-multisensor platform and using the BC to share information between segregators [23]. Also sensor-based, the approach of *Chaudhari et al.* employs Decision Trees (DT), Latent Dirichlet Allocation (LDA), Support Vector Machines (SVM) and perception algorithms to identify malicious data packages from water quality sensors, improving the integrity of data stored on a BC to protect water treatment facilities [24]. Sensors in combination with AI and BC can help not only in an ecologic context, but also in delivering social Due Diligence. *Pan et al.* developed a system that collects and combines sensor data and manual inspection records from construction equipment for safety and hazard analysis, which are then processed by a Convolutional Neural Network (CNN)-based text classification model to improve information retrieval performance when the data is stored on a BC [25]. Data integrity is not only paramount in monitoring applications alone, but also for sharing between companies. *Bist et al.* proposed a model of AI-enabled BCs for information exchange in an agricultural setting, to the benefit of not only sellers and buyers for efficiency reasons, but also for other stakeholders [26]. Federated Learning (FL) in conjunction with BC consensus mechanism can improve computing resource usage and the efficiency of data sharing [27]. Systems combining BC and AI do not serve only a single company but the entire SC, solving issues in ecological and social contexts and also generating visibility.

4.3 Data Privacy, Ownership and Anomaly Detection

With data sharing being a major factor in 4.2, data privacy, security and ownership questions become prominent. *Durrant et al.* discussed advantages and barriers to employ a combination of BC and FL approaches to ensure data privacy and engage ownership issues to reach sustainability goals [28]. Other security-related applications are referred to in *Pan et al.* [24], where *Chen et al.* designed a decentralized Stochastic Gradient Descent (SGD) algorithm to learn a general predictive model over the BC, to protect participating parties' data privacy [29]. Another important method is the detection of anomalies in datasets, *Hassan et al.* presented a comprehensive overview on how different AI methods may help to identify anomalies in different layers of BC implementations [30]. *Tsoukas et al.* employed a BC backend and combined it with an AI-driven anomaly detection based on TinyML to detect data manipulation attempts [31]. BC in this review is mostly discussed as a database, but it is commonly a platform for digital currency and funding opportunities, which may be vulnerable for malicious attacks. Relying on AIs anomaly detection capabilities, *Zkik et al.* utilized Graph Neural Network (GNN) models to defend BC-based crowdfunding platforms from smart contract-based attacks [32]. Thus, AIs multiple realms of application can enhance BCs security and help solve data privacy and ownership issues.

5 Discussion and Implications

Given the limited search results and the very narrow scope of this review, we rather identified a research gap around the usage of BC and AI in sustainability reporting in SCM, than addressing it. Still, the findings present insides, which are directly or can possibly be linked to sustainability reporting. It is noteworthy that most applications and scientific areas for a combination of BC and AI are in the realms of cyber security, healthcare, food and agriculture, which is in line with the results of other reviews [5, 18]. In regard of research question a), we found several applications, which could be important for sustainability reporting, but were not explicitly mentioned in a reporting environment. Combining sensor data, IoT, BC and AI, as shown in [23], is a common approach, which could be particularly useful for reporting, as waste management is part of the ESRS-reporting framework. This could also be utilized in managing data in water consumption, gas or other resources, which are also reporting-KPIs. As shown in [25], the security of machines can be documented and immutably saved on the BC, which is a useful way of securing data for social sustainability issues, fostering work safety approaches. FL is a promising method for analyzing sustainability data in supply chains, regulations not only demand data from OEMs, but also from their direct and indirect suppliers. In FL approaches, BC may solve privacy issues in data sharing; a similar approach can be seen in *Unal et al.*, where the integrity of a trained model was maintained by suppressing manipulation attempts through anomaly detection [33]. Regarding research question b), we also found eligible applications which might be transferable for sustainability reporting: AI may help classify large or unstructured sustainability datasets, which creates the possibility to only store relevant data, possibly encrypted, on the BC, which can also be seen in healthcare applications [12, 13]. Cyber security and data privacy measures can become important for sustainability reporting. Sharing data with several direct or indirect suppliers in AI-learning approaches may present the necessity to encrypt or protect supplier data. There is also a compelling case for anomaly detection in datasets saved on the BC, which may help address the Garbage-In-Garbage-Out principle in sustainability reporting, rendering data from suppliers more viable.

6 Conclusion

In this paper, we conducted a SLR to analyze the combination of BC and AI technologies in sustainability reporting. Relevant due to European regulatory measures, companies are prompted to extend their non-financial reporting. Given the immutability and distributed character of the BC and the large data analysis capabilities of AI, these technologies seem to be a perfect fit for this application. Based on the findings of this paper, it is safe to say that the synergetic use of BC and AI for sustainability reporting is still in its infancy, but the general use cases for the combination of both technologies inspire multiple applications. We explored the use of IoT-generated and AI-processed sensor data, which improves the quality and retrieval times of data from BCs. FL can be used to train AI models on data from different entities, while the BC addresses current concerns in respect to data privacy, security and viability. Sustainability reporting seems to be a green field for synergetic use of BC and AI, but our project is eager to cultivate on these unproven grounds, which are fertilized by the many diverse approaches and realms of application academia already has to offer.

Funding and Acknowledgement. The content of this article was created as part of the "Scalable AI and Blockchain Solutions for Automation and Autonomization in Value Networks (SKALA)" project. The SKALA research project is funded by the German Federal Ministry for Digital and Transport (BMDV) (funding reference: 45KI26B021).

References

1. Bose, S., Kumar Dey, S., Bhattacharjee, S.: Big data, data analytics and artificial intelligence in accounting: an overview. In: Akter, S., Wamba, S.F. (eds.) Handbook of Big Data Research Methods. Research Handbooks in Information Systems, pp. 32–51. Edward Elgar Publishing, Cheltenham, UK, Northampton, MA (2023)
2. Thayyib, P.V., et al.: State-of-the-art of artificial intelligence and big data analytics reviews in five different domains: a bibliometric summary. Sustainability (2023)
3. Mustapha, I., Khan, N., Qureshi, M.I., Sikandar, H., Hassan, M., Simarmata, J.: Revolutionizing supply chain processes: harnessing the power of IoT and blockchain technology to enhance opportunities. In: Ismail, A., Zulkipli, F.N., Mahat, R., Mohd Daril, M.A., Öchsner, A. (eds.) Innovative Technologies for Enhancing Experiences and Engagement. SpringerBriefs in Applied Sciences and Technology, 1st edn, pp. 87–94. Springer, Cham (2024). https://doi.org/10.1007/978-3-031-55558-9_10
4. Madumidha, S., Ranjani, P.S., Varsinee, S.S., Sundari, P.S.: Transparency and traceability: in food supply chain system using blockchain technology with internet of things. In: 2019 3rd International Conference on Trends in Electronics and Informatics (ICOEI). 2019 3rd International Conference on Trends in Electronics and Informatics (ICOEI), Tirunelveli, India, pp. 983–987. IEEE, Piscataway (2019)
5. Charles, V., Emrouznejad, A., Gherman, T.: A critical analysis of the integration of blockchain and artificial intelligence for supply chain. Ann. Oper. Res. (2023)
6. Wang, R., Luo, M., Wen, Y., Wang, L., Raymond Choo, K.-K., He, D.: The Applications of Blockchain in Artificial Intelligence. Security and Communication Networks (2021)
7. Hussain, A.A., Al-Turjman, F.: Artificial intelligence and blockchain: a review. Trans. Emerg. Tel. Tech. (2021)

8. Sarpatwar, K., Vaculin, R., Min, H., Su, G., Heath, T., Ganapavarapu, G., Dillenberger, D.: Towards enabling trusted artificial intelligence via blockchain. In: Calo, S., Bertino, E., Verma, D. (eds.) Policy-Based Autonomic Data Governance. LNCS, vol. 11550, pp. 137–153. Springer, Cham (2019). https://doi.org/10.1007/978-3-030-17277-0_8
9. Dinh, T.N., Thai, M.T.: AI and blockchain: a disruptive integration. Computer (2018)
10. Tang, H., Jiao, Y., Huang, B., Lin, C., Goyal, S., Wang, B.: Learning to classify blockchain peers according to their behavior sequences. IEEE Access (2018)
11. Dey, S.: A proof of work: securing majority-attack in blockchain using machine learning and algorithmic game theory. IJWMT (2018)
12. Zheng, X., Mukkamala, R.R., Vatrapu, R., Ordieres-Mere, J.: Blockchain-based personal health data sharing system using cloud storage. In: Proceedings of the 2018 IEEE 20th International Conference on e-Health Networking, Applications and Services (Healthcom). 2018 IEEE 20th International Conference on e-Health Networking, Applications and Services (Healthcom), Ostrava, pp. 1–6. IEEE (2018)
13. Juneja, A., Marefat, M.: Leveraging blockchain for retraining deep learning architecture in patient-specific arrhythmia classification. In: 2018 IEEE EMBS International Conference on Biomedical & Health Informatics (BHI). 2018 IEEE EMBS International Conference on Biomedical & Health Informatics (BHI), Las Vegas, NV, USA, pp. 393–397. IEEE (2018)
14. Finance: Corporate sustainability reporting (2025). https://finance.ec.europa.eu/capital-markets-union-and-financial-markets/company-reporting-and-auditing/company-reporting/corporate-sustainability-reporting_en. Accessed 18 Apr 2025
15. Ekramifard, A., Amintoosi, H., Seno, A.H., Dehghantanha, A., Parizi, R.M.: A systematic literature review of integration of blockchain and artificial intelligence. In: Choo, K.-K., Dehghantanha, A., Parizi, R.M. (eds.) Blockchain Cybersecurity, Trust and Privacy. AIS, vol. 79, pp. 147–160. Springer, Cham (2020). https://doi.org/10.1007/978-3-030-38181-3_8
16. Cai, T., Hong, Z.: Exploring the structure of the digital economy through blockchain technology and mitigating adverse environmental effects with the aid of artificial neural networks. Front. Environ. Sci. (2024)
17. Vărzaru, A.A.: An empirical framework for assessment of the effects of digital technologies on sustainability accounting and reporting in the European Union. Electronics (2022)
18. Ressi, D., Romanello, R., Piazza, C., Rossi, S.: AI-enhanced blockchain technology: a review of advancements and opportunities. J. Netw. Comput. Appl. (2024)
19. Kitchenham, B., Charters, S.: Guidelines for performing Systematic Literature Reviews in Software Engineering 2 (2007)
20. Pimenidis, E., Patsavellas, J., Tonkin, M.: Blockchain and artificial intelligence managing a secure and sustainable supply chain. In: Jahankhani, H., Jamal, A., Lawson, S. (eds.) Cybersecurity, Privacy and Freedom Protection in the Connected World. ASTSA, pp. 367–377. Springer, Cham (2021). https://doi.org/10.1007/978-3-030-68534-8_23
21. Shinde, R., Nilakhe, O., Pondkule, P., Karche, D., Shendage, P.: Enhanced road construction process with machine learning and blockchain technology. In: 2020 International Conference on Industry 4.0 Technology (I4Tech). 2020 International Conference on Industry 4.0 Technology (I4Tech), Pune, India, pp. 207–210. IEEE, Piscataway (2020)
22. Khan, A.A., Laghari, A.A., Rashid, M., Li, H., Javed, A.R., Gadekallu, T.R.: Artificial intelligence and blockchain technology for secure smart grid and power distribution automation: a state-of-the-art review. Sustain. Energy Technol. Assess. (2023)
23. Chidepatil, A., Bindra, P., Kulkarni, D., Qazi, M., Kshirsagar, M., Sankaran, K.: From trash to cash: how blockchain and multi-sensor-driven artificial intelligence can transform circular economy of plastic waste? Adm. Sci. (2020)
24. Chaudhari, R., Kumar Jadav, N., Tanwar, S., Garg, D.: AI and blockchain-based collaborative architecture for safeguarding IoT-driven water treatment plants. In: 2024 IEEE International

Conference on Contemporary Computing and Communications (InC4). 2024 IEEE International Conference on Contemporary Computing and Communications (InC4), Bangalore, India, pp. 1–7. IEEE (2024)
25. Pan, X., Zhong, B., Sheng, D., Yuan, X., Wang, Y.: Blockchain and deep learning technologies for construction equipment security information management. Autom. Constr. (2022)
26. Bist, A.S., Rawat, B., Prawiyogi, A.G., Wahyuningsih, Septiani, N., Fakhrezzy, M., Saputra, D.B.: AI-enabled blockchain for supply chain in agriculture. In: 2022 IEEE Creative Communication and Innovative Technology (ICCIT). 2022 IEEE Creative Communication and Innovative Technology (ICCIT), Tangerang, Indonesia, pp. 1–5. IEEE (2022)
27. Lu, Y., Huang, X., Dai, Y., Maharjan, S., Zhang, Y.: Blockchain and federated learning for privacy-preserved data sharing in industrial IoT. IEEE Trans. Ind. Inf. (2020)
28. Durrant, A., Markovic, M., Matthews, D., May, D., Leontidis, G., Enright, J.: How might technology rise to the challenge of data sharing in agri-food? Global Food Secur. (2021)
29. Chen, X., Ji, J., Luo, C., Liao, W., Li, P.: When machine learning meets blockchain: a decentralized, privacy-preserving and secure design. In: 2018 IEEE International Conference on Big Data (Big Data). 2018 IEEE International Conference on Big Data (Big Data), Seattle, WA, USA, pp. 1178–1187. IEEE (2018)
30. Ul Hassan, M., Rehmani, M.H., Chen, J.: Anomaly detection in blockchain networks: a comprehensive survey. IEEE Commun. Surv. Tutor. (2023)
31. Tsoukas, V., Gkogkidis, A., Kampa, A., Spathoulas, G., Kakarountas, A.: Enhancing food supply chain security through the use of blockchain and TinyML. Information (2022)
32. Zkik, K., Sebbar, A., Fadi, O., Kamble, S., Belhadi, A.: Securing blockchain-based crowdfunding platforms: an integrated graph neural networks and machine learning approach. Electron. Commer. Res. (2024)
33. Unal, D., Hammoudeh, M., Khan, M.A., Abuarqoub, A., Epiphaniou, G., Hamila, R.: Integration of federated machine learning and blockchain for the provision of secure big data analytics for internet of things. Comput. Secur. (2021)

Optimizing Reusable Food Packaging Flows Within a Circular Economy Framework

G. Bartolotti(✉), E. Ferrari, B. Guidani, R. Manzini, M. Ronzoni, and R. Accorsi

Department of Industrial Engineering, Alma Mater Studiorum Università di Bologna, Viale Risorgimento, 2, 40136 Bologna, Italy
giorgia.bartolotti2@unibo.it

Abstract. The agri-food industry is one of the most packaging-intensive sectors. Sustainability goals clearly indicate reusable packaging as a solution to reduce the environmental burdens of the food supply chain. While circular economy schemes are increasingly adopted for secondary and tertiary packaging, their large-scale implementation for primary packaging remains limited due to logistical and behavioral complexities. Closed-loop systems for reusable primary food packaging require coordinated direct and reverse flows, including filling, distribution, consumer return, sorting, and washing. Modeling these systems requires accounting for uncertainties such as return and breakage rates. This paper relies on a Mixed Integer Linear Optimization model to support the location-allocation of facilities and collection points in a reusable packaging network. The model integrates uncertainty in consumer behavior and packaging durability, quantifying their impact on cost, emissions, and system configuration. Results show that upstream nodes are highly sensitive to return and scrap rates, while downstream infrastructure remains largely stable across scenarios. This paper addresses return and breakage rate uncertainties, assessing their effects on supply chain configuration. The findings support resilient system design and guide policy incentives for scaling reusable packaging.

Keywords: Reusable primary packaging · Circular economy · Consumer behavior · Risk analysis · Logistic optimization

1 Introduction

As global environmental awareness increases, the food sector is progressively exploring reusable packaging as a key strategy to counteract the negative effects of single-use materials. While single-use packaging remains widespread for its convenience, it contributes significantly to plastic waste and resource depletion. Recent studies highlight that reuse should become a cornerstone of sustainable food systems [1]. Aligning with circular economy principles, reusable packaging systems are increasingly recognized for their potential to reduce environmental burdens [2, 3]. However, existing research and industrial implementations predominantly target secondary and tertiary packaging, often overlooking primary packaging (i.e., packaging in direct contact with food) in the

H. Kohl et al. (Eds.): GCSM 2025, LNME, pp. 215–223, 2026.
https://doi.org/10.1007/978-3-032-21154-5_24

food supply chain [4, 5]. The existing literature provides some examples of reusable primary packaging applications, but these are limited to specific product types and have not been implemented at scale across retail supply chains [6]. This gap persists despite growing recognition of the critical role that primary packaging could play in enabling circular food systems [7]. The shift toward reusable primary packaging introduces several structural and behavioral challenges that must be addressed to ensure system viability [8]. In particular, reverse logistics, including the processes of collection, cleaning, and redistribution, introduces operational complexities and cost implications, especially when not carefully optimized. These difficulties are further exacerbated by uncertainties in consumer return behavior.

The *ReturnRate*, defined as the proportion of packaging returned by the consumer into the reuse system, represents a major source of variability. When return rates are low or inconsistent, additional new packaging must be introduced into the network, thereby undermining the environmental and economic benefits of reuse. Barriers such as hygiene concerns, perceived inconvenience, and low engagement continue to limit consumer participation [9]. Socio-demographic factors, such as age, income, and environmental awareness, also influence adoption patterns [10], while educational initiatives and clear labeling have been shown to improve acceptance [11]. In addition to behavioral uncertainty, the *ScrapRate* introduces further risk. *ScrapRate* represents the share of packaging removed from the system due to breakage, contamination, or loss of integrity. This issue is relevant in the food and beverage sector, where hygiene and quality standards are mandatory. Packaging that fails to meet safety or visual requirements must be discarded. Optimization and operations research approaches are increasingly used to complement Life Cycle Assessment (LCA), especially in the context of Returnable Transport Items (RTIs) and Reusable Plastic Containers (RPCs) in B2B environments [12, 13]. However, these models have rarely been adapted to primary packaging systems in B2C contexts, which involve greater variability due to consumer behavior. In this context, the study addresses the following research question: (RQ.1) How do key uncertainty parameters in reusable primary packaging systems influence the optimal design of the logistics network? This study advances the field by exploiting a risk-informed optimization model, presented in detail in [14] which explicitly assesses the effects of *ReturnRate* and *ScrapRate* variability on the design and performance of reusable packaging networks. This paper's contribution lies in the implementation of a multi-scenario analysis framework specifically designed to assess how variations in *ReturnRate* and *ScrapRate* impact the final configuration of the reusable primary packaging logistics network. This allows for the evaluation of trade-offs across different configurations and uncertainty levels, offering actionable insights for the robust design of reuse systems. By simulating the full space of feasible behavioral and operational outcomes, the model enables data-driven decisions that account for variability in user behavior and system performance.

2 Methodology

This section details the methodology followed to perform the analysis and the optimization framework it falls into. The decision-supporting sensitivity analysis is based on an optimization model that aims at designing the optimal network configuration for a reusable primary packaging system at national level as presented in [14], analyzing the trade-off between economic performance and environmental impacts. The proposed optimization model adopts a bi-objective structure, simultaneously minimizing total system costs and associated greenhouse gas emissions. The model includes a comprehensive set of parameters describing the physical, economic, and environmental attributes of products, packaging, and logistics operations. For each product, relevant parameters include unit weight, volumetric saturation, and palletization efficiency. Packaging parameters include material type (e.g., glass or plastic), stackability, and volume, all of which directly affect compatibility with washing infrastructure and transport efficiency. Alongside products, the main entities of the system, defining the sets of the model, are the actors involved in the reusable primary packaging supply chain.

Figure 1 outlines the network actors and their relationship, highlighting in different colors packaging flows of different natures. The logistic parameters of the network accounted for in the optimization model include distances between actors, available transport modes, and associated unit costs and emissions. This data was compiled from industrial partners' interviews and companies' ERPs.

Each facility is characterized by processing throughput and capacity constraints. For instance, Packaging Producers (PPs) are defined by their ability to produce reusable containers, while Food Packagers (FPs) are evaluated based on their filling and sealing capacity. Distribution Centers (DCs) handle product flows and storage, while retail Shops (STs) are defined by local demand levels. Used packaging is returned by consumers into collection points (Reverse Vending Machines, RVMs). Sorting facilities (SOs) collect used packaging from RVMs and sort it according to the requirements of the Washing centers (Ws), some of which are specialized for glass, others for plastic. Ws sanitize the packaging, ensuring the absence of food traces or dirt.

The underlying optimization model evaluates only differential cost and emission components that arise when transitioning from single-use to reusable packaging systems. This dual-objective approach enables the identification of trade-offs between economic and environmental performance across different network configurations. For the cost objective function, the installation of RVMs and Washer facilities involves fixed investment, which are amortized over time assuming a one-year operational horizon. No opening costs are considered for the other nodes, as they are existing facilities that do not require new construction but only need to be adapted for the adoption of reusable packaging systems. These are complemented by unit operational costs (e.g., per-unit washing, filling, and sorting costs). Environmental impacts are parameterized as unit emissions and are specific for each facility typology.

The model's decision variables fall into two main categories. First, installation variables determine where infrastructure such as RVMs should be deployed and at what capacity. These decisions are discrete and define the structure of the reverse logistics system. Second, flow variables represent continuous logistics decisions, such as the quantity of packaging or products transported between actors. For each pair of interacting nodes (e.g., FP to DC, SO to W), a flow variable quantifies the optimized movement. These variables are used by the solver to minimize total cost and environmental impact across the network. Among the model parameters, the key uncertainty factors are ReturnRate and ScrapRate. The ReturnRate reflects the % of packaging returned by consumers, while the ScrapRate quantifies packaging % losses due to breakage or contamination. These two variables are the focus of the scenario-based sensitivity analysis. In the optimization model the ScrapRate and the ReturnRate are input parameters, and all their combinations were tested across the full range of interest: specifically, as illustrated in Fig. 1, ReturnRate was varied from 0 to 100 in steps of 10, while ScrapRate ranged from 5 to 55 in steps of 10. By modeling both infrastructure setup and operational flows, the decision-support tool enables a comprehensive assessment of strategic configurations for packaging reuse networks, responsive to varying levels of system uncertainty.

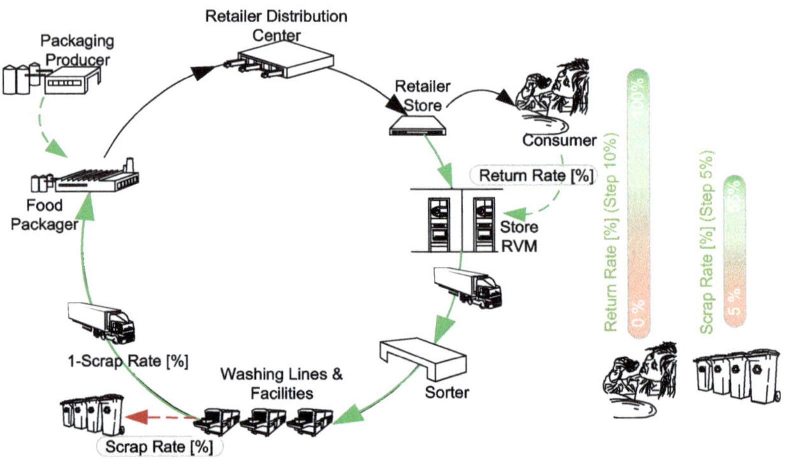

Fig. 1. Stakeholders and flows in the reusable packaging network

3 Case Study

The case study covers France, Belgium, and Luxembourg, and is built on real data collected through the European project R3PACK. This national-scale scope ensures the model reflects real operational complexity and allows for meaningful sensitivity analyses on uncertainty factors like *ReturnRate* and *ScrapRate*. The network includes all nodes involved in the reusable packaging loop, from production to return. Table 1 summarizes the number of nodes considered. These nodes correspond to real and existing facilities currently involved in packaging production that can be integrated into reusable systems. They include sites already engaged in the washing of certain types of packaging (particularly glass bottles and secondary packaging). The food packagers, distribution centers, and retail shop are actual sites belonging to two of the main French retail chains. Regarding sorting facilities, only one existing center is currently in operation. Therefore, sorting centers are assumed to coincide with distribution centers (DCs), based on logistical efficiency considerations and in alignment with suggestions from industrial partners within the R3PACK project.

Table 1. Nodes selection

Node Type	Packaging Producers	Food Packagers	Distribution Centers	Retail Shop	Sorting Centers	Washing Facilities
N°	29	47	145	2 366	151	30

In the case study, 20 distinct food products are modeled, each associated with a unique packaging format. These products span various food categories, including prepared salads, yoghurt, chips, savory and sweet biscuits, juice, beverages, condiments, and oil. Each product is paired with a single reusable packaging type, establishing a one-to-one relationship between the product and its packaging. A total of 11 packaging types are utilized, primarily divided into two main categories: rigid plastic packaging, such as trays, buckets, and pots made of polypropylene (PP), and glass packaging, including bottles and jars. These packaging types were selected for their suitability for reuse, durability, and compatibility with existing production processes.

4 Results and Discussions

This section presents the model's recommendations regarding the utilization of key supply chain nodes across different uncertainty scenarios. Six reference configurations were selected to analyze the system behavior: combinations of two *ScrapRate* values (5% and 15%) and four *ReturnRate* values (0%, 40%, and 70%). The output of the optimization model is visualized using bubble plots, where the bubble size reflects the quantity of packaging units handled by each facility. Across all scenarios, the model consistently selects sites that are geographically central to the areas of highest demand, particularly where retail shops (STs) are densely located. This spatial centralization of flows enables the minimization of both transport distances and the number of trips required to satisfy demand. Figure 2 illustrates the model's choice of PPs across different scenarios. PPs show the greater variability in network decisions across the different values of *ScrapRate* and *ReturnRate.* As expected, the number of required packaging producers declines as *ReturnRate* increases. This is because higher levels of reuse reduce the need for introducing new packaging into the system. For instance, at a 70% *ReturnRate*, increasing the *ScrapRate* from 5% to 15% leads to the activation of additional PP nodes, compensating for increased losses. This outcome confirms the logical correlation between system losses and new packaging demand. The results show high stability in the configuration of FPs, DCs, SOs, and Ws across all scenarios. This is largely due to structural constraints: FPs have limited feasible locations due to product-specific capabilities and dependence on nearby washing infrastructure, while DCs are already optimized for proximity to retail demand. SOs and Ws, either modeled with unlimited capacity or already well-positioned relative to return flows, remain unchanged even as *ReturnRate* and *ScrapRate* vary. These findings indicate that the downstream and reverse logistics nodes of the network are robust to uncertainty and do not require major reconfiguration. This provides critical insight for long-term planning: investment in flexibility should be prioritized at the upstream end of the supply chain, while the existing downstream configuration is adequate to support a wide range of operational conditions. To evaluate logistical efficiency, the model analyzes the probability distribution of distances traveled between each pair of actors, derived from the number of full truckloads. An example of this analysis for *ReurnRate* = 70% and *ScrapRate* = 15% is outlined in Fig. 3. A Weibull distribution is fitted to each empirical distribution to estimate the likelihood of specific transport distances. The shape and scale parameters of the Weibull function offer insights into distance variability.

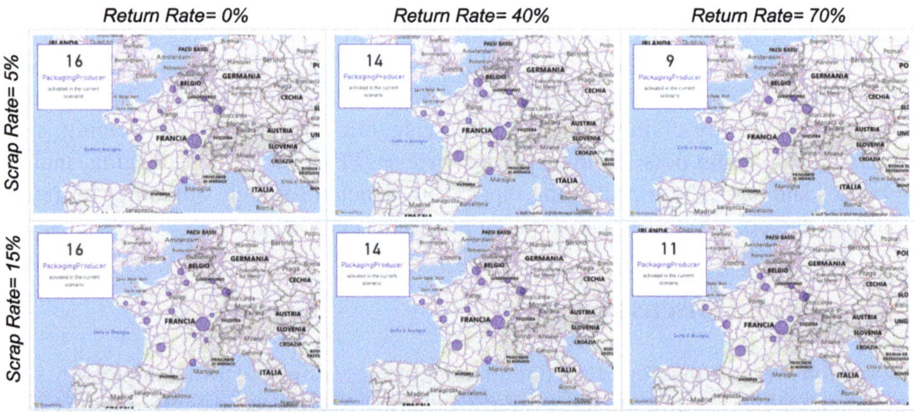

Fig. 2. PPs' location throughout different scenarios

Flatter curves indicate high variability and uncertainty, while sharper peaks reflect more consistent and optimized routes. The spatial distribution of logistical flows in the modeled network reveals distinct patterns across actor pairs. DC-ST distances display a Poisson-like distribution, sharply peaking around 50 km, a result of both strategic co-location of distribution centers and stores by retailers and the model's objective to minimize transportation costs and emissions. Similarly, the SO-W flows exhibit a negative exponential distribution, indicating consistently short routes within the reuse network, which enhances both economic and environmental efficiency. In contrast, flows such as PP-FP, FP-DC, and W-FP show flat Gaussian distributions, highlighting significant variability and limited optimization due to a sparse facility network. This inefficiency suggests the need for redesigning the network, particularly through the strategic relocation of PP and FP nodes, to improve accessibility, reduce transport distances, and enhance overall system performance. These results confirm that certain segments of the supply chain are resilient and well-optimized, while others, particularly those dependent on packaging production, are more sensitive to network design and uncertainty parameters. Strategic expansion or relocation of PP and FP facilities could improve system performance, reducing travel distances and emissions. The findings underscore the importance of rethinking upstream logistics to balance the network and enhance sustainability.

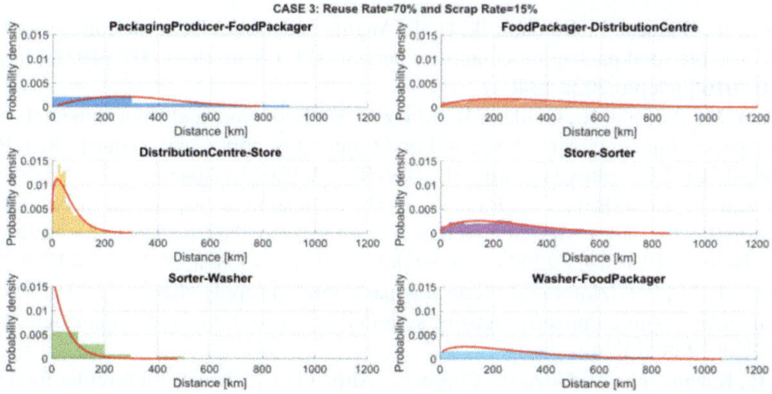

Fig. 3. Weibull distribution of network distances per stage. ReturnRate = 70% ScrapRate = 15%

5 Conclusions

This study highlights the critical role of uncertainty in the strategic design of reusable primary packaging systems. A scenario-based optimization model was used to assess the influence of varying consumer *ReturnRates* and packaging *ScrapRates* on supply chain performance. Results indicate that downstream and reverse logistics nodes, such as food packagers, distribution centers, sorting, and washing facilities, maintain a stable configuration across scenarios. In contrast, the number and location of packaging producers are highly sensitive to changes in *ReturnRates* and *ScrapRates*. Higher *ReturnRates* reduce the need for new packaging production, while higher *ScrapRates* increase it, creating a clear trade-off. The findings suggest that flexibility and investment should be prioritized in the upstream portion of the supply chain, where uncertainty has the greatest impact. Conversely, the stability of downstream nodes confirms that existing infrastructure is generally adequate to support large-scale reuse, provided sufficient consumer participation and packaging durability are achieved. These insights inform both strategic planning and policy development. Incentives that enhance return behavior and reduce breakage, through design, education, or regulation, can significantly improve system performance and sustainability. The results offer a robust basis for scaling up reusable packaging systems in the agri-food sector.

Acknowledgement. This project has received funding from the European Union's Horizon Europe Research (R3PACK) and Innovation program under Grant Agreement No 101060806.

References

1. Yadav, P., Silvenius, F., Katajajuuri, J.M., Leinonen, I.: Life cycle assessment of reusable plastic food packaging. J. Clean. Prod. **448**, 141529 (2024). https://doi.org/10.1016/J.JCLEPRO.2024.141529
2. Lu, Z., Hasselström, L., Finnveden, G., Johansson, N.: Cost-benefit analysis of two possible deposit-refund systems for reuse and recycling of plastic packaging in Sweden. Clean. Waste Syst. **3**, 100048 (2022). https://doi.org/10.1016/J.CLWAS.2022.100048

3. Accorsi, R., Battarra, I., Guidani, B., et al.: Augmented spatial LCA for comparing reusable and recyclable food packaging containers networks. J. Clean. Prod. **375** (2022). https://doi.org/10.1016/j.jclepro.2022.134027
4. Ronzoni, M., Accorsi, R., Guidani, B., Manzini, R.: Economic and environmental optimization of packaging containers choice in Food Catering Supply Chain. Transp. Res. Procedia **67**, 163–171 (2022). https://doi.org/10.1016/J.TRPRO.2022.12.047
5. Sazdovski, I., Batlle-Bayer, L., Bala, A., et al.: Comparative assessment of two circularity indicators for the case of reusable versus single-use secondary packages for fresh foods in Spain. Heliyon **10**, e27922 (2024). https://doi.org/10.1016/J.HELIYON.2024.E27922
6. Mahmoudi, M., Parviziomran, I.: Reusable packaging in supply chains: a review of environmental and economic impacts, logistics system designs, and operations management. Int. J. Prod. Econ. **228** (2020)
7. Ada, E., Kazancoglu, Y., Gozacan-Chase, N., Altin, O.: Challenges for circular food packaging: circular resources utilization. Appl. Food Res. **3** (2023). https://doi.org/10.1016/j.afres.2023.100310
8. Bartolotti, G., Guidani, B., Ronzoni, M., et al.: Data architecture framework for reverse logistics management in reusable food packaging networks. Procedia Comput. Sci. **253**, 1343–1351 (2025). https://doi.org/10.1016/J.PROCS.2025.01.196
9. Hoseini, M., Greenwood, S.C., Eman, S., et al.: Integrating behavioural, material and environmental science to inform the design and evaluation of a reuse system for takeaway food. Resour. Conserv. Recycl. **209**, 107815 (2024). https://doi.org/10.1016/J.RESCONREC.2024.107815
10. Schuermann, H., Woo, J.R.: Estimating consumers' willing-ness to pay for reusable food containers when ordering delivery food: a contingent valuation approach. J. Clean. Prod. **366** (2022). https://doi.org/10.1016/j.jclepro.2022.133012
11. Norton, V., Alexi, N., Contente, A., Lignou, S.: The 3Is: let's INVOLVE, INFORM and INSPIRE the next generation on disposing food packaging sustainably. J. Clean. Prod. **424**, 138807 (2023). https://doi.org/10.1016/j.jclepro.2023.138807
12. Accorsi, R., Cascini, A., Cholette, S., et al.: Economic and environmental assessment of reusable plastic containers: a food catering supply chain case study. Int. J. Prod. Econ. **152**, 88–101 (2014). https://doi.org/10.1016/J.IJPE.2013.12.014
13. Bortolini, M., Galizia, F.G., Mora, C., et al.: Bi-objective design of fresh food supply chain networks with reusable and disposable packaging containers. J. Clean. Prod. **184**, 375–388 (2018). https://doi.org/10.1016/j.jclepro.2018.02.231
14. Accorsi, R., Bartolotti, G., Guidani, B., et al.: Reusable food primary packaging in retail supply chain: a multi-objective optimization framework. Sustain. Prod. Consum. **56**, 364–384 (2025). https://doi.org/10.1016/J.SPC.2025.04.003

Simulation-Based Optimization for Supply Chains with Multi-criteria Target Systems - A Comparison of Multiple Algorithms

Lucas Schreiber(✉) and Luca Bernstiel

Fraunhofer Institute for Material Flow and Logistics IML, Joseph-von Fraunhofer-Str. 2-4, 44227 Dortmund, Germany
{lucas.schreiber,luca.bernstiel}@iml.fraunhofer.de

Abstract. The increasing importance of sustainability in industrial operations necessitates the integration of environmental considerations alongside traditional economic metrics in supply chain optimization. This study presents a comprehensive comparison of three metaheuristic approaches for multi-criteria supply chain optimization: Non-Dominated Sorting Genetic Algorithm II (NSGA-II), Particle Swarm Optimization (PSO), and Surrogate-Assisted Grey Wolf Optimization (SAGWO). These methods are applied to optimize a three-tier supply chain network considering costs, energy consumption, and service level simultaneously. The research contributes to sustainability by demonstrating how environmental factors can be systematically integrated into supply chain decision-making processes, enabling companies to achieve eco-efficient operations while maintaining economic viability. Results show that SAGWO achieves superior performance by reducing computational time while maintaining solution quality, making it particularly suitable for complex real-world applications where sustainability and efficiency must be balanced.

Keywords: Simulation-based optimization · Supply Chains · Sustainability · Metaheuristics · Material Flow Simulation

1 Introduction

In modern industrial applications changes to production or logistics systems are meticulously planned before implementation. This allows unexpected additional costs to be identified at an early stage. A popular tool for this kind of planning is discrete-event material flow simulation of a system to test previously developed scenarios and configurations. With this technique, complex projects can be tested and validated in a more cost-effective and sustainable digital environment [1].

In addition to the straightforward economic improvement of the present system regarding costs, energy consumption has a significant role. The concept of sustainability and energy-efficiency is becoming increasingly relevant not only in the private sector, but also in the industrial sector. Reasons for the increasing significance of sustainability

H. Kohl et al. (Eds.): GCSM 2025, LNME, pp. 224–231, 2026.
https://doi.org/10.1007/978-3-032-21154-5_25

for manufacturing companies and supply chains include the decreasing availability of fossil energy sources [2].

Accordingly, sustainable indicators such as energy consumption need to be included in corporate planning in addition to the classic target values. Within this integration, it is important that the consideration of sustainable indicators does not have a too severe negative impact on quality-determining targets, such as monetary expenses and on-time delivery of goods. In this study, energy consumption is chosen as sustainable indicator. The choice to focus on energy consumption as the environmental metric, rather than carbon emissions or comprehensive Life Cycle Assessment (LCA), deserves clarification. While LCA would provide a more holistic environmental assessment, energy consumption serves as a practical and measurable proxy for environmental impact that correlates strongly with carbon emissions in most industrial contexts. Energy consumption data is readily available in supply chain operations.

Material flow simulation can be used to investigate and optimize complex systems, such as supply chains. Specifically, configurations of supply chain parameters can be tested and validated, since these parameters have a direct impact on the output of the simulation. It is extremely difficult to set all supply chain parameters in an appropriate way to optimize the predefined target variables. Such an optimization quickly includes a large number of parameter combinations, which requires an enormous amount of computing power and cannot be optimized by testing every possible combination with today's technology [3]. Within this paper, the interaction between discrete-event simulation and optimization components is presented and the performance of the results is compared.

2 Literature Review

Simulation-based optimization for supply chains is already a widespread field of research. The contribution of Torcedilla et al. (2021) gives a comprehensive overview regarding simulation-based optimization for supply chains. It gathers papers presenting hybrid sim-opt methods that aim at improving supply chains through the interaction of simulation and optimization components. The majority of the presented contributions use exact optimization methods for this purpose, while a smaller fraction considers different metaheuristics [3]. Koo et al. address the optimization of safety stocks and capacity investments in an integrated refinery supply chain. In this use case, the simulation model and a Non-Dominated-Sorting Algorithm (NSGA-II) form the iterative optimization loop. A two-dimensional objective function consisting of profit margin and a customer satisfaction index is to be maximized is optimized [4]. The simulation-based application of Particle Swarm Optimization (PSO) in the context of supply chains is demonstrated by Kadadevaramath et al. A three-stage supply chain simulation model is being optimized using four different types of PSO algorithms [5].

A novel surrogate-based optimization (SBO) is presented by Dong & Dong. Here, a Grey Wolf Optimizer (GWO) is combined with the SBO approach, which saves a lot of computations of the objective functions. This results in the Surrogate Assisted Grey-Wolf Optimizer (SAGWO) [6]. Schreiber & Wilk then successfully applied this approach to a simulation model of a three-tier supply chain network to reduce the number of computationally intensive simulation runs and thus evaluate potentially multi-criteria solutions faster [7].

3 Supply Chain Use Case

The study examines a three-tier supply chain network comprising five suppliers, five distribution centers, and 25 customers, each representing geographically contiguous customer regions. Distribution centers maintain inventories of 21 distinct products sourced from suppliers over the simulation period. When customers place orders, the bill of materials calculates the required components and quantities, which are then deducted from the inventory. These components undergo a transformation process to produce the final product, which is subsequently delivered to customers via truck.

The network operates under an (s,S) inventory policy at the distribution centers. Components are replenished to a target stock level S as soon as the inventory level falls below the reorder point s [8].

Key data, including transportation distances between entities, transportation and production costs, truck capacities, and component weights, are incorporated into the simulation model. Furthermore, the model records energy expenditures associated with these costs, providing a comprehensive assessment of resource utilization. Costs and energy consumption occur at various stages of the supply chain. Transportation costs and associated energy expenditures arise during the movement of components from suppliers to distribution centers and from distribution centers to customers. Production costs and energy usage are incurred during the transformation process of components into finished products. The simulation model meticulously records these costs and energy expenditures, enabling accurate calculation of the target function values and facilitating a holistic evaluation of the system's performance.

A comprehensive elucidation of the system, including the implementation of the simulation model using Tecnomatix Plant Simulation© by Siemens, is provided in a related study by Schreiber and Wilk [7]. This detailed analysis offers further insights into the operational dynamics, parameter configurations, and constraints that underpin the supply chain network, thereby supporting the validity and applicability of the current investigation. As an assumption the reorder points (s) can vary from 200 to 1600, while target stocks (S) can take values between 1600 and 2400. These margins were declared to be economic in a sensitivity analysis with an identical use case scenario [9].

The optimization problem is highly complex, as it considers three decision variables for each product at each facility: the reorder point (s), the target stock level (S), and the decision whether a particular product is produced as a specific facility. This results in 315 (3 variables $\cdot$ 21 products $\cdot$ 5 locations) parameters per possible solution. For this 315-dimensional decision space, $7.34 \cdot 10^{666}$ unique combinations arise - a 667-digit number.

4 Selection of Algorithms and Methodology

Since the use case discussed in this paper has a high complexity, exact optimization methods are hard to apply because of the huge computation effort. Heuristic methods promise to obtain results faster with a trade-off concerning optimality and computation time.

An interesting approach to multicriteria problems is to use algorithms that follow a Pareto-efficient approach. Thus, optimization can be performed in the direction of any objective function without having to perform a previous weighting. In the area of supply chain optimization using Pareto-efficient optimization methods, Jadaan et al. present the Non-dominated Ranking Genetic Algorithm (NRGA) and simultaneously also mention the NSGA-II [10]. When comparing the methods, Alizadeh Afrouzy et al. conclude that the NRGA dominates the NSGA-II algorithm by about two percent in a direct comparison of CPU computation time. However, if this efficiency loss can be tolerated, the NSGA-II can improve the generated solutions by almost nine percent [11]. Thus, in this application, the *NSGA-II* is used for closer examination. A detailed explanation of the NSGA-II algorithm can be found in Deb et al. [12].

In a different application, but still multi-critical optimization problem, Song & Chen draw a comparison between Genetic Algorithms (GA) and PSO. Here, the weak point of the GA is determined to be a minimal degree of variation in the proposed solutions. The GA finds better solutions, but the PSO approach shows larger variations in the solutions [13]. This is seen as an interesting potential and accordingly the *PSO* is also considered as an additional method. Detailed explanation of the PSO algorithm can be found in Liang et al. [14].

Potential savings in computation time can be realized with surrogate-assisted algorithms by abstracting the simulation model. Minimizing the number of iterations also increases the savings potential. Such a method is presented by Dong & Dong. Here, a Grey Wolf Optimizer (GWO) is used together with a surrogate model to solve the problem. The resulting Surrogate-Assisted Grey Wolf Optimizer (SAGWO) shows good benchmark results for high dimensional optimization problems [15]. Further insights into the functioning of the Grey Wolf Optimizer (GWO) algorithm are provided by Mirjalili et al. [16].

5 Results

The previously presented optimization procedures are now applied to the use case to compare the generated solutions of these algorithms. The compared KPIs are the aggregated costs and energy consumption of the supply chain as well as the obtained service level so that there is a three-dimensional targe system. In addition, by generating a comparable set of random solutions, it is determined whether the application of an optimization method adds value. The statistical significance is confirmed by means of several t-tests. The comparison of the solutions is done by scalarizing the weighted score of the three target values of the best solutions in each case. This is done by a min-max normalization, where the best solutions over the comparable algorithms are then used as the best and worst solutions. In the following, the best solutions in each case are listed together with the algorithms that generated them. Three different objective function weightings are presented in this paper as an excerpt of the experiments conducted. The selected combinations each underrepresent one of the three target functions in the weighting to be able to check the sensitivity of individual KPIs (Table 1).

To be able to perform a transparent comparison, similar framework conditions are set for each algorithm, so that the available computing time respectively the maximum number of simulation runs to be performed in the optimization cycle is identical. As described in Sect. 3, the number of possible parameter combinations is extremely high.

Table 1. Weight combinations for objective functions

Combination	Costs	Energy	Service Level
COM1	0,4	0,2	0,4
COM2	0,4	0,4	0,2
COM3	0,2	0,4	0,4

Accordingly, the randomly chosen starting position has a great influence on the performance of the applied algorithm. Thus, if the algorithms start randomly in already very promising data regions, they converge faster and find good solutions already in early iterations. For this reason, the starting positions of the algorithms are standardized so that all have the same chance to find optimized solutions. The results derived from COM1 are presented in Table 2.

Table 2. Results for COM1

Algorithm	Costs [€]	Energy [MJ]	Service Level [%]	Score
SAGWO	17.329.159	74.901.082	87,95	0,8215
PSO	17.398.231	75.771.700	88,52	0,6201
NSGA-II	17.278.235	75.012.443	87,39	0,7748
Random	17.540.101	76.673.246	87,75	0,0637

It can be clearly observed that the randomly generated solutions are far behind the solutions optimized by algorithms. This is already the first indication that the algorithms show a convergence. The PSO approach achieved a comparatively poor result. Although the solution exhibits the best service level, it accepts a high energy consumption. The SAGWO and the NSGA-II stand out. The result of the NSGA-II is more cost-efficient, but the SAGWO has less energy consumption and a better service level and thus achieves the highest score. The following Fig. 1 shows all generated data points of the SAGWO for COM1 with the best ones in the top left corner with low energy consumption, low costs and a high service level.

Table 3 below provides the results for the second weighting configuration COM2.

SAGWO provides the solution with the best service level. However, since this is not weighted high in this combination, this has a minor effect on the score. Thus, both the PSO and the NSGA-II have found better solutions in the area of costs and energy expenditures. In comparison, the NSGA-II performed best. Figure 2 all generated data points of the NSGA-II for COM2 with good results in the middle with a high service level but also higher costs and energy consumption and the bottom left corner with low costs and energy consumption but a lower service level. Table 4 shows that also in the final combination COM3, the randomly generated solutions do not find a comparable quality of solutions. In general, the scores of the algorithms are similar to the results from COM1. Again, the NSGA-II achieves the best solution followed by PSO and SAGWO.

All data points w_1=0.4, w_2=0.2, w_3=0.4

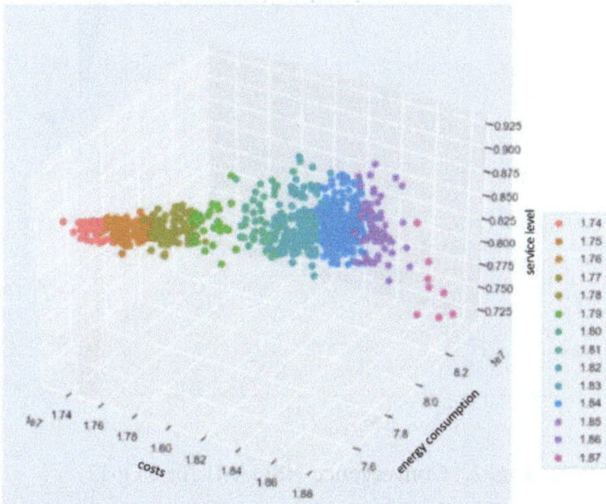

Fig. 1. Convergence of SAGWO for COM1

Table 3. Results for COM2

Algorithm	Costs [€]	Energy [MJ]	Service Level [%]	Score
SAGWO	17.456.285	75.990.126	88,37	0,4930
PSO	17.304.727	75.273.461	86,75	0,6966
NSGA-II	17.278.235	75.012.443	87,39	0,8790
Random	17.540.101	76.673.246	87,75	0,1234

Table 4. Results for COM3

Algorithm	Costs [€]	Energy [MJ]	Service Level [%]	Score
SAGWO	17.490.165	75.668.163	88,88	0,5147
PSO	17.322.538	75.546.545	87,67	0,6849
NSGA-II	17.322.679	74.867.334	87,41	0,7995
Random	17.540.101	76.673.246	87,75	0,0462

Examining the rankings, the NSGA-II achieves the best solutions most frequently. There is hardly any difference between the PSO and the SAGWO in the mean of the rankings. However, the SAGWO stands out with a few excellent results. From this it can be interpreted that the SAGWO has the potential to find very good solutions in multiple runs but is not as consistent as the PSO. Here, a fine-tuning between exploration and exploitation in interaction with the surrogate model and simulation could still reveal potential.

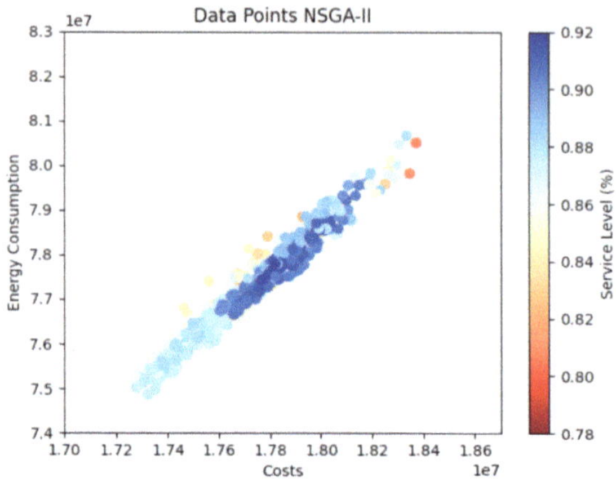

Fig. 2. Convergence NSGA-II for COM2

To ensure the statistical significance of the different methods a t-test with $\alpha = 0{,}05$ was performed regarding the comparison to the random results for all combinations. For all potential permutations the null hypothesis could be negated. There is a statistically significant difference between the mean result of the randomized solution and the algorithms.

6 Conclusion

The interaction of discrete-event simulation and optimization algorithms has led to good solutions for the parameterization of the inventory units of the distribution centers. The performance of the supply chain, with an extended view on eco-efficiency, measured by costs, energy consumption and service level, could be improved. Compared to the mere testing of randomly generated input parameter combinations, far better results are achieved by means of the optimization methods.

The comparison between the presented metaheuristics shows that the NSGA-II achieves the best solutions on average. At the same time, only a single optimization run has to be performed, since the NSGA-II converges in parallel in all directions of the objective functions. The SAGWO is still assumed to have potential in which the exploration scope of the algorithm is extended.

References

1. März, L., Krug, W., Rose, O., Weigert, G.: Simulation und Optimierung in Produktion und Logistik. Springer, Heidelberg (2011)
2. Centobelli, P., Cerchione, R., Esposito, E.: Environmental sustainability and energy-efficient supply chain management: a review of research trends and proposed guidelines. Energies **11**, 275 (2018)
3. Tordecilla, R.D., Juan, A.A., Montoya-Torres, J.R., Quintero-Araujo, C.L., Panadero, J.: Simulation-optimization methods for designing and assessing resilient supply chain networks under uncertainty scenarios: a review. Simul. Model. Pract. Theory **106**, 102166 (2021)

4. Koo, L.Y., Adhitya, A., Srinivasan, R., Karimi, I.A.: Decision support for integrated refinery supply chains. Comput. Chem. Eng. **32**, 2787–2800 (2008)
5. Kadadevaramath, R.S., Chen, J.C., Latha Shankar, B.: Application of particle swarm intelligence algorithms in supply chain network architecture optimization. Expert Syst. Appl. **39**, 10160–10176 (2012)
6. Gong, W., et al.: Multiobjective adaptive surrogate modeling-based optimization for parameter estimation of large, complex geophysical models. WRR **52**, 1984–2008 (2016)
7. Schreiber, L., Wilk, D.: Surrogate assisted simulation-based grey wolf optimization for energy efficient supply chain configuration. EUROSIM (2023)
8. Tempelmeier, H.: Analytics im Bestandsmanagement. Norderstedt (2020)
9. Schreiber, L., Niehus, C., Moroff, N.: Simulation-based optimization for eco-efficient supply chain configuration. In: Franke, J., Schuderer, P. (eds.) Simulation in Produktion und Logistik 2021, pp. 61–74. Cuvillier Verlag, Göttingen (2021)
10. Jadaan, O.A., Rao, C.R., Rajamani, L.: Solving constrained multi-objective optimization problems using non-dominated ranked genetic algorithm. In: Asia International Conference on Modelling & Simulation, pp. 113–118. IEEE (2009)
11. Alizadeh Afrouzy, Z., Paydar, M.M., Nasseri, S.H., Mahdavi, I.: A meta-heuristic approach supported by NSGA-II for the design and plan of supply chain networks considering new product development. J. Ind. Eng. Int. **14**, 95–109 (2018)
12. Deb, K., Pratap, A., Agarwal, S., Meyarivan, T.: A fast and elitist multiobjective genetic algorithm: NSGA-II. IEEE Evol. Comput. **6**, 182–197 (2002)
13. Song, M., Chen, D.: A comparison of three heuristic optimization algorithms for solving the multi-objective land allocation problem. GIS **24**, 19–31 (2018)
14. Liang, X., Li, X., Ercan, M.F.: A PSO – line search hybrid algorithm. In: Gervasi, O., Taniar, D., Murgante, B., Laganà, A., Mun, Y., Gavrilova, M.L. (eds.) Computational Science and Its Applications. Lecture Notes in Computer Science, vol. 5593, pp. 547–556. Springer, Heidelberg (2009)
15. Dong, H., Dong, Z.: Surrogate-assisted grey wolf optimization for high-dimensional, computationally expensive black-box problems. Swarm Evol. Comput. **57**, 100713 (2020)
16. Mirjalili, S., Mirjalili, S.M., Lewis, A.: Grey wolf optimizer. Adv. Eng. Softw. **69**, 46–61 (2014)

Analysis of Local Mitigation Strategies for Material Procurement Risks for Sustainable and Resilient Manufacturing: A Simulation-Based Approach for Linear and Non-linear Production Layouts

Florian Sinn(✉), Benjamin Gorgas, and Frank Straube

Chair of Logistics, Berlin University of Technology, 10623 Berlin, Germany
{sinn,gorgas,sekretariat}@logistik.tu-berlin.de

Abstract. This paper presents a simulation-based analysis of local mitigation strategies for managing material procurement risks in manufacturing environments. Focusing on factory-level mitigation strategies, the study investigates how different types of procurement-related disruptions affect production efficiency and identifies and evaluates mitigation strategies - such as safety stock adjustments, procurement volume control and variable processing - that enhance resilience and sustainability. Two archetypal production layouts are considered: a linear assembly line and a non-linear fractal configuration. By systematically modelling risk scenarios across these layouts, the study evaluates the effectiveness of individual and combined strategies in maintaining stable and efficient production. The results highlight key interdependencies between risk type, factory layout and mitigation approach, offering actionable insights for production planners. The analysis reveals that the effectiveness of mitigation strategies strongly depends on the production layout, with flexible configurations showing significantly greater potential to absorb disruptions. Linear systems, by contrast, exhibit more limited adaptability, reinforcing the need for layout-sensitive resilience planning. These findings support the development of adaptive, locally resilient manufacturing systems capable of responding to supply volatility while contributing to broader sustainability objectives.

Keywords: Procurement Risk Management · Simulation-based Analysis · Resilient Manufacturing · Linear and Non-linear Production · Local Mitigation Strategies · Multifactorial Experiments

1 Introduction

In recent years, global supply chains have been exposed to an increasing number of external shocks that have grown significantly in intensity, frequency and unpredictability. Natural disasters, pandemics, geopolitical conflicts, economic downturns and regulatory interventions in international trade relations have a serious impact on the availability of

H. Kohl et al. (Eds.): GCSM 2025, LNME, pp. 232–240, 2026.
https://doi.org/10.1007/978-3-032-21154-5_26

critical materials and components [1–3]. As a result, material procurement is increasingly becoming the focus of corporate resilience strategies as a key bottleneck factor. The just-in-time philosophy, long regarded as an efficiency paradigm, is becoming increasingly vulnerable to volatile conditions [4]. Companies worldwide are faced with the challenge of not only protecting their procurement and production systems against short-term risks, but also making them more sustainable and resilient in the long term [5]. Sustainability must be understood in a multidimensional way: Ecological responsibility, economic resilience and social stability must be harmonised in order to ensure the long-term competitiveness of industrial manufacturing [6, 7].

The majority of scientific and practical approaches to risk minimisation to date have focused on global strategies, for example through the geographical diversification of suppliers, nearshoring or multi-sourcing models [8]. These measures can undoubtedly contribute to increased robustness in upstream supply chains, but often fall short when it comes to direct operational security within the production facility itself. Especially at the local level, there are decisive levers for risk reduction that have not been systematically researched to date. These include, for example, adjustments to order quantities, the targeted management of safety stocks, flexible production planning or the organisation of buffers and alternative processes within production [9, 10]. These internal, intralogistical measures not only offer a short-term response option to disruptions, but also contribute to increasing operational resilience in the long term [11]. The need to systematically analyse local risk strategies and dovetail them with global measures is therefore a key challenge for holistically resilient and sustainable production.

Against this background, this study pursues three research objectives: Firstly, it analyses which different types of risks affect the efficiency of production systems (RO1). Secondly, specific local risk mitigation strategies will be analysed in terms of their effectiveness under different framework conditions, particularly with regard to flexibility potential, inventory strategies and planning buffers (RO2). Thirdly, the influence of different factory layout types - linear versus non-linear (i.e. fractal) layouts - on the resilience and effectiveness of these strategies will be analysed (RO3). The analysis is carried out using simulation-based modelling in order to depict complex dynamic interactions within the production system under realistic disruption scenarios. The aim is to formulate evidence-based recommendations for the design of sustainable and resilient procurement and mitigation strategies at local level.

2 Methodology

An explorative, sequential mixed-methods design was chosen to systematically analyse local risk strategies in material procurement for sustainable and resilient production systems. This research design allows qualitative and quantitative procedures to be methodically built upon one another, whereby a structured body of knowledge is first generated through qualitative research and then analysed quantitatively in greater depth [12]. The combination of both methodological approaches is particularly suitable for complex, previously underdeveloped fields of research, such as in the context of local procurement risks and their connection with the factory layout. The exploratory nature of the study makes it possible to openly examine correlations and mechanisms of action without preconceived hypotheses in order to derive robust, empirically sound findings.

The first phase of the research comprises a qualitative systematic literature review (SLR) in order to gain a structured understanding of the relevant risk types and local intralogistics and production-related countermeasures. Methodologically, this phase is based on the SLR approach according to Durach, which is characterised by a high degree of transparency, replicability and methodological rigour [13]. The literature analysis was conducted in two stages. In a first step, systematic search strategies (search strings, inclusion/exclusion criteria, databases such as Scopus, Web of Science) were used to identify key risks that are relevant in the context of local production systems, especially those that are not based on global but on internal causes. In a second step, suitable intralogistics and production-internal measures were recorded that are discussed in the literature as responses to these risks (Table 1, 2).

The qualitative phase led to theoretical saturation, as recurring measures were identified in the literature. Based on these findings, specific risk-measure pairs were formed for the quantitative analysis and transferred to experimental designs.

In the second phase of the study, a quantitative, simulation-based research design was implemented to analyse the effectiveness of the identified local risk strategies under realistic conditions. Simulations are particularly suitable for analysing dynamic systems with complex interactions, such as those prevalent in production systems [14]. The modelling was carried out using *Tecnomatix Plant Simulation* software, a proven tool for discrete-event modelling of production and logistics systems [15]. The core of the simulation study was the matching of risk types with corresponding countermeasures in two different layout configurations: a linear production layout, which is characterised by a sequential process structure with little networking and a non-linear, fractal layout, which is based on modular self-organisation and decentralised control [16].

The simulated factory layouts and parameters are based on an example model created by *Tecnomatix* to ensure realistic processing times [17]. The corresponding model was modified to contain the following general setup: The factory produces two main products on seven production steps followed by a chance-based quality check. The two main products inherit different parameters for processing times, demand of additional supply (eight supply types in total) and required production steps. Main products arrive in the factory twice a day, while supply orders are based on safety stock levels. While the linear layout follows the seven production steps, the factorial layout allows a variable order in which the production steps are finished for each part.

For each factory, three different risk factors were implemented that either influence the incoming products or the incoming supply and vary in level of impact (low, medium, high) resulting in 18 different risk setups to be observed individually. Risks were calibrated such that the low, medium and high levels of impact resulted in a drop of the output per day of 25%, 50% and 75% respectively.

In addition to aforementioned risks, six possible mitigation strategies based on the results of the SLR were implemented into the model. Mitigation strategies are meant to be implemented after the risk started occurring and therefore focus on operational rather than a strategic level.

During the simulation we conducted two sets of multifactorial experiments, systematically testing all 18 risk types against every possible combination of mitigation

Table 1. Risk Types and modelling description

Risk Type	Description
Delay Product/Supply	Lead times for incoming parts are determined by normal distribution. The Delay-Risk increases the corresponding average by a fixed amount to increase lead time. The level of impact determines the amount.
Reduce Product/Supply	The number of incoming parts is determined by a normal distribution. The Reduce-Risk decrease the average of the distribution to determine the actual number of parts. Level of impact determines the decrease.
Quality Product/Supply	Incoming parts have a low chance of being defective. The Quality-Risk increases this chance based on the level of impact. Defective parts are discarded during quality checks and do not count towards the total output.

Table 2. List of mitigation strategies and descriptions

Abbr.	Mitigation Strategy	Description
M1	Parallel Stations	Stations with the longest processing times are doubled. Processing time on parallel stations is longer than usual to account for inefficiencies due to the short notice of implementation.
M2	Flexible Station	A station is introduced which can conduct every processing step at an increased processing time.
M3	Increase Buffer Capacity	Small buffers close to stations are increased in size.
M4	Increase Product Order	The number of ordered products is increased. In the status quo the order size is close to the maximum throughput of the factory without any risks or mitigation strategies.
M5	Initial Quality Check	A quality check is introduced for incoming goods to reduce defective parts going into production.
M6	Increase Workforce	A two-shift system is used to limit the possibility to increase production facilities. This strategy adds workers to each shift on short notice.

strategies, applied across both factory layouts. I.e. one risk type with one level of intensity was activated within the experiment. The results act as benchmark for this risk. Afterwards one simulation was conducted for each possible combination of activated mitigation strategy resulting in a total of $2^6 = 64$ experiments per risk including the benchmark. This was done for the Status Quo with no risk or mitigation, each of the 18 risk types, and both factory layouts resulting in a total of 2306 experiments. Each set was simulated three times to account for outrunners. Simulated time was seven working days. The controlled variable was the output of non-defective parts per day.

The results offer insights on the correlation between specific risk types and the most efficient short term mitigation strategies, for both, linear and non-linear production environments. The results serve as an empirical basis for deriving context-sensitive recommendations for the design of local resilience strategies in manufacturing companies. For the correlation analysis, a multiple linear regression was conducted separately for each risk type and across both factory layouts (Table 3).

3 Results

Mitigation strategies exhibit significantly different effects across both layouts regarding the resulting output per day. While the linear layout demonstrates more moderate impacts, the fractal layout exhibits more extreme variations, with both pronounced positive and negative outcomes. In the fractal layout, specific measures are markedly more impactful, indicating a higher sensitivity of the fractal system to systemic changes (Fig. 1).

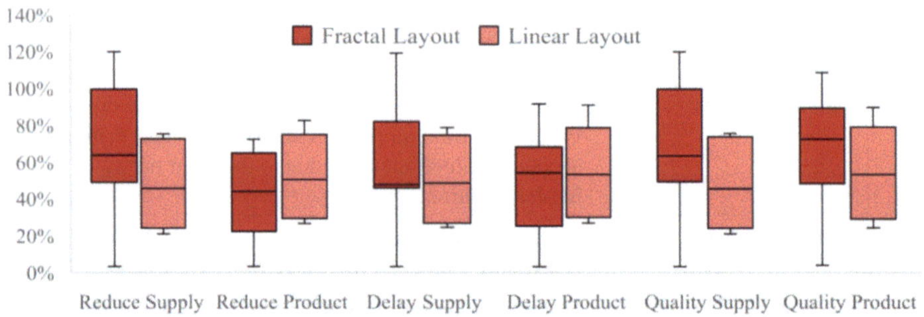

Fig. 1. Box-Plot of output per day relative to the status quo across all experiments

Each measure's effect on the controlled variable was calculated relative to the benchmark specific to each risk category and level of intensity. The regression coefficients were then normalised and presented as percentage changes compared to the respective benchmark and presented in Table 3.

The regression analysis revealed distinct patterns across factory layouts and risk types. E.g. in a *Linear Layout* and during the risk *Quality Product* with *Low* intensity, implementing the measure *Parallel Stations* (M1) results in an average increase of 1,76% (highlighted in Table 3) of the output per day compared to the output when no countermeasure is implemented (Table 3, for a description of each strategy see Table 2).

For the fractal layout, implementing *Parallel Stations* (M1) for the slowest production steps consistently demonstrated substantial positive effects on output, achieving improvements of up to approximately + 20% compared to the benchmark. Additionally, mitigation strategies such as the introduction of a *Flexible Station* (M2) and increasing *Buffer Sizes* (M3) in front of stations resulted in increases in output of up to 10%. Notably, the measure *Initial Quality Check* (M5) exhibited severely negative impacts in most cases. The only exception were the mid- and high-levels of the risk *Quality Product*, where 50% - 75% of incoming products are defective. In all other cases, particularly within scenarios involving delayed deliveries, the strategy reduced the output

Table 3. Relative change of output for mitigation strategies in given risk scenario in %

		Fractal Layout						Linear Layout					
Risk	**Level**	**M1**	**M2**	**M3**	**M4**	**M5**	**M6**	**M1**	**M2**	**M3**	**M4**	**M5**	**M6**
Reduced Supply	**Low**	20,74	8,01	8,13	1,57	-17,84	2,11	-0,04	0,18	0,00	-0,06	0,17	-0,25
	Mid	12,31	4,07	5,39	-2,03	-23,94	1,14	-0,59	-0,58	0,09	1,05	-0,06	-0,82
	High	13,07	5,50	5,76	-0,70	-15,84	1,12	-1,82	-1,36	-0,09	1,43	0,23	-1,66
Reduced Product	**Low**	17,56	7,01	5,72	4,26	-16,27	0,25	-0,01	0,11	0,00	4,96	0,01	0,00
	Mid	11,44	6,00	6,39	2,25	-11,00	0,48	-0,04	0,07	0,01	4,90	-0,02	0,00
	High	4,42	4,36	4,35	4,29	-4,19	0,48	-0,12	-0,04	0,01	5,08	0,01	-0,05
Delayed Supply	**Low**	16,90	6,69	8,23	-2,05	-23,66	1,47	0,72	1,00	0,41	0,22	-0,06	0,67
	Mid	8,69	2,55	4,51	-1,20	-29,16	1,63	1,24	0,89	0,44	-1,27	0,22	1,35
	High	14,30	5,84	5,93	-0,47	-15,76	1,33	2,88	1,04	0,33	-2,18	-0,54	3,75
Delayed Product	**Low**	18,53	9,56	8,18	3,19	-17,00	1,73	2,94	2,55	0,01	3,17	-0,48	1,64
	Mid	19,11	8,05	6,90	3,66	-15,20	2,11	2,84	2,67	0,02	3,58	-0,50	1,15
	High	14,37	9,63	3,34	3,66	-10,62	1,48	2,58	1,95	0,09	4,06	-0,48	2,88
Quality Supply	**Low**	20,94	7,83	8,11	1,91	-17,53	2,13	-0,10	-0,03	-0,34	0,40	-0,14	-0,17
	Mid	12,10	4,24	5,21	-1,76	-24,19	1,15	-0,76	-0,58	-0,31	0,82	-0,14	-0,73
	High	13,16	5,35	5,78	-0,78	-16,30	1,16	-1,84	-1,25	-0,01	1,43	0,22	-1,55
Quality Product	**Low**	20,75	14,35	5,85	3,72	-9,43	2,07	1,76	2,04	0,03	4,66	1,30	0,43
	Mid	12,87	10,93	11,51	2,49	6,74	4,24	1,22	1,36	0,01	5,25	1,83	0,51
	High	6,55	2,86	7,29	5,27	29,49	3,55	1,06	1,53	0,20	5,10	2,11	0,49

relative to the benchmark even further, suggesting that doing nothing would be the better choice. Other measures displayed inconsistent effects, generally minimal compared to the significant outcome of the mentioned strategies.

Within the linear layout, increases in outcome were significantly smaller. *Increase Product Order* (M4) showed notable positive effects, though more moderate - typically between + 1% and + 5%. In contrast, measures such as *Parallel Stations* (M1) and *Flexible Station* (M2) yielded only small positive effects if any, indicating a more stable production environment compared to the fractal layout.

Regarding the different levels of risk intensity (*Low/Mid/High*), no difference in the best choice of mitigation strategy can be observed. The model suggests, that while the level of intensity influences the overall impact of countermeasures – mitigation strategies have a higher absolute and relative impact on the output per day when the risk intensity is low - only the risk type and factory layout influence which mitigation strategy is best suited to counter the given risk type.

Overall, the linear layout was characterised by moderate effect sizes and lower variance among measures, whereas the fractal layout exhibited larger potential gains accompanied by increased sensitivity and pronounced negative effects from specific measures.

4 Discussion

The analysis demonstrates that different factory layouts distinctly influence the suitability and effectiveness of specific mitigation strategies for a given risk type. The model suggests, that implementing parallel and flexible production capacities greatly increase the ability to counteract risks in a highly variable factory setting. Our model indicates a much smaller effect for a strictly linear factory setting (Table 3). For risks involving reduced delivery volumes the measure *Increase Product Order* (M4) consistently emerged as effective, significantly improving production outputs across both factory layouts. However, the mitigation strategies mentioned above (M1 and M2) were significantly more effective for the fractal factory. In contrast, risks associated with quality issues also benefited from increased order quantities, though careful consideration is necessary due to potential accumulation of defective inventory when not combined with early quality checks.

In cases of delayed deliveries, mitigation strategies that increase buffer capacities or workforce flexibility generally provided minor improvements in the linear layout, suggesting their suitability as supplementary measures. Conversely, the fractal layout was significantly more sensitive: activating *Increase Product Order* (M4) while explicitly avoiding measures such as an initial quality check that counteract the flexible inflow of parts is critical to preventing a severe output reduction.

Risks characterised by volume and delay issues strongly favour increased ordering practices, supplemented by buffer and flexibility measures. Quality-related risks require balanced strategies that emphasise proactive supplier management rather than solely relying on increased inventories or stringent quality controls.

For both layouts, strategies had better effects when the risk affected the main part of the product rather than the supply of support goods, indicating that the chosen mitigation strategies might not be best suited to address risks associated with the supply. Future research should include more strategies and variations of their implementation. In addition, the interplay of different strategies shows the importance of combination effects. While some strategies show only marginal effects on their own, their combination with other measures (e.g. increased buffers with increased order quantities) can offer synergies, especially with the more responsive fractal arrangement. This emphasises the value of integrated, rather than isolated, planning of remedial actions.

5 Conclusion

In light of increasing supply chain disruption, this study responds to the need for localised, factory-level mitigation strategies that strengthen manufacturing resilience and sustainability. Addressing this gap, the research delivers three main contributions. (1) It classifies important procurement-related risks that occur within the production system. (2) It evaluates six local mitigation strategies through simulation experiments under controlled, multifactorial conditions. (3) It highlights how strategy effectiveness is significantly influenced by production layout and risk type (e.g. *Parallel* and *Flexible Stations*), and on the other hand, with the *Increase Product Order* strategy emerging as particularly impactful across scenarios.

These findings underline the importance of layout-sensitive, operational resilience planning and provide production planners with simulation-based decision support for managing short-term disruptions. However, several limitations must be noted. The simulation approach, while suitable for capturing dynamic interactions, relies on idealised factory models and fixed parameter settings. Thus, the results should be interpreted as indicative rather than prescriptive and future research should include hybrid factory layouts. Their direct applicability to real-world factories requires careful adaptation to specific structural and operational conditions. Furthermore, the study focuses on a limited set of risks and mitigation strategies; broader sets of disruptions - e.g., cascading failures or interdependent risks - and more complex adaptive countermeasures should be considered in future research. Strategies are also influenced by the effort of implementation and being possible at all. This was not part of this study but will be discussed in future research.

Despite these constraints, the study advances the understanding of how localised operational strategies can complement global sourcing measures. It supports the core proposition that sustainable manufacturing resilience is not only a question of structural diversification but also context-aware, layout-specific adaptability at the factory level.

References

1. Straube, F., Nitsche, B.: Heading into "the new normal": potential development paths of international logistics networks in the Wake of the Coronavirus Pandemic. Internationales Verkehrswesen **72**(3), 31–35 (2020)
2. Ivanov, D.: Supply chain viability and the COVID-19 pandemic: a conceptual and formal generalisation of four major adaptation strategies. Int. J. Prod. Res. **59**(12), 3535–3552 (2021)
3. Queiroz, M.M., Ivanov, D., Dolgui, A., Fosso, W.S.: Impacts of epidemic outbreaks on supply chains: mapping a research agenda amid the COVID-19 pandemic through a structured literature review. Ann. Oper. Res. **319**(1), 1159–1196 (2022)
4. Chopra, S., Sodhi, M.S.: Reducing the risk of supply chain disruptions. MIT Sloan Manag. Rev. **55**(3), 73–80 (2014)
5. Wieland, A.: Dancing the supply chain: toward transformative supply chain management. J. Supply Chain Manag. **57**(1), 58–73 (2021)
6. Elkington, J.: Cannibals with Forks: The Triple Bottom Line of 21st Century Business. Capstone Publishing Ltd. (1997)
7. Silvestre, B.S.: Sustainable supply chain management in emerging economies: environmental turbulence, institutional voids and sustainability trajectories. Int. J. Prod. Econ. **167**(C), 156–169 (2015)
8. Kamalahmadi, M., Parast, M.M.: A review of the literature on the principles of enterprise and supply chain resilience: major findings and directions for future research. Int. J. Prod. Econ. **171**(P1), 116–133 (2016)
9. Tang, C.S.: Perspectives in supply chain risk management. Int. J. Prod. Econ. **103**(2), 451–488 (2006)
10. Bode, C., Wagner, S.M.: Structural drivers of upstream supply chain complexity and the frequency of supply chain disruptions. J. Oper. Manag. **36**, 215–228 (2015)
11. Straube, F., Gorgas, B., Coll, A.: Leveraging digital logistics twins to harness sustainability and resilience in manufacturing networks. Lecture Notes in Mechanical Engineering (2024)
12. Mayring, P.: Qualitative Content Analysis - Theoretical Foundation, Basic Procedures and Software Solution. Klagenfurt (2014)
13. Durach, C.F., Kembro, J., Wieland, A.: A new paradigm for systematic literature reviews in supply chain management. J. Supply Chain Manag. **53**(4), 67–85 (2017)
14. Law, A.M., Kelton, W.D.: Simulation Modeling and Analysis. McGraw-Hill, New York (2000)
15. Siemens AG: Plant Simulation 2404.0013 (Software) (2024). https://plm.sw.siemens.com/en-US/tecnomatix/plant-simulation-software/
16. Wiendahl, H.-P., Reichardt, J., Nyhuis, P.: Handbuch Fabrikplanung: Konzept, Gestaltung und Umsetzung wandlungsfähiger Produktionsstätten Springer (2014)

17. Tecnomatix: Turbo Charger Assembly Example Model - Tecnomatix Plant Simulation (2019). community.sw.siemens.com/s/article/turbo-charger-assembly-example-model-tecnomatix-plant-simulation

Risk Mitigation and Sustainability in Oil Supply Networks: A Multi-objective Optimization Approach

Alperen Bal(✉), Hakan Tozan, Oumayma Hamlaoui, and Mohamed Karim Hajji

College of Engineering and Technology, American University of the Middle East, 54200 Egaila, Kuwait

alperen.bal@aum.edu.kw

Abstract. Oil supply chains are a critical infrastructure for both nation's economies and global energy markets. However, these supply chains face disruptive risks from sudden refinery outages and pipeline capacity losses to geopolitical events that can interrupt oil flows. Simultaneously, stricter environmental regulations, particularly on sulfur emissions, demand cleaner fuel production and operations. This study develops a multi-objective optimization model to enhance the resilience of oil supply chains in the Gulf region while balancing economic and environmental objectives. The model minimizes total supply chain cost and sulfur emissions, and maximizes resilience, under three disruption scenarios: (1) refinery shutdowns, (2) pipeline capacity reductions, and (3) geopolitical supply constraints. The formulation incorporates sulfur emission regulations and broader environmental policies as constraints. Scenario analyses are conducted to evaluate trade-offs among cost efficiency, compliance with sulfur caps, and the ability to sustain supply during disruptions. The results are expected to provide decision-makers with a set of Pareto-optimal strategies that reinforce supply chain resilience against disruptions, ensure regulatory compliance, and manage costs effectively.

Keywords: Supply Chain Resilience · Risk Mitigation · Multi-objective Optimization · Oil Supply Networks · Low-sulfur Production

1 Introduction and Background

Oil supply chains (OSC) are critical infrastructures that face mounting pressure from both operational disruptions and environmental regulations [1–3]. Recent disruptions such as refinery outages, pipeline failures, and geopolitical supply shocks have highlighted the importance of integrating resilience into OSC planning [4,5]. Simultaneously, environmental constraints—especially those targeting sulfur and CO_2 emissions—require models that account for regulatory compliance [6]. While earlier studies focused on cost-minimization under deterministic conditions [7], more recent research has emphasized stochastic and robust

H. Kohl et al. (Eds.): GCSM 2025, LNME, pp. 241–249, 2026.
https://doi.org/10.1007/978-3-032-21154-5_27

approaches to manage risk [8]. Multi-objective optimization (MOO) methods have emerged to address trade-offs among economic performance, environmental impact, and supply chain resilience [9,10]. However, few models offer integrated decision support that captures all three dimensions simultaneously [11,12].

This study develops a tri-objective optimization framework that integrates economic, environmental, and resilience objectives. It highlights the role of newly operational refinery, which enhances refining capacity and supports compliance with low-sulfur fuel standards, thereby advancing environmentally responsible manufacturing within the oil supply network. The model incorporates real-world constraints, including refinery outages, pipeline disruptions, and geopolitical supply risks. Scenario-based analyses are conducted to inform resilience investments and sustainability planning in energy logistics systems.

2 Methodology

2.1 Problem Description

Kuwait's oil supply chain forms a strategic backbone of its national economy, covering all supply chain operations including upstream extraction, midstream transport, and downstream refining and distribution. The country's upstream infrastructure consists of five major crude oil production zones: Greater Burgan, Raudhatain, Sabriyah, Minagish & Umm Gudair, and Wafra. As summarized in Table 1, these fields collectively enable daily production capacities exceeding two million barrels per day (bpd), with Greater Burgan alone contributing up to 1.3 million bpd.

Table 1. Oil field production capacities in Kuwait.

Oil Field	Region	Approximate Production (bpd)
Greater Burgan	Southeastern Kuwait	1,100,000–1,300,000
Raudhatain	Northern Kuwait	Up to 450,000
Sabriyah	Northern Kuwait	~100,000
Minagish & Umm Gudair	Western Kuwait	~200,000 (combined)
Wafra (Neutral Zone)	Southern Kuwait	170,000–180,000

Crude oil extracted from these fields is transported via an extensive pipeline network to three major refineries: Mina Al-Ahmadi, Mina Abdullah, and Al-Zour. As shown in Table 2, these facilities differ significantly in capacity and strategic role. Al-Zour, the newest and largest facility, plays a central role in increasing refining capacity and enabling compliance with low-sulfur fuel regulations.

On the demand side, Kuwait meets both domestic and international requirements. Domestic consumption is relatively stable at around 450,000 to 500,000

Table 2. Refinery capacities in Kuwait.

Refinery Name	Location	Capacity (bpd)
Mina Al-Ahmadi	∼40 km south of Kuwait City	466,000
Mina Abdullah	∼46 km south of Kuwait City	280,000
Al-Zour	Southern Kuwait (coastal)	615,000

bpd [13], while the majority of refined products are exported. Crude exports have declined slightly over time, whereas refined product exports have grown steadily from 650,000 bpd in 2022 to 968,000 bpd in 2024 [14]. This shows the country's strategic shift toward capturing more value in the downstream segment.

2.2 Model Assumptions

- Weekly planning horizon over 52 periods; demand disaggregated from monthly data with stochastic variation.
- Fixed demand shares: 32% crude, 68% refined products; product-level split based on 2024 data.
- Well and refinery capacities are predefined; 10% refining gain applied to account for output expansion.
- Emissions calculated using sulfur content and refinery-specific CO_2/SOx factors; annual emission caps enforced.

2.3 Mathematical Model

The model formulation is given below.
Sets:

W Set of wells
R Set of refineries
P Set of refined products
M Set of markets
T Set of time periods

Parameters:

cap_w^{wt}	Production capacity of well w in period t
cap_{wr}^{wrt}	Pipeline transfer capacity from w to r in period t
cap_r^{rt}	Refinery production capacity at r in period t
D_t^{crude}	Crude oil demand in period t
D_{pt}^{prod}	Demand for product p in period t
$\delta_w^{wt}, \gamma_{wr}^{wrt}, \theta_r^{rt}$	Disruption factors for wells, pipelines, and refineries
$p_t^{\text{crude}}, p_{pmt}$	Prices of crude and refined products in period t
$c^{\text{prod}}, c^{\text{pipe}}, c_p^{\text{ref}}, c^{\text{inv}}, c^{\text{pen}}$	Production, transportation, inventory holding costs, and penalties for unmet demand
σ_w	Sulfur content from well w
ϕ_r, ψ_r, η_r	CO_2, SOx emission factors and SRU efficiency at refinery r

Decision Variables:

x_{wt}	Crude oil produced at well w in period t
y_{wrt}	Crude transported from w to r in period t
e_{rt}	Crude exported from refinery r in period t
z_{prt}	Refined product p produced at refinery r in period t
s_{prmt}	Product p shipped from r to market m in period t
I_{rt}^{crude}	Crude inventory at refinery r in period t
I_{prt}^{prod}	Inventory of product p at refinery r in period t
u_t^{crude}	Unmet crude oil demand in period t
u_{pt}^{prod}	Unmet product p demand in period t
$\text{CO2}_{rt}, \text{SOx}_{rt}$	Emissions at refinery r in period t

The model integrates three conflicting objectives—economic profit, environmental sustainability, and supply chain resilience—using a weighted-sum formulation: $Z = \alpha f_1 - \beta f_2 + \gamma f_3$. The economic component is defined as $f_1 = \text{Revenue} - \text{Operational Costs} - \text{Inventory Costs} - \text{Unmet Demand Penalties}$, capturing net profit from exports minus all associated costs. The environmental term is $f_2 = \text{CO}_2 + \text{SOx}$ emissions, based on refinery throughput and sulfur content. The resilience objective is $f_3 = \text{Fulfilled Shipments} + \text{Crude Exports} - \text{Unmet Demand}$, promoting service continuity during disruptions.

Constraints:

$$x_{wt} \leq \text{cap}_w^{wt} \cdot \delta_w^{wt}, \quad \forall w, t \tag{1}$$

$$\sum_r y_{wrt} = x_{wt}, \quad \forall w, t \tag{2}$$

$$y_{wrt} \leq \text{cap}_{wr}^{wrt} \cdot \gamma_{wr}^{wrt}, \quad \forall w, r, t \tag{3}$$

$$I_{rt}^{\text{crude}} = I_{r,t-1}^{\text{crude}} + \sum_w y_{wrt} - \frac{1}{1+g} \sum_p z_{prt} - e_{rt}, \quad \forall r, t \tag{4}$$

$$\sum_p z_{prt} \leq \text{cap}_r^{rt} \cdot (1+g) \cdot \theta_r^{rt}, \quad \forall r, t \tag{5}$$

$$I_{prt}^{\text{prod}} = I_{pr,t-1}^{\text{prod}} + z_{prt} - \sum_m s_{prmt}, \quad \forall r, p, t \tag{6}$$

$$\sum_r e_{rt} + u_t^{\text{crude}} = D_t^{\text{crude}}, \quad \forall w, r, t \tag{7}$$

$$\sum_{r,m} s_{prmt} + u_{pt}^{\text{prod}} = D_{pt}^{\text{prod}}, \quad \forall p, t \tag{8}$$

$$\sum_r e_{rt} \geq E_t, \quad \forall t \tag{9}$$

$$\text{SOx}^{\text{load}} = \sum_w \sigma_w \cdot y_{wrt}, \quad \forall t \tag{10}$$

$$\mathrm{CO2}_{rt} = \phi_r \sum_p z_{prt}, \quad \forall r, t \tag{11}$$

$$\mathrm{SOx}_{rt} = \left(\sum_w \sigma_w \cdot y_{wrt} \right) (1 - \eta_r), \quad \forall r, t \tag{12}$$

$$\sum_{r,t} \mathrm{CO2}_{rt} \leq \mathrm{CO2}_{\mathrm{cap}}, \quad \sum_{r,t} \mathrm{SOx}_{rt} \leq \mathrm{SOx}_{\mathrm{cap}} \tag{13}$$

$$x_{wt}, y_{wrt}, e_{rt}, z_{prt}, s_{prmt}, I_{rt}^{\mathrm{crude}}, I_{prt}^{\mathrm{prod}}, u_t^{\mathrm{crude}}, u_{pt}^{\mathrm{prod}}, \mathrm{CO2}_{rt}, \mathrm{SOx}_{rt} \geq 0 \tag{14}$$

Constraints (1) and (2) ensure that the crude oil production at each well w in period t does not exceed its maximum available capacity, adjusted for potential geopolitical or technical disruptions in the disruption scenarios (δ_w^{wt}). Additionally, all produced crude must be fully allocated to the downstream transportation network, preserving mass balance. Constraint (3) limits pipeline capacity from wells to refineries, adjusted dynamically by the disruption factor γ_{wr}^{wrt}. Constraint (4) tracks crude inventory at each refinery by balancing incoming flows, prior inventory, refining requirements (adjusted by gain factor g), and exports. The factor g reflects volume expansion during refining. Constraint (5) limits refinery throughput based on nominal capacity, adjusted for refining gain and disruption factor θ_r^{rt}. Constraint (6) maintains accurate product-level inventories at each refinery. Inventories evolve over time by accounting for newly refined output and outgoing shipments to domestic and export markets. Constraints (7) and (8) guarantee the satisfaction of external crude and product demand across all markets and periods. Slack variables u_t^{crude} and u_{pt}^{prod} are introduced to quantify unmet demand and are penalized in the objective function to discourage shortfalls.

Another set of constraints are developed for policy requirements and sustainability targets. Constraint (9) enforces a minimum crude export level, reflecting policy commitments and OPEC quota obligations. Constraint (10) computes sulfur inflow to each refinery based on crude volumes received and well-specific sulfur content σ_w. Constraints (11) and (12) calculate emissions: CO_2 is proportional to throughput via ϕ_r, while SOx depends on crude sulfur content and SRU efficiency η_r. Constraints (13) enforce annual CO_2 and SOx emission caps to meet environmental regulations. Constraint (14) enforces non-negativity of all variables to ensure physical and operational feasibility.

3 Results and Discussion

The weighted-sum method was chosen for its intuitive interpretation of trade-offs among conflicting objectives. A grid search was conducted using the scalarized objective function in Eq. 15, where the weights α, β, and γ satisfy:

$$\alpha + \beta + \gamma = 1, \quad \alpha, \beta, \gamma \in [0, 1] \tag{15}$$

The model was implemented and solved using the Python-based PuLP optimization package within the Google Colab environment. The resulting model consists of 8,372 decision variables and 5,098 constraints, with weekly time resolution, multiple product types, and detailed disruption scenarios. Due to its linear programming formulation, the model was able to generate Pareto-optimal solutions in less than 2 s in each iteration.

3.1 Pareto Frontier Analysis

The Pareto frontier analysis yields 54 non-dominated solutions illustrating trade-offs among profit, emissions, and resilience, as shown in Fig. 1. Each solution reflects a different combination of weights in the scalarized objective function. Assigning greater weight to the economic objective increases profit but often leads to higher emissions and unmet demand. In contrast, prioritizing resilience reduces shortfalls and stabilizes flows, with a modest trade-off in economic performance.

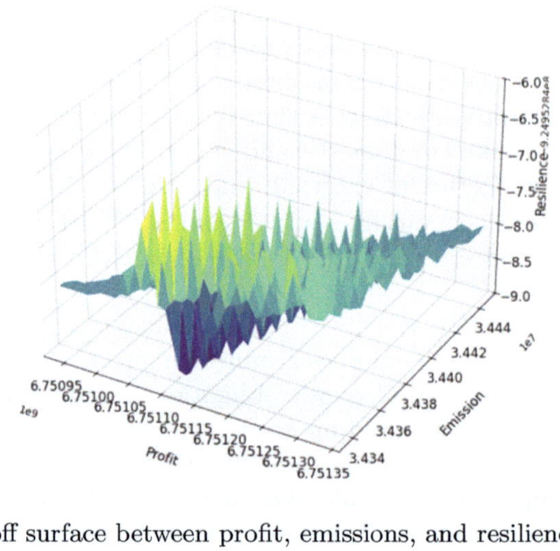

Fig. 1. Payoff surface between profit, emissions, and resilience objectives.

3.2 Scenario Comparison and Managerial Insights

The scenario analysis presented in Table 3 highlights how different types of disruptions impact key performance indicators (KPIs) of the oil supply chain model. Each scenario varies in its effect on economic outcomes, environmental performance, and operational reliability.

Table 3. Scenario performance comparison for baseline and disruption cases.

Scenario	Profit	Unmet Demand (%)	Export Fulfill. (%)	Inv. Util. (%)	Ref. Util. (%)	Emission Intensity
Baseline (α=0.6, β=0.2, γ=0.2)	7.23	1.98	100.00	5.94	94.13	0.4527
Refinery Disruption – Moderate	7.22	2.04	100.00	6.66	99.89	0.5032
Refinery Disruption – Severe	6.97	4.94	100.00	7.90	99.88	0.5147
Pipeline Disruption – Moderate	7.23	1.98	100.00	1.78	94.12	0.5030
Pipeline Disruption – Severe	7.21	2.28	98.99	5.50	94.12	0.5021
Geopolitical Supply – Moderate	7.10	4.02	93.20	4.11	94.13	0.5028
Geopolitical Supply – Severe	6.97	5.90	86.94	4.14	94.13	0.4987

Refinery disruptions, particularly those affecting Al-Zour—the newest and most environmentally efficient facility—have a relatively modest impact on total profit (less than 3% deviation from the baseline). However, they significantly increase emission intensity, as production is redistributed to older refineries with higher sulfur and carbon output per barrel. This emphasizes the strategic importance of protecting Al-Zour from operational downtime through prioritized maintenance schedules, rapid-repair protocols, and targeted investments in critical spare parts. In addition, maintaining buffer inventories of low-sulfur crude oil can help moderate environmental penalties when cleaner refining capacity is temporarily offline.

Pipeline disruptions are shown to be largely absorbable by the network. Even in severe cases, where multiple routes are constrained, the model identifies alternate transport paths, thereby limiting profit loss and unmet demand to negligible levels. This result suggests that the current pipeline infrastructure already contains sufficient physical redundancy.

Geopolitical disruptions at the well level, particularly involving the Burgan and Raudhatain fields, produce the most severe deterioration in supply chain performance. These scenarios reduce export fulfillment to 81.6% and significantly increase unmet domestic demand. Particularly, refinery utilization remains high, enabled by inventory drawdowns. These findings imply that upstream disruptions are less manageable. To prevent any lost sales engaging with OPEC and export partners to allow temporary adjustments to export obligations could provide further flexibility during such shocks.

Furthermore, weight sensitivity analysis confirms that increasing the relative importance of the resilience objective (γ) during disruptions enhances system robustness. Re-weighting in favor of resilience leads to significant improvements in unmet demand and service levels, with only marginal reductions in profit.

4 Conclusion and Future Work

This study presents a multi-objective optimization model for oil supply chain planning that integrates economic, environmental, and resilience objectives under disruption scenarios. The model investigates trade-offs across competing goals through Pareto frontier analysis. While this study focuses on economic and environmental sustainability in extraction and refining operations, it acknowledges the broader dimensions of sustainability. Future research could extend the model to capture downstream environmental impacts and incorporate social aspects like community health, workforce safety, and socioeconomic outcomes. In addition, the analysis could be enhanced by integrating disruption recovery strategies and demand-side uncertainty. Although a deterministic weighted-sum approach enables transparent trade-off analysis, future work may benefit from stochastic or robust formulations to explicitly address uncertainty.

References

1. Carvalho, H., Azevedo, S.G., Cruz-Machado, V.: Agile and resilient approaches to supply chain management: Influence on performance and competitiveness. Logist. Res. **4**(1–2), 49–62 (2012)
2. Ivanov, D.: Viable supply chain model: integrating agility, resilience and sustainability perspectives—lessons from and thinking beyond the COVID-19 pandemic. Ann. Oper. Res., 1–21 (2020). https://doi.org/10.1007/s10479-020-03640-6
3. Azadegan, A., Mellat Parast, M., Lucianetti, L., Nishant, R., Blackhurst, J.: Supply chain disruptions and business continuity: an empirical assessment. Decis. Sci. **51**(1), 38–73 (2020)
4. Urciuoli, L., Mohanty, S., Hintsa, J., Boekesteijn, E.G.: The resilience of energy supply chains: a multiple case study approach on oil and gas supply chains to Europe. Supply Chain Manag. **19**(1), 46–63 (2014)
5. Christopher, M., Peck, H.: Building the resilient supply chain. Int. J. Logist. Manag. **15**(2), 1–14 (2004)
6. Sarkis, J., Dhavale, D.G.: Supplier selection for sustainable operations: a triple-bottom-line approach using a Bayesian framework. Int. J. Prod. Econ. **166**, 177–191 (2015)
7. Gilani, H., Sahebi, H.: A multi-objective robust optimization model to design sustainable sugarcane-to-biofuel supply network: the case of study. Biomass Conv. Biorefinery **11**, 2521–2542 (2021)
8. Behzadi, G., O'Sullivan, M.J., Olsen, T.L., Zhang, A.: Agribusiness supply chain risk management: a review of quantitative decision models. Omega **79**, 21–42 (2018)
9. Govindan, K., Soleimani, H., Kannan, D.: Reverse logistics and closed-loop supply chain: a comprehensive review. J. Clean. Prod. **96**, 132–147 (2015)
10. Fahimnia, B., Jabbarzadeh, A.: Marrying supply chain sustainability and resilience: a match made in heaven. Transp. Res. Part E Logistics Transp. Rev. **91**, 306–324 (2016)
11. Dolgui, A., Ivanov, D., Sokolov, B.: Ripple effect in the supply chain: an analysis and recent literature. Int. J. Prod. Res. **56**(1–2), 414–430 (2018)
12. Ivanov, D., Dolgui, A.: A digital supply chain twin for managing the disruption risks and resilience in the era of Industry 4.0. Prod. Plann. Control **32**(9), 775–788 (2021)
13. Energy Information Administration.: Kuwait – International - Analysis (2024). https://www.eia.gov/international/overview/country/KWT, Retrieved 27 April 2025
14. WorldData.info.: Energy consumption in Kuwait (2024). https://www.worlddata.info/asia/kuwait/energy-consumption.php, Retrieved 27 April 2025

Case Study of Bochum and Chofu City in Installing E-Moped Sharing Service for Accessibility, Greenhouse Gas Emissions, and Cost

Seigo Takahashi[1], Yuki Kinoshita[1], Nora Schelte[2], Semih Severengiz[2(✉)], and Tetsuo Yamada[1]

[1] Department of Informatics, The University of Electro-Communications, 1-5-1 Chofugaoka, Chofu, Tokyo 182-8585, Japan
y-kinoshita@uec.ac.jp

[2] Sustainable Technologies Laboratory, Department of Electrical Engineering and Computer Sciences, Bochum University of Applied Sciences, Am Hochschulcampus 1, 44801 Bochum, Germany
semih.severengiz@hs-bochum.de

Abstract. Recently, there has been a call to transition toward low greenhouse gas (GHG) emission transportation to address global warming. One of the ecofriendly transportations is an electric moped scooter (e-moped) sharing service, which does not emit GHG during operation. Planning the location of charging stations and managing material procurement for e-moped manufacturing is essential to reduce cost and GHG emissions, while also enhancing service accessibility. In this study, two-stage design on the e-moped sharing services was adopted using integer programming to allocate charging stations and select material suppliers for e-mopeds. The method to determine suitable charging station locations and sizes, and supplier selection were presented. Numerical experiments were conducted to illustrate the proposed design and analysis method, assuming the installation of the Kumpan 1954 model in Bochum and Chofu cities. In the numerical experiments, the set covering and maximal covering location problems with a small coverage radius for charging stations yielded better results when accessibility, GHG emissions, and cost were evaluated comprehensively for Bochum and Chofu cities, respectively. In supplier selection, a higher GHG reduction target, such as 70% reduction, is desirable in both cases. The difference in the optimal scenarios for Bochum and Chofu can be attributed to the distinct population distribution in the two cities.

Keywords: Carbon Neutrality · Product Life Cycle · E-moped Sharing Service · Life Cycle Assessment (LCA) · Facility Location Problem · Integer Programming · Supplier Selection

1 Introduction

According to the International Energy Agency, road transportation accounts for 15% of global CO_2 emissions [1]. Decarbonizing the transportation section must be achieved by introducing zero-emission vehicles, which currently include battery electric vehicles

H. Kohl et al. (Eds.): GCSM 2025, LNME, pp. 250–258, 2026.
https://doi.org/10.1007/978-3-032-21154-5_28

(EVs) [2]. An electric moped scooter (e-moped) sharing services are expected to reduce greenhouse gas (GHG) emissions for short-distance travel [3]. The number of sharing scooters increased globally by 58% from 66,000 in 2019 to 104,000 in 2020 [4]. For e-mopeds sharing services, determining the locations and number of charging stations is a critical decision, as it directly affects accessibility, such as the distance from demand point to charging station and the availability of e-mopeds [5]. Charging station allocation can be formulated as a facility location problem, which involves identifying optimal locations for facilities using mathematical models [6].

By charging batteries at solar photovoltaic (PV) stations, e-mopeds do not emit GHG during the usage phase. However, GHG emissions in production phase remains unavoidable, and must be reduced to achieve carbon neutrality. Especially, material-based GHG emissions of assembly products can account for larger percentage against forward supply chain, including material production, manufacturing, and distribution [7]. As material-based GHG emissions and procurement cost differ among countries owing to different energy mix and economic situations, switching materials suppliers can reduce material-based GHG emissions with an affordable cost [7].

Csiszár et al. [8] developed a two-stage method for locating EV charging stations, aimed at assessing the potential for EV adoption and evaluating the effectiveness of charging stations. He et al. [7] proposed EV charging station location-allocation model that carefully incorporates supply-and-demand estimation of EV charging. Vazifeh et al. [9] presented a data-driven EV station allocation model to minimize the total number of stations and average distance drivers travel. Bai et al. [10] proposed a bi-objective model for total cost and service quality to determine locations, capacity options, and service types for EV charging stations. Those previous studies focused solely on EV charging station allocation and did not account for GHG emissions and costs associated with the material production phase. Takahashi et al. [5] proposed a two-stage design method for implementing e-moped sharing services, consisting of charging station allocation and supplier selection, using mathematical programming, and validated using Bochum city case. However, potential user demand vary across countries or regions.

This study adopts the two-stage design method proposed by Takahashi et al. [5] to Chofu city to further validate the approach and examine key insight and practical implications when applied to a city with characteristics different from those of Bochum. Specifically, Bochum city (reference case study) [5] has a population of 360,000, whereas Chofu city in Japan has a population of 240,000 and a population density of 11,200 people per square kilometer [11]. Although both cities have relatively large populations and potential for e-mopeds sharing service, their population densities differ.

2 2-Stage Design of the e-Moped Sharing Service Installation

2.1 Overview

The proposed method consisted of two stages, namely, first-stage for charging stations allocation and second-stage for material supplier selection [5]. Decision makings of both stages are supported using a mathematical model. Figure 1 shows the overview of the two-stage design of e-moped sharing service installation. The method employs two different types of data, including city data and product data. The first-stage allocates stations based on predefined demands and coverage radius, determining the required number of e-mopeds, accessibility, and equipment cost. The second stage selects material suppliers to minimize material procurement cost while satisfying predetermined GHG reduction target. Both procurement cost and GHG emissions are determined at this stage. The main reason why first-stage: station allocation is conducted before second-stage: supplier selection is needs of the required number of e-mopeds to conduct second-stage. To conduct supplier selection at second-stage, the required number of e-mopeds is a vital input parameter. Moreover, it depends on the size and number of charging stations. For those reasons, first-stage allocates charging stations, consequently, the required number of e-mopeds is determined.

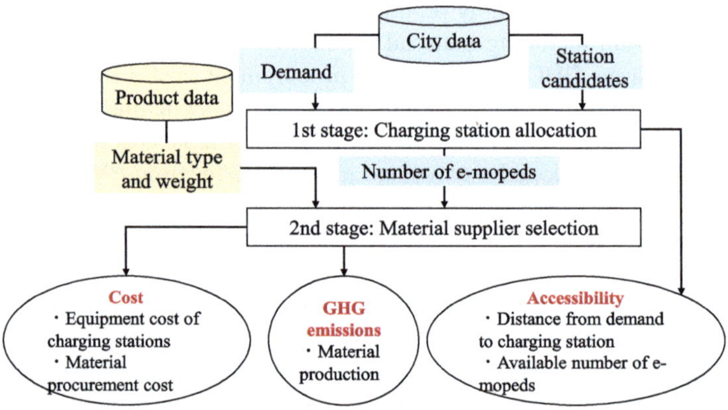

Fig. 1. Overview of the two-stage design of e-moped sharing service installation for accessibility, GHG emissions and cost proposed in Takahashi et al. [5]

2.2 Formulation of Two-Stage e-method Sharing Service Installation

2.2.1 First-Stage of Charging Station Allocation

Notation is listed as follows:

I, J: Set of demand/charging station candidate points indexed by i, j
A_{ij}: 1 if the demand point i can be covered by candidate j, otherwise 0.
W_i: User demand at demand point i.
SC: Cost to install a charging station.
NS: Number of charging stations.
CD: Cover demands.
x_j: 1 if the charging station is installed at demand point j, otherwise 0.
z_i: 1 if the demand point i is covered, otherwise 0.
EC: Equipment cost of charging stations.

The first stage applies set covering location and maximal covering location problems. The set covering location problem minimizes the number of facilities required to cover all demand points within a specified station coverage radius, as shown in Eqs. (1) to (3) [12]. Meanwhile, the maximal covering location problem aims to maximize demand coverage using a fixed number of stations [12] as shown in Eqs. (4) to (7). Contrarily to the set covering problem, the maximal covering problem does not ensure coverage of all demand points.

Set Covering Location Problem

$$NS = \sum_{j \in J} x_j \to min \tag{1}$$

$$EC = SC \times NS \tag{2}$$

$$\sum_{j \in J} A_{ij} x_j \geq 1 \forall i \in I \tag{3}$$

Maximal Covering Location Problem

$$CD = \sum_{i \in I} W_i z_i \to max \tag{4}$$

$$EC = SC \times NS \tag{5}$$

$$\sum_{j \in J} x_j = NS \tag{6}$$

$$\sum_{j \in J} A_{ij} x_j \geq z_i \forall i \in I \tag{7}$$

2.2.2 Second-Stage of Material Supplier Selection

Notation is listed as follows:

L:	Set of suppliers.
K:	Set of materials.
PC_{lk}/ME_{lk}:	Procurement unit cost/GHG emissions of material k in supplier l.
N_k:	Required volume of materials k to produce one unit.
NP:	Required number of e-mopeds.
E_{Max}/ε:	GHG emissions in the baseline/GHG reduction target.
f_{lk}:	Volume of material k procured from supplier l

$$MC = \sum_{l \in L} \sum_{k \in K} PC_{lk} f_{lk} \to min \tag{8}$$

$$E = \sum_{l \in L} \sum_{k \in K} ME_{lk} f_{lk} \tag{9}$$

$$E \leq (1 - \varepsilon) E_{Max} \tag{10}$$

$$\sum_{l \in L} f_{lk} = N_k NP \forall k \in K \tag{11}$$

$$f_{lk} \geq 0 \forall l \in L, \forall k \in K \tag{12}$$

The material supplier selection minimizes material procurement cost under GHG reduction target ε, as shown in Eqs. (8) to (12). Total cost (*TC*) is the sum of *EC* and *MC*.

3 Numerical Experiments

3.1 Input Data and Assumptions

Similar to Bochum city case [5], three different station size scenarios are considered at the first stage. S1, S2 and S3 indicate Small, Middle, and Large sizes of station scenarios, respectively, for the set covering location problem. Similarly, scenarios based on max coverage are represented as M1, M2, and M3. Larger stations can accommodate more operational e-mopeds but incur higher installation cost per station [5], as shown in Table 1. In the maximal covering location problems, the number of installed charging stations is predetermined and set as a parameter, as shown in Table 1. Note operational e-mopeds per 1 charging stations is estimated based on electrical production from solar charging station per day (kWh/day) and electrical demand per e-moped in a day (kWh/day). As both station production and e-mopeds consumption are fluctuation and the operational e-mopeds per 1 charging station affects the total number of e-mopeds strongly, the operational e-mopeds is set as continuous value to keep original estimated values.

Kumpan 1954 i model [3] was used as the reference e-moped, comprising 21 types of materials. Table 2 shows part of bill-of-materials (BOM) for e-moped. Each material can be procured from suppliers in China, Japan, and Korea. The procurement cost and GHG emission for each material of e-moped are estimated based on BOM [3], Japanese census of manufacturing, and life cycle inventory (LCI) database with the Asian International Input– Output Table as well as [5]. The LCI database can estimate GHG emissions by inputting the associated cost for each country.

First, Japanese material procurement cost is estimated using material weight and unit price obtained from BOM and Japanese census of manufacturing, respectively. Second, Chinese and Korean procurement cost is estimated using price levels. Finally, GHG emissions for each country is estimated using LCI database. These cost and GHG emissions include raw material manufacturing, material logistics, and material production only. Moreover, three GHG reduction targets ε were prepared as 26%, 50%, and 70% for each scenario. Consequently, 18 results of material supplier selection were obtained.

Table 1. Input data for each scenario at first-stage

Scenario	S1	S2	S3	M1	M2	M3
Operational e-mopeds per 1 charging station	90.3	181.5	27.8	90.3	180.5	270.8
Cost to install a charging station [thousand USD]	29	58	87	29	58	87
Coverage radius [km]	1	2	3	1	1	1
Installed number of stations	–	–	–	6	4	2

Table 2. Part of BOM for e-moped

Material name (weight [g])	Procurement cost [USD]			GHG emissions [kg-CO_2eq]		
	China	Japan	Korea	China	Japan	Korea
Lithium-ion battery (6,440)	7,906	18,386	12,503	19,725.0	5,707.1	8,301.3

3.2 Chofu City Case

Table 3 represents the results of the accessibility for each scenario obtained at the first-stage. Operational e-mopeds were calculated based on the number of charging stations *NS*. Cover rate *CR* represents ratio of covered potential demands against all potential demands. The e-mopeds per user were determined by dividing the number of operational e-mopeds by covered potential demands. Average distance *AD* and maximal distance *MD* indicate distances between demand points and their covered charging stations. From an accessibility perspective, a higher number of e-mopeds per user is desirable.

Table 3. Sation allocation for six scenarios at first-stage in Chofu city

Scenario	S1	S2	S3	M1	M2	M3
NS	9	5	2	6	4	2
Operational e-mopeds	812	903	542	542	722	542
CR [%]	100	100	100	95.9	80.1	50.0
E-mopeds per user [$\times 10^{-3}$]	3.4	3.8	2.3	2.4	3.8	4.5
AD [km]	0.49	0.77	1.58	0.52	0.58	0.62
MD [km]	0.97	1.71	2.90	0.97	0.96	0.92

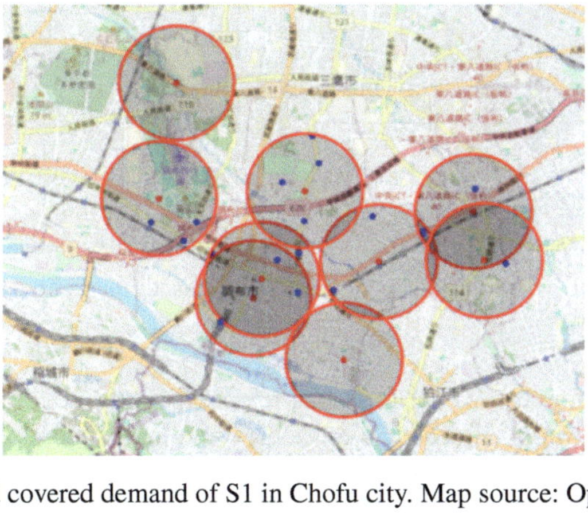

Fig. 2. Selected and covered demand of S1 in Chofu city. Map source: OpenStreetMap ®[13]

In the set covering location problem, all demands are satisfied; thus, the number of e-mopeds per user depends solely on the number of operational e-mopeds. *AD* and *MD* increase as station size grows. In the maximal covering location problem, the set of users varied across scenarios. The number of e-mopeds per user increased with larger station sizes, while *AD* slightly increased as the number of stations decreased. Over 80% demands were covered except the least charging station scenario M3. Considering accessibility only, greater e-mopeds per user and shorter *AD* and *MD* are desirable. Therefore, scenario S1 is preferable owing to its shortest *AD* and *MD*. Figure 2 shows the results of scenario S1. Red points are demands, where the charging station was allocated, whereas Blue points are the demand points, where the charging station was NOT allocated. Grey circles represent coverage radius. In scenario S1, only two charging station were allocated in the central area of Chofu city, which contains six demand points. Despite the high potential demand, this limited allocation may result in a missed opportunity. Remark Bochum case map is in [5].

3.3 Bochum and Chofu Comparisons for Accessibility, GHG Emissions, Cost

This subsection compares Bochum [5] and Chofu cases in terms of the selected locations of charging stations and selected suppliers. Note that, despite applying the same scenarios, the number of operational e-mopeds and cover radius differ between the Bochum and Chofu cases. The detailed input data of Bochum city is in [5]. Figure 3 presents 36 results from the first and second stages, showing *AD* relative to total costs per user (CpU) and GHG emissions per user (EpU) for Bochum and Chofu cities, respectively. Each shape and color identifies city and scenario, and colored patterns indicate GHG reduction targets in second-stage. For example, red solid square mean Bochum city case in scenario M2 with 26% GHG reduction target. As difference of CpU among GHG reduction targets were small, markers were overlapped as shown in Fig. 3.

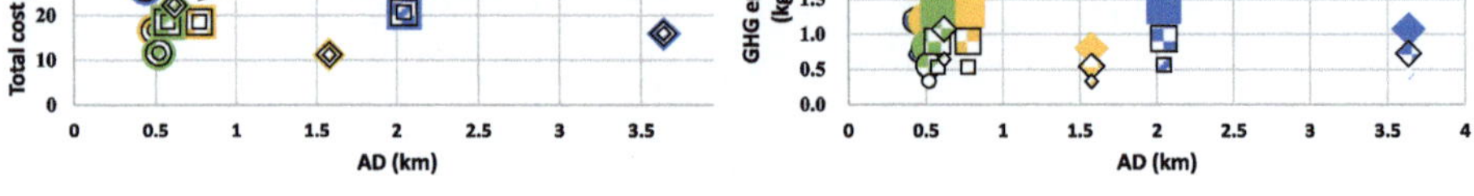

Fig. 3. Comparison of accessibility and GHG emissions through first and second stages in Bochum and Chofu cities

In case of Bochum city, scenario S1 had lower CpU and EpU than that of scenarios M1, M2, and M3. Moreover, scenario S1 achieved a larger reduction in *AD* compared to scenarios S2 and S3, with a slight increase of CpU and EpU. Therefore, scenario S1 should be adopted in Bochum city. Meanwhile, scenario M1 had lower CpU and EpU than that of scenarios S1, S2, M2 and M3 in case of Chofu city. Moreover, scenario M1 achieved a larger reduction in *AD* compared to scenario S3, with a slight increase in CpU and EpU. Therefore, scenario M1 should be adopted in Chofu city. A 70% GHG reduction target is desirable for both cases in the second stage as cost difference compared to 26% and 50% reduction targets in the same scenario were relatively small.

4 Conclusions

This study adopted the two-stage installation method of the e-moped sharing service proposed by Takahashi et al. [5] to Chofu city, and conducted a comprehensive comparison between the Bochum and Chofu cases in terms of accessibility, GHG emissions, and cost. It was found that suitable station size scenarios differed between the Bochum and Chofu cases. Moreover, a 70% GHG reduction target was considered desirable for both cases. The model would contribute to decisions of e-mopeds replacement period with new one. To achieve carbon neutrality, the reduced GHG emissions by e-mopeds has to be greater than GHG emissions of its production. The proposed model could be helpful to grasp the required volume of GHG reduction using e-mopeds for carbon neutrality. However, the model has two limitations about station allocations. One is only one charging station size can be addressed in each scenario. The other is not consideration of land cost for charging stations.

Future studies should incorporate manufacturing cost of e-mopeds and the land cost of charging stations. GHG emissions associated with e-moped production and station infrastructure should be evaluated. Furthermore, the facility location problem should be extended to allow for the simultaneous installation of charging stations of varying sizes, enabling station capacity to be tailored to the demand at each point.

Acknowledgements. This research was partially supported by the Japan Society for the Promotion of Science (JSPS) KAKENHI, Grant-in-Aid for Scientific Research (B), JP23K26329 from 2023 to 2025.

References

1. Ritchie, H.: Cars, planes, trains where do CO_2 emissions from transport come from?. https://ourworldindata.org/co2-emissions-from-transport. Accessed 20 May 2025
2. EU: Regulation - 2023/851 - EN - EUR-Lex. https://eur-lex.europa.eu/eli/reg/2023/851/oj. Accessed 20 May 2025
3. Schelte, N., Severengiz, S., Schünemann, J., Finke, S., Bauer, O., Metzen, M.: Life cycle assessment on electric moped scooter sharing. Sustainability **13**, 8297 (2021)
4. UNU: Unu launches global moped sharing market report 2020. https://unumotors.com/en-de/unu-launches-global-moped-sharing-market-report-2020/. Accessed 21 May 2025
5. Takahashi, S., Kinoshita, Y., Schelte, N., Severengiz, S., Yamada, T.: 2-stage design of e-moped-sharing service for accessibility, greenhouse gas emissions, and cost through station and supplier selections. Energies **18**, 1644 (2025)
6. He, S.Y., Kuo, Y.H., Sun, K.K.: The spatial planning of public electric vehicle charging infrastructure in a high-density city using a contextualised location-allocation model. Transp. Res. Part A: Policy Pract. **160**, 21–44 (2022)

7. Kotegawa, M., Kinoshita, Y., Yamada, T.: Global supply chain network design under cap-and-trade system and analysis of carbon price and emission allowance. Green Low-Carbon Econ. **3**(2), 159–170 (2025)
8. Csiszár, C., Csonka, B., Földes, D., Wirth, E., Lovas, T.: Urban public charging station locating method for electric vehicles based on land use approach. J. Transp. Geogr. **74**, 173–180 (2019)
9. Vazifeh, M.M., Zhanga, H., Santia, P., Ratti, C.: Optimizing the deployment of electric vehicle charging stations using pervasive mobility data. Transp. Res. Part A: Policy Pract. **121**, 75–91 (2019)
10. Bai, X., Chin, K.S., Zhou, Z.: A bi-objective model for location planning of electric vehicle charging stations with GPS trajectory data. Comput. Industr. Eng. **128**, 591–604 (2019)
11. City Population. https://www.citypopulation.de/. Accessed 21 May 2025
12. Daskin, M.S.: What you should know about location modeling. Naval Res. Logist. **55**, 283–294 (2008)
13. OpenStreetMap ®. https://www.openstreetmap.org/copyright/en. Accessed 21 May 2025

Digitalization

Case Study Analyses of Benefits and Barriers of Circular Manufacturing Systems: A Step Towards a Simulation-Based Approach

Jorge Francisco Cabello Oqueña(✉), Shun Yang, and Sebastian Thiede

University of Twente, Drienerlolaan 5, 7522 NB Enschede, The Netherlands
j.f.cabello@utwente.nl

Abstract. The industry's circularity transition is a critical but challenging issue in Circular Manufacturing Systems (CMS) mainly due to increased process complexity and high uncertainties, particularly but not limited to the quantity and quality of returned products. This study applies the PRISMA framework to identify the most significant drivers and barriers on industry-based case studies. A systematic review of case studies indicates that while circular practices increase process complexity, they also offer other benefits, such as cost savings. To address these challenges, this research proposes a simulation-based analysis focused on uncertainty management. The outcomes offer practical recommendations for companies to address these challenges through structured steps in a discrete simulation environment.

Keywords: Sustainability · Circular Economy · Circular Manufacturing Systems · Uncertainty Management · Simulation-Based Analysis

1 Introduction

Circular Manufacturing Systems (CMS) are manufacturing systems that promote companies to shift towards a Circular Economy (CE). In this case, companies are increasingly exploring sustainability programs, particularly with the regulations that were started by the European Commission [1]. CMS as seen in Fig. 1, attempts to integrate both manufacturing and Re-X processes (including assembly and disassembly) in such a way that the closed-loop system retains product value. Although this integration looks straightforward in theory, practical implementation is thwarted by various limitations that hinder extended usage, like the complexity of reverse logistics, consumer sentiment, and inadequate capacity, among others.

Although related research, such as [2], introduces the concept of Circulation Factories—a system-oriented approach designed to enable hybrid production environments where processes are balanced between physical and digital domains—it relies on data analysis and feedback loops for optimization retrofitting, and most existing work remains narrowly focused. Traditionally, research has been assembly- or disassembly-focused, rarely addressing their empirical integration. Assembly, as a result of the linear economy,

H. Kohl et al. (Eds.): GCSM 2025, LNME, pp. 261–269, 2026.
https://doi.org/10.1007/978-3-032-21154-5_29

is already optimized as far as possible [3]. Disassembly can be optimized independently, but when combined with assembly, uncertainty still exists [4]. This resembles the Texas sharpshooter fallacy that considering assembly or disassembly alone but not their net impact on CMS, certain optimization methods or models look like resolving the problem but are actually not viable in real-world production.

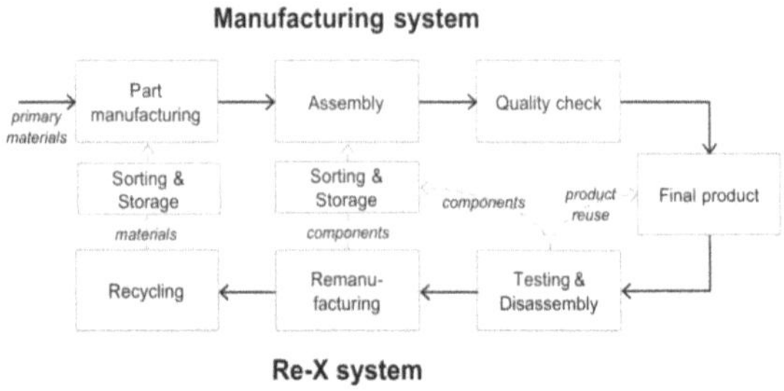

Fig. 1. Functional view on circular manufacturing systems (CMS) for learning factories [5].

The proposed research seeks to reduce this gap, first by systematically reviewing empirical-based case studies to identify interactions among encountered benefits and barriers. Then, a conceptual simulation-based approach is proposed to address the identified constraints by managing uncertainties and considering their impact on economic, environmental, and social aspects. This approach must be adaptable across industries, processes, and products.

2 State of Art

The first critical question this research addressed is whether disassembly can even be integrated into manufacturing systems. The real challenge lies in determining how to achieve this integration in a way that is both efficient and practically viable for real-world applications. To answer this, it is necessary to examine previous research addressing similar gaps, although with a more theoretical approach.

While the literature on CE concepts and associated technologies such as disassembly, remanufacturing, and recycling is extensive, the integration of CMS has received less attention. For instance, [6] provides an overview of green manufacturing and product recovery, addressing issues beyond disassembly, such as product design, closed-loop supply chain, and remanufacturing. For the situation in disassembly, sequencing and scheduling have been widely studied. However, these are only one aspect of the overall issue. For example, reverse logistics planning, where inflows are extremely variable, can lead to defining a system as capable, yet not necessarily doing well on received products under real conditions. Additionally, most industry-based case studies included in this research are focused on specific, context-dependent issues and not on presenting generalized frameworks applicable across industries. Different perspectives can be found in [7] and [8], emphasizing design for disassembly. Though their study indicates that this strategy improves disassembly flow performance, it remains challenging to most industries, particularly the automotive, electronics, and battery sectors.

Although earlier theoretical research is extensive, it is not universal across industries and does not address the overall structural problems. Certain empirical works, such as [9] and [10], examine specific aspects such as scheduling via optimization algorithms—fruit fly optimization and knowledge-driven artificial bee colony algorithms, respectively. While these approaches provide solutions to specific industrial problems, they fail to address further disadvantages due to the narrow nature of their strategies.

To bridge this gap, the methodology investigates empirical case studies in which companies develop CMS, aiming to identify the strengths and weaknesses of their strategies. The collected information then supports the development of a simulation-based solution designed to address these weaknesses in manufacturing systems.

3 Methodology

There was a systematic review of the literature based on the PRISMA approach, as shown by Fig. 2. The process entailed three phases: identification, screening, and inclusion. [11] posit that PRISMA methodology provides a logical framework; however, in the case of the current research, a further activity—highlighted in red—was included on the inclusion phase. It involved the inclusion of 15 company website case studies. These cases were not mentioned in research articles, because companies do not publish long detailed content due to competitiveness. Nevertheless, the information found on their websites provided satisfactory estimates of their circularity processes [12].

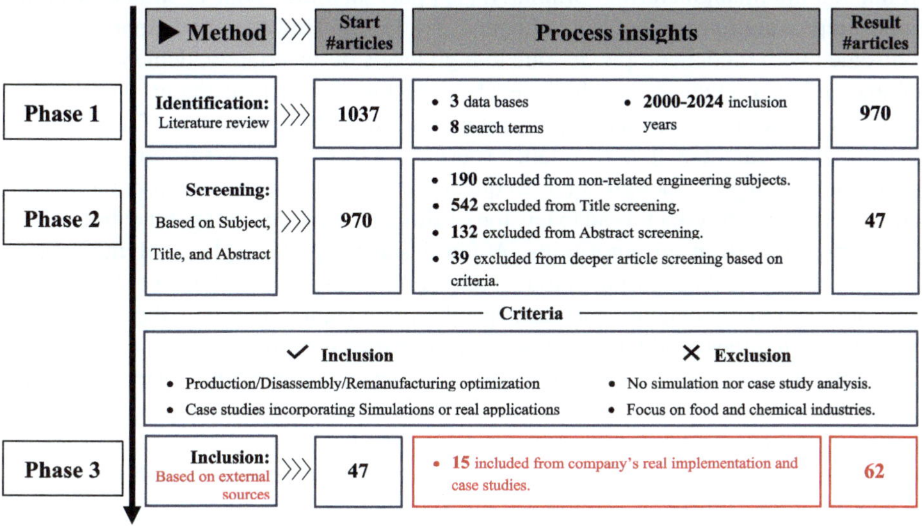

Fig. 2. Adapted PRISMA flow diagram—additions highlighted in red.

The outcome of the methodology, which consists of 62 case studies [13], will work as the basis for establishing the interconnections among agents in CMS. This foundation justifies the inclusion criteria based on practical application or simulations and will later support the simulation-based approach by defining different levels (i.e., system, process, and product), uncertainty of variables (e.g., product condition and quantity), and CE impact dimensions (i.e., environmental, economic, and social) to systematically identify effects, allowing an efficient optimization transition towards circularity.

4 Literature Analysis

An overall classification of the selected case studies recognizes three general categories in accordance with their strategy and emphasis: (1) pure disassembly, (2) production modeling, and (3) optimization algorithms.

The first category includes cases where disassembly is realized as an isolated activity for CMS integration. For example, [12] depicts a remanufacturing program with both material flows within the same manufacturing and remanufacturing system, incorporating reverse logistics, disassembly operations, and a closed-loop system [14]. Similarly, [15] presents a remanufacturing program with a divided product flow after quality inspection, dividing frames and individual parts. This method focuses on disassembly and prior operations as significant decision nodes, through which, in turn, capacity-based decision-making can be facilitated.

The second theme, production modeling, seems to be a response to the integration of case studies with simulation-based methods. These investigations apply single- or multi-criteria simulations as decision tools for disassembly operations optimization. For instance, [16] proposes an effective approach to manage uncontrollable incoming production in remanufacturing. Similarly, [17] puts forward a control strategy with the aim of improving remanufacturing efficiency by 15.3%. However, both approaches are confronted with model complexity due to high levels of uncertainty, demands for large volumes of data analysis, and high levels of technological investment. The latter is particularly relevant, since the implementation of new tracking devices for measuring variables and guaranteeing performance is usually a significant cost factor.

The third category, though not as frequent, includes case studies incorporating optimization techniques, either as an extension of simulation models or as standalone strategies leveraging advanced algorithms. A few instances include [18], solving specific problems like cycle time minimization for remanufacturing, and [19]. However, they usually entail extremely detailed and case-specific optimization, and thus their immediate applicability to other products, processes, or industries is narrow.

Figure 3(a) shows the most frequent benefits and constraints identified in the case studies. These results reveal a strong tendency to adopt strategies that locally optimize production [19], along with context-suitable solutions that improve assembly/disassembly processes [20] or shorten time frames [21]. Even though the benefits display an improvement in economics, environment, and operations—cost efficiency and environmental impact—these strategies add extra layers of complexity to processes (Fig. 3(b)), limiting their scalability and further application in diverse contexts while keeping them locally flexible within their own product range.

Figures 3(c) and 3(d) show that cost efficiency, process complexity, and improved efficiency are the most consistently high-impact factors across the case studies. Moreover, by looking at them in Fig. 3(b), benefits are relevantly present. This is due to the local focus on improved efficiency, which requires additional resources and contributes to capacity limitations [22]. As a result, although these cases are highly cost-efficient, they also involve high complexity. When this complexity combined with capacity limitations such as limited data and parts, it can pull back economic viability due to the high investment required.

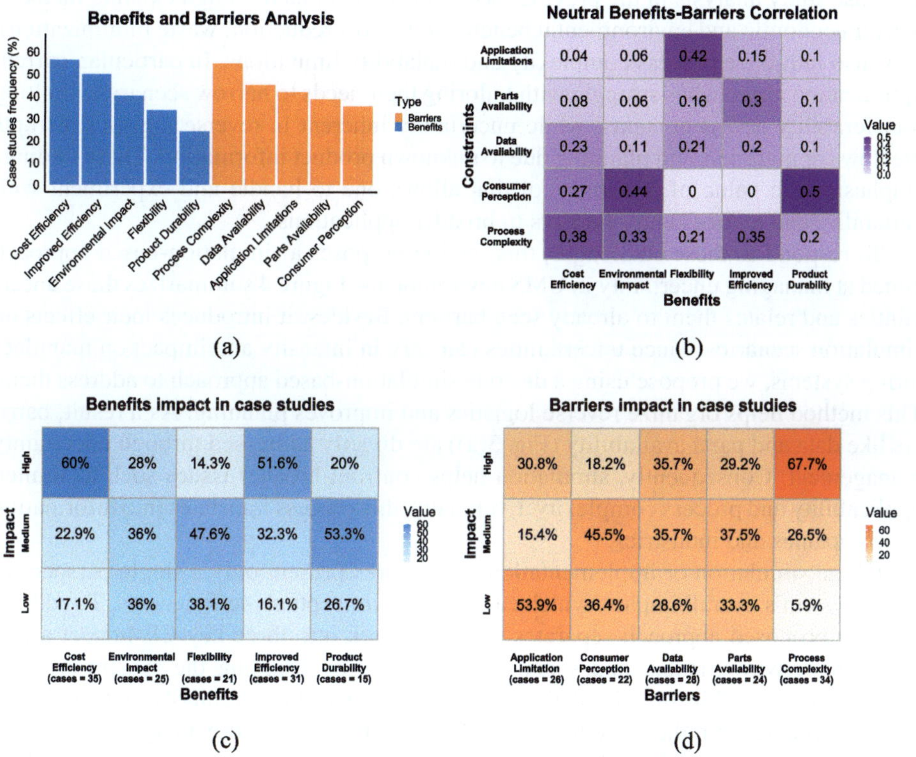

Fig. 3. (a) Benefits and barriers frequency in case studies. (b) Neutral benefits-barriers correlation in case studies. (c) Benefits impact in case studies. (d) Barriers impact in case studies.

This lack of viability stems not only from insufficient parts, data, and capacity but also from consumer perception. This accounts for almost 35% of cases, as CE products often provide no added value beyond appealing to conscious consumers and offering minor price reductions. As pictured in Fig. 3(b), the main benefits lie in environmental impact and extended product durability. Besides, [23] highlights that consumers are even more reluctant when dealing with high-value remanufactured items, such as car engines. Despite its sociological dimension, consumer perception should also play a key role for CMS when evaluating the feasibility of circularity at different production stages.

These findings indicate that companies have a tendency to invest in solutions that will give an immediate return on investment. The key challenge, however, is not whether circularity should be pursued but how to determine its appropriate scope for certain processes or products and how to effectively put it into practice. This refers to addressing identified barriers to allow broader industry applications.

5 Uncertainty Management Using Simulation-Based Approach

The case study analysis in the previous section revealed that although existing methods deliver economic and environmental benefits (e.g., cost reduction, waste minimization), they also introduce process complexity and scalability limitations. In particular, current optimization approaches struggle with tailoring their needs to narrow scenarios, limiting transferability across contexts, while uncertainty inherent in reverse logistics disrupts the flows of materials and planning due to unknown product information. These findings emphasize the value of a visual tool that allows one to handle and experiment with variability and reduces entry barriers to broader applications.

To respond to these challenges, this section proposes a simulation-based approach aimed at managing uncertainty in CMS environments. Figure 4 summarizes these uncertainties and relates them to already seen barriers. Besides, it introduces their effects on simulation scenarios. Since uncertainties can vary in intensity and impact on manufacturing systems, we propose using a discrete simulation-based approach to address them. This method helps organize reverse logistics and improves planning. As a result, barriers like data and parts availability (Fig. 3(a)) are directly addressed through uncertainty management. Consequently, simulation helps confront broader issues such as limited applicability and process complexity. Eventually, this process translates into information for companies and industries.

Often, simulation or implementation frameworks present only a single perspective of the system's overall impact, which can result in conceptual inefficiencies. To address this, the proposed approach separates the framework into three key CE impact areas: economic, environmental, and social. For instance, by adapting the system-process-product layers within the simulation, one can evaluate the circularity potential of a product/process from the three impact area angles. However, these analyses should not be confused into a single holistic parameter, since independent calculations like Life Cycle Costs (LCC) or Life Cycle Assessment (LCA) already provide that information.

After establishing the conceptual framework for simulation, its applicability appears in Fig. 4, where it is tailored for discrete engineering simulation software such as Visual Components, FlexSim, etc. Starting from the product layer, it requires a forecasted range of percentages of uncertainties in returned products, allowing capacity adjustments to meet incoming demand. In cases where uncertainty data is unavailable—highlighting the need for improved reverse logistics and data management—hypotheses can serve as a basis for simulations. To operationalize uncertainty, parameters such as the quantity, condition, and processing time of returned products are modeled using normal distributions, although the choice of distribution can be customized based on the system context or data availability. Therefore, the expected outcome should include not only the layout arrangement but also an estimation of the system's performance through indicators such as makespan, utilization, and throughput, ultimately leading to a reduction of constraints listed in Fig. 3(a), such as process complexity and application limitations, and increasing adaptability for broader industrial applications. This approach ranks different scenarios

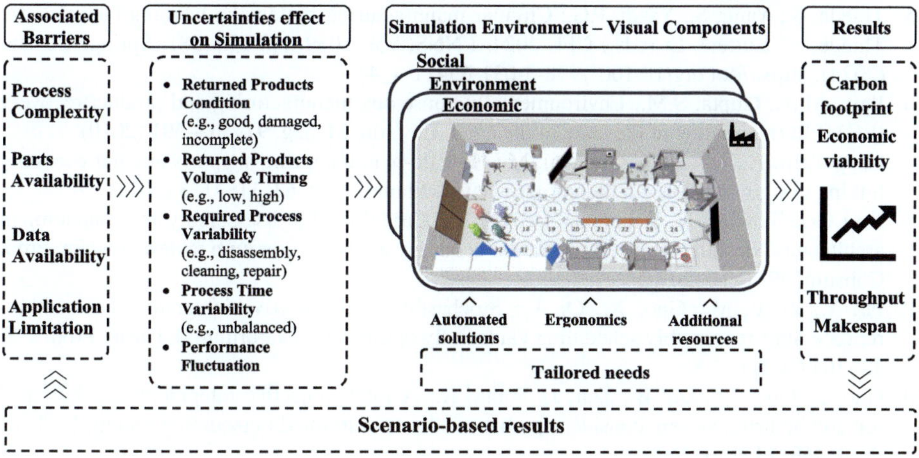

Fig. 4. Addressed barriers in an uncertainty-based simulation environment.

based on KPIs and validates them through bootstrap resampling and cross-validation techniques to ensure reliability across scenario subsets.

6 Conclusion

This research analyzes the benefits and barriers of implementing circular manufacturing systems based on case studies. By linking these barriers to underlying uncertainties, a discrete event simulation approach has been proposed to manage the potential uncertainties. Furthermore, the simulation analysis concept is designed considering the analysis of the CE impact dimensions separately only after formulating the entire model.

One of the pitfalls of these industry-oriented case studies is the Fallacy of Composition [24], that isolated improvements (e.g., optimizing the disassembly or material recovery sequence) will necessarily improve the system as a whole. Lacking an overall perspective, such optimizations may introduce inefficiencies elsewhere. By addressing uncertainty management at different levels and assessing interdependencies across the entire system, the proposed analysis seeks a more industry-oriented applicable model for the development of circular manufacturing.

References

1. European Commission: European green deal – Delivering on our targets. Publications Office of the European Union, Luxembourg (2021)
2. Cerdas, F., et al.: Defining circulation factories – a pathway towards factories of the future. Procedia CIRP **29**, 627–632 (2015)
3. Linder, M., Sarasini, S., van Loon, P.: A metric for quantifying product-level circularity. J. Ind. Ecol. **21**, 545–558 (2017)
4. Cioffi, R., Travaglioni, M., Piscitelli, G., Petrillo, A., Parmentola, A.: Smart manufacturing systems and applied industrial technologies for a sustainable industry: a systematic literature review. Appl. Sci. **10**(8), 2897 (2020)

5. Thiede, S., Yang, S., Yazdi, P.G.: Circular manufacturing systems in learning factories. In: Thiede, S., Lutters, E. (eds.) CFL 2024. LNNS, vol. 1059, pp. 374–381. Springer, Cham (2024). https://doi.org/10.1007/978-3-031-65411-4_44
6. Ilgin, M.A., Gupta, S.M.: Environmentally conscious manufacturing and product recovery (ECMPRO): a review of the state of the art. J. Environ. Manag. **91**, 563–591 (2010)
7. Cappelletti, F., Rossi, M., Germani, M.: How de-manufacturing supports circular economy linking design and EoL – a literature review. J. Manuf. Syst. **63**, 118–133 (2022)
8. Kręt-Grześkowiak, A., Baborska-Narożny, M.: Guidelines for disassembly and adaptation in architectural design compared to circular economy goals – a literature review. Sustain. Prod. Consum. **39**, 1–12 (2023)
9. Fu, Y., Zhou, M., Guo, X., Qi, L.: Stochastic multi-objective integrated disassembly-reprocessing-reassembly scheduling via fruit fly optimization algorithm. J. Clean. Prod. **278**, 123364 (2021)
10. Guo, J., Guo, W., Du, B., Zou, J., Wang, K.: A multi-objective integrated scheduling of remanufacturing system considering time window constrained outsourcing option. J. Clean. Prod. **468**, 142916 (2024)
11. Moher, D., Liberati, A., Tetzlaff, J., Altman, D.G.: Preferred reporting items for systematic reviews and meta-analyses: the PRISMA statement. BMJ **339**, b2535 (2009)
12. REMAN JD: Remanufactured Engines, Electronics, and Parts. https://www.deere.com/en/parts-and-service/parts/remanufactured/. Accessed 31 Mar 2025
13. Cabello Oqueña, J.F.: Data underlying the publication: case study analyses of benefits and barriers of circular manufacturing systems: a step towards a simulation-based approach. 4TU.ResearchData. Dataset. https://doi.org/10.4121/92cc0b7b-d709-421c-8eb9-06d661cff5b9. Accessed 23 June 2025
14. Geissdoerfer, M., Savaget, P., Bocken, N.M.P., Hultink, E.J.: The circular economy – a new sustainability paradigm? J. Clean. Prod. **143**, 757–768 (2017)
15. King, A., Barker, S., Cosgrove, A.: Remanufacturing at Xerox: evaluating the process to establish principles for better design (2007)
16. Golinska-Dawson, P., Pawlewski, P.: Multimodal approach for modelling of the materials flow in remanufacturing process. IFAC-PapersOnLine **48**(3), 2133–2138 (2015)
17. Liu, C., et al.: An integrated optimization control method for remanufacturing assembly system. J. Clean. Prod. **248**, 119261 (2020)
18. Su, C., Shi, Y., Dou, J.: Multi-objective optimization of buffer allocation for remanufacturing system based on TS-NSGAII hybrid algorithm. J. Clean. Prod. **166**, 756–770 (2017)
19. Gal, B., Zeh, C., Szaller, Á.: How grey wolf optimization affects remanufacturing. Procedia CIRP **126**, 927–932 (2024)
20. Romero-Silva, R., Shaaban, S., Marsillac, E., Laarraf, Z.: The impact of unequal processing time variability on reliable and unreliable merging line performance. Int. J. Prod. Econ. **235**, 108108 (2021)
21. Sarkar, M., Park, K.S.: Reduction of makespan through flexible production and remanufacturing to maintain the multi-stage automated complex production system. Comput. Ind. Eng. **177**, 108993 (2023)
22. Schimanek, R., Bilge, P., Dietrich, F.: Automated remanufacturing of lithium-ion batteries with shear separation, ultrasonic separation, and ultrasonic welding. Procedia CIRP **116**, 227–232 (2023)
23. Cummins Inc.: Cummins ReCon. https://www.cummins.com/parts/recon. Accessed 31 Mar 2025
24. Finocchiaro, M.A.: The fallacy of composition: guiding concepts, historical cases, and research problems. J. Appl. Log. **13**, 24–43 (2015)

Digitization in Serial Timber Construction Manufacturing: A Model-Based Approach for Sustainable and Efficient Production

Nicole Oertwig[1(✉)], Valentin Eingartner[2], Patrick Gering[1], and Holger Kohl[1,2]

[1] Fraunhofer Institute for Production Systems and Design Technology IPK, Pascalstraße 8-9, 10587 Berlin, Germany
nicole.oertwig@ipk.fraunhofer.de
[2] Technische Universität Berlin, Pascalstraße 8-9, 10587 Berlin, Germany

Abstract. Progressing climate change and growing resource scarcity call for innovative and sustainable manufacturing concepts in serial timber construction. A central challenge is the seamless integration of design and material data into production to enhance material efficiency, reduce waste, and ensure environmentally responsible manufacturing. Integrated modelling techniques enable the incorporation of design and logistics information into production, supporting optimized material use, minimal offcuts, and the targeted selection of sustainable resources. Real-time data and intelligent information and communication technologies (ICT) further improve process control, lower energy consumption, and prevent rework. Achieving these efficiency and sustainability gains requires comprehensive digitalization across the value chain. This paper presents a systematic, model-based framework for assessing the digitalization level in serial timber manufacturing, identifying existing gaps, and providing targeted recommendations to support a data-driven, resource-efficient, and low-emission production process.

Keywords: Serial Timber Construction · Digitalization Assessment · Reference Model

1 Introduction

The construction sector is under increasing pressure to reduce emissions, conserve resources, and adopt more sustainable practices [1]. In this context, serial timber construction has gained attention as a viable alternative to conventional building methods due to its potential for prefabrication, climate performance, and circularity [2]. However, fully realizing this potential requires the systematic digitalization of design, production, and logistics processes [3]. Existing digitalization approaches, ranging from BIM-based design coordination [4, 5] to process automation [6, 7], remain fragmented and lack semantic alignment between design and production. The proposed framework addresses these gaps by integrating process-oriented enterprise modelling, IFC-based semantic mapping, and sustainability metrics into a unified approach. Such integration enables consistent, traceable, and transparent workflows across the value chain, distinguishing it from existing methods.

H. Kohl et al. (Eds.): GCSM 2025, LNME, pp. 270–277, 2026.
https://doi.org/10.1007/978-3-032-21154-5_30

Here, an integrated modelling framework that links process-oriented enterprise modelling with standardized building data structures is presented. The core is a domain-specific reference model [8] developed by using the Integrated Enterprise Modelling (IEM) methodology in the MO2GO modelling environment [9]. It provides a process-based representation of serial timber construction workflows, extended through semantic mapping to the IFC 4.3.2.0 standard [10] to capture building elements, material attributes, and process logic in a consistent digital form. The framework aims to enable open, vendor-neutral BIM workflows across the built environment, allowing stakeholders to exchange, visualize, and analyze data seamlessly throughout the asset life cycle. Additionally, a Process Assistant (PA) is presented that evaluates the digital maturity of individual process steps. It identifies gaps in digital integration and provides recommendations to enhance data continuity and quality across the production system. The framework is based on three core components: the reference model, which structures serial timber construction workflows, the IFC standard, which ensures semantic interoperability, and the Process Assistant, which assesses digital maturity and identifies integration gaps. Together, these components form the framework's operational structure, organized into four interconnected layers: the Process Layer, Interoperability Layer, Sustainability Layer, and Assessment Layer. These layers represent distinct yet interrelated dimensions, enabling a comprehensive approach to digitalization, sustainability, and process optimization. The framework's modular architecture ensures adaptability and scalability across diverse applications. The paper is structured as follows: Sect. 2 describes the operational structure, Sect. 3 introduces the demonstrator implementation, and Sect. 4 concludes with a discussion and outlook on future research.

2 Integration Framework for Sustainable and Efficient Production

To enhance sustainability, efficiency, and information flow in serial timber construction, we propose a multi-layer integration framework comprising four interconnected layers. The Process Layer formalizes workflows, roles, and data flows across planning, production, and logistics phases using the IEM, while identifying critical digital touchpoints where structured data is generated, transformed, or utilized. The Interoperability Layer ensures semantic consistency by linking process elements to IFC 4.3.2.0 class attributes, enabling traceable and reliable data exchange among stakeholders. The Sustainability Layer integrates lifecycle and environmental data, facilitating the assessment of material efficiency, emissions, and other sustainability metrics. These three layers provide the foundation for the Assessment Layer, which implements a structured self-assessment to evaluate digital maturity in terms of data availability, automation, and quality, thereby highlighting gaps and guiding targeted improvements. Collectively, the layers establish a coherent framework that supports data-driven decision-making, optimizes resource efficiency, and advances sustainability across the serial timber construction value chain.

2.1 Process Layer

The Process Layer provides a structured reference model of the value chain in serial timber construction, encompassing the key phases of planning, production, and logistics.

Each phase is detailed into sub-processes, such as CNC preparation, material assignment, assembly, and quality inspection, modeled using the IEM to capture relationships between activities, roles, and resources. Each sub-process is linked to specific roles (e.g., production staff, quality controllers, logistics coordinators) and supporting IT tools such as CNC software or ERP systems. Particular emphasis is placed on identifying digital touchpoints where structured data is created, transformed, or used. These touchpoints ensure seamless data integration across the value chain and connect the Process Layer with other layers of the framework. For example, data from CNC preparation is mapped to IFC entities in the Interoperability Layer, while material usage data informs the Sustainability Layer. In this way, the Process Layer forms the foundation for assessing digital maturity and optimizing workflows.

2.2 Interoperability Layer

The process elements modeled in MO2GO were subsequently linked to corresponding entities from the IFC 4.3.2.0 standard. The mapping was implemented by representing the IFC schema through the class structure within MO2GO, where IFC classes were modeled and assigned as parts of the process elements (Table 1). While the role of BIM as an enabler for integration has been introduced in [11], the table introduces the specific implementation. It covers the integration of IFC 4.3.2.0 class attributes with the IEM-based process model to ensure semantic alignment and traceability across the value chain.

2.3 Sustainability Layer

Sustainability is a critical focus in timber-based prefabrication due to timber's renewable nature and low embodied energy. This framework integrates sustainability data within the IEM–IFC modelling approach to enable comprehensive environmental assessment and decision-making across the value chain. Qualitative attributes, such as material sourcing policies, are translated into model-relevant data by mapping them to corresponding IFC property sets (e.g., *Pset_MaterialEnvironmentalProperties*) or metadata fields. These attributes are then linked to IEM objects, ensuring traceability and alignment with sustainability goals (Table 2).

2.4 Assessment Layer

The framework incorporates a digital maturity assessment designed to evaluate the availability and quality of data for each process element and its corresponding IFC entities. The assessment is structured around three core criteria: information availability, which examines the existence of digital data, automation level, which determines whether data acquisition is system-based or manual, and data quality, assessed through the sub-criteria of completeness, consistency, and accuracy. Applied within the specific process context, the self-assessment ensures that domain-specific knowledge is adequately considered. Each process element is subsequently assigned a digital maturity level that reflects its readiness for digital integration. Identified deficiencies, such as incomplete or inconsistent data, are addressed through targeted interventions, including the automation of data collection and the implementation of validation mechanisms.

Table 1. Mapping Structure for IEM Object with IFC Entities.

IEM Object	Corresponding IFC Entities	Explanation
Order	*IfcControl* (for controlling/informational objects)	Orders are modeled as controlling or informational objects influencing processes, e.g., instructions, approvals, or regulatory documents using *IfcControl*
Resource	*IfcConstructionResource, IfcEquipment, IfcPerson*	Resources include machines, tools, personnel, and optionally transport means. These are modeled via *IfcConstructionResource* or specialized subclasses
Action	*IfcTask, IfcProcess*	Process steps or actions are represented by *IfcTask* or *IfcProcess* *IfcTask* especially models defined work packages or activities
Product	*IfcProduct (e.g., IfcWall, IfcBeam, IfcBuilding), IfcMaterialLayerSet, IfcMaterial*	The model covers all physical objects along the value chain. Raw materials, such as tree logs, are represented as products using *IfcMaterial*. Semi-finished and finished products, including wall or ceiling elements, are modeled as instances of *IfcProduct* (e.g., *IfcWall, IfcBeam*) and linked to material definitions through *IfcRelAssociatesMaterial* and *IfcMaterialLayerSet*. The final building is represented by *IfcBuilding* or *IfcBuildingElementProxy*, incorporating a hierarchical structure that includes floors and rooms

Table 2. Assignment of IEM Objects and Sustainability Attributes.

IEM Object	Description	Examples of Sustainability Attributes
Order	Models sustainability rules, certifications, and guidelines as *IfcControl* objects	• Material sourcing policies • Emission limits • Certification requirements (e.g., FSC) • Attributes such as Purpose, Status, and Identification for traceability

(*continued*)

Table 2. *(continued)*

IEM Object	Description	Examples of Sustainability Attributes
Resource	Represents resources as *IfcConstructionResource, IfcEquipment, IfcLabor*	• Energy consumption (kWh/use) • Emissions (GHG, dust, noise) • Global Warming Potential • Fuel type and consumption • Worker qualifications • Lifecycle data (service, maintenance)
Action	Maps process steps as *IfcTask* or *IfcProcess* with sustainability indicators	• Environmental classification (e.g., low-emission) • Waste generation potential (via *IfcControl*) • Process duration and efficiency
Product	Models physical components (raw materials to finished elements) as IEM Products linked to e.g. *IfcMaterial* and *IfcProduct*	• Embodied carbon (kg CO_2-eq/unit) • Renewable content (e.g., FSC) • Recyclability • Origin & supply chain transparency • Transport emissions

3 Implementation for Prefabrication of Timber Frame Walls

In the context of the proposed multi-layer integration framework, the prefabrication of interior timber frame wall elements serves as a representative use case to demonstrate the digital linkage of process, building data, sustainability, and assessment layers.

3.1 Use Case Scenario

The core prefabrication workflow, including CNC preparation, material assignment, component assembly, and quality inspection, is modeled using IEM. Each process step is defined as an action and mapped to corresponding IFC entities such as *IfcTask* or *IfcProcess*. Within the IEM framework, IFC entities are embedded directly into the class structure: classes like *IfcWall*, *IfcMaterial*, *IfcTask*, and *IfcConstructionResource* are represented as subclasses and associated objects within the process model. This ensures seamless semantic integration, allowing building and process data to coexist within a unified, domain-specific model. Roles (e.g., production staff, quality controllers) and resources (e.g., CNC machines, tools, personnel) are modeled as *IfcConstructionResource* or specialized subclasses such as *IfcEquipment* and *IfcLaborResource*. Inputs and outputs, including materials and semi-finished products, are linked through the IEM part structure, ensuring traceability and consistent data flows without relying on explicit IFC relationship classes. The semantic linkage between process and building data is achieved by integrating IFC entities as core elements of the IEM model. For instance, prefabricated wall elements are represented as *IfcWall* or *IfcBuildingElementProxy* and linked via the part structure to their associated materials and processes.

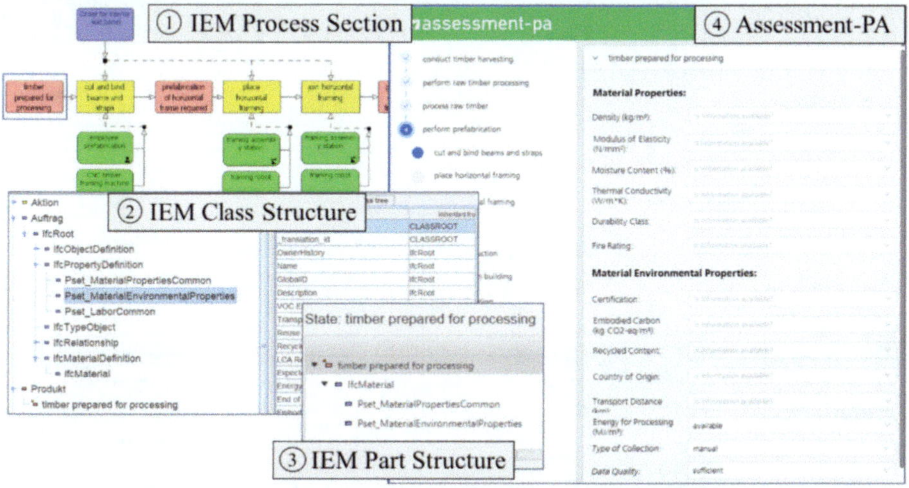

Fig. 1. Major Elements of Demonstrator for IEM Product "timber prepared for processing".

As shown in Fig. 1, input materials, particularly timber components, are represented as *IfcMaterial* entities enriched with property sets *Pset_MaterialPropertiesCommon* and *Pset_MaterialEnvironmentalProperties*. These integrate key physical (e.g., density, thermal conductivity) and environmental attributes (e.g., embodied carbon, recyclability, certification status) into the IEM model, ensuring traceability and alignment with sustainability objectives across the value chain. Sustainability indicators at the process level, such as waste generation and emissions, are linked to *IfcTask* or *IfcProcess* entities, while resource-related data, including energy consumption and worker qualifications, are modeled via subclasses of *IfcConstructionResource* within the same structure. With the PA, the user evaluates each process step in the IEM model using digital maturity criteria, data availability, automation level, and data quality. The evaluation allows experts to judge data completeness, consistency, and accuracy. As illustrated in Fig. 2, the results are visualized to highlight improvement areas, such as limited automation or data quality issues. Beyond the example of interior timber wall prefabrication, the integrated modelling approach is applicable throughout the entire serial timber construction value chain. By combining process modelling, IFC-based semantic mapping, and sustainability data, it enables the automated generation of process instructions and resource allocations by directly linking building information (e.g., material properties, dimensions) to production workflows. This reduces manual effort and errors during, for example, CNC preparation. The framework also enhances interoperability across design and supply chain partners through standardized IFC data structures, supporting consistent data exchange and minimizing miscommunication. Furthermore, process data can be mapped to MES and ERP systems, while automated validation tools help resolve data inconsistencies. Collectively, these features enable targeted digitalization initiatives, allowing companies to systematically optimize material efficiency, environmental performance, and digital maturity.

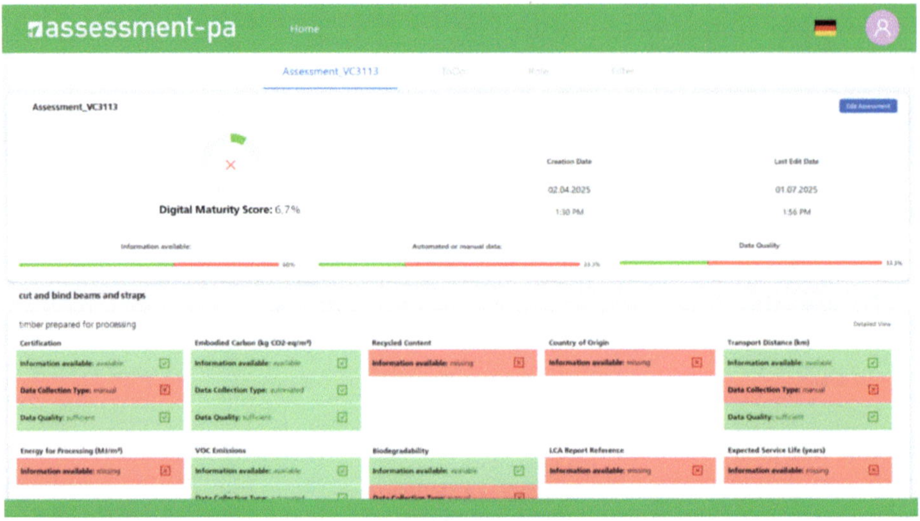

Fig. 2. Assessment-View for "cut and bind beams and straps".

4 Conclusion and Outlook

This paper presents a modular integration framework that combines process-oriented enterprise modelling with standardized building information representations to address fragmented data flows and insufficient digital integration in serial timber construction. Unlike existing approaches that focus solely on process modelling or BIM-based data exchange, the framework integrates process models, IFC-based semantic structures, and sustainability data within a unified environment. This enables consistent, traceable, and transparent representation of production workflows and building elements. The demonstrator for timber frame wall prefabrication illustrates how the framework can be operationalized to assess digital maturity across process steps, highlighting its potential to support a systematic transition toward data-driven and resource-efficient production. By enabling evaluation and visualization of digitalization levels, integrating environmental performance indicators, and embedding IFC-based structures directly into enterprise models, the framework fosters optimization and collaboration across the construction value chain. Future research will extend the approach to additional prefabrication processes and building systems, incorporate dynamic lifecycle assessment, and develop interfaces to upstream planning and downstream logistics. Validation in industrial settings and quantification of impacts on productivity, resource efficiency, and data quality will be essential to demonstrate practical benefits. Overall, the framework provides a foundation for more transparent, interoperable, and sustainable timber construction workflows, promoting broader adoption of digital methods in the construction industry.

Acknowledgements. We thank the Federal Ministry of Food, Agriculture and Consumer Protection, Berlin, Germany, as well as the Federal Agency of Renewable Resources (FNR) for funding of the research project DiKieHo (Funding Code 2221HV082A-B).

References

1. 2023 Global Status Report for Buildings and Construction: Beyond foundations - Mainstreaming sustainable solutions to cut emissions from the buildings sector. United Nations Environment Programme (2024)
2. Moustafa, Z., Asif, M., Wuni, I.Y.: Circular economy in the building sector: a systematic review of environmental, economic, and social dimensions. Sustain. Futures **9**, 100690 (2025)
3. Lu, W., et al.: Digital technologies for construction sustainability: status quo, challenges, and future prospects. NPJ Mater. Sustain. **2**(1) (2024)
4. Sepasgozar, S., et al.: BIM and digital twin for developing convergence technologies as future of digital construction. Buildings **13**(2), 441 (2023)
5. Tan, Y.L., Mohamed, S., Ahmodu, A.A.: Usage of BIM for improvement on design coordination in construction industry. RMTB **5**(1):1165–1178 (2024)
6. Dindorf, R., Wos, P.: Challenges of robotic technology in sustainable construction practice. Sustainability **16**(13), 5500 (2024)
7. Shareef, M.: A review on automation and robotic technology in construction industry. IJSR **12**(1), 855–859 (2023)
8. Orozco, S., Oertwig, N., Feldhaus, A., Eingarnter, V., Mintchev, M., Kohl, H.: Planning and control of value creation networks in timber construction. In: Kohl, H., Seliger, G., Dietrich, F., Mur, S. (eds.) Sustainable Manufacturing as a Driver for Growth, pp. 437–444. Springer, Cham (2025)
9. Mertins, K., Jochem, R., Jäkel, F.-W.: A tool for object-oriented modelling and analysis of business processes. Comput. Ind. **33**(2–3), 345–356 (1997)
10. buildingSMART International. IFC 4.3.2.0 Documentation (IFC 4.3 ADD2) (2024). https://technical.buildingsmart.org/standards/ifc/ifc-schema-specifications/. Accessed 12 June 2025
11. Schramm, N., Oertwig, N., Kohl, H.: Conceptual approach for a digital value creation chain within the timber construction industry – potentials and requirements. In: Kohl, H., Seliger, G., Dietrich, F. (eds.) Manufacturing Driving Circular Economy, pp. 595–603. Springer, Cham (2023)

Digitalizing Circular Construction: Operationalizing Circularity Metrics Through BIM and BI–Integrated Workflows in Timber Prefabrication

Johannes Reinders[1(✉)], Rosa Pulido[1], Sebastian Orozco[1,2], Yaoning Yang[3], and Ignacio Borrego[1]

[1] Technical University of Berlin, 10623 Berlin, Germany
reinders@tu-berlin.de
[2] Costarican Institute of Technology, 10101 San José, Costa Rica
[3] Yunnan University, Kunming 650504, Yunnan, China

Abstract. The built environment significantly contributes to Greenhouse Gas emissions, resource consumption, and waste generation. Circular Construction (CC) and Timber Construction with Prefabricated Components (TCPC) are seen as key contributors to curb these environmental hazards, with digital design workflows as major enablers. However, the integration of effective assessment procedures for CC into digital design workflows remains insufficient. This study addresses the gap by proposing a prototypical workflow for real-time circularity assessment through the operationalization of selected metrics. Utilizing Building Information Modelling (BIM) and Business Intelligence (BI) technologies, two metrics from the German Sustainable Building Council (DGNB) Circularity Index (CI), *Material Origin* and *Recyclability*, were operationalized. The four-stage workflow, comprised of semantic BIM modelling, Life-Cycle-Assessment (LCA) data mapping, circularity metric computation, and BI visualization, was implemented on an exemplary component catalogue. The prototype demonstrates the feasibility of integrating circularity metrics into digital workflows, enabling real-time, component-level evaluation. Despite the limited scope, focusing on two metrics and a small component catalogue, the workflow exhibits considerable potential in replicability. Future developments should expand the metric library, improve automation, and enable broader applicability. Ultimately, the approach supports low-effort, continuous assessments of environmental trade-offs during design, contributing to data-driven sustainable planning.

Keywords: Circular Construction · Timber Prefabrication · Building Information Modeling (BIM) · Circularity Metrics · Sustainable Building Design

1 Introduction

The Built Environment (BE) plays a fundamental role in global warming and environmental impact, accounting for ca. 37% of global energy-related carbon emissions [1] and generating an estimated nearly one-third of the world's total solid waste [2].

H. Kohl et al. (Eds.): GCSM 2025, LNME, pp. 278–286, 2026.
https://doi.org/10.1007/978-3-032-21154-5_31

Circular Construction (CC) is a key component in achieving climate neutrality and resource conservation, as it helps reduce raw material consumption and minimize waste. This leads to greater material efficiency, potentially saving up to 80% of the greenhouse gas (GHG) emissions generated by the BE [3]. Additionally, Timber Construction with Prefabricated Components (TCPC) offers numerous benefits in this matter, from reducing on-site waste by up to 65% to improving deconstruction processes [4]. Recent studies highlight the outstanding properties of timber for modular and prefabricated construction, as well as the high potential of these technologies to provide practical solutions for designing constructions with a circular approach [5].

Digital technologies are emerging as essential enablers of circularity in the context of TCPC, where digital- and factory-based processes facilitate precise control over material inputs and augment disassembly and reuse potential [6]. However, insufficient standardized specifications or tools for assessing circularity in TCPC still represent a challenge [3, 7]. This research proposes a prototypical workflow that leverages Building Information Modelling (BIM) and Business Intelligence (BI), to operationalize circularity metrics for TCPC, aiming for a real-time evaluation of circularity and Life-Cycle-Assessment (LCA) as integrated parts of the digital design process.

2 Background and Related Work

To address the need for circular performance quantification, numerous indicators such as the Urban Mining Index (UMI) [8], Madaster Circularity Indicator (MCI) [9], and German Sustainable Building Council's (DGNB) Circularity Index (CI) [10] have emerged. These systems evaluate factors of a building's life cycle, which represent present and future contribution to circularity. Furthermore, with the aim of translating complex criteria into operational standards, a variety of sustainability certification systems have been developed, including DGNB, BREEAM (Building Research Establishment Environmental Assessment Method), LEED (Leadership in Energy and Environmental Design), and BNB (Assessment System for Sustainable Building). These systems have become widely adopted frameworks for assessing environmental performance in the built environment. However, their integration into digital design workflows remains limited, and the evaluation of circularity factors lacks harmonization [11].

Digital tools such as BIM and BI play a pivotal role in this context. BIM consolidates building data into 3D models known as digital twins, enabling a synchronized database, transparent material tracking, and life cycle assessments [3]. BI is being progressively used in the building sector to analyse, process, and visualize building data [12]. By means of interactive dashboards that facilitate real-time analysis and benchmarking, BI technologies complement BIM to support data-driven decision-making [13].

However, these technologies are still underutilized in supporting the integration of circularity metrics [7]. This highlights a broader absence of operational, standardized workflows, and reference implementations that can translate circularity indicators into actionable decision criteria within the digital design and construction processes. To address this gap, this research proposes a prototypical digital workflow that operationalizes two selected circularity metrics in the context of TCPC using BIM and BI technologies. By means of a prototypical implementation of the proposed workflow, this research seeks to answer the following questions:

1. How can circularity metrics be integrated into digital design workflows for TCPC?
2. Which data inputs and data flows could enable such integration?

3 Methodology

This study aims to operationalize two selected circularity metrics. Operationalization in this context refers to defining abstract concepts in measurable terms, with the aim to empirically observe and analyse these concepts [14], which was translated in this study to the transformation of abstract circularity indicators into quantifiable, computable variables integrated into a digital workflow.

Two metrics that are most representative for the pre-use and post-use life cycle phases were chosen based on the DGNB CI: Material Origin (pre-use) and Recyclability (post-use). These were selected due to their frequent occurrence in circularity measuring systems and their ability to reflect both input resource quality and end-of-life material potential [10].

To contextualize circularity performance, the CI metrics were assessed alongside Embedded Carbon (EC). LCA data was obtained from the ÖKOBAUDAT database [15] and linked to material quantities using the Carbo Life Calculator plug-in for Revit. This enables EC assessments within the same workflow presented hereafter.

3.1 Workflow Overview

The prototypical workflow consists of four replicable stages: (1) BIM-based component modelling and semantic enrichment with parameters that facilitate data categorization, (2) environmental data mapping to enrich BIM data via LCA tools, (3) enriched BIM and circularity data integration, (4) real-time assessments of EC and CI through dynamic data visualization for user interaction. For stage four, an interactive user interface was developed using Power BI, allowing component-level performance assessments by filtering design options and visualising real-time outputs on EC and CI (Fig. 1).

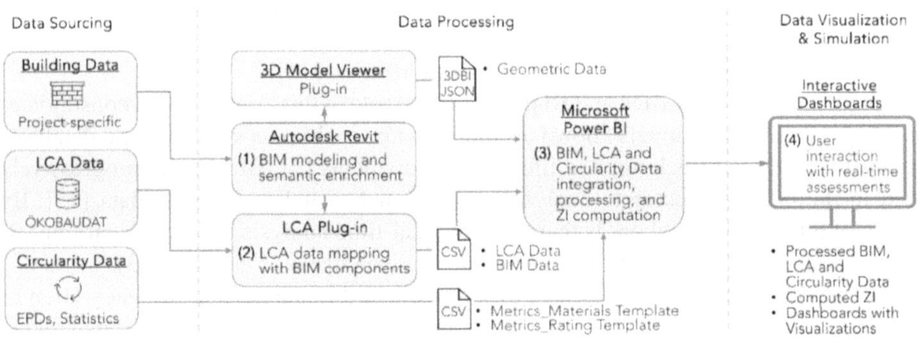

Fig. 1. Workflow overview with tools and data flow

3.2 Metric Operationalization

To compute component level circularity scores, two structured templates were developed. The first, a rating template, which is illustrated in Table 1, defines the *circularity classes* of the metrics *Material Origin* and *Recyclability* and assigns a numerical rating factor to each class based on the DGNB CI methodology [10]. The second, the material classification template, listed the materials and assigned them to their respective circularity classes for each metric. Together, these templates established a mapping from material type to circularity class to rating factor, which forms the basis for computing component level circularity.

Table 1. Snippet of rating template with a selection of circularity classes.

Material Origin		Recyclability	
Circularity Class	Rating Factor	Circularity Class	Rating Factor
Reused	1	Reuse	1
Closed-Loop-Recycled	0.75	Open-Loop-Recycling	0.6
Primary, renewable	0.5	Downcycling	0.4
Primary, non-renewable	0	Landfilling	0.2

Circularity computation was carried out in Power BI in three steps, as summarized in Table 2: total mass aggregation, classification-based mass percentage calculation, and weighted average using rating factors for each material class.

Table 2. CI Calculation procedure.

Step	Description	Inputs	Formula
1	Compute total mass of all elements in selected component	BIM Data, material masses	$M_t = \sum m_i$
2	Compute mass shares of the component's circularity classes	material masses, material classification template, Total mass	$P_i = m_{ci}/M_t$
3.a	Compute CI for Material Origin	Rating template: Material Origin	$CI_{Origin} = \sum P_i \times r_{i,origin}$
3.b	Compute CI for Recyclability	Rating template: Recyclability	$CI_{Recycling} = \sum P_i \times r_{i,recycling}$

m_i = mass of material i; m_{ci} = material mass of circularity class i; r_i = rating factor for circularity class i; P_i = mass share of circularity class; M_t = Total Mass of component

For the first step, the total material mass was computed in BI from imported BIM data. Materials were matched to circularity classes via the 'material type' attribute, which links BIM data to the material classification and rating templates downstream. The second step uses the material classification template to map each material to its circularity class. Here, the mass share of each class is calculated relative to the component's total mass. In the third step, the CI score for each metric was computed as a mass-weighted average of the circularity class ratings using the rating templates.

The calculation steps described were implemented using DAX (Data Analysis Expressions) language. The total mass per component and circularity class mass shares were computed using calculated columns and DAX measures. CI scores were calculated as dynamic outputs, allowing real-time updates based on user-defined filters or component selections.

3.3 Implementation Scope

To demonstrate the feasibility of the proposed workflow, a modular digital component catalogue was developed using Autodesk Revit. It is composed of four components, two walls and two floors, and three structural elements, two beams and one column. Together, these elements form a representative module equivalent to a typical room unit, as illustrated in Fig. 2.

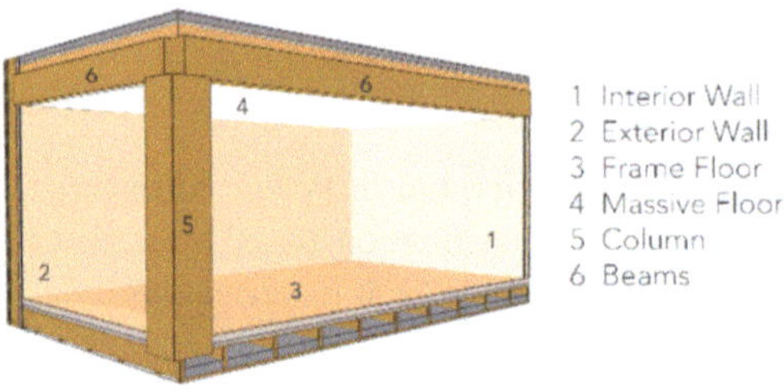

Fig. 2. Assembled components in 3D View

The layout for the floor panels was provided by a regional timber manufacturer and a regional architect. A suspended ceiling in the massive floor is complemented by the authors to facilitate service integration. The configurations of all other components and elements are based on standard regional layouts.

Moreover, all elements within a given component were modelled individually and enriched with custom parameters (e.g. component type, material type, classification code). These parameters were managed using a shared parameters file, enabling structured semantic enrichment and downstream data extraction for analysis. The modelling was conducted at a Level of Development (LOD) 300, ensuring sufficient detail.

3.4 Limitations and Assumptions

The implementation is subject to several limitations and assumptions.

- Scope of Metrics: this workflow does not bring together several circularity metrics to calculate a global CI. Instead, it concentrates exclusively on two selected metrics, one representative of each phase (pre-use and post-use).
- LCA Data Quality: ÖKOBAUDAT provides reliable LCA data, yet static and mostly generic, limiting the accuracy for current project-specifications.
- User Interface Prototype: the user interface is a proof-of-concept implementation.
- It does not yet fully meet the usability standards for end-user applications. Furthermore, integration between BIM and BI remains partial.

- Lack of Circularity Data Standards: unlike LCA, circularity metrics lack standardized, centralized datasets. This entailed manual assignment of circularity classes and rating factors, based on assumptions drawn from DGNB methodologies.

4 Results and Discussion

The proposed workflow was successfully implemented across all four stages described in Sect. 3.1. The resulting dashboard enables the real-time quantification and visualization of EC and CI at component, material, and system levels. Key elements of the process, such as material classification, mass aggregation, and index calculation, are fully automated within the BI environment. Users can explore results through interactive views, supporting performance-based design comparisons. Figure 3 illustrates one such view, showing a component-level performance comparison.

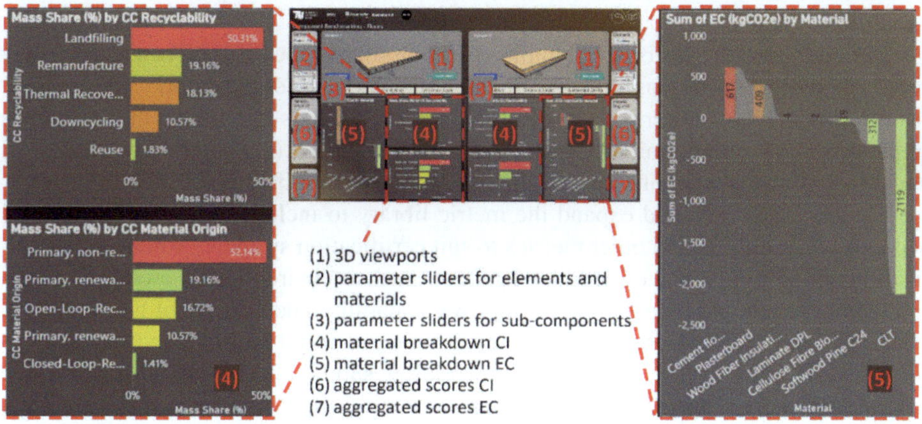

Fig. 3. Example dashboard for component benchmarking (centre), with zoomed-in snippets (left and right) and red annotative overlays.

Two key observations emerged from the prototype. First, components with lower EC did not necessarily achieve higher circularity scores. For instance, comparing two ceiling panels, a wooden beam ceiling (laminated veneer lumber) and a massive timber ceiling (cross-laminated timber), revealed such a discrepancy. While the massive timber ceiling scored better in EC with -1.440 kgCO2e in comparison to 276 kgCO2e, the wooden beam ceiling scored higher in both, Pre-Use CI with 0.29 in comparison to 0.21 and 0.39 in comparison to 0.36 for Post-Use CI.

Second, a notable divergence between Pre-Use and Post-Use circularity performance. This was evident in the comparison of alternative insulation materials for a partition wall using the material benchmarking view. Cellulose fibre blowing insulation achieved a high Pre-Use CI of 0.75 but a significantly lower Post-Use CI of 0.40. In contrast, expanded cork scored lower in the Pre-Use CI with a value of 0.50 yet outperformed in the Post-Use CI with a value of 0.60. These differences were additionally reflected in their respective EC values, with cellulose fibre resulting in −62.47 kgCO2e and expanded cork resulting in 13.33 kgCO2e. These divergences indicate that designers relying solely on LCA may overlook key material life cycle considerations. Additionally, the observed differences between pre-use and post-use metrics highlight the importance of balancing both in circular design strategies.

These observations were made possible by the workflow's ability to embed real-time, structured performance analysis directly into the digital design process. This integration allows for rapid and nuanced comparison of material and design options. This dynamic feedback supports early-stage planning decisions and provides designers with operational access to circularity performance during design development.

Ultimately, the developed workflow constitutes a methodological advancement in operationalizing circularity metrics in digital design environments. The workflow significantly streamlines performance evaluation, reducing processing times from several hours or days to just a few minutes. Compared to conventional post-hoc or document-based approaches, the workflow supports iterative, data-driven design exploration, empowering planners to balance trade-offs between embedded carbon and circularity in a structured and efficient manner.

Limitations of the prototypical implementation such as the focus on two metrics may overlook synergistic or contradictory interactions across other indicators. Additionally, insufficient technical expertise among planners and designers may hinder the workflows implementation. Other limitations persist due to the manual uploading from BIM to BI and a lack of integration of standard BIM viewers into dashboards.

Future research should expand the metric library to include disassembly, material separability, pollution and other metrics to suit certification systems. Automated classification and scoring based on product databases such as environmental product declaration databases should be explored to improve scalability and transfer of the workflow. BIM-BI integration requires further development and could be improved through tailormade APIs or plug-ins. Moreover, the workflow should also be validated and tested across construction projects and design processes to assess efficiency, robustness and transferability.

5 Conclusion

This study presented a prototypical BIM-BI-based workflow for embedding circularity metrics into digital design procedures. Through indicator operationalization, the workflow enables real-time, component-level circularity and LCA assessment. The presented workflow contributes to industry wide efforts of implementing data-driven design within planning processes. While the here introduced implementation remains limited in scope, future developments of the suggested workflow could bridge the gap between sustainability certification frameworks and day-to-day design practices through digital integration. Future work should focus on metric expansion, automation, and validation to support mainstream implementation. To maximize practical relevance, the data requirements and evaluation criteria should be aligned with major sustainability certification systems. This would greatly increase the applicability of the workflow in real-world practices.

Acknowledgements. We thank the Federal Ministry of Food, Agriculture and Consumer Protection, Berlin, Germany, as well as the Federal Agency of Renewable Resources (FNR) for funding of the research project DiKieHo (Funding Code 2221HV082B).

References

1. United Nations Environment Programme. Building Materials and the Climate: Constructing a new Future (2023)
2. Soto-Paz, J., Arroyo, O., Torres-Guevara, L.E., Parra-Orobio, B.A., Casallas-Ojeda, M.: The circular economy in the construction and demolition waste management: a comparative analysis in emerging and developed countries. J. Build. Eng. **78** (2023)
3. Landesanstalt für Umwelt Baden-Württemberg. Ein praxisnaher Leitfaden für Entscheidungstragende, Bauverantwortliche und Planende. Karlsruhe (2024)
4. Akinade, O.O., et al.: Design for deconstruction (DfD): critical success factors for diverting end-of-life waste from landfills. Waste Manage. **60**, 3–13 (2017). https://doi.org/10.1016/j.wasman.2016.08.017
5. Gutiérrez, N., Negrão, J., Dias, A., Guindos, P.: Bibliometric review of prefabricated and modular timber construction from 1990 to 2023: evolution, trends, and current challenges. Sustain. (Switz.) **16** (2024)
6. Van der Lugt, P., Harsta, A.: Tomorrow's Timber: Towards the next Building Revolution, 1st edn. MaterialDistrict, The Netherlands (2020)
7. Al-Qazzaz, I., Osorio-Sandoval, C.A., Tokbolat, S., Thermou G.: Integration of building information modeling into building circularity assessment: a systematic review. Built. Environ. Proj. Asset Manage. (2024)
8. Rosen, A.: Entwicklung einer Systematik zur quantitativen Bewertung der Kreislaufkonsistenz von Baukonstruktionen in der Neubauplanung (2020)
9. Madaster. Madaster Circularity Indicator (2021)
10. DGNB e.V. Qualitätsstandard für Zirkularitätsindizes für Bauwerke (2024)
11. Klein, M., Osterhage, T., Müller, D., Kosanović, S., Hildebrand, L., On Erc, E.: KLABS | sustainable and resilient building design _ approaches, methods and tools building certification systems and processes building certification systems and processes (2018)
12. Mahmudin, A., Mohsin, A.M.A., Islam, M.S.: A metadata review on business intelligence in the construction industry. Int. J. Bus. Innov. Res. **32**, 147–168 (2023). https://doi.org/10.1504/IJBIR.2023.134656
13. Rodrigues, F., Alves, A.D., Matos, R.: Construction management supported by BIM and a business intelligence tool. Energies (Basel) **15** (2022). https://doi.org/10.3390/en15093412
14. Enrique Pérez-Martínez, J., Díaz, J., González-Prieto, Á., Gil, S., Pérez, J., Gil-Borrás, S.: Theory building for empirical software engineering in qualitative research: operationalization (2024)
15. Federal Ministry for Housing, Urban Development and Building. ÖKOBAUDAT (n.d.). https://www.oekobaudat.de/. Accessed 23 May 2025

Utilizing Manufacturing Digital Twins for Sustainability Reporting

Yujia Luo[1], Juan Ramon Candia[2], and Peter Ball[3](✉)

[1] York Business School, York St John University, York, UK
[2] Priestley Centre for Climate Futures, University of Leeds, Leeds, UK
[3] School for Business and Society, University of York, York, UK
peter.ball@york.ac.uk

Abstract. The advance of digital technologies and the societal imperative to achieve sustainability exposes new pathways for businesses. The environmental and social impacts of industrial operations are well known and there are increasing demands to report impact through voluntary and mandatory corporate stakeholder reporting. The extensive live operations reporting is disconnected from corporate reports built retrospectively from aggregate enterprise data. This paper argues that the disconnect between operations and corporate reporting is a missed opportunity hindered by data integration, absence of frameworks to guide adoption and the missed value that comes from integrated reporting. By considering digital twin data and those of reporting standards such as from the EU's Corporate Sustainability Reporting Directive, the two could be mapped. Once mapped, a digital twin could be created to generate reporting data and enable improvement evaluation. This paper uses literature and expert interviews to report the requirements for linking digital twins to sustainability reporting. In turn it proposes a framework to guide data mapping and model configuration to use to evaluate the business level impact of operational changes. A discussion on the utility of the framework precedes the conclusion with directions for future research.

Keywords: Digital twins · sustainability reporting · manufacturing systems

1 Introduction

The impact of manufacturing is growing [1] and affecting ecosystems [2]. New practice implementations are needed [3]. To address the sustainability imperative [4], digital twins are promoted to address the evaluation of new environmental practices [5] as well as the long-standing productivity focus. As a core technology of Industry 4.0 [6], they can model efficiency, environmental and possibly social improvement scenarios. Digital twins, based on simulation, can capture the dynamic, time-varying interactions between materials and other resource flows through the stages of production system [7]. However, there are gaps in addressing environmental and social sustainability challenges using digital twins in operations [8] and the wider supply chain [9].

Communicating company activity voluntarily against selected Sustainable Development Goals (SDGs) [10] has become commonplace for organizations. Mandatory

H. Kohl et al. (Eds.): GCSM 2025, LNME, pp. 287–295, 2026.
https://doi.org/10.1007/978-3-032-21154-5_32

reporting requirements are being established, e.g. EU Corporate Sustainability Reporting Directive (CSRD). Until recently, they have been voluntary [11] with no single international standard [12].

Despite manufacturing operations having a significant role in a company's impact and that manufacturing has utilized extensive measurement for centuries, there is a disconnect between manufacturing operations and compulsory sustainability reporting. Irrespective of the use of digital twins, granular, frequent manufacturing reporting is not the source of sustainability reporting data. Instead, aggregate data is drawn from site meters, billing, etc. This represents a knowledge gap and an industrial opportunity.

Mandatory sustainability reporting standards have granularity that could be mapped to digital twin outputs and wider sources. The benefits would be twofold. First, corporate stakeholder reporting could be efficiently rolled up from operational metrics for traceability. This could be done regularly rather than annually. Second, operations improvement could be driven by business scenarios triggered by the desire to improve reportable corporate performance. The objective of this research is therefore to establish if and how digital twins can bridge between operations and sustainability reporting.

In this paper, we first examine the literature on digital twins and compulsory sustainability reporting and the potential for links between them. Based on a mixed methodology we then conduct empirical data collection with experts from across industry to consider the findings from the literature and the prevailing visions in manufacturing. Next, we develop a framework that combines the operations and simulation modeling domain with the domain of corporate sustainability reporting. We consider the applicability to a SME in the UK that has advanced thinking and practices in the sustainability arena. Finally, we discuss the findings, conclude and propose research.

2 Literature

Digital twins [6] have been used for sustainability related analysis such as energy [5] but in general are limited in their scope [8] and other techniques have been used to engage in broader environmental and social analysis [13].

There is a weakness in using digital twins for sustainability performance tracking [14] perhaps because environmental benefits are often being consequential rather than the prime focus [15]. With external reporting pressures requiring a manufacturing response [16] there is potential for digital twins to be implemented to respond to corporate, stakeholder-facing imperatives.

Stakeholder accountability necessitates effective measuring, monitoring and reporting for reasons of societal awareness and associated demand for insight and transparency [17] as well as industry expectations and emerging legislation. Against a backdrop of stakeholder dissatisfaction [17] standards are developing.

Such reporting standards span the breadth of environmental, social and economic metrics [18] and include ISO standards, UN Global Compact, Global Reporting Initiative (GRI), Task Force on Climate-related Financial Disclosures (TCFD), Task Force on Nature-related Financial Disclosures (TNFD). One such standard is the European Sustainability Reporting Standard (ESRS) (based on CSRD) framework that incorporates cross-cutting standards (1–2), and specific (topical) standards for environment (E1–E5), social (S1–S4) and general (G1). The five environmental standards (E1–E5) of Climate change, Pollution, Water and marine resources, Biodiversity and ecosystems and Resource use and circular economy are the focus of this paper.

3 Approach

Figure 1 presents the research approach used. Step 1 was the literature review to establish knowledge available in the academic literature on digital twins (DTs) and sustainability performance reporting (SPR). This sought to confirm the guiding research objective and formed the basis for subsequent interviews and framework development. The interviews in Step 2 were conducted to refine the established knowledge from the literature by incorporating the prevailing thinking and practice from industry. It served to confirm the validity of the research endeavor in case academic publications were not indicative of current practice. Interviewee questions posed included: frameworks or guidance they used, the scale and scope of analysis they considered, criteria for including/excluding data, their data sources, how unavailable data is addressed, how they consider manufacturing and sustainability together and whether they were aware of any companies working on such integration. Step 3 used literature and empirical evidence gained from industry to develop a conceptual framework to guide a process to align digital manufacturing with sustainability reporting standards. Finally, Step 4 was the high-level illustration of the framework application based on a UK company.

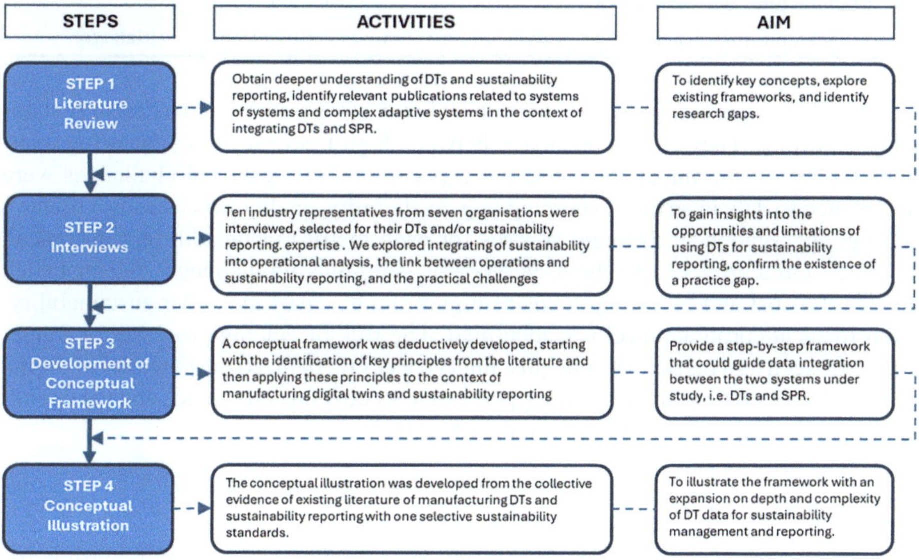

Fig. 1. The research approach used.

4 Results

4.1 Interview Insights

Table 1 summarizes the 10 interviewees from seven companies. All companies were operating in the manufacturing industry, i.e. each consultancy directly served manufacturing companies. Between one and three interviewees took part from each company. Each interviewee was invited for their depth of expertise in one of the three domains (indicated by the table text) of simulation modeling/digital twins, sustainability or manufacturing. All had depth of knowledge of sustainability or manufacturing (indicated by the table shading) even if their dominant role was in modeling. All were at least mid-career stage so had significant experience.

Table 1. Summary of the company type and main expertise of each interviewee.

#	Type \ Expertise	Digital twins	Sustainability	Manufacturing
1	Consultancy *sustainability*		Director	
2	Consultancy *sustainability*	Consultant	Consultant	
3	Consultancy *simulation*	Consultant	Consultant	Bus. development
4	Consultancy *simulation*	Director		
5	Consultancy *digital*		Innovation lead	
6	Manufacturer *automotive*		Director	
7	Manufacturer *aerospace*			Manager

All interviewees confirmed the disconnect between operations and external sustainability reporting. They cited a mismatch between high frequency manufacturing operations reporting and annual sustainability reporting. Many technical challenges were cited related to data in terms of sourcing (e.g. availability, quality, consistency, volume), use (e.g. aggregation of different sources for each metric) or breadth (e.g. technical, workforce, circularity, supply chain, etc.) For decision-making challenges, the mismatch between the cost and efficiency focus of digital twins versus the wider sustainability-driven focus at corporate level was identified. The lack of guiding processes or frameworks was seen as a barrier. Despite this, interviewees recognized the potential of digital twin enabled operations analysis to support progress beyond compliance and the exploit the utility of multi-scale strategic sustainability decision making.

4.2 Framework Development

The development of a conceptual framework to bring the sustainability reporting and the manufacturing domains together took a data flow view through standard approaches to sourcing for both domains. Figure 2 illustrates the data flows used in each of the two domains split across input and output categories. Each side/domain in the figure shows (once scope has been established) the flow of data from what is available and recorded to its use in performance reporting. In the planning stage of simulation methodologies, the performance reporting requirement drives the information collection plan and in turn the sourcing of the data items from the available inventory of data.

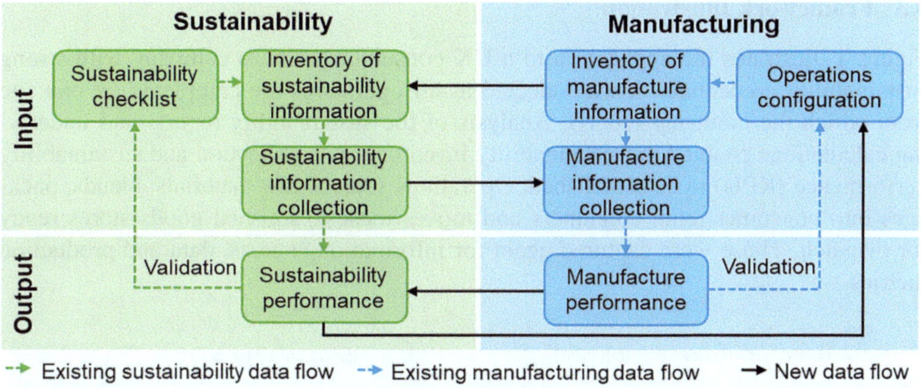

Fig. 2. Foundation of the information flows in sustainability reporting and manufacturing domains and their interlinking

Using Fig. 2 as the conceptual data flow representation, a conceptual framework was developed to operationalize the integration between the two domains shown in Fig. 3.

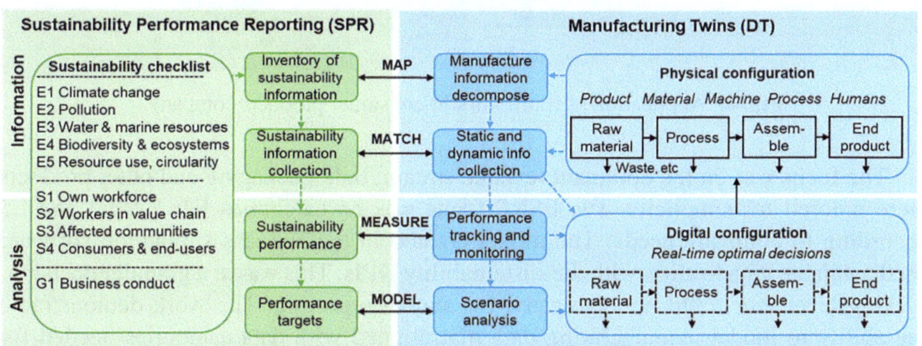

Fig. 3. Conceptual framework to operationalize the integration between sustainability performance reporting (SPR) with the manufacturing digital twin (DT).

The left-hand side of the figure is drawn from the standard steps within the selected EU standard, the European Sustainability Reporting Standard (ESRS) based on CSRD. This is mirrored on the right-hand side by the typical representations in the manufacturing domain. In the top right is a generic production process for the provision of information and this is mirrored in the bottom right by a corresponding digital twin for the analysis. Four steps of mapping, matching, measuring and modeling are used to align data collection, analysis and reporting between the two domains. For example, energy consumption in the manufacturing system is mapped and matched to the sustainability reporting requirements, i.e. to ESRS E1 Climate change. For example, manufacturing water consumption is matched to ESRS E3 Water and marine resources.

4.3 Framework Illustration

Figure 4 illustrates the application to a UK consumer products company with strong sustainability credentials. It was selected as all operations are completed on one site from which the company reports. Analysis of the sustainability reports and underlying calculations enabled the sustainability inventory, data collection and sustainability performance (KPIs) to be established. Operations sources raw materials, blends, packages into consumer retail containers and moves these to finished goods stores ready for dispatch. These were captured again for information sources, data and production metrics.

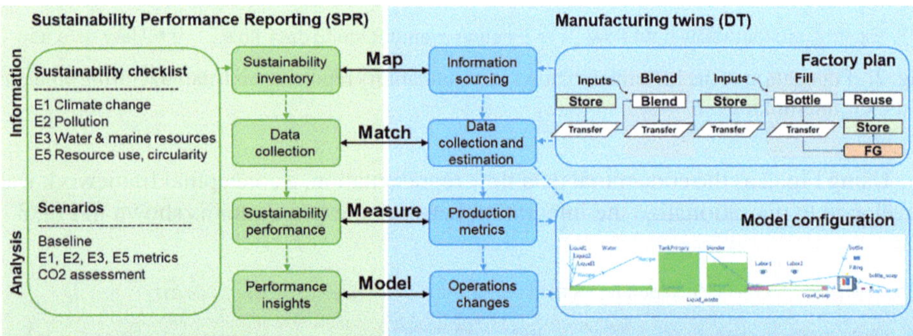

Fig. 4. Application of framework to consumer products company.

The focus was on the dominant revenue stream; bulk packaging and other products were ignored for simplicity. The ESRS focus was on categories E1, E2, E3 and E5 according to company needs. The model created in the Witness simulation software used available data to align with the sustainability KPIs. This was at a granular decision-making level that could be aggregated for annual reporting. The work demonstrated the ability to model manufacturing data that aligned with data categories needed for performance reporting.

5 Discussion

This research has sought to connect operations, enabled by digital twins, with a recognized sustainability reporting standard by creating an integrated mapping framework. It takes the language and standards from manufacturing and corporate reporting standards and guides the process of mapping, matching, measuring and modeling.

The research objective was to examine the bridge between operations and standard sustainability reporting, i.e. how granular operations reporting can support aggregate external reporting. The difference being the dynamic, granular data used from day-to-day production operation rather than aggregate sourcing from meter readings, billing, etc. Akin to dynamic Lifecycle Analysis (LCA), this has the advantage of granular insight for decision making but is variable according to prevailing conditions rather than the average, annualized reporting. The question of when to 'roll-up' the operations reporting to corporate level for effective decision making needs investigation.

The language of operations analysis and corporate reporting is suggestive of a one-way process in which improvements are inspired, investigated and implemented in operations and realized at corporate level. It is difficult to work from corporate external stakeholder reporting to inspire operational changes given the inability to breakdown the aggregate reporting data. However, if corporate reporting is built up from digital twin data it then becomes possible to inspire and investigate from a corporate reporting start point and use digital twins to game different improvement scenarios.

6 Conclusion

This exploration into the feasibility of connecting manufacturing operations, in particular, digital twins with compulsory sustainability reporting has demonstrated value and potential. Motivated by the significant environmental impact of operations and the imperatives on external sustainability reporting, this research established an academic and practice-oriented baseline before developing and demonstrating a framework.

From a theoretical perspective, no work exists nor do frameworks or supporting guidance exist in other fields. From a practice perspective, although the value of the connection was seen, there are no industrial instances of such connections from interviewing leaders in the simulation, manufacturing and sustainability fields.

This paper identified the untapped synergies between digital twin data and that of compulsory sustainability reporting. By aligning existing ways to build simulation models with those of a sustainability standard, a novel framework is proposed to enable operations data to be linked with sustainability reports. Our contribution is the previously unseen demonstration of how to connect day-to-day operations data with previously only annualized reporting that did not use operations data. This gives potential for operations data to feed into frequent assembly of sustainability reports and for the review of sustainability reports to drive focused operations practice change.

This work extends practice change from operational to strategic through the demonstration of corporate level improvements. The work shows potential for bridging between operations and sustainability domains and presented a means to achieve this. Future research lies in extending the digital twin scope beyond main shop floor operations, reviewing whether this bridging of enterprise activities improves decision making and understanding whether there are changes to decision maker behavior as a result.

Acknowledgements. We thank the UK's Engineering and Physical Sciences Research Council (EPSRC) for funding the EP/T024844/1 Multiscale Digital Twin-Driven Smart Manufacturing System for High Value-Added Products project.

References

1. Li, L., et al.: Sustainability assessment of intelligent manufacturing supported by digital twin. IEEE Access: Pract. Innov. Open Solut. **8**, 174988–175008 (2020)

2. Garetti, M., Taisch, M.: Sustainable manufacturing: trends and research challenges. Prod. Plann. Control **23**(2–3), 83–104 (2012)
3. Larreina, J., Gontarz, A., Giannoulis, C., Nguyen, V.K., Stavropoulos, P., Sinceri, B.: Smart manufacturing execution system (SMES): the possibilities of evaluating the sustainability of a production process. In: Global Conference Sustainable Manufacturing, pp. 517–522 (2013)
4. Karwowski, W., et al.: Grand challenges in industrial and systems engineering. Int. J. Prod. Res. 1–46 (2025)
5. Tagliabue, L., Cecconi, F., Maltese, S., Rinaldi, S., Ciribini, A.: Leveraging digital twin for sustainability assessment of an educational building. Sustainability **13**(2), 480 (2021)
6. Shojaeinasab, A., et al.: Intelligent manufacturing execution systems: a systematic review. J. Manuf. Syst. **62**, 503–522 (2022)
7. Oates, M.R., et al.: Design of sustainable industrial systems by integrated modelling of factory building and manufacturing processes. In: Global Conference on Sustainable Manufacturing (GCSM), pp. 742–747 (2012)
8. Jiang, H., Qin, S., Fu, J., Zhang, J., Ding, G.: How to model and implement connections between physical and virtual models for digital twin application. J. Manuf. Syst. **58**, 36–51 (2021)
9. Ivanov, D.: Intelligent digital twin (iDT) for supply chain stress-testing, resilience, and viability. Int. J. Prod. Econ. **263**(108938) (2023)
10. United Nations. Transforming our world: The 2030 agenda for sustainable development. Resolution adopted by the general assembly on 25 September 2015 (2015). https://sustainabledevelopment.un.org/post2015/transformingourworld
11. OICOU-IOSCO. Report on Sustainability-related Issuer Disclosures Final Report (2021)
12. Silva, S., Nuzum, A.K., Schaltegger, S.: Stakeholder expectations on sustainability performance measurement and assessment. J. Clean. Prod. **217**, 204–215 (2019)
13. Juvvala, R., Sangle, S., Tiwari, M.K.: Post-Covid challenges and opportunities: rethinking ESG performance in the logistics sector. Int. J. Prod. Res. **63**(4), 1–19 (2024)
14. Ma, X., Qi, Q., Tao, F.: An ontology-based data-model coupling approach for digital twin. Robot. Comput.-Integr. Manuf. **86**(102649) (2024)
15. Tao, F., Xiao, B., Qi, Q., Cheng, J., Ji, P.: Digital twin modeling. J. Manuf. Syst. **64**, 372–389 (2022)
16. Perotti, S., Cannava, L., Ries, J., Grosse, E. Reviewing and conceptualising the role of 4.0 technologies for sustainable warehousing. Int. J. Prod. Res. **63**(6), 2305–2337 (2024)
17. Osobajo, O., Oke, A., Lawani, A., Omotayo, T., Ndubuka-McCallum, N., Obi, L.: A bibliometric lit review of sustainability performance reporting. Sustainability **14**, 8523 (2022)
18. Abeysekera, I.: A framework for sustainability reporting. Sustain. Account. Manage. Policy J. **13**(6), 1386–1409 (2022)

Exploring the Role of DPPs in Smart Circular Economy: A Conceptual Synthesis

Luana Coelho de Morais[1], Carlos Eduardo Silveira[1], Maxim Mintchev[2], Ana Carolina Bertassini[4], Meire Ramalho de Oliveira[5], Valentin Eingartner[2], Holger Kohl[2], and Mateus Cecílio Gerolamo[1,3](✉)

[1] São Carlos School of Engineering at the University of São Paulo, São Carlos, Brazil
mateus.gerolamo@usp.br
[2] Institute for Machine Tools and Factory Management, Technische Universität Berlin, Berlin, Germany
[3] Luiz de Queiroz College of Agriculture, University of São Paulo, Piracicaba, Brazil
[4] Chalmers University of Technology, Gothenburg, Sweden
[5] Federal Institute of Sao Paulo, Capivari, Brazil

Abstract. This article presents a systematic literature review (SLR) that examines the strategic role of Digital Product Passports (DPPs) in facilitating transitions to Smart Circular Economies (SCE). DPPs are conceptualized as dynamic, interoperable digital records linked to unique product identifiers, facilitating multidimensional traceability across product lifecycles. The review synthesizes recent research on the conceptual foundations, regulatory drivers, technological enablers, life cycle integration, and organizational challenges. While European policies drive progress, barriers such as digital infrastructure, governance and stakeholder engagement persist, particularly in Global South contexts. The proposed conceptual synthesis integrates technological, organizational, market, and regulatory dimensions, with a focus on change management and collaborative governance. Key research gaps include a limited focus on downstream actors and behavioral factors. Findings suggest that realizing the full transformative potential of DPPs requires the coevolution of technology, organizational culture, and inclusive policy frameworks, supported by future studies in diverse economic and institutional settings.

Keywords: Digital Product Passport · Smart Circular Economy · Traceability · Organizational Change · Socio-Technical Systems

1 Introduction

The transition to a circular economy (CE) has emerged as a strategic response to the growing environmental, social, and economic concerns associated with the linear production model [1, 2]. Advancing this transition requires changes in technology, structure, culture, and information management across product lifecycles [3]. Digital Product Passports (DPPs) gained prominence as tools to enhance traceability, support sustainable

H. Kohl et al. (Eds.): GCSM 2025, LNME, pp. 296–304, 2026.
https://doi.org/10.1007/978-3-032-21154-5_33

decisions, and enable smart circular economy (SCE) models [4]. Driven by digital innovations and policies such as the CE Action Plan and the ESPR (Eco-design for Sustainable Products Regulation), DPPs are primarily applied in the electronics, construction, and textiles sectors, particularly in Europe [5, 6].

However, many DPP frameworks remain conceptual and disconnected from organizational and behavioral realities [7–9]. Barriers such as low digital maturity, fragmented governance, and institutional resistance limit their adoption, especially in emerging economies [10, 11]. Effective implementation requires a combination of technical infrastructure, institutional capacity, stakeholder trust, and adaptive leadership [4, 11, 12].

This article asks: how can a conceptual framework integrate technical and organizational functions to position DPPs as strategic tools for the SCE? The central proposition is that their effectiveness depends on embedding DPP within a socio-technical ecosystem that encompasses infrastructure, data standards, governance, culture, and stakeholder roles.

The main objective is to propose a conceptual synthesis for DPPs that represents their strategic role in the SCE by integrating functions, barriers, impacts, and actors across product life cycle stages, structured in a tabular format. Based on a systematic literature review (SLR), this study analyses five core contributions, encompassing conceptual foundations, applications, enabling technologies, and organizational challenges. The article is structured as follows: Sect. 2 details the SLR methodology; Sect. 3 presents findings and the framework; Sect. 4 discusses applicability and limitations; Sect. 5 concludes and suggests future research directions.

2 Methodology

This is a qualitative and exploratory study aiming to understand the role of Digital Product Passports (DPPs) in enabling Smart Circular Economies (SCE). A systematic literature review (SLR) was adopted as the method, following the PRISMA protocol [13] to ensure rigor and reproducibility.

Data collection was conducted in April 2025 using the Web of Science and Scopus databases. The search terms included: "Digital Product Passport*", "circular econom*", "smart circular econom*", "design", "sustainability", and "secure design". The publication period was limited to 2021–2025. A total of 105 records were identified, with 30 duplicates removed, resulting in 75 documents screened. After assessing relevance and availability, the final sample comprised 57 articles.

Content analysis followed the procedures of Whittemore and Knafl [14], including: (i) exploratory reading; (ii) data extraction based on analytical units; and (iii) categorization and subcategorization of findings.

Beyond theoretical and methodological criteria, the analysis focused on identifying stakeholder roles throughout the product lifecycle, organizational barriers, and behavioral drivers—key elements for constructing the proposed socio-technical framework.

While the SLR encompassed 57 peer-reviewed studies published between 2021 and 2025, we explored five of these contributions in greater depth, based on their theoretical richness, sectoral diversity, and relevance to the research question. These studies offered complementary perspectives on conceptual foundations, sectoral applications, traceability technologies, and organizational challenges, providing a robust foundation for synthesis. The remaining studies are not inconsequential, but their insights are integrated into the results and discussed to reinforce thematic coherence and expand contextual understanding. This two-tiered strategy balances breadth and depth: while the full SLR informs thematic trends and general patterns, the close reading of selected studies allows for deeper conceptual integration. Selection criteria were grounded in relevance to the research question, sectoral variety, and analytical clarity, rather than subjective preference.

3 Results and Discussions

3.1 An Overview on DPPs for SCE

The systematic literature review (SLR) defines DPPs as structured digital repositories linked to unique identifiers, containing product lifecycle data [15]. EU regulatory frameworks are key enablers of DPP development across sectors such as textiles, batteries, electronics, and construction [5, 16, 17]. Wan and Jiang [9] identified four core components: data sources, support, security, and access verification, noting that 64% of initiatives remain conceptual and only 20% have been implemented. Common data carriers include Quick Response (QR) codes and Radio-Frequency Identification (RFID), shaped by ESPR requirements, while blockchain is explored to ensure data integrity despite limitations in updating data during use.

Significant barriers persist, especially for SMEs, due to limited IT infrastructure, budgets, and expertise, challenges amplified in the Global South [11]. Beyond technical design, DPPs foster sustainable products by embedding lifecycle criteria such as modularity and recyclability early in design [9, 18] and guiding material selection [5, 19]. They enhance traceability, compliance, transparency, circular business models, reverse logistics, and impact assessments [15, 16, 20]. However, organizational barriers remain, especially where SMEs face low digital maturity and institutional fragility [8, 11, 12].

The literature stresses the need for inclusive, interoperable infrastructures. Future directions include using AI to estimate product longevity and integrating social sustainability indicators [21]. Addressing digital inequities [11] and broadening sustainability metrics supports a more inclusive DPP agenda, especially for underrepresented regions in the Global South.

3.2 Application of DPP in the Electro-Electronic Sector

DPPs have been applied in various sectors, including textiles, batteries, electronics, construction, and automotive [15, 18, 22]. In electronics, Koppelaar et al. [8] proposed a circular supply chain using DPPs to recover critical raw materials (CRMs) from electronic waste. A key barrier is the lack of integrated CRM traceability and coordination systems. Using literature and CircThread project data, the authors developed an architecture with product identifiers and a secure digital repository to track lifecycle data and end-of-life CRM flows.

Challenges include weak manufacturer incentives, product complexity, high costs of CRM technology, and a lack of standardized material declarations, reflecting technical and governance gaps that require validation and deeper policy analysis [5, 19, 23].

In countries like Brazil, where Waste of Electrical and Electronic Equipment (WEEE) management is evolving, existing regulations and a growing recycling sector support the adoption of DPP. Progress depends on actor coordination, overcoming data-sharing barriers, and building trust among supply chain stakeholders [11, 24].

3.3 Practical Implementation of DPPs

Çetin et al. [9] developed a Material Passport (MP) framework for European social housing, addressing existing buildings that often lack material and circularity data. Through literature review, stakeholder interviews, and a Dutch case study, they created a stakeholder map and data template.

The pilot revealed fragmented data, exposing challenges in retrofitting digital tools. To overcome this, the study proposed a digital framework that integrates Building Information Modelling (BIM), computer vision, laser scanning, web scraping, and artificial intelligence (AI) to consolidate material data and assess conditions. The model supports four CE strategies: narrow (reduce resource use), slow (extend lifespan), close (reuse/recycle), and regenerate (restore ecosystems).

AI plays a crucial role in enhancing data quality and traceability [4, 15, 25]. The integration of BIM and digital twins aligns with circularity trends and supports approaches like those by Koppelaar et al. [8]. Though only piloted, the model shows scalability. In contexts like Brazil, low digitization and informal construction highlight t the need for public investment, and organizational change.

3.4 Challenges in the Implementation of DPPs

DPP adoption faces significant challenges [5, 16, 20, 22]. Domskienė and Gaidule [12] studied the Baltic textile sector, assessing DPP familiarity, practices, and technologies. QR codes emerged as the most feasible entry point due to low cost and simplicity, while RFID, Near Field Communication (NFC), and blockchain offer greater robustness and scalability but still have limitations.

Barriers include a lack of standardization and integration, fragmented policies, poor coordination, high implementation costs, and privacy concerns. The authors recommend adjusting DPP complexity by product value, at the batch level for low-value items and at the item level for premium products. These challenges mirror those in construction [6], highlighting the need for systemic coordination In emerging economies, DPP adoption relies heavily on trust, cultural change, and governance balancing privacy, transparency, and inclusion [10].

3.5 Traceability Technologies and Structural Barriers to DPPs

Jakubowicz and Yarahmadi [12] reviewed traceability technologies for ABS plastic, a standard but hard-to-recycle material. Due to limited research, they examined both digital and physical tracking approaches. DPPs have emerged as promising tools for creating digital twins that link material composition to lifecycle data through QR codes, barcodes, and blockchain. Physical marking methods, such as photoluminescence, laser etching, and digital labels, were also considered.

The study identified four main barriers: technological (interoperability, data standardization, data quality); governance (ownership, transparency); regulatory (gaps), economic (high costs, limited SME resources); and cultural/organizational (low awareness, limited coordination, and weak cross-sector collaboration).

Although not empirically tested, the authors proposed hybrid systems that combine physical and digital tools, tailored to product type, value, and regulatory context, to address the transparency and implementation challenges discussed by Koppelaar et al. [8] and Jansen et al. [5].

The study concludes that institutional readiness is crucial for the success of DPP. In developing regions like Brazil, systemic interventions are necessary to overcome structural limitations.

3.6 Conceptual Framework: DPP for SCE

Based on the synthesis of the SLR, Table 1 summarizes the conceptual dimensions that define DPPs as enablers of smart circular economy (SCE) ecosystems.

The structure was built by triangulating empirical evidence, theoretical foundations, and gaps identified in the reviewed studies.

The Digital Product Passport (DPP) framework highlights its role as a dynamic, interoperable digital record essential for the smart circular economy. The table summarizes its strategic functions, lifecycle scope, involved stakeholders, benefits, and implementation challenges.

The DPP operates across the entire product lifecycle, engaging multiple stakeholders and technologies. Organizational culture and change management are critical, especially for SMEs and in the Global South. Despite clear benefits, overcoming technical, regulatory, and organizational challenges is key to effective adoption.

4 Conclusions

This study examined the strategic role of Digital Product Passports (DPPs) in advancing smart circular economies (SCE) through a systematic literature review, exploring how DPPs have been conceptualized, applied, and challenged across sectors, institutions and governments, to develop a guiding theoretical framework.

Findings indicate growing adoption despite persistent barriers in standardization, interoperability, governance, and stakeholder engagement. The conceptual synthesis (Table 1) emphasizes that DPPs 'effectiveness depends on integration with organizational behavior, collaboration, and inclusive policies, beyond technical design.

The central proposition - that DPPs act are socio-technical tools supporting circular practices, was confirmed. However, gaps remain in change management, the inclusion of downstream actors (consumers, repairers), and attention to organizational culture.

Table 1. Conceptual Framework of the Digital Product Passport (DPP) Ecosystem

Aspect	Details
Core	Dynamic, interoperable, and accessible digital record linked to physical products via unique identifiers
Strategic Functions	Transparency and traceability (origin, composition, manufacturing)
	Support for circular practices (repair, reuse, remanufacturing, recycling)
	Regulatory compliance and automated reporting
	Lifecycle-informed decision-making (e.g., ecodesign)
	Standardized data for interoperability and collaboration
	Consumer engagement and trust by reducing information asymmetry
Traceability Dimensions	Material traceability
	Process traceability
	Environmental traceability
Lifecycle Scope	Design, raw material extraction, use, maintenance, end-of-life
Priority Sectors	Batteries, electronics, textiles, packaging, construction [15, 16, 26–28]
Stakeholder Roles	Producers (data input), consumers (information access), regulators (verification/compliance), recyclers
Contextual Layers	Shared digital infrastructure (blockchain, IoT, cloud), interoperability, data standards
Foundational Layer	Organizational culture and change management, especially critical for SMEs and Global South contexts
Expected Benefits	Environmental (waste reduction, resource conservation)
	Economic (resilience, circular business models)
	Social (trust, safety, improved labor conditions)
Feedback Loops	Continuous improvement, regulatory learning, strategic refinement
Implementation Challenges	Technical (data standardization, platform scalability), regulatory (norm harmonization), organizational (governance, costs, training), market-related (sustainable business models)

Limitations include the qualitative nature, reliance on secondary sources, and focus on literature from 2021–2025, which may exclude emerging trends. Future research should explore empirical applications in Global South contexts, especially SMEs; assess governance and trust mechanisms; investigate organizational adoption; and develop indicators for circular and social impact measurement.

DPPs are not merely digital tools, but institutional enablers that require coevolution of technology, culture, and policy. Achieving their full potential requires interdisciplinary, locally adapted strategies, particularly in unequal contexts such as the Global South.

Acknowledgements. This study is financed by the Brazilian Coordination for the Improvement of Higher Education Personnel (CAPES) - Finance Code 001, the German Research Foundation (Deutsche Forschungsgemeinschaft - DFG) and the Brazilian National Council for Scientific and Technological Development (CNPq).

References

1. Geissdoerfer, M., et al.: The circular economy – a new sustainability paradigm? J. Clean. Prod. **143**, 757–768 (2017)
2. Valencia, M., et al.: The social contribution of the circular economy. J. Clean. Prod. **408**, 137082 (2023)
3. Voulgaridis, K., et al.: Digital product passports as enablers of digital circular economy: a framework based on technological perspective. Telecommun. Syst. **85**(4), 699–715 (2024)
4. Langley, D.J., et al.: Orchestrating a smart circular economy: guiding principles for digital product passports. J. Bus. Res. **169**, 114259 (2023)
5. Jansen, M., et al.: Stop guessing in the dark: Identified requirements for digital product passport systems. Systems **11**(3), 123 (2023)
6. Adisorn, T., Tholen, L., Götz, T.: Towards a digital product passport fit for contributing to a circular economy. Energies **14**(8), 2289 (2021)
7. Wan, P.K., Jiang, S.: Enabling a dynamic information flow in digital product passports during product use phase: a literature review and proposed framework. Sustain. Prod. Consum. (2025)
8. Koppelaar, R.H., et al.: A digital product passport for critical raw materials reuse and recycling. Sustainability **15**(2), 1405 (2023)
9. Çetin, S., et al.: Data requirements and availability for material passports: a digitally enabled framework for improving the circularity of existing buildings. Sustain. Prod. Consum. **40**, 422–437 (2023)
10. Jakubowicz, I., Yarahmadi, N.: Review and assessment of existing and future techniques for traceability with particular focus on applicability to ABS plastics. Polymers **16**(10), 1343 (2024)
11. Costa, A., Crupi, A., De Marco, C.E., Di Minin, A.: SMEs and open innovation: challenges and costs of engagement. Technol. Forecast. Soc. Chang. **194**, 122731 (2023)
12. Domskienė, J., Gaidule, E.: An overview of technological challenges in implementing the digital product passport in the textile and clothing industry. AUTEX Res. J. **24**(1), 20240002 (2024)
13. Page, M.J., et al.: The PRISMA 2020 statement: an updated guideline for reporting systematic reviews. BMJ **372** (2021)
14. Whittemore, R., Knafl, K.: The integrative review: Updated methodology. J. Adv. Nurs. **52**(5), 546–553 (2005)
15. Zhang, A., Seuring, S.: Digital product passport for sustainable and circular supply chain management: a structured review of use cases. Int. J. Log. Res. Appl. **27**(12), 2513–2540 (2024)
16. Barwasser, A., Schuseil, F., Werner, A., Jung, M., Zimmermann, N.: The digital product passport: scenario-based recommendations for the manufacturing industry. IFAC-PapersOnLine **58**(19), 1186–1191 (2024)

17. Franke, M., et al.: Towards a service marketplace to empower circular economy transition: an example application in the supply chain of textile industry. In: IFIP International Conference on Advances in Production Management Systems. Springer, Cham (2024)
18. Pohlmann, A., Popowicz, M., Schöggl, J.P., Baumgartner, R.J.: Digital product pass-ports for electric vehicle batteries: stakeholder requirements for sustainability and circularity. Clean. Prod. Lett. **8**, 100090 (2025)
19. Kebede, R., Moscati, A., Tan, H., Johansson, P.: A modular ontology modeling approach to developing digital product passports to promote circular economy in the built environment. Sustain. Prod. Consum. **48**, 248–268 (2024)
20. Adhikari, S.: The AAS and its Submodels: driving the DPP for a sustainable circular economy. ATP MAGAZINE (2025)
21. Cattaneo, M., Elia, S., Zaninotto, E.: From traceability to responsibility: integrating social impact indicators in digital product passports. J. Clean. Prod. **408**, 137186 (2023)
22. Kim, J., Moon, J., Do, Y., Kim, H., Jeong, J.: Research on a SaaS (software as a service)-based digital product passport system model for the EV battery industry value chain. WSEAS Trans. Bus. Econ. **20**, 2466–2475 (2023)
23. King, M.R.N., Timms, P.D., Mountney, S.: A proposed universal definition of a digital product passport ecosystem (DPPE). J. Clean. Prod. **384**, 135538 (2023)
24. Jensen, S.F., et al.: An ecosystem orchestration framework for the design of digital product passports in a circular economy. Bus. Strateg. Environ. **33**(7), 7100–7117 (2024)
25. Morganti, L., et al.: Data-driven and LCA-based framework for environmental and circular assessment of modular curtain walls. J. Facade Design Eng. **12**(1), 9–42 (2024)
26. Nowacki, S., Sisik, G.M., Angelopoulos, C.M.: Digital product passports: use cases framework and technical architecture using DLT and smart contracts. In: 2023 19th International Conference on Distributed Computing in Smart Systems and the Internet of Things (DCOSS-IoT), pp. 373–380. IEEE (2023)
27. Ferrante, M., Vitti, M., Sassanelli, C.: The evolution of circular economy performance assessment: a systematic literature review. Renew. Sustain. Energy Rev. **217**, 115757 (2025)
28. Lopes, C., Barata, J.: Digital product passport: a review and research agenda. Procedia Comput. Sci. **246**, 981–990 (2024)
29. Berger, K., Schöggl, J.-P., Baumgartner, R.J.: Digital battery passports to enable circular and sustainable value chains: conceptualization and use cases. J. Clean. Prod. **353**, 131492 (2022)
30. Geissdoerfer, M., Morioka, S.N., de Carvalho, M.M., Evans, S.: Business models and supply chains for the circular economy. J. Clean. Prod. **190**, 712–721 (2018)
31. Schick-Makaroff, K., MacDonald, M., Plummer, M., Burgess, J., Neander, W.: What synthesis methodology should I use? A review and analysis of approaches to research synthesis. AIMS Public Health **3**(1), 172 (2016)
32. Hulea, M., Miron, R., Muresan, V.: Digital product passport implementation based on multi-blockchain approach with decentralized identifier provider. Appl. Sci. **14**(11), 4874 (2024)
33. Boukhatmi, Ä., Nyffenegger, R., Größer, S.N.: Designing a digital platform to foster data-enhanced circular practices in the European solar industry. J. Clean. Prod. **418**, 137992 (2023)
34. Mendes, J.A.J., Carvalho, N.G.P., Mourarias, M.N., Careta, C.B., Zuin, V.G., Gerolamo, M.C.: Dimensions of digital transformation in the context of modern agriculture. Sustain. Prod. Consum. **34**, 613–637 (2022)
35. Campos, C.J.G.: Método de análise de conteúdo: ferramenta para a análise de dados qualitativos no campo da saúde. Rev. Bras. Enferm. **57**, 611–614 (2004)
36. Gallego-Schmid, A., et al.: Links between circular economy and climate change mitigation in the built environment. J. Clean. Prod. **260**, 121115 (2020)
37. Kirchherr, J., Reike, D., Hekkert, M.: Conceptualizing the circular economy: an analysis of 114 definitions. Resour. Conserv. Recycl. **127**, 221–232 (2017)

Artificial Intelligence

Application of Generative AI in Industry 5.0: A Fictional SME Case Study

Tim van Erp[1(✉)], Bruno Pereira[2], and Adela McMurray[2]

[1] College of Science and Engineering, Flinders University, Bedford Park, Australia
tim.vanerp@flinders.edu.au

[2] College of Business, Government and Law, Flinders University, Bedford Park, Australia

Abstract. Novel generative Artificial Intelligence (genAI) tools are increasingly being introduced to the marketplace, offering new opportunities for manufacturing companies. Often, Small and Medium-Sized Enterprises (SMEs) struggle to navigate the landscape of newly released software solutions due to a lack of AI competencies and limited resources for technology scouting. This qualitative research discusses a novel and fictional SME case study for combining multiple genAI tools to reduce the time-to-market of new products significantly. The case study covers ideation, computer-aided design (CAD), and computer-aided manufacturing (CAM) as relevant manufacturing phases for SMEs engaged in New Product Development. Based on the case study, a critical discussion of Industry 5.0 benefits and barriers to using genAI in manufacturing is presented. Ultimately, the research contributes to fostering a culture of innovation by demonstrating a novel combination of genAI tools to support the design-for-manufacturing process. This can significantly decrease the time-to-market and improve the return-on-investment, competitive advantage, and potentially a transition towards Industry 5.0 if the barriers to genAI adoption are overcome.

Keywords: Sustainable Manufacturing · New Product Development · Innovation

1 Introduction

Novel genAI tools are increasingly being introduced to the marketplace and are changing how organisations create value. While the use of genAI is rapidly increasing among organisations, larger companies are outperforming smaller companies in adopting genAI [1]. It seems to be difficult for SMEs to navigate and adopt new genAI tools for useful applications as they engage in the Industry 4.0 and 5.0 era. Our study contributes to bridging this gap by highlighting a novel and fictional case study on how manufacturing SMEs can substantially reduce their time-to-market by using genAI tools throughout their engineering design processes without the need for extensive AI domain knowledge or financial budget, which are often scarce resources for SMEs. Further, potential benefits and barriers of manufacturing SMEs for using genAI are discussed from an Industry 5.0 perspective following the three dimensions of Industry 5.0: resilience, sustainability, and human-centricity [2].

H. Kohl et al. (Eds.): GCSM 2025, LNME, pp. 307–314, 2026.
https://doi.org/10.1007/978-3-032-21154-5_34

2 Literature Review

The literature review presents an overview of the state-of-the-art in genAI applications across manufacturing systems and in SMEs. GenAI in this sense "[…] refers to computational techniques that are capable of generating seemingly new, meaningful content such as text, images, or audio from training data" [3]. This includes software solutions from companies such as OpenAi's ChatGPT, to generate text, or Dall-E, to generate images, but in a manufacturing context, also from companies such as Sloyd or Tripo AI, which offer software tools for generating 3D models.

2.1 Application of GenAI in Manufacturing Systems

Doanh et al. discuss that genAI can support engineering design, quality control, predictive maintenance, demand forecasting, and marketing strategy, among others [4]. Kusiak adds application potentials of genAI for software development, work instructions and user manuals, and motion planning and control of robots [5]. Westphal and Seitz highlight the application of genAI specifically for Additive Manufacturing [6]. Mata et al. discuss the application potential for the digital twin design [7]. Shafiq et al. address the smart materials design based on genAI [8] while Badini et al. add a perspective for enhancing mechanical and biobased materials by integrating genAI into its design and manufacturing processes [9]. Wang and Zhang discuss genAI in the context of supporting sustainable business model innovations [10]. Haridasan and Jawale stress the long-term benefits of genAI, such as supporting sustainability, personalised production and autonomous systems [11]. Yang et al. discuss the relevance of genAI for a future transition towards Manufacturing 5.0 and Industry 5.0, while Sai et al. also highlight the importance of ethics for a future integration of genAI [12, 13].

2.2 Application of GenAI in SMEs

Hussain and Rizwan present a framework that discusses the role of leadership, as well as the integration of general purpose, task-specific, and in-house developed genAI tools to support the genAI adoption in SMEs [14]. Walke et al. present a grounded theory approach for characterising genAI usage in SMEs [15]. Rajaram and Tinguely list promises and challenges for the application of genAI in SMEs [16]. Shore et al. argue that genAI can significantly enhance entrepreneurial resilience in SMEs [17]. Ahmad et al.'s research supports these findings by arguing that genAI in decision-making enhances supply chain resilience for SMEs [18].

2.3 Research Gap

While current studies discuss the usage of genAI across manufacturing systems, the literature seems to insufficiently address the perspective of manufacturing SMEs. This includes application cases of genAI for manufacturing SMEs as well as a more holistic discussion of potential benefits and barriers of using genAI from an Industry 5.0 perspective with its three dimensions: resilience, sustainability, and human-centricity [2].

3 Research Aims and Method

Our study aims to explore the opportunities for manufacturing SMEs to use genAI tools to substantially reduce the time-to-market. Additionally, the study discusses and highlights the benefits and barriers of genAI adoption from an Industry 5.0 perspective. Hence, our study is loosely following the idea of fictional-critical writing and analysis as a qualitative research method [19]. For this purpose, we firstly present the fictional case study, enhanced by real tool applications, and secondly provide a critical analysis and discussion of the case in a broader context from an Industry 5.0 perspective.

4 Case Study

The case study presents a fictional small toy manufacturing company that uses additive manufacturing as well as laser cutting as manufacturing technologies to produce simple toys and sets of wooden toy puzzles. The fictional toy company covers the engineering design phases, 'ideation, Computer-Aided Design (CAD), and Computer-aided Manufacturing (CAM)', to eventually create a CAM model for other contract manufacturers that provide manufacturing-as-a-service. These contract manufacturers manufacture the toys according to the CAM models and send them back to the case company for final assembly, commissioning and packaging. The case study intends to explore how this fictitious small toy manufacturing company can apply different genAI tools to substantially reduce the time-to-market. Easy replication of the case study is important to facilitate application across the manufacturing SME sector with its typical limitations.

4.1 Ideation

Creativity informs ideation, which is the first phase of the company's value creation process for developing new designs of toy sets. Ideation is supported by market research and customer development to identify new market trends for the target group, which is younger children aged 4 to 8. The toy company uses genAI tools, and more specifically, text-to-image generation, to support the ideation process. The genAI tool translates generated ideas for new toy sets into specific images of how the new toy set could look like. For example, Microsoft Copilot® [20] can be used to create an image of a toy castle based on the prompt "create a virtual image of a simple toy castle for children aged 5, made from one piece of material. The image should have no background to serve as a foundation for a 3D model". The usage of the genAI tool allows the company to quickly visualise different design variants for a new toy set. This AI-generated image can further facilitate the discussion process with different external and internal stakeholders, such as focus groups, engineers or production planners.

4.2 Computer-Aided Design (CAD)

The next phase is to create a CAD model of the actual toy set. For this purpose, the toy company uses image-to-3D-model generation as a suitable genAI tool to quickly create different 3D models of the toy set. This reduces substantial CAD time as the genAI tool can create typical digital engineering files, such as OBJ or STL files. For example, Tripo AI [21] can be used to AI-generate a 3D model based on the AI-generated image from the previous ideation phase. Consequently, the company realises a direct connection between the output of the ideation phase and the output of the CAD phase. This streamlining of processes reduces the time-to-market. The engineer from the company now spends most of the time on post-processing the output generated from the 3D model, as the 3D mesh, as output from the genAI tool, usually requires further optimisation.

4.3 Computer-Aided Manufacturing (CAM)

The last phase covers the creation of a CAM model and G-code of the 3D toy model to be sent to the contract manufacturers for mass production. For this purpose, the toy company directly feeds the STL file from the 3D model generation or the optimised post-processed STL file into a CAM software tool. The fictional company can use the Autodesk Slicer for Fusion 360 [22] to automatically generate the laser cutting patterns for the wooden toy puzzle sets, as well as the Prusa Slicer [23] to automatically generate a 3D print model (G-code) of the toy. As the creation of these types of CAM models might not be based on computational algorithms using training data, this CAM phase as part of the engineering design process might not be directly supported by genAI in this case. However, real genAI CAM tools such as 'AI-NC.com' [24] already exist.

5 Critical Discussion

The case study covers the value creation of a fictional toy manufacturing SME. Hereby, the application opportunity of genAI for two relevant phases of the value creation process, ideation and CAD, is demonstrated (Fig. 1). The fictional case study shows a substantial reduction in time-to-market by applying genAI tools compared to traditional engineering tools, as ideation and CAD modelling can be facilitated and sped up considerably. For example, genAI can help to quickly generate images of a new product, and the CAD department can outsource complex design tasks to the genAI tool, which performs the tasks that previously took hours within minutes. Through this fast generation of quality outputs, which are also interconnected, genAI supports very quick iterations and testing of different product variants.

Fig. 1. Value creation of the fictional toy manufacturing SME

Limitations of the case study include the fictionality of the case as well as the toy sector. Both aspects lead to a constructed and unique case, which does not reflect the reality of the whole manufacturing industry and manufacturing SMEs. However, we believe that the case, even though fictional, can demonstrate important impacts of genAI on typical engineering tasks as well as future opportunities for reducing the time-to-market in manufacturing SMEs.

The following sections discuss potential Industry 5.0 benefits and barriers to using genAI in manufacturing based on the study findings. The critical discussion follows the resilience, sustainability and human-centric dimension of an Industry 5.0 [2].

5.1 Potential Resilience Benefits of Using GenAI Tools in Manufacturing

Resilience benefits of using genAI tools in manufacturing might include:

- Reducing the time-to-market for new products and product variations by decreasing engineering time for ideation and CAD modelling.
- Offering new opportunities for open innovation and co-creation with stakeholders by simply creating different ideas and a variety of designs for new products.
- Facilitating the exploration and easy integration of new market trends by quickly testing different design features, materials, etc.
- Enabling agile and lean manufacturing methods by adopting genAI to simplify and accelerate new product development and validation processes.
- Supporting asset-less and manufacturing-as-a-service business models by creating CAM files for quick and automated exchange with contract manufacturers and other supply chain actors, which can facilitate the integration of the regional and national manufacturing sector into the global market.

5.2 Potential Sustainability Benefits of Using GenAI Tools in Manufacturing

Sustainability benefits of using genAI tools in manufacturing might include:

- Facilitating novel sustainable product designs by exploring designs with new geometries for reduced material consumption or bio-based, reused, remanufactured, recycled materials and components.
- Supporting the co-creation of products for reduced material and energy consumption by supporting the discussion of different designs and manufacturing options with relevant stakeholders, such as contract manufacturers.
- Improving the generation of machining-toolpaths and instructions for improved material and energy consumption by optimising manufacturing parameters.
- Contributing to the elimination of waste along the engineering and manufacturing process chain by reducing the overall time-to-market.
- Supporting the environmental competency and awareness building of engineers by guiding the engineering process.

5.3 Potential Human-Centric Benefits of Using GenAI Tools in Manufacturing

Human-centric benefits of using genAI tools in manufacturing might include:

- Supporting the training and overall competency development of engineers and other professionals by guiding the engineering process and allowing for easy experimentation during the digital design and manufacturing processes.

- Facilitating the meaningful integration of stakeholders into the engineering design and manufacturing process by fostering easy demonstration of design and manufacturing results without requiring specific domain knowledge.
- Improving the possibility for considering and testing relevant stakeholder perspectives and human-centric features during the product design by allowing quick iterations and experimentations in a digital environment.
- Lowering the entry barriers for engineering design and manufacturing and thus contributes to democratising industrial value creation by offering simple-to-use toolkits for end-to-end engineering and manufacturing.
- Supporting the exploration of novel and more socially relevant products by opening the solution space for manufacturing solutions.
- Facilitating human and AI interaction by offering easy-to-use software tools that can often be guided through prompts in natural language.

5.4 Potential Barriers to Using AI Tools in Manufacturing for SMEs

Barriers to using genAI tools in manufacturing might include:

- Finding the right tools for the right design and manufacturing applications might be challenging, as new tools are released frequently.
- Lacking data interoperability between different genAI software tools makes streamlining engineering processes difficult.
- Reworking, adjusting, and converting genAI outputs are often required, for example, to create assemblies of multiple parts and components or to make genAI designs suitable for the intended purpose.
- A reluctance to use genAI tools by established and experienced engineers might prevail in some manufacturing companies.
- Integrating genAI tools into existing IT infrastructure and formal engineering best practices and procedures might be difficult as there could be a trend towards "patchwork solutions" with a variety of different genAI tools that must be used and maintained.
- GenAI tools often require fine-tuning of parameters and some degree of domain knowledge to generate suitable outputs for manufacturing.
- For external co-development, data privacy protection methods are required to protect against cyber threats, ensuring responsible genAI implementation.
- Ethics- and overreliance-related risks of using genAI tools seems not well understood yet for using these tools in manufacturing.

6 Conclusion

This research demonstrated a fictional SME case study for combining multiple genAI tools throughout the engineering design and manufacturing process to significantly reduce the time-to-market of new products. Subsequently, a critical discussion of the case in a broader context from an Industry 5.0 perspective has been presented. Multiple potential benefits in the resilience, sustainability, and human-centricity dimensions of Industry 5.0 have been laid out, and barriers for genAI adoption in manufacturing have been addressed. Manufacturing SMEs should try to overcome these barriers for an efficacious implementation of genAI in manufacturing value creation and to realise a genAI-supported transition towards Industry 5.0. Future research should focus on shaping complex manufacturing value creation using a combination of genAI tools. A streamlined and interoperable data exchange between the genAI tools and existing digital engineering processes seems to be a crucial research field in this context. Further, ethics- and overreliance-related risks to genAI adoption in manufacturing should be explored.

References

1. Singla, A., Sukharevsky, A., Yee, L., et al.: The state of AI: How organizations are rewiring to capture value. McKinsey (2025). https://www.mckinsey.com/capabilities/quantumblack/our-insights/the-state-of-ai. Accessed 06 May 2025
2. Directorate-General for Research and Innovation, Breque, M., De Nul, L., Petridis, A.: Industry 5.0 – Towards a sustainable, human-centric and resilient European industry, Publications Office of the European Union, p. 13 (2021)
3. Feuerriegel, S., Hartmann, J., Janiesch, C., et al.: Generative AI. Bus. Inf. Syst. Eng. **66**, 111 (2024). (2024). https://doi.org/10.1007/s12599-023-00834-7
4. Doanh, D.C., Dufek, Z., Ejdys, J., et al.: Generative AI in the manufacturing process: theoretical considerations. Eng. Manage. Prod. Serv. **15**(4), 76 (2023). https://doi.org/10.2478/emj-2023-0029
5. Kusiak, A.: Generative artificial intelligence in smart manufacturing. J. Intell. Manuf. **36**, 1–3 (2025). https://doi.org/10.1007/s10845-024-02480-6
6. Westphal, E., Seitz, H.: Generative artificial intelligence: analyzing its future applications in additive manufacturing. Big Data Cogn. Comput. **8**(7), 74 (2024). https://doi.org/10.3390/bdcc8070074
7. Mata, O., Ponce, P., Perez, C., et al.: Digital twin designs with generative AI: crafting a comprehensive framework for manufacturing systems. J. Intell. Manuf. **37**, 1049–1072 (2026). https://doi.org/10.1007/s10845-025-02583-8
8. Shafiq, M., Thakre, K., Pandurangan, R., et al.: Generative AI designs the next generation of smart materials from pixels to products. Int. J. Adv. Manuf. Technol. (2025). https://doi.org/10.1007/s00170-025-14999-w
9. Badini, S., Regondi, S., Pugliese, R.: Enhancing mechanical and bioinspired materials through generative AI approaches. Next Mater. **6**, 100275 (2025). https://doi.org/10.1016/j.nxmate.2024.100275
10. Wang, S., Zhang, H.: Leveraging generative artificial intelligence for sustainable business model innovation in production systems. Int. J. Prod. Res. 63(18), 6732–6757 (2025). https://doi.org/10.1080/00207543.2025.2485318
11. Haridasan, P.K., Jawale, H.: Generative AI in manufacturing: a review of innovations, challenges and future prospects. J. Artif. Intell. Mach. Learn. Data Sci. **2**(2), 1418 (2024). https://doi.org/10.51219/jaimld%2Fpraveen-haridasan%2F321
12. Yang, J., Wang, Y., Wang, X., et al.: Generative AI empowering parallel manufacturing: building a "6S" collaborative production ecology for manufacturing 5.0. IEEE Trans. Syst. Man Cybern.: Syst. **54**(11), 6525–6533 (2024). https://doi.org/10.1109/TSMC.2024.3349555
13. Sai, S., Sai, R., Chamola, V.: Generative AI for industry 5.0: analyzing the Impact of ChatGPT, DALLE, and other models. IEEE Open J. Commun. Soc. **6**, 3064 (2025). https://doi.org/10.1109/OJCOMS.2024.3400161
14. Hussain, A., Rizwan, R.: Strategic AI adoption in SMEs: a prescriptive framework. arXiv preprint arXiv:2408.11825, pp. 3–10 (2024)

15. Walke, F., Klopfers, L., Winkler, T.J.: Leveraging generative AI and ChatGPT in SMEs: a grounded model. In: Proceedings of the 58th Hawaii International Conference on System Sciences, pp. 5479–5488 (2025). https://doi.org/10.24251/HICSS.2025.658
16. Rajaram, K., Tinguely, P.N.: Generative artificial intelligence in small and medium enterprises: navigating its promises and challenges. Bus. Horiz. **67**(5), 631–633 (2024). https://doi.org/10.1016/j.bushor.2024.05.008
17. Shore, A., Tiwari, M., Tandon, P., Foropon, C.: Building entrepreneurial resilience during crisis using generative AI: an empirical study on SMEs. Technovation **135** (2024). https://doi.org/10.1016/j.technovation.2024.103063
18. Ahmad, K., Rozhok, A., Revetria, R.: Supply chain resilience in SMEs: integration of generative AI in decision-making framework. In 2024 International Conference on Machine Intelligence and Smart Innovation (ICMISI), pp. 295–299. IEEE (2024). https://doi.org/10.1109/ICMISI61517.2024.10580495
19. Bolton, G.: Stories at work. Fictional-critical writing as a means of professional development. Br. Educ. Res. J. **20**(1), 55–68 (1994)
20. Microsoft: Microsoft Copilot®. https://copilot.microsoft.com/
21. Tripo: Tripo AI. https://studio.tripo3d.ai/
22. Autodesk: Slicer for Fusion 360. https://www.autodesk.com/support/technical/article/caas/tsarticles/ts/3yg7zznS94MHNDG7KMV8Qg.html
23. Prusa: PrusaSlicer. https://www.prusa3d.com/page/prusaslicer_424/
24. AI-NC: Manufact. https://www.ai-nc.com/

Implementing Artificial Intelligence to Enhance Sustainable Value Creation of Palm Date Supply Chain: Case Study of Palestine

Ahmed Abu Hanieh(✉) and Afif Akel Hasan

Birzeit University, Al Marj Street 1, P.O. Box 14, Birzeit, Palestine
ahanieh@birzeit.edu

Abstract. This research concentrates on involving modern Artificial Intelligence (AI) technologies in the different stages of palm date production in Palestine. Palm date farms are now increasing and accelerating to compete with olive sector as one of the most important income sources, knowing that olive sector was the main contributor in GDP in Palestine since a long time ago. This diversification in agricultural income helps in improving the Palestinian economy which is climbing slowly due to political volatile situation. These sectors suffer from many problems and obstacles related to high weakness in implementing modern agricultural technologies which influences the production's quality and quantity. Farmers in Palestine still use traditional means in cultivation and post-harvest industries. Using modern AI technologies is becoming a stringent need to improve production and increase the value creation of these products. Simulation shows that implementing AI techniques can double the added value increasing the value creation index from 1.4 to 2.3. To implement AI while enhancing sustainability, energy sources used in agricultural processes can be transferred into renewable ones, mainly solar energy. Using high efficiency Photovoltaic panels in irrigation and cultivation processes can enhance quality and productivity of dates significantly.

Keywords: Added Value · Artificial Intelligence · Palm Dates · Sustainability · Value Creation Index

1 Introduction

Palm tree is a well rooted plant in the Middle East and North Africa MENA and it was mentioned in the holy books. Date fruit is a healthy food that is consumed extensively in the area due to its nutrient values being rich in copper, magnesium, potassium minerals and fibers [1, 2]. Dates and their products are consumed in large amounts in Ramadan, Muslims fasting month, and in condolence occasions. The ten top producing countries in dates in 2022 as shown in Table 1 are mostly from the MENA region.

Palm tree although has been in Palestine since biblical times it has been revived in the last 20 years [3], and becoming one of the important agriculture products in Palestine. Figure 1 shows production yields per hectare for the last few years, this figure indicates that the yield per hectare in Palestine is relatively lower than other countries like Egypt.

H. Kohl et al. (Eds.): GCSM 2025, LNME, pp. 315–324, 2026.
https://doi.org/10.1007/978-3-032-21154-5_35

Table 1. Top ten dates producing countries in 2022 [2]. (https://worldpopulationreview.com/country-rankings/date-production-by-country retrieved Feb 1 2025).

Country	Production Yield 2022 (Tons)	Area Harvested 2022 (Hectare)	Yield Per Hectare 2022 (100g/Hectare)
Egypt	1,733,433	58,557	296,027
Saudi Arabia	1,610,731	156,460	102,948
Algeria	1,247,404	176,044	70,858
Iran	1,030,460	122,332	84,235
Pakistan	732,936	102,081	71,800
Iraq	715,293	278,159	25,715
Sudan	442,667	37,046	119,492
United Arab Emirates	397,329	40,739	97,531
Oman	376,980	26,355	143,037
Tunisia	369,000	77,805	47,426

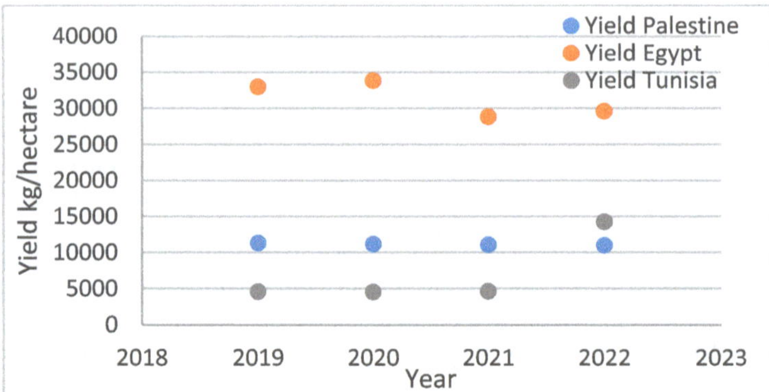

Fig. 1. Production yield kg/hectare in some MENA countries.

Pandey et al. 2024 carried out a comprehensive review of advanced biotechnology including genetic engineering, the incorporation of AI, and machine learning. The genetic engineering could lead to date cultivars resilient to biotic stresses with enhanced disease resistant and high productivity [4]. They concluded that employing AI can ensure long term sustainability. They also concluded that optimizing irrigation requirements, and improving cold storage using AI and IoT can help farmers to reach solutions that improve date palm farming. Invoking IoT based subsurface irrigation can lead to a 70% saving in water usage compared to conventional irrigation was found by Maged et al. 2021 in [5].

Implementing AI for the prediction and pest management solutions are investigated by Hessane et al. 2024 [6], the developed model employs climatological parameters for the prediction of the nymphs and the adult females in order to destroy them in time and provide effective pest management.

A comprehensive review is given by Fotis et al. 2024 of AI applications in the various stages of date palm supply chain. Review covered selecting crop, managing resources, detecting disease, production prediction, and market integration. They pointed out that to exploit AI potential in supply chain technical, economic, as well as ethical problems need to resolved [7].

Hossein et al. 2023 simulated the employment of anti-coronavirus optimization algorithm (ACVO) and an adaptive neuro-fuzzy inference system (ANFIS) in the cultivar of dates and found that good performance in estimating the production yield [8].

Khriji et al. 2020 used Machine-Learning (ML) methods for classification of date fruits, based on color, shape, size, and texture features. Tested methods included; Artificial Neural Networks (ANN), Support Vector Machine (SVM), and K-Nearest Neighbor (KNN). They found that ANN classifier gives better performance than SVM and KNN with classification accuracy of 99.2% [9]. Also Ali et al. 2024 Developed a convolutional neural network (CNN) method to classify date based on size, and color characteristics using machine learning and deep learning. They proved that the CNN method achieved 98.08% accuracy for dates classification [10].

Saudi Arabia as the second world producer of date with a moderate productivity per planted area would benefit greatly from implementing AI technologies in the date palm supply chain. Application of AI and IoT in the date palm cultivation in Saudi Arabia was investigated by Mohammed et al. 2023[11], they presented some experimental results and verification of developed models in the various stages of the date palm cultivation. Models included the use of sensors, and ML algorithms, for control of irrigation processes, pest management, detection of disease, quality and readiness for fruits for harvesting.

Value creation cycles according to Heyer and Seliger in [12], "constituting consecutive multiple phases of usage and remanufacturing enable for increased use productivity of resources.". They explained that an engineering value creation model would consist of planning, scheduling of manufacturing and all recycling processes. According to Pinar et al. in [13], the ratio between output and input during production processes leads to value creation. They propose a model based approach for value creation that incorporates profitability and sustainability. The attributes are specified and then assigned a value, then all values are weighted to obtain an overall value for all processes involved.

2 Methodology and Implementation

The main research methodology in this paper is based on studying the possibilities of integrating the AI technologies into the process and supply chain of the date palm sector. This can be assessed by presenting the benefits and drawbacks of the implementation in each stage of the process. In the following subsections, AI techniques are suggested for each step:

2.1 Integrating AI in Palm Date Supply Chain

As mentioned in the previous section, AI is one of the main technologies that lead to improving value creation and sustainability of the products. In this section, AI techniques will be integrated into the different stages of palm date production. Palm date is produced through four stages; palm tree cultivation, date harvesting, product treatment and post-harvest industries.

Figure 2 shows that irrigation of palm trees can be modified by smart dripping irrigation that save 60–70% of consumed water in comparison to flooding irrigation. Using fertilizers can be governed by chemical sensors while common diseases can be controlled specific sensors depending on the infection. Pollination can be done using smart extractors to exploit the maximum amount of pollens to ensure the best fertility. Automatic guided vehicles can be used to simplify the pruning and thinning processes during cultivation to keep workers away from the palm harming thorns.

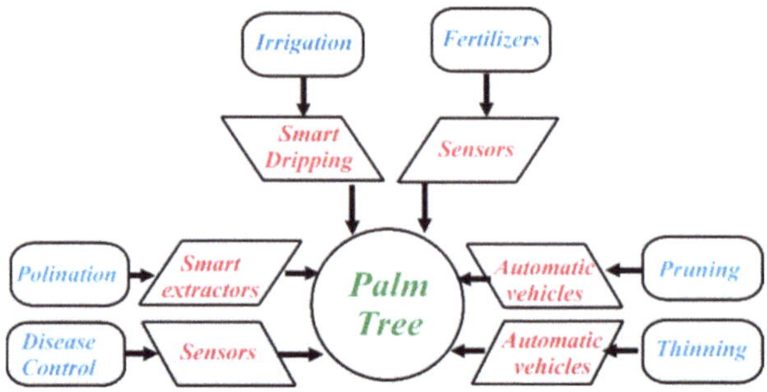

Fig. 2. Implementing AI technologies in palm tree cultivation

The second stage in palm date production is the harvest stage. Labor, machinery and energy are widely used in this stage. Figure 3 shows that it is possible to use automatic guided vehicles and smart tools in the harvest process. Using these tools and machines help in keeping farmers away from the big thorns and needles existing between the palm leaves especially in the Medjoul date type.

After harvest, dates are taken to special storage places with controlled temperature and humidity. Humidity and temperature sensors are used for this purpose and for the processes of hydration and dehydration necessary to avoid damage of fruits. Fumigation is another process implemented to protect date fruits from germs and insects and here it is necessary to use acidity and vapor sensors as shown in Fig. 4.

The final stage in post-harvest is the industrial applications of date fruits. Sorting, packaging and transporting palm dates need to be done in smart warehouses and flexible conveying systems as shown in Fig. 5. Date fruits are then targeted either directly to table for eating or to specific industries like bakery sweets and pharmaceuticals. The palm leaves and date seeds are used widely as a protein-rich animal feed.

Application of suggested AI and automation technologies would improve the sustainability of the sector and contribute to achieving the SDGs. During the cultivation stage application of AI for irrigation and pest control results in improving material efficiency by reducing water consumption for irrigation and pesticide use for pest management. Using automations and energy efficient machinery for harvesting, reduces fuel consumption leading to less CO2 emissions. In addition, less water and energy will be used once AI is applied in the dates treatment stage. Application of AI and automation in the date supply chain would contribute to SDG goals including SDG 12 Responsible Consumption and Production through resource efficiency, SDG 13 Climate Action through energy reduction and GHG emission reduction, and SDG 2 Zero hunger through promoting sustainable agriculture.

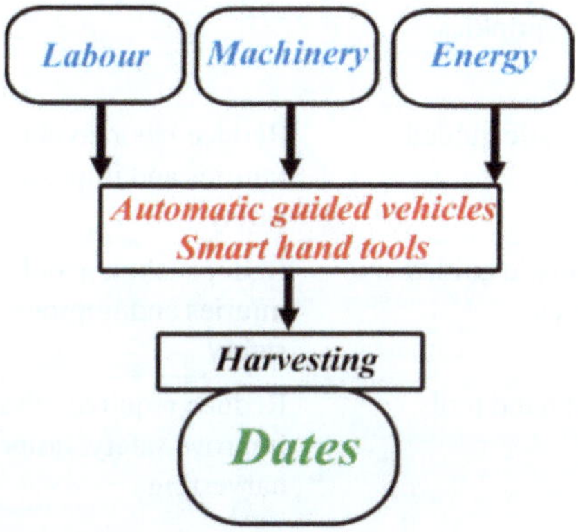

Fig. 3. Implementing AI to palm date harvesting stage.

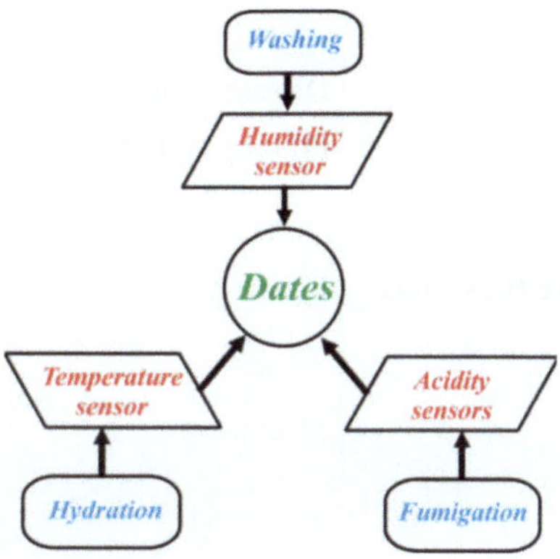

Fig. 4. AI techniques in date treatment stage

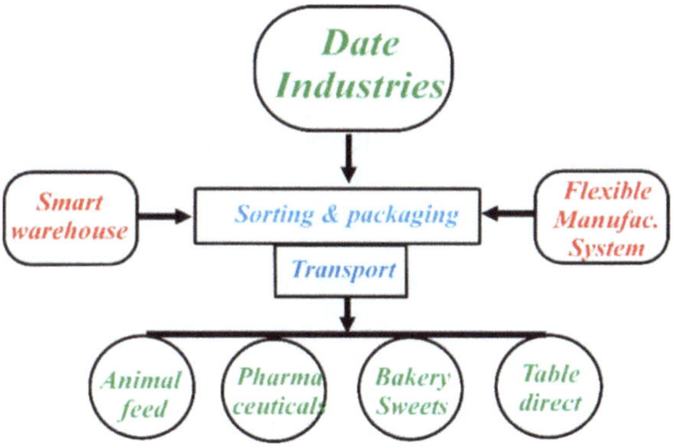

Fig. 5. Implementing AI technologies in palm date after-harvest industries.

2.2 Limitations and Values of Implementing AI Techniques

Table 2 summarizes the different AI techniques used in palm date sector to improve the added value and develop the local economy. The different AI techniques for each process with the potential values and limits is shown in this table.

Table 2. Details of the AI techniques used in palm date sector

Phase	Process	AI Technologies	Limits of AI technologies
Cultivation	Pollination	Smart extractors	Reduce required labor
	Fertilizing	Nitrogen level sensor	Exact amount of fertilizer
	Irrigation	Temperature sensors Humidity sensors Smart sprinkles	Use just the required water at the right time
	Disease control	sensors	
	Pruning and thinning	Automatic guided vehicles	Reduce labor, avoid injuries and improve safety
Date Harvesting	Labor	Automatic guided vehicles	Reduce labor, avoid injuries and improve safety
	Machinery	Smart hand tools	Reduce required labor improve safety, faster harvesting
	Energy		
Date Treatment	Transport	Automatic Guided Vehicle	Reduce labor, avoid injuries and improve safety
	Washing	Humidity sensor	Reduce consumed water
	Hydration	Temperature sensor	Improved quality of dates
	Dehydration	Temperature sensor	Improved quality of dates
	Fumigation	Acidity sensor	Improved quality of dates
Date Industries	Pharmaceuticals	Smart controlled machines	Improve product quality, larger and faster production
	Table direct	Direct Numerical Control	Improve product quality, larger and faster production
	Bakery sweets	Computer Numerical Control	Improve product quality, larger and faster production

(*continued*)

Table 2. (*continued*)

Phase	Process	AI Technologies	Limits of AI technologies
	Animal feed	Flexible manufacturing system	Improve product quality, larger and faster production
	Food		Improve product quality, larger and faster production

3 Implementing AI to Enhance Palm Date Value Creation

Any product needs an added value to the investment to be feasible. The added value can be defined as the difference between costs and revenues of a product.

$$AV = R - C \tag{1}$$

where, *AV* is the added value, *R* is the revenue and *C* is the cost. The revenue can be calculated by multiplying the price of palm date per kilogram by the amount of production kilograms. While the cost is calculated as the summation of costs of labor, water, machinery and energy. Simulation on MATLAB showed that the added value for the production of one palm tree without implementing AI techniques was about 200 USD per tree per year. While implementing AI technologies in irrigation, energy and labor proved to increase the added value to more than 400 USD per tree per year during the effective life period of the palm tree from 10 to 35 years, as shown in Fig. 6.

One of the most important indicators of the added value is the value creation index. Value creation index can be calculated from the equation:

$$VCI = R/C \tag{2}$$

where *VCI* is the value creation index, *R* is the revenue and *C* is the cost. Figure 7 shows that without AI, the value creation index is about 1.4. While with implementing AI technologies, the accumulated VCI is also presented in the same figure. The value creation index reached about 2.3 during the effective life period of the palm tree from 10 to 35 years.

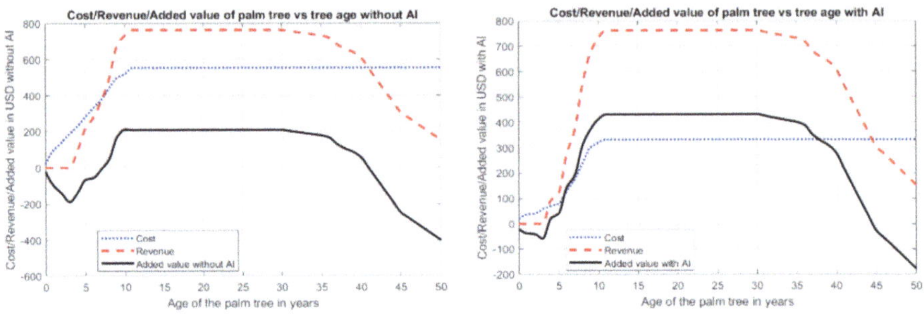

(a) without AI *(b) With AI*

Fig. 6. Palm date cost, revenue and added value without using AI

Fig. 7.Palm date VCI and accumulated VCI with and without using AI

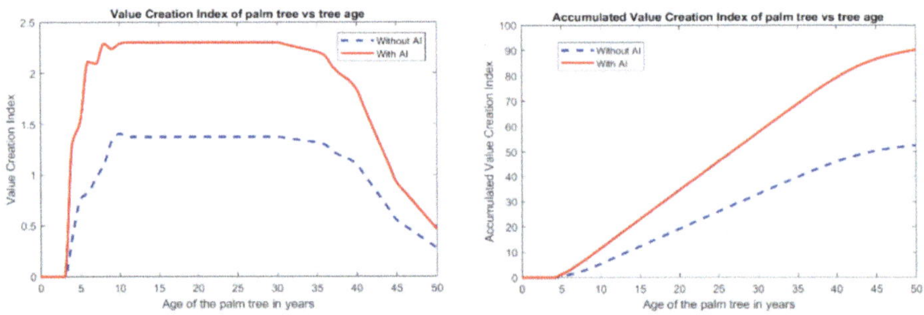

(a) without AI *(b) With AI*

Fig. 7. Palm date VCI and accumulated VCI with and without using AI

4 Conclusions

The foregoing study focused on the implementation of AI and modern technologies to develop the palm date sector in Palestine. The implementation of AI technologies in palm date is investigated toward achieving the United Nations SDGs. Different automation and AI techniques are considered for implementation in each of the four stages of date production; cultivation, harvesting, treatment and industrial applications.

Simulation on Matlab shows that using modern AI technologies can improve the added value of the date sector significantly. The added value proved to be doubled from 200 USD to 400 USD per tree per year. In addition, the simulation showed that the value creation index increased from 1.4 to 2.3 by applying AI technologies to the various stages and processes of palm date supply chain. Although using AI technologies increases the initial costs in the beginning of the palm lifecycle but it gives a better production and added value in the long run.

References

1. Dreher, K.: Whole fruits and fruit fiber emerging health effects. Nutrients **10**(12), 1833 (2018). https://doi.org/10.3390/nu10121833. https://health.clevelandclinic.org/benefits-of-dates. Accessed 30 Jan 2025
2. https://worldpopulationreview.com/country-rankings/date-production-by-country. Accessed 1 Feb 2025
3. Hanieh, A.A., Hasan, A., Assi, M.: Date palm trees supply chain and sustainable model. J. Clean. Prod. **258**, 120951 (2020)
4. Pandey, N., Tripathi, P., Pandey, N., Nakum, H., Vala, Y.S.: The sustainable improvement and precision agriculture in the date palm cultivation-a review. J. Agric. Ecol. **19**, 35–48 (2024). https://doi.org/10.58628/JAE-2419-203
5. Maged, M., Riad, K., Alqahtani, N.: Efficient IoT-based control for a smart subsurface irrigation system to enhance irrigation management of Date palm. Sensors **21**(12), 3942 (2021)
6. Hessane, A., El Youssefi, A., Farhaoui, Y., Aghoutane, B., Alaoui, E.A.A., Nayyar, A.: Artificial intelligence-driven prediction system for efficient management of Parlatoria Blanchardi in date palms. Multimed. Tools Appl. **84**, 15293–15329 (2024). https://doi.org/10.1007/s11042-024-19635-5
7. Assimakopoulos, F., Vassilakis, C., Margaris, D., Kotis, K., Spiliotopoulos, D.: Artificial intelligence tools for the agriculture value chain: status and prospects. Electronics **13**(22): 4362 (2024). https://doi.org/10.3390/electronics13224362
8. Dehghanisanij, H., Salamati, N., Emami, S., Emami, H., Fujimaki, H.: An intelligent approach to improve date palm crop yield and water productivity under different irrigation and climate scenarios. Appl Water Sci **13**, 56 (2023). https://doi.org/10.1007/s13201-022-01836-8
9. Khriji, L., Ammari, A.C., Awadalla, M.: Artificial intelligent techniques for palm date varieties classification. **11**(9) (2020)
10. Ali, J., Ramzan, M., Rafiq, A.: Deep palm: deep learning classification model for date palm varieties. J. Comput. Biomed. Inform. ICASET. 2710 - 1606 (2024)
11. Mohammed, M., Alqahtani, N.K., Munir, M., Eltawil, M.A.: Applications of AI and IoT for advancing date palm cultivation in Saudi Arabia. In: Internet of Things-New Insights. IntechOpen (2023)
12. Heyer, S., Seliger, G.: Open manufacturing for value creation cycles. In: Matsumoto, M., et al. (eds.) Design for Innovative Value Towards a Sustainable Society (2012). https://doi.org/10.1007/978-94-007-3010-6_23, © Springer Science+Business Media Dordrecht 2012
13. Bilgea, P., Badurdeenb, F., Seliger, G., Jawahir, I.S.: Model-based approach for assessing value creation to enhance sustainability in manufacturing. Procedia CIRP **17**, 106–111 (2014)

The Resource Friendly Experiment Agent

Enno Müller(✉) and Thomas Knothe

Fraunhofer IPK, 10587 Berlin, Germany
enno.mueller@ipk.fraunhofer.com

Abstract. Augmenting large language models (LLMs) with external tools has proven to be effective for producing consistent, deterministic results. While capabilities of agentic AI systems are growing rapidly and the integration into various domain application speeds up, their resource consumption finds little attention in literature. With continuously more capable models requiring more energy, coupled with ever increasing numbers of users, there is a need for alternative solutions. We present an approach of using small-scale open-source models locally powering an agentic AI system in the context of production planning. By defining and iteratively executing experiments, the system can optimize discrete-event simulation models – while running entirely on-device. With significant drawbacks in speed and performance, the local configuration consumes only a fraction of the estimated resources of state-of-the-art models. By offering further benefits in data sovereignty, operational independence and cost control, the concept of locally deployed models presents an attractive alternative for industry-grade manufacturing applications.

Keywords: Discrete Event Simulation · Agentic AI · Resource Consumption

1 Motivation

Conducting simulation studies in the context of Production Planning and Control (PPC) can be a time-consuming task. Testing different configurations virtually is a crucial part in finding the right system setup. A classic example from the field of material flow simulations is the optimization of buffer sizes to maximize target values such as overall throughput. Different experiments must be planned and performed to define the most influencing parameters for reaching the target values. The analysis of experiments requires a minimum set of knowledge in the respective domain for making conclusions and interpret the results accordingly. Large Language Models (LLM) are trained on a vast majority of publicly available data and shared knowledge across different fields [1]. Thus, the use of a LLM to coordinate simulation studies in the context of PPC represents a promising approach. With the fast adoption of LLMs throughout different application domains, their size and capabilities are constantly improving. While constrains in data privacy are a widespread argument against the integration of models hosted by external companies [2], the environmental impact during operation is rarely mentioned. Most published paper in this regard are focused rather on the training phase than the so-called inference phase during operation [3]. The deployment of state-of-the-art models

H. Kohl et al. (Eds.): GCSM 2025, LNME, pp. 325–333, 2026.
https://doi.org/10.1007/978-3-032-21154-5_36

at an industrial scale requires a considerable amount of hardware. A common alternative is to host smaller open-source models locally, which is feasible on business laptops even without a dedicated Graphics Processing Unit (GPU) [4]. Although using minimal resources, being independent from request limits, and allowing full control over ingested data, locally deployed models suffer significant drawbacks in output quality, consistency, and inference time. This paper aims to investigate the potential of a locally deployed LLM-powered system in the context of simulation studies.

2 Background

2.1 Agentic AI

Although LLMs are in the center of the widespread integration of artificial intelligence (AI) into domain applications, it is important to recognize their limitations when being confronted with tasks requiring deterministic output. Examples include resolving calculations, chemistry reactions or conducting simple simulations. Based on design, LLMs are trained to predict output based on the user's input, often delivering different results to identic prompts. A popular solution to overcome this problem, is to equip a model with external resources like plugins, databases or tools, expanding the model's capabilities beyond their training data. Using an external calculator tool enables the model to return exactly calculated answers consistently [5]. While there is no exact definition of an AI agent, a common understanding is that such systems make at least one autonomous decision based on the input prompted by the user. Typical decisions are tool selection, routing tasks or terminating a workflow when certain conditions are met. Other definitions state that an AI agent is capable of independent task planning, which translates to transferring a user input into a set of separate steps and actions [6]. A LLM equipped with a calculator tool is a simple form of an AI agent, also described as "the augmented LLM" [7]. The communication between LLM and external resource is usually realized with structured input and output formats. Recently, this development experienced rapid growth through the introduction of the Model-Context-Protocol (MCP). This protocol enables LLMs to discover existing tools, supports their execution process and adds error handling capabilities [8]. Today, there is a wide range of pre-built, ready-to-use MCP-servers available on GitHub [9] providing external tools. A variety of frameworks designed for developers and non-technical users greatly simplify the integration process, making it easy to adopt LLM-powered agentic systems. With high-level frameworks it is possible to build an agentic workflow system solving tasks by drag and drop without writing a single line of code [10]. Typical workflow patterns contain prompt chaining, routing, parallelization, orchestration and the evaluator-optimizer loop. More complex systems can consist of several independent agents each equipped with specialized external resources and potentially further nested workflow structures [7].

2.2 Resource Consumption

It is crucial to emphasize the implications on resource consumption when building agentic systems, especially when using high-level frameworks. Users have limited control over the communication process between agentic system and the LLM-provider. Every interaction in the workflow requires prompting a model. Text is broken into tokens, substring units that are used to process input and to concatenate the output [1]. Complex systems consisting of several agents can lead to a high number of token usage. When using state-of-the-art models, this can lead to an extraordinary energy consumption. The electricity demand of the main cloud-computing providers in America more than doubled between 2017 and 2021, a time before the rapid adoption of LLMs at the forefront of generative AI [11]. Often, the cooling is based on water, contributing to the environmental regional impact of data centers [12]. Recent studies suggest an average energy consumption of 0.14 kWh and 519 ml of cooling water to generate a mail consisting of 100 words when using ChatGPT-4 [13, 14]. The example of OpenAI surpassing one billion daily queries and 400 million weekly users at the beginning of 2025 demonstrates the need for alternative, more sustainable solutions [15]. Due to missing transparency of technology companies, exact numbers can only be retrieved by experiments. Luccioni et. al performed detailed experiments in their study comparing different models and different tasks. One of their key findings is that reasoning tasks consume energy on an utterly different scale [16] compared to simple queries such as the creation of a mail. Describing a general move towards more multi-purpose models, alternative resource-saving concepts are becoming increasingly important in the context of agentic AI.

3 Related Work

Coding agents are one of the fastest growing fields of agentic AI. With the "AI pair programmer", GitHub Copilot released their first version of a coding assistant in 2021 – enabling generation and auto-completion of code. Since then, coding assistants have seen a massive improvement in capability and are integrated into most Integrated Development Environments (IDE) [17]. The introduction of agentic AI coding tools marks the latest development in this field. Compared to early coding assistants, these systems can work solely with input in plain English text and are capable of building fully functional applications in seconds. Enabling non-technical people to prototype websites and mobile apps, the worldwide user adoption of these tools is exploding, setting new records in business figures such as company growths and annual recurring revenue [18]. Several other fields are addressed by agentic AI including research. A prominent example is "Deep research" introduced by OpenAI in 2024. A system capable of detailed web search, analysis and interpretation of vast amounts of data, images, files and text powered by a multi-purpose reasoning model. Reasoning involves a chain-of-thought process, where complex problems are broken down into subtasks and tackled one by one [19]. When prompted with a research task, it can take several minutes for the system to come back with a result. Another example in this field is the "AI Scientist-V2", introduced by Yamada et. al in 2025 [20]. This system was able to produce the first AI-generated paper passing a peer-review process prior to a research conference in machine learning. Based on a complex workflow including different phases like idea generation, experimentation and the actual paper generation including graphics, the system can provide a result within several hours up to a default maximum of 15 h. While no indications on energy consumption or system costs are included, the use of state-of-the-art models like GPT-4o [21] or Claude 3.5 Sonnet (v2) [22] suggests a significant environmental impact – mapped to 15 h and a maximum token length of 8,192 per request. In comparison – the generation of a 100-word mail consuming 0.14 kwh and half a liter of cooling water are equivalent to 133 output tokens. In the wide field of simulation, there has been a great utilization of LLMs enhancing the simulation capabilities. Due to the limited domain knowledge of LLMs, simulation represents an ideal approach of leveraging the LLMs capabilities and gain access to high-quality data. Power systems [23], war games [24], patients [25] and blockchain [26] are just a few examples, where simulation was used to augment the capabilities of LLMs. Gao et. al give a detailed overview on existing applications in various sectors [27]. LLM-powered agentic systems that conduct simulation experiments in the field of production planning have not been described yet in literature. The presented concepts above purely rely on externally hosted state-of-the-art LLMs emphasizing the lack of awareness of their environmental impact.

4 Experiment-Agent

4.1 System Design

This paper describes a concept for an agent-based AI system driven by small open-source LLMs hosted locally on a business laptop. The use case of the developed system is the investigation and optimization of existing simulation models in the context of production planning. Supplemented by a set of tools, the system processes a user request and performs simulation experiments, iteratively refining the model configuration until a certain target value is reached or no further improvement can be detected. The overall workflow is outlined in Fig. 1. In the first step, the user request is processed and categorized. With the execution of single experiments and the optimisation of specific model parameters, there are currently two different use cases supported. Based on the request, relevant model resources and attributes are being identified through interaction with a database. For the system evaluation, we will focus on the more complex optimisation use case, meaning the process of finding an ideal system configuration through iterative experiments. Once the relevant parameters have been narrowed down, a sequential experiment loop is triggered. In this loop, the parameters are first tested for general influence on the target value in short experiments. After each experiment, the results are analysed. For parameters with influence on the target value, more detailed experiments are performed. Thus, an ideal system configuration can be specified efficiently.

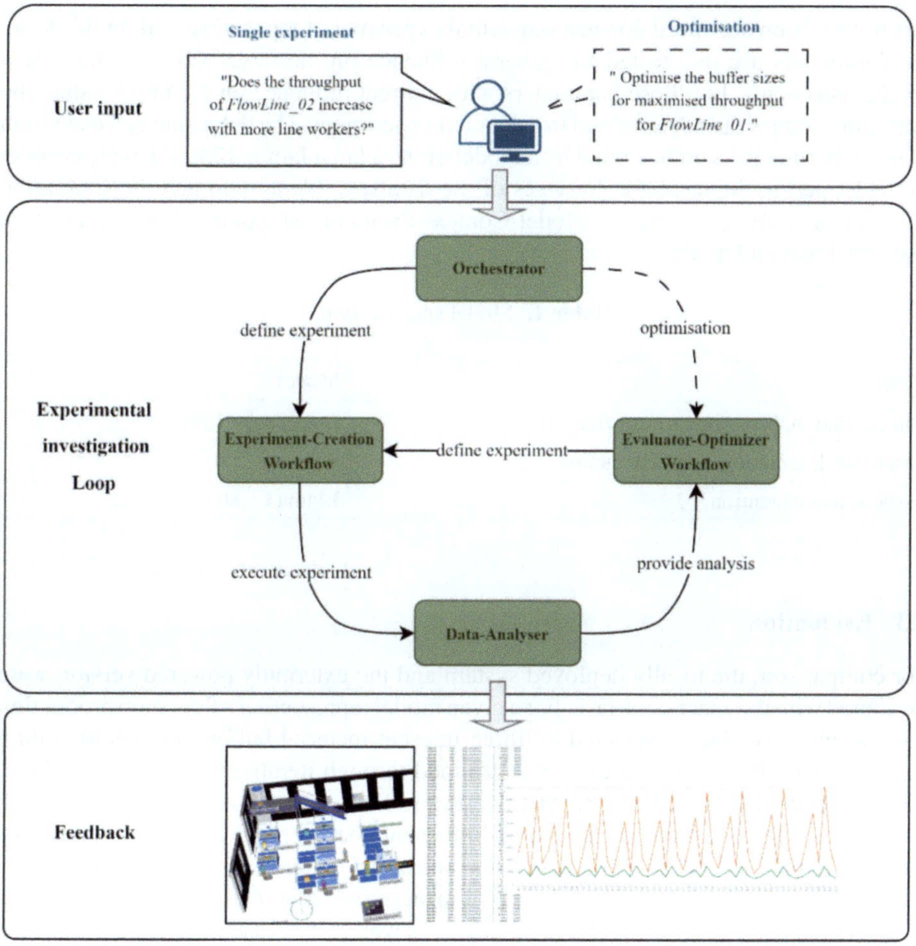

Fig. 1. System workflow

4.2 Implementation

As benchmark, a first proof-of-concept was built with OpenAI's GPT-4o. Subsequently, all interactions with external models were replaced by smaller local models, requiring the complexity of individual function calls to be split into multiple simpler calls. For the initial input processing and categorization, a distilled version of the DeepSeek R1 [28] model is used, leveraging its reasoning capabilities. Once, the request is understood and extracted from the user input, it must be translated into attributes of resources and parameters of the existing model. This is managed by interacting with a database containing specification and corresponding descriptions, enabling a LLM to understand the meaning of specific model parameters. The model specification and the complete task allocation of this application are shown in Table 1. For the database interaction, a model from the Granite-3.2 family [28] developed by IBM is applied, making use of its function calling capabilities. Once the relevant parameters for the optimization target have been narrowed down, a sequential experiment loop is triggered. In this loop, the parameters are first tested for general influence on the target value in short two-level experiments. For those parameters with a direct influence on the target value, the optimum value is then determined in a detailed experiment. The definition and execution of the experiments is orchestrated by a model of the Llama family [28]. Each experiment run is logged in the database. Services of the database, file system and the simulation software are exposed using the Model-Context-Protocol, wrapping all communication between tools and models.

Table 1. Model specification

Task	Model
Orchestration, Experiment data analysis	DeepSeek-r1:8b
Database-Interaction, Json-Extraction	Granite3.2:8b
Experiment execution	Llama3.1:8b

4.3 Evaluation

For comparison, the locally deployed system and the externally powered version were prompted with the exact same task. For a given model representing a flow shop production with several machines connected by three internal material buffers, the configuration for maximum throughput should be determined through iterative experiments. Table 2 displays the average results of successful test runs. Realizing the same functionality with smaller, less capable models results in a higher number of required LLM-interactions. While external models can extract the full scope of information at once, this process had to be split into manageable parts locally. Furthermore, functions requiring input data structures such as maps or lists were separated as well to decrease complexity. Thus, the integration of less capable models generates a considerable amount of additional work on the architecture side. Calculating out any programmatic delays necessary due to request limitations by external LLM-providers, the average cycle time differs drastically between both setups. The per-request energy consumption of the remote setup is calculated based on publicly available information. The energy demand of the local setup is roughly approximated by observing the total battery drainage. A typical application run consumes about 25% of the battery capacity, which corresponds to roughly 18,75 watthours. Since all components of the application run locally, the energy consumption is likely to be less than 18,75 Wh. The Chat-GPT based version requires fewer LLM calls overall. Based on 2,9 Wh per request [11] and a total of 35 requests per run, its energy use reaches approximately 102 Wh – over five times higher than the local setup. This estimate does not include other resource demands like cooling water used in the data center.

Table 2. System comparison

	LLM interactions	Cycle time (min)	Est. Energy (Wh)
Remote	35	3	101,5[a]
Local	56	21	<18,75

[a]Based on 2.9 Wh per request [11].

5 Discussion

The presented paper shows the potential resource savings when an agent-based AI system runs entirely on local consumer hardware. The proof-of-concept runs on a business laptop - without an internet connection or external services - and still performs complex tasks such as optimizing a simulation model. However, performance measurements show a clear trade-off. The time for inference and the consistency of the output differ drastically. While the remote configuration delivers reliable results in 3 min, local configuration takes up to 21 min with no guarantee of success. Due to the lack of reliability, the smaller distilled models with 8 billion parameters are not suitable for productive use. Nevertheless, for applications, where time is secondary to data integrity, cost control and the environmental impact, local deployment can be attractive. Most of the tasks do not require huge foundational models (and corresponding hardware demands). For local systems, finding an optimum between performance and hardware requirements and thus the resulting resource consumption depends on balancing the size of the model against the complexity of the task to be accomplished. Furthermore, it depends on the time one is willing to invest in the system architecture to break down complexity and adapt it to the model capabilities. The concept shows that small-scale LLMs can power agentic workflows entirely on-device in the context of production planning, serving as a proof of concept rather than a deployable solution. Future research will focus on integrating more powerful models and refining the balance between environmental impact, computational speed and solution quality to achieve an optimal system design suitable for use in production without being dependent on external actors.

References

1. Zimmerman, J.W., Hudon, D., Cramer, K., et al.: Tokens, the oft-overlooked appetizer: large language models, the distributional hypothesis, and meaning. arXiv (2024)
2. Yan, B., Li, K., Xu, M., et al.: On protecting the data privacy of large language models (LLMs): a survey. arXiv (2024)
3. de Vries, A.: The growing energy footprint of artificial intelligence, vol. 7 (2023)
4. IBM. What is a GPU? | IBM (2025). https://www.ibm.com/think/topics/gpu. Accessed 23 Apr 2025
5. Bran, A.M., Cox, S., Schilter, O., et al.: ChemCrow: augmenting large-language models with chemistry tools. arXiv (2023)
6. GitHub. What are AI agents? (2025). https://github.com/resources/articles/ai/what-are-ai-agents. Accessed 24 Apr 2025

7. anthropic. Building effective agents (2024). https://www.anthropic.com/engineering/building-effective-agents
8. Hou, X., Zhao, Y., Wang, S., et al.: Model Context Protocol (MCP): Landscape, Security Threats, and Future Research Directions (2025)
9. GitHub. GitHub - modelcontextprotocol/servers: Model Context Protocol Servers (2025). Accessed 10 May 2025
10. Pajo, P.: Vibe Coding: Revolutionizing Software Development with AI-Generated Code (2025, Unpublished)
11. Aljabour, J., Wilson, T.: Powering Intelligence: Analyzing Artificial Intelligence and Data Center Energy Consumption, Palo Alto, California (2025)
12. Verma, P., Tan, S.: A bottle of water per email: the hidden environmental costs of using AI chatbots. The Washington Post (2024)
13. Wright, I.: ChatGPT Energy Consumption Visualized - BEUK. Business Energy UK (2025)
14. OpenAI, Achiam, J., Adler, S., et al.: GPT-4 Technical report. arXiv (2023)
15. Reuters. OpenAI's weekly active users surpass 400 million. Reuters Media (2025)
16. Luccioni, S., Jernite, Y., Strubell, E.: Power Hungry Processing: Watts Driving the Cost of AI Deployment? (2024)
17. Zhang, B., Liang, P., Zhou, X., et al.: Practices and Challenges of Using GitHub Copilot: An Empirical Study, vol. 2023 (2023)
18. Hillion, P.: How Lovable grew to $17M ARR in 3 months. The Growth Mind (2025)
19. OpenAI. Deep Research System Card (2025)
20. Yamada, Y., Lange, R.T., Lu, C., et al.: The AI Scientist-v2: workshop-level automated scientific discovery via agentic tree search. arXiv (2025)
21. OpenAI. GPT-4o System Card (2024)
22. anthropic. Claude 3.5 Sonnet Model Card Addendum (2024)
23. Jia, M., Cui, Z., Hug, G.: Enhancing LLMs for Power System Simulations: A Feedback-driven Multi-agent Framework (2024)
24. Hua, W., Fan, L., Li, L., et al.: War and peace (WarAgent): large language model-based multi-agent simulation of world wars. arXiv (2023)
25. Yu, H., Zhou, J., Li, L., et al.: AIPatient: simulating patients with EHRs and LLM powered agentic workflow. arXiv (2024)
26. Diamatopoulos, G., Theodoropoulos, G., Tziritas, N., et al.: Towards LLM augmented discrete event simulation of blockchain systems. In: Loper, M., Pellegrini, A. (eds.) Proceedings of the 38th ACM SIGSIM Conference on Principles of Advanced Discrete Simulation, pp. 75–76. ACM, New York (2024)
27. Gao, C., Lan, X., Li, N., et al.: Large language models empowered agent-based modeling and simulation: a survey and perspectives, vol. 11 (2024)
28. Ollama (2025). Ollama. https://ollama.com/library/. Accessed 11 May 2025

The Path to a More Efficient Circular Economy by Integrating Deep Learning into Robotic Sorting Systems

Alexander Schlosser[1](✉), Jonas Walter[1], Lukas Roming[2], Jörg Franke[1], and Sebastian Reitelshöfer[1]

[1] Institute for Factory Automation and Production Systems, Friedrich-Alexander-Universität Erlangen-Nürnberg, Erlangen, Germany
alexander.schlosser@faps.fau.de

[2] Fraunhofer Institute of Optronics, System Technologies and Image Exploitation, Karlsruhe, Germany

Abstract. Automated robotic sorting plays a central role in boosting material recovery and reducing environmental impact in the circular economy. Lightweight packaging waste consists of heterogeneous and visually complex materials. This presents persistent challenges for object-level detection and classification. Deep learning–based instance segmentation offers a scalable and precise approach for identifying target materials and contaminants under dynamic real-world conditions. A newly developed image processing pipeline integrates structured preprocessing with state-of-the-art segmentation models. These models are trained on synthetic and real datasets designed to reflect both idealized and challenging scenarios, including occlusions, clutter, and variable lighting. In controlled environments, the models achieve high segmentation accuracy and real-time performance. However, generalization drops significantly when applied to real data, especially for visually ambiguous objects. These results highlight the need for domain-adaptive training and realistic variation in dataset design. Integrating such segmentation pipelines into industrial systems improves classification, reliability, and operational efficiency. Continued development of these technologies is essential to meet regulatory goals, reduce resource consumption, and promote sustainable practices in recycling.

Keywords: Recycling · Sorting · Automation

1 Introduction

Sustainable resource management and the reduction of packaging waste are central objectives of global environmental policy and contribute significantly to the achievement of the United Nations Sustainable Development Goals (SDGs) [1]. In this context, corporate environmental responsibility is gaining increasing relevance. The European Sustainability Reporting Standards (ESRS) require companies to transparently disclose their environmental impacts and to implement concrete measures for resource conservation and waste prevention (Fig. 1) [2].

H. Kohl et al. (Eds.): GCSM 2025, LNME, pp. 334–342, 2026.
https://doi.org/10.1007/978-3-032-21154-5_37

Within the broader category of packaging waste, lightweight packaging (LWP) represents a particularly critical fraction. This category includes used, empty sales packaging made from materials such as plastic, aluminum, tinplate, or composite materials [3].

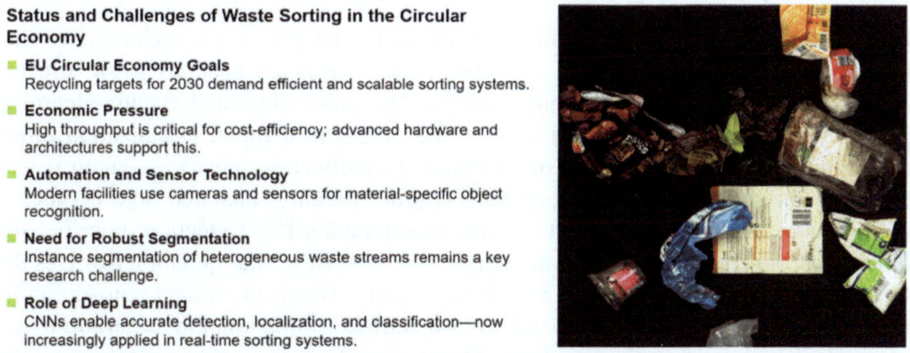

Fig. 1. Achieving the EU's circular economy targets requires precise, AI-driven sorting technologies with high scalability.

In 2021, Germany generated approximately 19.7 million metric tons of packaging waste—an increase of 4.9% compared to the previous year. While 67.9% of this waste was recycled and 96.3% was subject to some form of recovery, a considerable share was not diverted into material recycling but instead incinerated or landfilled [3].

In response to these challenges, the European Union has launched several initiatives aimed at accelerating the transition toward a circular economy. Among them is the Circular Economy Action Plan, which sets specific recycling targets to be achieved by 2030 [4]. A core prerequisite for meeting these targets is the efficiency and accuracy of waste sorting systems. Especially for LWP, high-quality recycling depends on the precise separation of materials at the sorting stage [5].

Integrating AI-based image recognition into robotic sorting processes offers a promising strategy for improving the automated classification of lightweight packaging waste. Deep learning–based instance segmentation models enable object-level detection and differentiation within complex and heterogeneous material streams. These models not only identify the presence of objects but also delineate their precise boundaries, allowing multiple overlapping items to be separated within a single image. Unlike semantic segmentation, which assigns a class label to each pixel, instance segmentation distinguishes between individual objects of the same class—crucial for accurate sorting in automated systems. By enhancing the speed and precision of material separation, these technologies lay the groundwork for more efficient and sustainable sorting operations. Improved classification accuracy not only supports higher recycling rates and the recovery of high-quality secondary raw materials but also contributes to reducing environmental impacts and advancing key objectives of the circular economy [6].

2 Methods

Here, several state-of-the-art instance segmentation models are evaluated. The study examines their suitability for automated recycling processes. It also assesses their potential to improve sorting efficiency and enable higher levels of process automation.

Object Detection. Automated sorting relies on robust perception technologies that allow accurate object localization and classification in heterogeneous waste streams [7]. Perception technologies combine sensors, data processing, and machine learning to extract meaningful information from visual or physical input. Localization refers to determining the exact position of an object. Classification means assigning it to a specific material category. Current systems require spatially separated objects on conveyor belts to ensure reliable detection and manipulation [8]. Detection describes the ability to recognize objects within the sensor's field of view. Manipulation involves the physical handling of these objects by robotic systems. However, optical sensors can be affected by environmental factors such as dust, surface contamination, or airborne particles, which reduce detection accuracy. Maintaining high throughput is also essential from an operational standpoint [6, 9].

Camera-based systems are typically used for object detection. These are often supplemented by material-sensitive sensors to identify specific waste fractions [8]. In most cases, objects are identified in real-time camera images using machine learning algorithms [10]. Advances in deep learning and computer vision have significantly enhanced recognition capabilities in industrial settings. Convolutional neural networks (CNNs), enable precise object detection, localization, and classification. While modern networks are computationally intensive, recent optimizations in model architecture and hardware now enable real-time deployment in sorting systems [11]. Studies confirm that deep learning-based systems offer high flexibility and robustness, even under complex input conditions such as mixed waste streams [7, 10].

Instance Segmentation. Instance segmentation plays a critical role in distinguishing material types and enabling object-specific manipulation. As a subtask of computer vision, instance segmentation involves pixel-accurate detection and delineation of individual object instances within an image. This is particularly relevant in domains such as autonomous robotics and recycling automation.

Modern segmentation models combine object detection with mask prediction. These masks provide binary, pixel-level representations of an object's shape and position within the image. High-performing architectures such as YOLOv8, YOLOv9, YOLOv11 and RTMDet offer both high accuracy and low inference latency, making them suitable for real-time applications [12, 13]. One-stage models like YOLACT++ achieve high processing speeds through parallel detection and segmentation but may struggle with overlapping objects [14]. In contrast, two-stage models such as Mask R-CNN separate detection and segmentation into distinct modules, achieving higher accuracy in complex scenes [15]. Multistage architectures like Cascade Mask R-CNN further refine the output masks through successive iterations [16].

Models such as SOLOv2 and YOLOv9-Seg are based on CNNs and offer different design paradigms: SOLOv2 employs a direct pixel-to-instance mapping approach, while YOLOv9-Seg integrates detection and segmentation into a unified framework, favoring speed over fine-grained precision [12]. Vision Transformers (ViTs) provide an alternative, non-convolutional method. By dividing an image into fixed-size patches and processing them through transformer blocks, ViTs generate classification tokens used in downstream prediction tasks [17].

Training Datasets. Accurate instance segmentation requires extensive and diverse training datasets. Deep neural networks depend on such datasets to learn object boundaries and generalize to unseen input conditions. However, generating and annotating real-world datasets is labor-intensive and cost prohibitive. As a result, synthetic data generation is widely used [6].

A commonly employed method is the cut-and-paste technique, in which real object crops are inserted into new image contexts. This approach enables the creation of diverse training scenarios with varying levels of occlusion and object positioning while allowing for automated annotation. Due to its scalability and minimal manual effort, it is particularly well-suited for industrial applications in instance segmentation [18].

Three self-generated datasets are used in the following analyses. The first, ALPHA-Dataset, is a synthetic dataset created under controlled laboratory conditions. It contains simple object geometries and is designed to assess the basic functionality of the segmentation pipeline and the mechanical performance of the grippers. ALPHA-Dataset serves as a reference for fundamental segmentation tasks. The second dataset, BETA-Dataset, is also synthetic but reflects more complex scenarios. It includes typical impurities and representative objects commonly found in recycling facilities, allowing the evaluation of model robustness under increased variability. The third dataset, GAMMA-Dataset, consists of real-world data collected using a proprietary test setup. It captures authentic contaminant types and realistic fault scenarios, providing a benchmark for performance under operational conditions.

3 Results and Discussion

The growing demand for automation in the recycling industry reflects the need for greater efficiency, higher precision, and increased throughput. These factors support the achievement of the United Nations Sustainable Development Goals (SDGs), particularly SDG 12 (responsible consumption and production) and SDG 9 (industry, innovation, and infrastructure) [1]. To address these goals, this study develops and evaluates a deep learning–based instance segmentation pipeline. The system detects and classifies foreign objects in heterogeneous lightweight packaging waste streams with high accuracy. It enables object-level differentiation in visually complex environments.

The segmentation model follows a practical taxonomy based on contaminants frequently observed in real-world sorting processes. Recognizable object classes include “battery”, “shoe”, “cartridge”, and “phone”. These reflect typical non-conforming materials found in industrial waste streams.

The following results present the segmentation performance of the system under different conditions and across multiple datasets. The pipeline uses state-of-the-art architectures, including YOLOv8-seg, YOLOv9-seg, RTMDet-Ins, and the latest version,

YOLO11. Performance is evaluated using mean average precision (mAP50:95), per-class AP50:95, and inference time. The analysis focuses on real-time applicability in high-throughput recycling environments (Fig. 2).

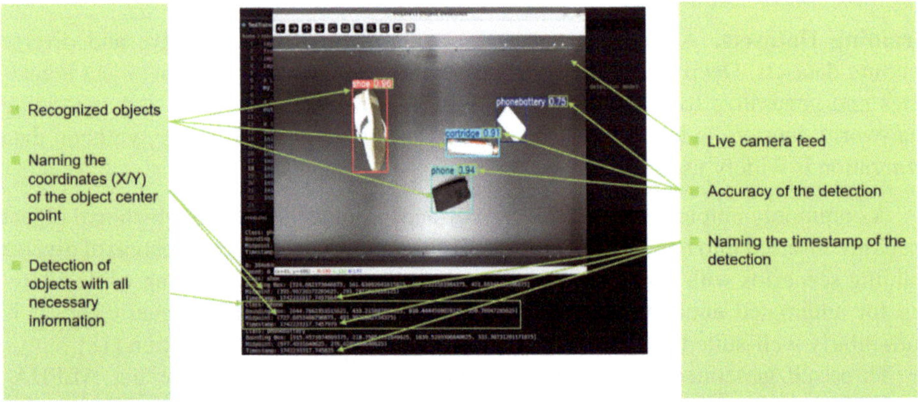

Fig. 2. Segmentation accuracy is evaluated using a live camera feed across both synthetic and real-world datasets representing heterogeneous lightweight packaging waste streams.

ALPHA-Dataset. The ALPHA dataset serves for pretraining and model evaluation. All tested models achieve consistently high mAP50:95 scores. At 1280 × 1280 resolution, YOLOv8x-seg reaches 94.6%, and RTMDet-Ins-x reaches 92.2%, with inference times of 17 ms and 23 ms, respectively. The controlled setup of ALPHA results in low inter-class variance and enables robust detection across all categories. However, this performance does not transfer reliably to real-world conditions. Accuracy drops significantly, especially for complex or visually ambiguous classes. Increasing input resolution does not consistently improve generalization, indicating overfitting in larger models. Despite this limitation, RTMDet-Ins-x maintains 59.9% mAP50:95 in real-world tests while keeping inference times below 33 ms. This identifies it as a strong candidate for real-time deployment.

BETA-Dataset. The BETA dataset serves as a reference for external validation. Although limited in scope, it reflects operationally relevant conditions and offers insight into model generalization. Most models achieve over 90% mAP50:95 on synthetic-like classes. However, the "battery" class reaches only 67.4%, highlighting persistent challenges in detecting small, low-contrast objects. Upscaling the input resolution to 1280 × 1280 improves performance for such items, underscoring the importance of input fidelity in fine-grained segmentation. On actual plant data, performance varies widely across classes. AP50:95 scores for categories like "notebook", "battery", "phone battery", and "shoe" drop substantially. Among all models, RTMDet-Ins-m (640 × 640) provides the best balance between speed and accuracy. It achieves 57.1% mAP50:95 with an inference time of just 3.7 ms. YOLO11 demonstrates strong real-time capabilities, supported by multi-label architectures and optimized feature extraction. YOLO11 reaches 90% mAP50:95 in under 50 training epochs. Still, classes such as "shoe" and "cartridge" remain difficult to detect, pointing to ongoing limitations with ambiguous and low-contrast backgrounds. Nonetheless, YOLO11's combination of speed and precision confirms its suitability for deployment in automated recycling systems (Fig. 3).

GAMMA-Dataset. The GAMMA dataset captures realistic error patterns and complex visual structures typical of operational recycling systems. When applied to GAMMA, YOLO11 improves recognition of small and variably lit objects. This is supported by enhanced image preprocessing, adaptive input resolutions (e.g., 1080p), and a stricter confidence threshold of 80%. These adjustments reduce the rate of false positives in a measurable way.

While all models perform well on synthetic data, their accuracy decreases considerably under real-world conditions (Fig. 3). This highlights the ongoing challenge of bridging the gap between laboratory performance and operational robustness. LWP waste streams present specific challenges. These include overlapping objects, cluttered backgrounds, and low-contrast materials. While synthetic pretraining proves effective, real-world deployment reveals the need for domain-specific data and stronger generalization mechanisms. YOLO11 and RTMDet-Ins-x perform well in this context. They offer a favorable balance between speed and precision, making them strong candidates for real-time industrial use.

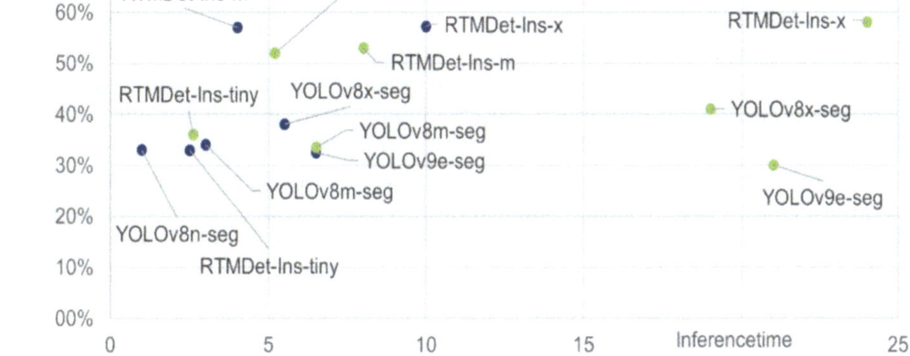

Fig. 3. Performance of YOLOv11x-seg and YOLOv12x on the BETA-Dataset (synthetic). YOLOv11x-seg and YOLOv12x (640 × 640, blue circle, 1280 × 1280, green circle) are compared to other models by mAP50:95 and inference time.

Deep learning segmentation architectures show significant potential for advancing automation in sustainable waste management. In LWP sorting, one of the most heterogeneous waste categories, high segmentation fidelity is essential to meet material purity standards. However, model performance depends strongly on architectural complexity and the realism of the training data. Larger models perform well on synthetic benchmarks but often generalize poorly. In contrast, smaller architectures sometimes deliver more stable results in practice. This challenges the assumption that increasing model size always leads to greater robustness.

Small, low-contrast, or partially obscured items remain major challenges. These cases expose structural weaknesses in segmentation models when confronted with visual uncertainty. YOLO11 addresses many of these limitations through architectural improvements and multi-label detection. It maintains both accuracy and speed across varying conditions which are essential qualities for real-world deployment.

However, benchmark performance is only one aspect of practical implementation. Success depends on system-level integration, including optimized camera setups, reduced latency, and alignment with robotic handling systems.

Future work should prioritize modular architectures. These must tightly connect training data, simulation environments, and live operations. Such integrated approaches are crucial to scaling resilient, intelligent sorting technologies that support global sustainability goals through more effective packaging waste recycling.

4 Conclusion

Deep learning–based instance segmentation models form a critical technological foundation for automating the sorting of lightweight packaging waste. Their relevance is particularly pronounced in the context of the circular economy, where scalability, precision, and throughput are essential. Models such as YOLOv8, YOLOv9, RTMDet, and YOLO11 show high segmentation accuracy and real-time performance under controlled conditions. They enable object-level classification of irregularly shaped, overlapping, and visually complex items, thereby enhancing the accuracy and efficiency of material recovery.

However, current findings also highlight a fundamental limitation. Models trained on synthetic or idealized datasets often fail to generalize under real-world conditions, where contamination, occlusion, and variable lighting introduce significant variability. This performance gap reveals a structural shortcoming in current modeling approaches.

Overcoming this challenge requires a shift from isolated model development to system-level integration. Robust segmentation must be embedded into modular pipelines that align model architecture, data diversity, and the operational context. Realistic training data, adaptive learning strategies, and close integration with sensor and robotic infrastructure are key to enabling reliable performance at scale.

Ultimately, only such integrated and context-aware approaches will allow instance segmentation to move beyond laboratory demonstrations. This is essential for deploying intelligent sorting systems that are not only technically robust but also economically viable and environmentally sustainable at an industrial scale.

References

1. Vereinte Nationen - Regionales Informationszentrum für Westeuropa. 17 Ziele (2024). https://unric.org/de/17ziele/. Accessed 07 Apr 2025

2. Finance. Corporate sustainability reporting (2025). https://finance.ec.europa.eu/capital-markets-union-and-financial-markets/company-reporting-and-auditing/company-reporting/corporate-sustainability-reporting_en. Accessed 07 Apr 2025
3. Verpackungswirtschaft.de. Leichtverpackung (2024). https://www.verpackungswirtschaft.de/leichtverpackung/?utm_source=chatgpt.com. Accessed 07 Apr 2025
4. Europäische Kommission. Ein neuer Aktionsplan für die Kreislaufwirtschaft Für ein saubereres und wettbewerbsfähigeres Europa: EUR-Lex - 52020DC0098 - DE (2020). https://eur-lex.europa.eu/legal-content/DE/TXT/PDF/?uri=CELEX:52020DC0098. Accessed 20 Mar 2025
5. Fischer, M., Foord, D., Frecè, J., et al.: Sustainable Business: Managing the Challenges of the 21st Century. SpringerBriefs in Business. Springer Nature, Cham (2023)
6. Kiyokawa, T., Takamatsu, J., Koyanaka, S.: Challenges for future robotic sorters of mixed industrial waste: a survey. IEEE Trans. Autom. Sci. Eng. **21**, 1023–1040 (2024). https://doi.org/10.1109/TASE.2022.3221969
7. Raptopoulos, F., Koskinopoulou, M., Maniadakis, M.: Robotic pick-and-toss facilitates urban waste sorting. 1149–1154 (2020). https://doi.org/10.1109/CASE48305.2020.9216746
8. Mao, W.-L., Chen, W.-C., Wang, C.-T., et al.: Recycling waste classification using optimized convolutional neural network. Resour. Conserv. Recycl. **164**, 105132 (2021). https://doi.org/10.1016/j.resconrec.2020.105132
9. Lubongo, C., Alexandridis, P.: Assessment of performance and challenges in use of commercial automated sorting technology for plastic waste. Recycling **7**, 11 (2022). https://doi.org/10.3390/recycling7020011
10. Mao, W.-L., Chen, W.-C., Fathurrahman, H.I.K., et al.: Deep learning networks for real-time regional domestic waste detection. J. Clean. Prod. **344**, 131096 (2022). https://doi.org/10.1016/j.jclepro.2022.131096
11. Manakitsa, N., Maraslidis, G.S., Moysis, L., et al.: A review of machine learning and deep learning for object detection, semantic segmentation, and human action recognition in machine and robotic vision. Technologies **12**, 15 (2024). https://doi.org/10.3390/technologies12020015
12. GitHub. ultralytics/docs/en/models (2025). https://github.com/ultralytics/ultralytics/blob/main/docs/en/models. Accessed 20 Mar 2025
13. Lyu, C., Zhang, W., Huang, H., et al.: RTMDet: An Empirical Study of Designing Real-Time Object Detectors (2022)
14. Bolya, D., Zhou, C., Xiao, F., et al.: YOLACT: real-time instance segmentation. In: 2018 IEEE/CVF Conference on Computer Vision and Pattern Recognition, pp. 9156–9165 (2018). https://doi.org/10.1109/ICCV.2019.00925
15. Ren, S., He, K., Girshick, R., et al.: Faster R-CNN: towards real-time object detection with region proposal networks. IEEE Trans. Pattern Anal. Mach. Intell. **39**, 1137–1149 (2017). https://doi.org/10.1109/TPAMI.2016.2577031
16. Cai, Z., Vasconcelos, N.: Cascade R-CNN: Delving into High Quality Object Detection (2017)
17. Dosovitskiy, A., Beyer, L., Kolesnikov, A., et al.: An Image is Worth 16 × 16 Words: Transformers for Image Recognition at Scale (2020)
18. Dirr, J., Bauer, J.C., Gebauer, D., et al.: Cut-paste image generation for instance segmentation for robotic picking of industrial parts. Int. J. Adv. Manuf. Technol. **130**, 191–201 (2024). https://doi.org/10.1007/s00170-023-12622-4

Wax Form-Finding Through Artificial Intelligence Assisted Robotic Arm for Sustainable Design and Manufacturing

Chor-Kheng Lim(✉)

Department of Art and Design, Yuan Ze University, Taoyuan City, Taiwan
kheng@saturn.yzu.edu.tw

Abstract. This research introduces a wax form-finding system integrating robotic arm control with artificial intelligence-based gesture recognition. Traditional form-finding methods often suffer from unpredictability and excessive material waste. In contrast, the proposed system combines intuitive, gesture-driven interaction with the precision of robotic manipulation, enabling designers to shape wax by hand gestures remotely. The designer's gestures are interpreted in real time through Artificial Intelligence-powered image recognition and transformed into commands for the following manufacturing stages. This method significantly reduces material waste, production time and energy consumption compared to conventional 3D printing. Preliminary experiments indicate a material utilisation efficiency of 90%, a reduced total production time, and an 88% reduction in energy use. Furthermore, material recovery rates are significantly high, between 95% and 98%. These findings underscore the environmental advantages of the proposed system. Overall, this research demonstrates the potential of integrating robotics, Artificial Intelligence, and a sustainable material (plant-based wax) in product design and manufacturing, providing a practical contribution to the circular economy.

Keywords: Gesture Recognition · Artificial Intelligence · Wax · Form-Finding Systems · Sustainability

1 Introduction

For centuries, architects have based the design process on their experience, intuition and personal visualisation. The form of the building has always been the design factor that highlights the individual style of an architect. Form-finding has a long-standing tradition in the design field, from Gaudí's catenary curve models to Frei Otto's experiments with tensile structures [1]. It is a process in which the shape or form of the building is determined through a logical process that can be inspired by nature and living organisms. The target is to achieve the optimal form that delivers dynamic stability, adaptability, and sustainability. Form-finding systems can be based on traditional technologies such as physical modelling using different materials (mainly wood, steel, timber, reinforced concrete and wax). Wax is commonly used in model-making due to its malleability,

H. Kohl et al. (Eds.): GCSM 2025, LNME, pp. 343–351, 2026.
https://doi.org/10.1007/978-3-032-21154-5_38

while its phase change properties offer a distinctive medium for artistic expression. Wax belongs to the Phase Change Materials, PCMs, which are substances that absorb or release significant amounts of latent heat without changing temperature during their phase transitions, typically between solid and liquid states. These properties of thermal energy storage and reusability make them highly valuable in different applications, including sustainable design [2, 3]. Among PCMs, bio-based variants offer superior sustainability benefits compared to paraffin-based alternatives, thanks to their affordability, safety, and lower environmental impact [4]. However, conventional wax form-finding techniques often lack precision, relying on natural flow dynamics to shape the material. This unpredictability makes it difficult for designers to anticipate outcomes, prolongs the creative process, and increases material waste [5]. Nordez M. investigates the formal potential of wax casting [6]. These applications have primarily emphasised aesthetic expression, with limited attention to sustainability considerations.

In the digital age, computational design methods have further expanded the possibilities for form exploration [7]. Digital design tools aid architects in exploring various building forms influenced by several factors such as functionality, site context, strategy, and aesthetics to realise the optimised output. In particular, 3D modelling software has replaced hand-drawing methods, offering architects a more innovative and holistic approach to the design process. It allows them to explore various building forms and choose the optimised output that responds to the design constraints.

As natural body movements, gestures have been increasingly utilised as input interfaces in 3D modelling environments [8]. Kang et al. developed a system that enables real-time generation and visualisation of 3D design concepts through gesture recognition, significantly enhancing the efficiency and intuitiveness of the design process [9]. Meanwhile, Artificial Intelligence (AI) technologies have been widely applied in manufacturing to boost production efficiency, reduce energy consumption, and optimise resource utilisation [10]. Viliunas et al. applied an AI-based tool to shape-finding in architecture [11]. However, research directly integrating AI-powered gesture recognition with sustainable manufacturing remains limited.

This study addresses that gap by proposing a novel approach that combines the mentioned technologies to support more sustainable design and production practices in architecture. In particular, this research proposes a wax form-finding system that integrates AI-based gesture recognition with robotic arm control, enabling intuitive and precise form generation while enhancing sustainability and resource efficiency.

2 System Implementation

The design process using a wax form-finding system is grounded in four core sustainability principles:

- Material Circularity. Wax is selected as the primary manufacturing material due to its ability to undergo repeated melting and solidification cycles, enabling nearly 100% material recycling.
- Energy Efficiency. The system minimises unnecessary energy consumption by precisely controlling the wax's melting temperature and cooling process. Localised heating and short-duration operations further reduce energy demands.

- Process Optimisation. Integrating AI-powered gesture recognition allows for intuitive and accurate control, enabling designers to predict outcomes more effectively and reduce trial-and-error iterations. The system also records successful parameter combinations, building a form-parameter database to support continuous process optimisation.
- Value Maximisation. The system supports rapid prototyping and creative form exploration, empowering designers to build prototypes with minimal material and energy input.

The sustainable wax form manufacturing system developed in this study consists of three main components: (1) the AI Gesture Recognition Module that captures user gestures as input commands; (2) the Robotic Arm Control System, which executes the form-generation operations; (3) the Wax Phase Change Control Unit that manages the thermal state of the wax material (see Fig. 1).

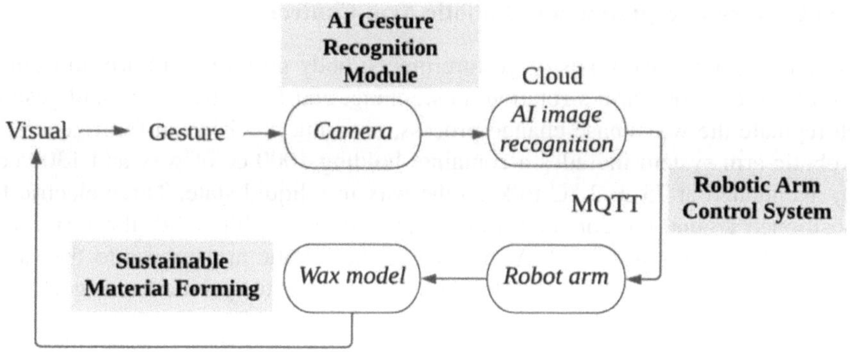

Fig. 1. Architecture of the sustainable wax form manufacturing system. Revised from [5].

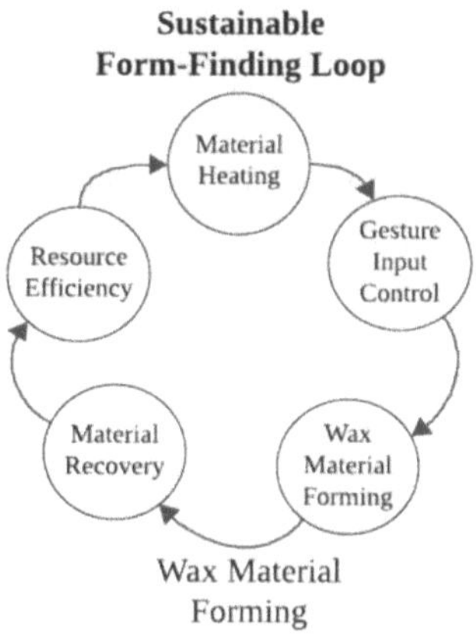

Fig. 2. A closed-loop framework of the sustainable wax-based form-finding system.

The system operates as follows: first, a camera captures user gestures, enabling intuitive interaction. Second, the AI recognition module interprets gestures and sends control signals to the robotic arm. Finally, the robotic arm manipulates the wax according to the control parameters. The system functions in a closed-loop configuration, allowing users to observe the form-generation process in real time and to make real-time modifications possible by adjusting gestures accordingly. This feedback loop significantly enhances success rates and reduces material waste. Figure 2 illustrates the closed-loop framework of the wax-based form-finding system.

The system uses plant-based wax, which is biodegradable, fully recyclable, non-toxic, and requires low energy for phase transitions, adopting a safe-and-sustainable-by-design approach. Used wax is fully recyclable through melting and reformation, enabling circular material use. This process improves design efficiency and maximises resource utilisation, achieving a near-zero-waste manufacturing process.

2.1 AI Gesture Recognition and Robotic Arm Control

The system supports two types of gesture inputs: body gestures, full-arm movements that control the robotic arm's rotation, positioning, and fan activation; hand gestures, which regulate the wax phase change process, ensuring precision in form generation. The robotic arm system includes a container holding 1000 cc of wax and 1300 cc of water, maintained at 75 ± 2 °C to keep the wax in a liquid state. Three electric fans are positioned around the container to manage cooling. A thin solid film forms when the wax surface cools to 60 °C. At this stage, the robotic arm moves to the surface and uses wax-coated elements to lift the film, shaping it into specific forms gently. Key control parameters include: temperature regulation: 75 ± 2 °C; surface cooling rate: 0.5 ÷ 2 °C/s; robotic arm angular rotation: 0 ÷ 360°; robotic arm movement range: 0 ÷ 150 mm; lifting speed: 2 ÷ 5 mm/s.

2.2 Sustainability Assessment Methods

To assess the process sustainability performance, the following indicators have been used and evaluated (see the following Sect. 3.3):

1. Material Use Efficiency. Ratio of the final product mass to the initial input material mass.
2. Material Recovery Rate. Ratio of the amount of recyclable or reusable material to the total waste generated.
3. Energy Consumption Indicators. Total Energy Consumption, Energy per mass.
4. Production Time Indicators. Total Production Time.
5. Comprehensive Sustainability Indices. Resource Efficiency Index, Circular Potential Assessment, Environmental Impact Factor.

Each metric has been compared against conventional 3D printing methods to evaluate the system's sustainability performance comprehensively.

3 Experimental Results

3.1 System Operation

The operation of the system unfolds through five main phases. It begins with the system preparation phase, during which the equipment undergoes a five-minute preheating process to ensure optimal working conditions for wax manipulation. This step is followed by the gesture input phase, where users interact with the system using hand movements. Seven users participated in the testing stage, and the system achieved a gesture recognition accuracy of 95%, thanks to its integration of AI-based image recognition.

Next is the form generation phase, where the interpreted gestures guide the robotic arm to shape the wax. This phase typically takes around ten minutes to complete. Once the form is generated, the process moves into the form completion phase, which finalises and stabilises the structure. The final step is the material recovery phase, which allows for the efficient reuse of leftover wax, minimising waste and enhancing sustainability.

Across 30 experimental trials, the system consistently produced seven distinct basic form types: *arched*, *loral-like strut*, *spiral line*, *perforated*, *funnel-shaped*, *cylindrical surface*, and *layered spiral surface* (see Fig. 3). These results highlight the system's reliability and versatility in generating complex geometries through intuitive, gesture-based interaction.

Fig. 3. Seven basic form types generated by the wax form-finding system. Revised from [5].

To contextualise the environmental benefits of the proposed system, a comparative evaluation was conducted against a standard Fused Deposition Modelling (FDM) 3D printing process under equivalent prototyping conditions. Both the wax-based and the FDM-printed samples were cylindrical models with a height of approximately 30 cm, a diameter of 10 cm, an estimated volume of about 2.36 L, and a mass of about 150 ÷ 200 g per unit. The comparison focused on key sustainability metrics.

The system demonstrated several sustainability advantages (see Tab. 1):

- Fast Formation. The total process time is 15 min (5 min preheating + 10 min forming), enhancing the design iteration speed.
- High Success Rate. 85%, significantly higher than the 50 ÷ 60% typical of first-time 3D printing attempts.
- Low Material Usage. Average of 150 ÷ 200 g of wax per form, with over 95% being recyclable

Table 1. Comparison between the proposed wax-based system versus 3D Printing (FDM)

Criteria	Wax-Based System	3D Printing (FDM)
Production Time	15 min per unit	85 min per unit
Material Waste	<10% (high reuse rate)	20 ÷ 30% (support structures & purge)
Recyclability	Up to 10 reuse cycles per unit	Typically non-recyclable
Energy Consumption	~88% less than FDM (for same size)	Higher (due to process duration and hot chamber)
Form Control	Intuitive, real-time gesture control	Predefined CAD models required

- Low Energy Consumption. Approximately 0.15 kWh per operation, just 12% of the energy used in comparable 3D printing processes
- Zero Waste. All materials used are either recyclable or biodegradable.

3.2 Material Use Efficiency and Recovery Rate

The sustainable wax form manufacturing system demonstrates an average Material Use Efficiency of 90%, compared to 65.3% for conventional 3D printing of similar complexity, an improvement of 37.8% (Table 2).

Material Recovery Rate tests indicate that 95 ÷ 98% of the system's waste can be fully recycled, with only 2 ÷ 5% requiring treatment due to contamination or oxidation. After 20 reuse cycles, performance tests on the wax material show a 92% retention of physical properties and a 96% retention of chemical properties.

Material losses account for approximately 10% of the input material. These losses arise from several sources:

- Process-related losses, such as dripping during lifting and surface oxidation.
- Operational losses, including evaporation during heating.
- System losses, such as residues left in containers and conduits.

These findings are reported and collated in Tab. 2. While the system is efficient, there is still room for improvement, particularly in minimising process and system-related losses to enhance material circulation.

Efficiency tests across different form complexities reveal that simple forms can achieve Material Use Efficiency values of up to 92%. In contrast, complex forms drop to 78%, highlighting a clear correlation between form complexity and material efficiency.

Table 2. Comparison of Sustainability Key Performance Indicators between the Wax Form manufacturing systems and traditional 3D Printing systems. Sustainability

Indicator Sustainable	Wax-Based System	3D Printing (FDM)	Differences [%]
Material Use Efficiency (%)	90.0	65.3	+37.8%
Material Recovery Rate (%)	95.0 ÷ 98.0	15.6	+518.6%
Total Energy Consumption (kWh)	0.15	1.25	−88.0%
Energy per mass (kWh/kg)	0.75	6.80	−89.0%
Total Production Time (min)	15	85	−82.4%
Resource Efficiency Index (%)	85.3	62.1	+37.4%
Circular Potential Assessment (%)	92.5	25.3	+265.6%
Environmental Impact Factor	0.24	1.00	−76.0%

3.3 Measurements of Sustainability Key Performance Indicators

Experimental tests showed that the system consumes a total of 0.15 kWh to produce a standard test form (30 cm height, 10 cm diameter), with energy use distributed as follows: heating system: 0.08 kWh (53.3%); robotic arm movement: 0.04 kWh (26.7%); control system and AI computation: 0.02 kWh (13.3%); fans and auxiliary equipment: 0.01 kWh (6.7%). In comparison, producing the same model using 3D printing techniques (FDM) requires: 1.25 kWh, 8.3 times more than this system, representing an 88% energy saving. When measured by energy consumption per unit mass (see Table 2), the system uses 0.75 kWh/kg, while 3D printing consumes 6.8 kWh/kg. Additionally, the system recovers approximately 15% of its total energy consumption through heat exchangers. This recovery rate can increase to 22% in batch production scenarios. The system's average total production time is 15 min (5 min of preparation and 10 min of forming).

In contrast, 3D printing the same model takes an average of 85 min, resulting in a reduction of 82.4% in production time.

From initial design concept to completed physical form, the system requires an average of 25 min, compared to 210 min for the whole 3D printing process.

Comprehensive sustainability assessment results indicate that the wax-forming system significantly outperforms traditional manufacturing methods in key areas such as material circularity, energy efficiency, production time, and environmental impact.

4 Final Remarks

4.1 System Advantages and Limitations

The main technical advantages of the sustainable wax form manufacturing system include: intuitive interaction interface, intelligent control of the phase change process, remote operation capability and circular material management. From the sustainability point of view, several advantages can be remarked: efficient material circulation, significant energy efficiency, production time optimisation, and waste minimisation.

Regarding user experience, the system provides design freedom, real-time feedback loops, low entry barriers, and minimised iteration costs. However, the system that has been developed at present still has several limitations, such as the limited control capability for high-precision geometric shapes or extremely detailed structures. A maximum form size of approximately 40 cm high, 20 cm wide. It is not as precise as 3D printing, and the form variation rate is about $5 \div 8\%$; the used material has limited physical strength and functionality.

Future development efforts are necessary to overcome present limitations and should be focused on the following areas:

1. Technology Expansion and Optimisation. Enhancing the robotic arm's degrees of freedom and motion precision, integrating hybrid manufacturing technologies, and improving control over form complexity and geometric accuracy.
2. Material Innovation. Developing a broader range of circular phase-change materials, particularly bio-based options derived from renewable or recycled sources, to expand the system's sustainability and application scope.
3. Intelligent Control Enhancement. Incorporating advanced sensing systems and AI-driven form generation algorithms to enable precise closed-loop control, improving both form prediction accuracy and production repeatability.
4. Sustainability Optimisation. Introducing features such as zero-energy standby modes, renewable energy-powered versions, and integrated carbon footprint tracking systems to reduce environmental impact further.
5. Distributed Network Development. Establishing a distributed manufacturing network based on the system, exploring new production models that combine global design with localised fabrication, and contributing to a lower-carbon, more resilient manufacturing ecosystem.

5 Conclusion

The sustainable wax form-finding system developed in this research achieves efficient and controllable form generation while significantly improving resource use efficiency by integrating AI gesture recognition and robotic arm control technologies. Experimental results prove that the system has significant advantages in material efficiency, energy consumption, production time, and overall environmental impact, highlighting its enormous potential as a sustainable manufacturing method.

The wax-based, gesture-controlled system demonstrates superior performance across all parameters, positioning it as a more sustainable alternative for early-stage shape and form exploration.

The system's main sustainable aspects include: (1) proposing a new material circulation model, achieving near-zero waste in the design iteration process; (2) optimising energy utilisation, simultaneously achieving high efficiency and low energy consumption; (3) developing intuitive sustainable design methods, merging creative freedom with resource efficiency.

This integrated system represents an innovative attempt to increase sustainability in the building design process.

References

1. Lopes, J.V., et al.: Parametric urban models based on Frei Otto's generative form-finding processes. In: Proceedings of the 19th Conference on Computer Aided Architectural Design Research in Asia (CAADRIA), pp. 595–604 (2014)
2. Osibodu, S.J., Adeyinka, A.M., Mbelu, O.V.: Phase change material integration in concrete for thermal energy storage: techniques and applications in sustainable building. Sustain. Energy Res. **11**(1), 45 (2024). https://doi.org/10.1186/s40807-024-00138-8
3. Chakraborty, T., et al.: Application of phase change materials for sustainable energy: a review. J. Energy Storage **45**, 103774 (2022)
4. Radhakrishnan, S., et al.: Review on the sustainability of phase-change materials used in buildings. Sustain. Mater. Technol. **32**, 100605 (2022)
5. Lim, C.K.: Gesture modeling: in between nature and control. In: Proceedings of the 28th International Conference of the Associations for Computer-Aided Architectural Design Research in Asia (CAADRIA), pp. 189–198 (2023)
6. Nordez, M.: Lost-wax casting: a widespread technique to produce copper alloy adornments in Atlantic Europe since the mid-second millennium BC. J. Archaeol. Sci. **168**, 106008 (2024). https://doi.org/10.1016/j.jas.2024.106008
7. Williams, K.: Form-finding, architecture and mathematics. Nexus Netw. J. **24**, 13 (2022)
8. Choul, D., Gao, Z., Huang, S., Fuh, C.-S.: Hand gesture recognition, tracking, and project to a 3D model. In: Proceedings of the Conference on Computer Vision, Graphics and Image Processing (IPPR), Taiwan (2013)
9. Kang, J., et al.: Instant 3D design concept generation and visualization by real-time hand gesture recognition. Comput. Ind. **64**(7), 785–797 (2013)
10. Wang, Y., et al.: Deep learning for smart manufacturing: methods and applications. J. Manuf. Syst. **48**, 144–156 (2018)
11. Viliunas, G., Grazuleviciute-Vileniske, I.: Shape-finding in biophilic architecture: application of AI-based tool. Archit. Urban Plan. **18**(1), 68–75 (2022)

Framework and Development of an AI-Based Application to Support Object Classification and Consumer Waste Management

Lisa Klatt(✉), Klaus Bolze, Thomas Potempa, and Max Patrick Ehleben

Institute of Recycling, Ostfalia University of Applied Sciences, Wolfsburg, Germany
lis.klatt@ostfalia.de

Abstract. The recycling of waste materials plays a key role in conserving natural resources. To promote material circularity, it is necessary to consider all processes along the value chain, from raw material extraction to waste management and processing. Effective collection and sorting of end-of-life products and packaging are crucial prerequisites to achieve high recycling rates, minimize energy consumption and reduce material losses. However, a significant proportion of post-consumer waste is not properly returned to established waste management systems by citizen.

To address this challenge, recent developments in AI-based object classification can support citizens in the form of a mobile app. This type of application can assist users in identifying items and suggests the most effective location-aware recycling pathways to facilitate waste separation by type. To raise these potentials, a framework is presented that combines two apps: one for citizen to optimize recycling routes and a second app for collecting data to train the machine learning models. A novel classification system was created, and data suitable for AI training were collected to ensure the effectiveness of the system. In future, this application can contribute to directing household material flows into specific recycling pathways. In this way, the application can strengthen the circular economy in Europe and support environmental protection efforts.

Keywords: AI-supported object classification · recycling · circular economy · material flows

1 Introduction

1.1 Motivation

The recycling of waste materials plays a key role in conserving natural resources by reducing dependence on fossil fuel extraction. Effective recycling requires consideration of all processes along the value chain, from resource extraction to waste management. One critical factor is the correct return of end-of-life objects and packaging into established waste management systems [1]. However, current recycling efforts are hindered by significant challenges, such as improper disposal ("mis-throwing") including material contamination and the incorrect separation of hazardous waste, undermines recycling

H. Kohl et al. (Eds.): GCSM 2025, LNME, pp. 352–360, 2026.
https://doi.org/10.1007/978-3-032-21154-5_39

efficiency and increases operational risks and costs [2]. These issues prevent the effective circulation of valuable materials which is reflected in a recycling rate of only 68% in Germany in 2022 [3].

To address this, it is crucial to improve the collection and sorting of waste at a very early stage. Achieving high recycling rates, minimizing energy demands, and reducing losses in recycling processes depend on the purity, quality, and quantity of materials [2]. However, the presorting performance of consumer waste with respect to often localized waste management system is currently not sufficient [4]. An AI-powered software application available as a public app for citizens offers a promising solution for early-stage intervention in the recycling process. This can support users in identifying waste objects, recommend the best and locally appropriate recycling pathways and help manufacturers and recyclers to recover recyclable materials.

1.2 Related Work

European regulations [5] implemented at a national level, as for instance in Germany [6], and municipal strategies form the basis for the regulations on how waste should be treated. In addition to household waste, municipalities often collect two to three types of recyclable materials (paper, biowaste and lightweight packaging). These services are supplemented by on-demand collections at the property (for bulky waste or electrical appliances) and drop-off facilities for used glass and clothing at locations in the residential areas, sometimes complemented by recycling depots with drop-off facilities for all relevant waste and recyclable materials. The EU packaging regulations implemented in Germany through the Packaging Act [7] oblige manufacturers to collect and recycle the sales packaging they place on the market. In some regions of Germany, however, non-packaging materials of the same type are also collected alongside lightweight packaging in order to integrate more material flows into the circular economy. The variety of regional collection systems demonstrates the diversity of implementations, resulting in inconsistent and often confusing recycling practices for the public. In addition, existing systems aimed at reducing mis-throwing primarily rely on centralized collection facilities and automated as well as manual sorting technologies. However, current research highlights that the efficiency of these systems is heavily influenced by citizen engagement and behavior, presenting a significant barrier to achieving higher recycling rates [8]. Mis-throwing not only increases waste management costs but also reduces the efficiency of recycling systems by causing the loss of high-value waste streams within the waste hierarchy [5]. They further pose safety hazards across the recycling chain, such as fire risks in bins, vehicles, and sorting facilities [9]. Addressing these issues is essential for improving recycling outcomes and ensuring safer material recovery.

Current education campaigns and interventions by the waste management industry lack integration and real-time solutions for citizens. Interventions, including bin labeling, collection refusal, and awareness campaigns, lack real-time interactivity and integration into everyday routines, limiting their impact on citizen behavior [10, 11]. In addition to awareness campaigns, digital solutions such as AI-supported systems in waste collection vehicles are already being used to improve waste separation. Companies such as Lixo and the Brantner Group focus on post-collection processes, utilising AI-based object classification to identify contaminants in biowaste during container emptying. Pre-trained models detect abstract features, such as shape, size, texture or pattern, enabling reliable object recognition without the need for symbolic markings. In practice, vehicle-mounted cameras capture the waste stream and analyze it in real time using machine learning algorithms to identify items such as plastic bags or hazardous materials, including nitrous oxide cartridges. The driver is then alerted prior to loading. While this approach mitigates the impact of incorrect disposal during collection, it offers no real-time guidance to citizens at the point of disposal.

Against this background, the aim of this work is to focus on an early stage of the process chain by proposing a framework for a software application that intervenes earlier in the process and provides citizens with a user-friendly, real-time tool to correctly separate waste at the point of disposal to improve citizen waste separation behavior.

2 Application Framework

2.1 Requirements and Methodology

The framework to address the beforementioned challenges must take critical requirements into account to ensure an effective and sustainable solution for waste management and recycling. It should provide real-time information on alternative recycling routes in order to support the collection and recovery of materials further up the waste hierarchy. This functionality promotes reuse and resource circulation while supporting the enforcement of concepts like Extended Producer Responsibility (EPR) [5] and its derivatives Collective Producer Responsibility (CPR) and Individual Producer Responsibility (IPR) [12]. Moreover, the framework should be translated into tools that facilitate high-quality resource recovery from urban waste, thus reducing operational costs and fostering a sustainable circular economy. To achieve these goals, the tool must be user-friendly, accessible and adaptable to individual localization and needs. It should provide citizens with efficient access to recycling information in collaboration with local authorities. Finally, it must take into account European, national and local recycling legislation and future material flows to ensure the adaptability of the system.

As the primary objective it was set to develop a legally compliant, adaptable classification system and implement AI components using diverse, structured image datasets. To manage the complexity of waste and object classification in a citizen-oriented context, the challenge was divided into manageable sub-tasks. Following early stakeholder involvement, the framework's requirements were defined, from which the system's core functions were derived. The architecture was then structured into four interdependent modules: classification, object data acquisition, system integration, and citizen interaction. This modular design ensures that all functional layers—from data collection to classification results and recycling recommendations—are seamlessly connected and transparently traceable throughout the entire process.

2.2 Application Framework for AI-Based Waste Object Classification

As illustrated in Fig. 1, the framework comprises eight interconnected modules. It begins with the regulatory analysis (1), which forms the basis for a hierarchical classification system (2). Subsequent components—from structured image acquisition (3), data curation (4–5), to dual application implementation (6–9)—ensure transparency and coherence throughout the system. To enable this, a system that links data with relevant classifications is essential for accurate object identification. The classification framework draws on an analysis of European, national, and local disposal and recycling regulations, ensuring legal compliance while addressing stakeholder needs (Fig. 1, number 1). This analysis identifies key elements from which the logical structure for the machine learning models is derived. Additionally, anticipated future material flows must be considered to ensure the system's long-term adaptability. Based on these requirements, a set of suitable waste categories can be established (2). The resulting classification should follow a hierarchical structure that reflects current recycling and disposal practices, while also accommodating regional differences and future waste streams.

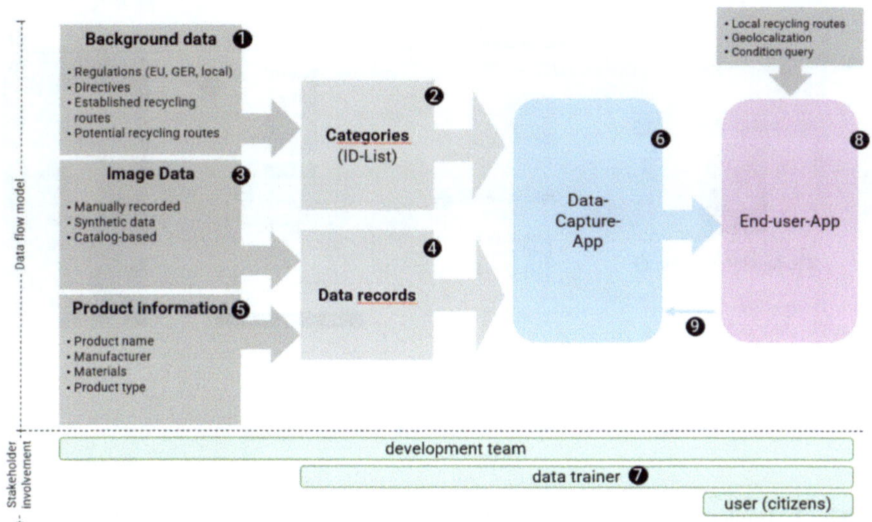

Fig. 1. Application framework with data flow model connecting the core modules (top) and stakeholder involvement (bottom)

Successful object classification of household waste items requires image data of these objects, which can be used to train machine learning models. The resulting database focus specifically on items to be recognized and classified later by the application. The image data can be captured manually, artificially generated or extracted from catalogs (3). A major challenge is the procurement of suitable and labeled data records (4) that are provided with the necessary object information (5) to ensure accurate classification within the respective categories. To guarantee the reliability and precision of the training data, two distinct applications are proposed in the framework. The first application serves as a tool for data collection (6) and is intended for use by trained personnel (7). The second application is designed to assist end-users by offering a user-friendly interface for recycling guidance by using the previously trained machine learning models (8). Following the training phase, the functionalities of the data collection application can be incorporated into the end-user application (9). This integration enables the system to gather more targeted insights into citizens' recycling challenges while facilitating refined training for the machine learning models.

3 Implementation Results and Case Study

Based on the presented framework, a tailored solution as a user-friendly mobile application called WERTIS was developed for the district of Peine, Germany, to provide citizens with accessible and efficient support for waste disposal using image classification technology. Figure 2 shows the implementation of the case study in this context.

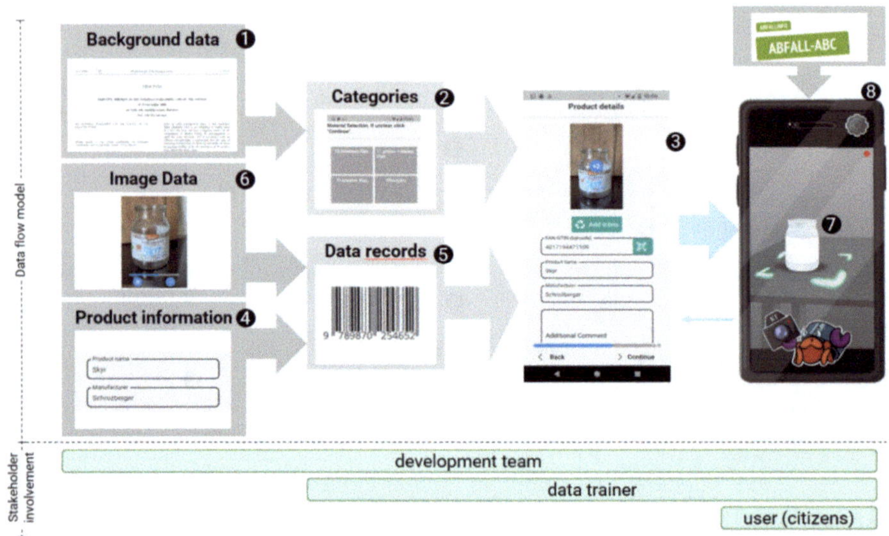

Fig. 2. Implementation of the framework in the case study towards the WERTIS app

The most important criteria for creating the categories for **classification** include compliance with European and German regulations (see Fig. 2, number 1) - such as separate collection, directive on waste of electrical and electronic equipment (WEEE) [13] and deposit systems, as well as adaptability to regional differences, such as the distinction in collection between packaging and non-packaging of the same material. In addition to the legal regulations, European classifications, such as the European waste catalogue (EWC) [14], were also used as a starting point for the development of the system and expanded to include the level of detail, such as the recycling symbols. As a result, a list of 185 categories was developed. Each category was assigned an ID number, which was then assigned to a local recycling pathway. Subdividing the assignment of objects to the correct recycling route into individual steps via the ID number makes it possible to adapt to local differences and enables regional adaptability. This is due to the fact that a universally usable framework has been created up to the identification numbers. It is only through the additional step of assignment that the utilization paths are defined that relate specifically to a region. Furthermore, the list had to be detailed enough to map new material flows as they emerged. The categories and numbers were structured hierarchically to allow detailed object classification while accommodating imprecise data. For example, plastic packaging could be categorized in detail according to plastic type (plastic packaging made of PP) or generally recorded as plastic packaging if no specific information was available. The 17 main waste categories at the first hierarchy level range from materials such as glass, metal and plastic to waste categories such as bulky waste, gases and liquids. Objects at lower hierarchy levels are classified according to their condition, for example as "used" or "damaged". This classification is relevant for further disposal routes. Textiles in perfect condition, such as T-shirts, are recycled in different ways than textiles that are already worn out, such as cleaning rags. This leads to recommendations for second-hand stores or landfill sites. This distinction at lower hierarchy levels enables objects to be managed at a higher level of the recycling hierarchy.

The implemented classification scheme facilitates the systematic and **efficient recording of data** and the interlinking of information. To ensure the creation of realistic and diverse data for AI training while preserving copyright, all data was manually collected using smartphones. A dedicated data capture app (3) was developed to streamline the recording and labeling of images, facilitating the systematic acquisition of data relevant to recycling and waste disposal. Important attributes are recorded as metadata for each object in order to create a comprehensive data record (5). A typical data record consists of the assignment to the matching ID number (2), one to four images of the object (6), up to three text fields (4) in which information such as product name and manufacturer are recorded, a number field for the barcode to uniquely assign items and optionally further images for more precise specification of existing symbols. Special consideration was given to recyclable materials and hazardous items such as batteries in order to improve waste separation and raise safety standards. An object is therefore considered classified and recorded (7) if it is linked to the photos and metadata as well as to a recycling path via the ID number. This logic links the object classification with the real waste infrastructure and enables granular adaptation to local regulations and collection systems. The labeled images from the dataset were used to train and test machine learning models. More than 4,000 labeled data records of labeled household objects were systematically recorded and stored in a structured way in a PostgreSQL database and an object storage solution.

To develop and evaluate object classification within the app, several **AI architectures** were explored. Several model architectures were evaluated, including classical classifiers, CLIP-based models, and Visual Question Answering (VQA) systems, to account for complex scenarios such as ambiguous product features or multipart objects.

The most robust results were achieved using a classification model with a Vision Transformer (ViT) backbone, fine-tuned on the recorded images and evaluated on a hold-out dataset. The ViT model, fine-tuned on the labeled data records, demonstrated robust classification performance – reaching ~75% accuracy at category level 2. Accuracy decreased for more granular categories, as expected, due to increased inter-class similarity.

The **WERTIS app** (8) is built upon the classification scheme, structured data collection, and machine learning model selection developed in earlier phases. Together, these components enable a customized, citizen-oriented solution that provides real-time, location-specific disposal recommendations through smartphone image input. By photographing items with their smartphone, users receive instant, location-aware disposal recommendations via the app, supported by geolocation and multilingual features for greater accessibility and personalized responses. The classification system enables regional adaptation and legal compliance through a scalable structure suited to diverse waste management systems. In addition, the application's modular architecture allows it to capture a wide range of household items and adapt to previously unknown items through continuous learning of models. Through the synthesis of AI-driven classification, localized regulatory data, and user-centered design, WERTIS exemplifies a scalable and sustainable approach to intelligent waste management.

4 Conclusion and Outlook

The WERTIS-KI project showcases the potential of AI-powered solutions in modern waste management to improve separation and collection of household waste. By integrating image classification, geolocation-based recommendations, and interactive functionalities, the WERTIS app empowers users to accurately identify and dispose of waste while adhering to waste hierarchy principles. This innovative tool addresses the challenges of early-stage recycling to support the reduction of mis-throwing, minimize contamination, and promote circular material flows. Additionally, its incorporation of alternative recycling channels, like second-hand stores, strengthens urban waste management and sustainable manufacturing systems, bridging gaps in existing processes and promoting informed citizen behavior. To improve the accuracy of the app's predictions and overcome the limitations of vision-only models when assessing semantic context, such as determining the condition of an object or its fill level, multimodal large language models (LLMs) are to be tested.

Developing real-time solutions is a major challenge, especially when developing classification systems that take into account the various recycling options and regulations. A crucial aspect is the acquisition of a high-quality dataset that represent real-world scenarios, especially smartphone-based inputs. This complexity requires adaptability, scalability and collaboration among stakeholders, including researchers, waste management experts, and policymakers, to ensure that public good-oriented solutions can be successfully established.

The proposed framework offers a scalable template for intelligent waste systems. Enhanced collaboration with producers enables tighter material cycles, paving the way for extended concepts such as Consumer2Manufacturer feedback.

Acknowledgements. This research was conducted within the project "Wertstoff-Informations-System mit Künstlicher Intelligenz (WERTIS-KI)" (Grant No. 67KI31009A) funded by the Federal Ministry for the Environment, Nature Conservation, Nuclear Safety and Consumer Protection (BMUV) on the basis of a resolution of the German Bundestag.

References

1. Kusch, A., Gasde, J., Deregowski, C., Woidasky, J., Lang-Koetz, C., Viere, T.: Sorting and recycling of lightweight packaging in Germany. - Climate impacts and options for increasing circularity using tracer-based-sorting. Mater. Circular Econ. **3**, 10
2. Hahladakis, J.N., Purnell, P., Iacovidou, E., Velis, C.A., Atseyinku, M.: Post-consumer plastic packaging waste in England: assessing the yield of multiple collection-recycling schemes. Waste Manage. (New York, N.Y.) (2018). https://doi.org/10.1016/j.wasman.2018.02.009
3. Statistisches Bundesamt: Recycling und Reduzierung von Siedlungsabfällen, Verpackungsmüll und Elektroschrott (2025). https://www.destatis.de/Europa/DE/Thema/Umwelt-Energie/Muell_und_Recycling.html. Accessed 28 Apr 2025
4. Hahladakis, J.N., Iacovidou, E.: Closing the loop on plastic packaging materials: what is quality and how does it affect their circularity? Sci. Total Environ. (2018). https://doi.org/10.1016/j.scitotenv.2018.02.330
5. European Union: DIRECTIVE 2008/98/EC on waste and repealing certain Directives. Abfallrahmenrichtlinie (2008)
6. Deutscher Bundestag: Gesetz zur Förderung der Kreislaufwirtschaft und Sicherung der umweltverträglichen Bewirtschaftung von Abfällen. Kreislaufwirtschaftsgesetz - KrWG (2012)
7. Deutscher Bundestag: Gesetz über das Inverkehrbringen, die Rücknahme und die hochwertige Verwertung von Verpackungen. Verpackungsgesetz - VerpackG (2017)
8. Mielinger, E., Weinrich, R.: Insights into plastic food packaging waste sorting behaviour: a focus group study among consumers in Germany. Waste Manage. (New York, N.Y.) (2024). https://doi.org/10.1016/j.wasman.2024.02.038
9. Ghiji, M., Edmonds, S., Moinuddin, K.: A review of experimental and numerical studies of lithium ion battery fires. Appl. Sci. (2021). https://doi.org/10.3390/app11031247
10. Deci, E.L., Koestner, R., Ryan, R.M.: A meta-analytic review of experiments examining the effects of extrinsic rewards on intrinsic motivation. Psychol. Bull. (1999). https://doi.org/10.1037/0033-2909.125.6.627
11. Lemon, K.N., Verhoef, P.C.: Understanding customer experience throughout the customer journey. J. Mark. (2016). https://doi.org/10.1509/jm.15.0420
12. Atasu, A., Subramanian, R.: Extended producer responsibility for E-waste: individual or collective producer responsibility? Prod. Oper. Manage. (2012). https://doi.org/10.1111/j.1937-5956.2012.01327.x
13. European Union: DIRECTIVE 2012/19/EU on waste electrical and electronic equipment. WEEE (2018)
14. European Commission: Decision 2000/532/EC on the List of Waste. 2000/532/EC (2023)

Industry 5.0

Slow Manufacturing: Its Conceptualisation, Application and Research Agenda

Hideki Kobayashi[1(✉)], Haruto Katayama[1], Hidenori Murata[1], Barbara Cimatti[2], and Giampaolo Campana[2]

[1] The University of Osaka, Suita 5650871, Japan
kobayashi@mech.eng.osaka-u.ac.jp
[2] University of Bologna, 40136 Bologna, Italy

Abstract. The manufacturing paradigm is transitioning from Industry 4.0 to Industry 5.0. While Industry 4.0 emphasised IoT technologies to enhance flexibility and efficiency, Industry 5.0 seeks to integrate these advancements with human-centric, sustainable, and resilient approaches. This paper introduces the concept of slow manufacturing, comprising three core elements rooted in craftsmanship, and explores its alignment with the goals of Industry 5.0. Practical and research-based examples from industrialised nations are presented, followed by an analysis of how these cases may be adapted for developing countries. A research agenda is then proposed and structured around the foundational elements of slow manufacturing. This approach suggests that adopting slowness in manufacturing can offer a viable pathway to realising the human-centred ideals of Industry 5.0.

Keywords: Slowness · Craftsmanship · Extra Value Creation · Industry 5.0

1 Introduction

Since the emergence of Industry 4.0 (I4.0) [1], numerous studies and practical applications have explored efficient and flexible manufacturing through technologies such as digital twins and Cyber-Physical Production Systems (CPPS) [2, 3]. However, a new paradigm - Industry 5.0 (I5.0) - is now evolving. Building upon the technological foundations of I4.0, I5.0 introduces concepts of sustainability, resilience, and human-centricity [4, 5]. Although I4.0-related efforts have addressed sustainability and resilience, for example, in decarbonising supply chains [6] and enhancing supply chain resilience [7], many of these approaches lack meaningful integration between human capabilities and autonomous technologies in the context of I5.0.

In this study, we focus on craftsmanship as a vital competence within manufacturing, encompassing not only technical skills but also cultural and social dimensions. In particular, we investigate the concept of slowness in manufacturing [8, 9], which is believed to enhance product value and user satisfaction while supporting worker well-being. The remainder of this paper is organised as follows: Sect. 2 reviews related work on craftsmanship in manufacturing; Sect. 3 introduces the proposed concept of slow manufacturing and its core elements; Sect. 4 presents industrial and academic examples;

H. Kohl et al. (Eds.): GCSM 2025, LNME, pp. 363–371, 2026.
https://doi.org/10.1007/978-3-032-21154-5_40

Sect. 5 discusses adaptation for developing countries; and Sect. 6 proposes a research agenda aligned with the slow manufacturing framework.

2 Related Work

Historically, the Arts and Crafts Movement emerged as a response to the rise of mass production during the first industrial revolution [10]. This movement emphasised using natural materials and traditional craftsmanship, aiming to enhance the enjoyment and artistic integrity of production work. However, it ultimately failed to establish workshop-based production as a mainstream model. In more recent times, the growth of craft-oriented e-commerce platforms, such as Etsy, has rekindled interest in craftsmanship. Yet, these platforms predominantly feature non-essential items, such as jewellery, rather than essential goods.

In the manufacturing industry, several cases have demonstrated the integration of craftsmanship. For example, the Japanese car manufacturer Mazda has incorporated the principles of craftsmanship into mass production through a concept known as "mass craftsmanship" [11]. This approach analyses and systematises the mechanisms behind expert manual skills. In academia, the Digital Triplet (D3) framework has been proposed to capture both explicit and tacit knowledge of craftsmen within digital manufacturing systems [12]. D3 aims to externalise and integrate artisanal skills into automated systems. However, such efforts often overlook the irreplaceable value generated through human touch in manufacturing processes, a value that modern technologies alone cannot replicate.

The earliest formal discussion of "slowness" in the context of factory operations was introduced by Cimatti and Campana, who classified slow manufacturing based on the types of technologies used and the speed of production [8]. In line with their framework, our own research has explored methods for evaluating craftsmanship as a business competence in manufacturing environments [13–16]. While a unified framework has yet to be established, we aim to qualitatively and quantitatively extract the intrinsic value of craftsmanship, not merely for the purpose of automation, but as an independent contributor to product and social value.

Industrial clusters or districts also play an essential role in promoting craftsmanship within manufacturing ecosystems. For instance, Japan's Tsubame-Sanjo region in Niigata Prefecture has a rich history in producing metal household goods and hand tools. At the same time, Sabae is renowned for its eyeglass frame manufacturing [17]. These clusters evolved through the accumulation of specialised artisanal skills and collaborative networks.

In Italy, several industrial districts originate in a distant past, based on the culture and traditions of a specific territory. For example, the popular glass district of Murano-Venice was born in 1291 when the local governor decided to move all the glassmakers of the City to the Murano Island to prevent Venice from the possible fires connected to this kind of activity. The concentration of glassmakers in the same site fostered excellence and the development of new products and techniques, such as the invention of crystal glass in 1450, from which two further Italian inventions had their origin: eyeglasses and telescopes.

One of the most famous Italian industrial districts is the one of sports vehicles located around Bologna and Modena. The "Motor Valley", as Emilia-Romagna's sports cars and motorbikes district is called, has its roots in this region's relevant mechanical industrial tradition. Indeed, in the XV century, the Bologna silk mill was one of the most advanced technologies in Europe, which gave a vocation towards mechanics even before the invention of the car at the end of the XIX century.

3 Concept of Slow Manufacturing

The concept of slow manufacturing is illustrated in Fig. 1 and is composed of three core elements: (1) the integration of craftsmanship and technology, (2) local characteristics, and (3) the creation of extra value.

The first element emphasises the synergistic use of craftsmanship and modern technology, including both traditional machinery and advanced automation systems such as those used in Industry 4.0. As discussed earlier, technologies developed under the Industry 4.0 paradigm aim to improve productivity and operational flexibility. However, many artisanal skills - particularly those involving tacit knowledge and sensory judgment - cannot be replaced by machines. Therefore, manufacturers are encouraged to blend technological efficiency with the human touch of skilled workers, not merely to reduce costs, but to enhance the value of their products. While combining traditional tools and manual labour represents one pattern of slow manufacturing, the concept is broader. In particular, craftsmanship also plays a vital role in promoting worker well-being and supporting the creation of emotionally resonant products in line with human-centred principles of Industry 5.0.

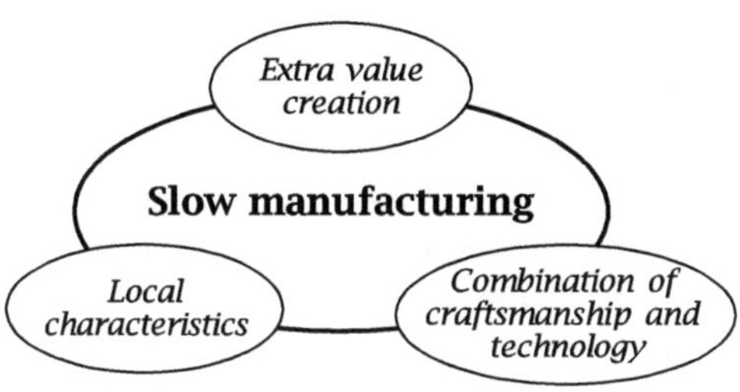

Fig. 1. A concept of slow manufacturing.

The second element concerns the consideration and utilisation of local characteristics. As highlighted in Sect. 2, industrial clusters and districts offer a viable framework for integrating artisanal skills into manufacturing systems. Critical success factors for these clusters include the coexistence of complementary firms, favourable historical and geographical conditions, strong subcontracting relationships, and supportive local government policies. Additionally, using indigenous materials - sourced from the local environment - can add authenticity and uniqueness to the final products.

The third element is creating extra value, which results from the optimal integration of the first two elements. In this context, "extra value" refers not only to a premium price, but also to intangible attributes such as brand prestige or product storytelling. Since such value varies greatly by context and product category, a flexible and case-based approach is essential. Empirical studies that document and analyse successful examples are especially valuable for understanding how slow manufacturing can be effectively implemented across different settings.

4 Industrial Practice and Academic Research

4.1 Industrial Practice

This section presents examples of Small and Medium-sized Enterprises (SMEs) in Japan and Italy that embody the principles of slow manufacturing. These companies demonstrate how craftsmanship can be effectively integrated with automation to enhance product value while preserving artisanal integrity.

Citizen, a leading Japanese watchmaker, exemplifies this dual approach. While automated processes are used for cost-effective mass production, the company reserves the creation of high-end models for skilled artisans, including nationally and regionally certified master craftsmen. This strategy positions craftsmanship as a top corporate priority and differentiates their premium product lines. The company's factory is also located within an industrial district known for precision instrument manufacturing, further reinforcing collaborative advantages.

Kaneko Optical, based in Sabae - a famous city for its eyeglass manufacturing - represents another notable example. The company has gained popularity among ethically conscious consumers for its handcrafted products and meticulous attention to detail. At the same time, Kaneko Optical invests in research and development to enhance its production capabilities, aiming to find the optimal balance between manual techniques and specialised automation. This includes the development of proprietary technologies tailored to their artisanal workflow.

Ferrari, Lamborghini, Maserati and Pagani are the most significant brands for sports cars and Ducati for motorbikes, located in the already mentioned Italian "Motor Valley". Vehicles are complex products that require high-quality materials and processes to be fabricated and assembled. In all these medium-sized enterprises, craftsmanship and advanced technologies join to manufacture excellent vehicles. Local vocational skills in mechanics, good technical schools, and a locally diffused entrepreneurial spirit have allowed the establishment of factories where vehicles are manufactured, prioritising attention and quality over time reduction. The craftwork component is expressed by the capability of workers to take care of all the details using an artisanal manual approach, which adds significant extra value, making each vehicle unique and integrated into the use of advanced materials and processes.

Another case of an Italian SME where the slow manufacturing approach is adopted is Babbi, a company producing chocolates and creams. Being nicknamed the "Ferrari of food", Babbi chooses slow processes to guarantee taste and the superior quality of its products. For example, in the case of the production of caramelised figs, the company has evaluated a modern pasteurisation system as a possible process innovation because it lasts a few hours. However, based on the expected product quality, the longer traditional cooking method lasting several hours has been finally preferred because this method assures the best taste of the figs.

All these examples reflect the three core elements of slow manufacturing outlined in Sect. 3: the integration of craftsmanship and modern technology, the influence of regional characteristics (industrial clustering), and the pursuit of extra value creation through authenticity and narrative-rich products.

4.2 Academic Research

In academic research, efforts have been made to identify which manufacturing processes should remain reliant on artisanal skills and which are more suitable for automation. One such method involves a structured classification process, consisting of the following four steps:

Step 1: Investigate the production processes of benchmark companies that successfully integrate craftsmanship. These processes are categorised into three types:

- Skill-dependent processes. Where the artisan's expertise significantly influences product quality and value.
- Skill-neutral processes. Where worker skill is used but has minimal impact on the final value.
- Automated processes. Already optimised through mechanisation or digital technologies.

Step 2: Evaluate each process using qualitative criteria, such as whether it involves handcrafting or produces visually distinct or emotionally appealing variations in the product finish.
Step 3: Apply Hayashi's Quantification Theory Type II [18] to the qualitative assessments. This technique maps each process into a discriminant space, where the centroid of each classification category is calculated.
Step 4: Classify a target process by locating its position in the discriminant space based on qualitative evaluation, then assign it to the closest category centroid.

As a practical example, Fig. 2 illustrates the results of applying this classification method to an eyeglass manufacturing company. Training data were derived from best practices in knife and glass bottle production. Three specific processes - frame shaping, engraving, and temple bending - were identified as differing from current company practices. When these results were presented to the factory manager, the proposal to automate engraving was deemed infeasible due to its complexity and craftsmanship requirements. However, the recommendations to automate shaping and bending processes were considered promising. For all other processes, the proposed classifications aligned well with current practices, validating the method's effectiveness.

Although this methodology shows potential, further verification and expansion across other industries and regions is needed to generalise its applicability. Nevertheless, it offers a valuable tool for strategically integrating craftsmanship and automation in a balanced, human-centred way, consistent with the ideals of Industry 5.0.

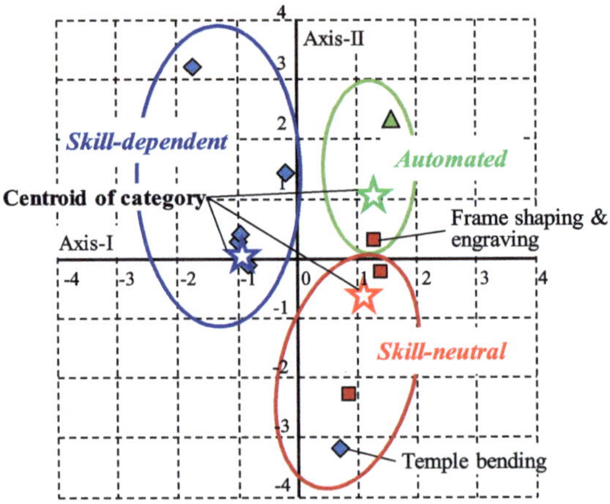

Fig. 2. Classification result of target manufacturing processes based on learning from the best practice of slow manufacturing.

5 Discussion

5.1 Extending the Network of Slow Manufacturing to Developing Countries

In many developing countries, particularly in Southeast Asia, large domestic manufacturing enterprises are scarce, and economic constraints often hinder investment in advanced manufacturing technologies. Consequently, SMEs have emerged as the primary actors in the manufacturing sector.

One viable approach for SMEs in such regions is the adoption of Appropriate Technology (AT). This approach focuses on producing original local products using naturally available materials such as wood, bamboo, and clay, typically in small production volumes. Workers are usually recruited from the surrounding community, resulting in the preservation of low-cost, indigenous, and culturally rooted technologies.

For instance, Mitti Cool in India manufactures clay-based, non-electric refrigerators designed for individuals who cannot afford conventional electric appliances [19]. This product addresses local challenges such as economic hardship and unstable electricity supply by utilising traditional materials, older machinery, and local labour. Similarly, Magno in Indonesia produces handcrafted wooden goods, with all employees sourced from the local area [20].

Slow manufacturing and AT share several features, including non-mass production principles, emphasis on worker satisfaction, and the creation of extra value through culturally embedded narratives. However, key differences also exist. Unlike AT, slow manufacturing aims to combine craftsmanship with advanced technologies, such as those developed under the Industry 4.0 paradigm, to enhance value creation and production efficiency. Moreover, industrial clusters that play a critical role in the slow manufacturing model are generally underdeveloped in emerging economies.

To enable the spread of slow manufacturing in these regions, it is essential to support the development of industrial districts and invest in engineering education and technical infrastructure. These measures would help bridge the technological gap and enable the integration of human skills with modern innovations in a contextually appropriate and sustainable manner.

5.2 Research Agenda

To advance the practical and theoretical foundation of slow manufacturing, we propose a research agenda focusing on three key areas:

Combination of Craftsmanship and Technology. As demonstrated in Sect. 4.1, further refinement of analysis methods is needed to identify better optimal combinations of manual skills and both traditional and modern technologies. Particular attention should be given to positively evaluating the time spent on artisanal processes that significantly contribute to value creation, rather than viewing such time as inefficiency.

Local Characteristics. Each successful industrial district evolves with its own unique configuration [19]. Therefore, it is crucial to identify and apply region-specific models that support slow manufacturing. This requires the continuous collection and analysis of case studies focused on artisanal skills, intending to identify compatible structural patterns for various local contexts.

Extra Value Creation. A deeper understanding of how slowness contributes to creating extra value is necessary. This value is twofold:

- On the worker side, indicators may include per-person sales or the subjective well-being derived from meaningful work.
- On the user side, value manifests as emotional attachment or a sense of timelessness toward products crafted by artisans. This emotional connection can extend product lifespan and reduce waste, thereby enhancing sustainability.

Moreover, fostering skill-centred product development is critical. If slow manufacturing enterprises can leverage craftsmanship to explore new product directions or niche markets, core artisanal competencies can be preserved and carried forward into the future.

It should be noted that these research areas are interrelated. Thus, a cyclical approach combining the development of new analytical methods with comprehensive case studies, spanning individual companies, regional clusters, and international comparisons, is recommended to achieve a more integrated understanding of slow manufacturing.

6 Conclusions

This paper proposed a concept of slow manufacturing applicable to the framework of Industry 5.0 and introduced examples of its application in industrialised countries, both in corporate practice and academic research. The discussion then extended to the potential for implementing this concept in developing countries, followed by the presentation of a future research agenda. Slow manufacturing offers a promising solution to the challenge of integrating human handwork with Industry 4.0 technologies, contributing to greater diversity in manufacturing approaches. Moving forward, we aim to pursue further studies that enhance the well-being of both workers within production systems and the users of the products they create.

Acknowledgement. A portion of this work was supported by Bilateral Collaborations of JSPS (JPJSBP120229925).

References

1. Kagermann, H., Wahlster, W.: Ten years of Industrie 4.0. Sci **4**(3), 26–35 (2022)
2. Schleich, B., et al.: Shaping the digital twin for design and production engineering. CIRP Ann. Manuf. Technol. **66**, 141–144 (2017)
3. Monostori, L., et al.: Cyber-physical systems in manufacturing. CIRP Ann. Manuf. Technol. **65**, 621–641 (2016)
4. Xu, X., et al.: Industry 4.0 and Industry 5.0 – inception, conception and perception. J. Manuf. Syst. **61**, 530–535 (2021)
5. Van Erp, T., et al.: Industry 5.0: a new strategy framework for sustainability management and beyond. J. Clean. Prod. **461**, 142271 (2024)
6. Sindhwani, R., Hasteer, N., Behl, A., Chatterjee, C., Hamzi, L.: Analysis of sustainable supply chain and Industry 4.0 enablers: a step towards decarbonization of supply chains. Ann. Oper. Res. (2023). https://doi.org/10.1007/s10479-023-05598-7
7. Qader, G., et al.: Industry 4.0 enables supply chain resilience and supply chain performance. Tech. Forecasting Soc. Change **185**, 122026 (2022)
8. Barbara, C., Campana, G.: The slow factory: a new paradigm for manufacturing. In: Proceedings of the 11th GCSM, pp. 273–277 (2013)
9. Kobayashi, H.: A concept proposition of slow-manufacturing. In: Proceeding of the JSME-MSD, pp. 136–138 (2022). (in Japanese)
10. Krugh, M.: Joy in labour: the politicization of craft from the arts and crafts movement to Etsy. Can. Rev. Am. Stud. **44**(2), 281–301 (2014)
11. Adachi, N.: Development of mass craftsmanship production system. Mazda Gihou **36**, 154–158 (2019). (in Japanese)
12. Umeda, Y., et al.: Digital triplet and its implementation on learning factory. IFAC PapersOnLine **55**(2), 1–6 (2022)
13. Campana, G., Cimatti, B.: Measures and methods for a new taxonomy in manufacturing enterprises. Procedia CIRP **26**, 287–292 (2015)
14. Campana, G., Cimatti, B., Melosi, F.: A proposal for the evaluation of craftsmanship in industry. Procedia CIRP **40**, 669–674 (2016)
15. Campana, G., Mele, M., De Bortoli, L., Cimatti, B., Kobayashi, H.: Evaluation of competencies for a sustainable industrial environment. In: Kohl, H., Seliger, G., Dietrich, F. (eds.) Manufacturing Driving Circular Economy. GCSM2022. LNME (2023). https://doi.org/10.1007/978-3-031-28839-5_93
16. Irino, Y., Murata, H., Kobayashi, H., Campana, G.: An analysis method considering manufacturing skills and technologies toward slow manufacturing. In: EcoDesign for Circular Value Creation, vol. I, pp. 547–561. Springer, Cham (2025)
17. Yamawaki, H.: The evolution and structure of industrial clusters in Japan. Small Bus. Econ. **18**, 121–140 (2002)
18. Hayashi, C.: On the quantification of qualitative data from the mathematico-statistical point of view. Ann. Inst. Stat. Math. **2**, 35–47 (1950)

19. Mitti Cool. https://mitticool.com. Accessed 15 Apr 2025
20. Magno. https://magno.works. Accessed 15 Apr 2025

Information Logistics as a Key Enabler for Biointelligent Value Creation Systems – Challenges and Requirements

Arber Shoshi[1,2,3](✉), Thomas Bauernhansl[2,3], and Robert Miehe[2,3]

[1] Graduate School of Excellence Advanced Manufacturing Engineering (GSaME), University of Stuttgart, Nobelstraße 12, 70569 Stuttgart, Germany
arber.shoshi@gsame.uni-stuttgart.de

[2] Fraunhofer Institute for Manufacturing Engineering and Automation IPA, Nobelstraße 12, 70569 Stuttgart, Germany

[3] Institute of Industrial Manufacturing and Management (IFF), University of Stuttgart, Allmandring 35, 70569 Stuttgart, Germany

Abstract. The concept of biointelligence has recently emerged as a convergence of biology, engineering and information technology. The use of biointelligent systems in decentralized, small-scale production units, designed for non-professionals, using local biological resources and involving different actors, constitute biointelligent value creation systems (BVC). The effective operation of BVCs requires an information system and the underlying system, namely information logistics, capable of providing the right information to the right actor at the right time and in the right quality. However, the inherent complexity of management, due among other things to biological components in the system and heterogeneous participants - including industrial and non-industrial actors - poses significant challenges to existing information systems. This paper addresses this gap by analyzing 19 expert interviews from industry and academia. It identifies key challenges such as a lack of semantic interoperability and unclear data ownership, as well as requirements for scalable, role-specific and adaptive information logistics systems. The findings provide actionable insights into how future information systems can be designed to support complex and dynamic BVC environments, including the importance of transparent governance structures and user-centered dashboards that address diverse stakeholder needs.

Keywords: Biointelligence · Information logistics · value creation systems · Sustainable production

1 Introduction

The biological transformation of industry represents a fundamental paradigm shift in how we perceive production and value creation. This transformation aims to integrate biological principles and processes into industrial systems, enabling more sustainable and efficient manufacturing methods [1]. Within this context, the concept of biointelligence has emerged as a promising approach, focusing on the convergence of biological,

H. Kohl et al. (Eds.): GCSM 2025, LNME, pp. 372–380, 2026.
https://doi.org/10.1007/978-3-032-21154-5_41

digital, and physical systems [2]. It is crucial to distinguish biointelligence from bioinspiration (using biological principles as models) and biointegration (embedding biological components). Biointelligence specifically refers to systems where biological, technical, and digital elements form adaptive, feedback-driven networks [3, 4]. These Biointelligent Systems (BIS) combine biological components (e.g., cells, enzymes) with technical structures (e.g., reactors, sensors) and digital control systems to create adaptive, self-optimizing production systems. Typical applications include cell-based manufacturing, 3D-printed food, or decentralized therapeutic production [1].

When scaled to the level of production networks, these systems form Biointelligent Value Creation Systems (BVC). A BVC is a decentralized, often modular, production or service network that integrates one or more BIS. They are characterized by the use of biological living components or resources, modular and decentralized production units, and the active participation of both industrial and non-industrial actors [5] BVC aim to reduce waste, minimize transport, and support closed-loop processes, positioning themselves as enablers of sustainable production [5, 6]. Furthermore, their integration of diverse actors—including industrial companies, small enterprises, community groups, and individual users—is particularly notable for democratizing production and promoting broad stakeholder participation [7].

The nature of biological systems introduces specific challenges: biological substrates behave non-deterministically. Cells or microorganisms may react unpredictably to stimuli, and their quality is difficult to measure and standardize. Often, suitable sensors for precise and continuous monitoring are lacking, complicating the control and modeling of processes. Moreover, the socio-technical structure of BVC adds further complexity, as stakeholders differ significantly in expertise, technological maturity, strategic goals, and available resources. While some require sophisticated data systems, others rely on intuitive, low-threshold interfaces. Consequently, BVC are characterized by high heterogeneity, dynamic systems, and an increased need for coordination and trust.

A representative example is the vision of decentralized CAR-T cell therapy mini-factories. In this scenario, biological components (cells), technical systems (automated bioreactors), and digital monitoring infrastructures interact across diverse actors such as medical staff and equipment providers [8, 9]. This localized, adaptive, and personalized production environment highlights the crucial need for robust Information Logistics (IL) across professional and organizational boundaries. IL refers to the systematic coordination of data flows to ensure that the right information reaches the right stakeholder at the right time, in the right format, and in the right context—between diverse actors and systems. To effectively realize BVC, Information Logistics Systems (ILS) are therefore essential [10, 11].

In BVC, the diversity of actors and decentralized operations significantly increase the complexity of IL. Seamless information flows are crucial to coordinate activities among biological processes, technical systems, and human users, manage uncertainties, and enable traceability and transparency throughout the network [12]. Despite the growing importance of biointelligent value creation, there is a recognized lack of structured knowledge on how information systems and particularly ILS must be designed to effectively support such complex environments. This paper addresses this gap by exploring the design of ILS for BVC based on 19 expert interviews from industry and academia. By systematically categorizing and analyzing the findings, we offer insights into the socio-technical requirements of IL in BVC and provide design-oriented implications. The following sections detail the methodological approach (Sect. 2), present the results and discuss their implications (Sect. 3), and cover the limitations of this research, followed by the conclusion in Sect. 4.

2 Methodology

To identify key challenges, requirements, and practices for ILS in BVC, a qualitative content analysis was conducted based on 19 expert interviews with participants from information systems, technical systems, and biological domains. Experts from both industry and academia were selected to ensure practical and conceptual relevance. The sample included engineering managers, product developers, and digital transformation leads from industry, as well as professors and postdoctoral researchers in bioengineering, manufacturing, and IT. Participants held mid- to senior-level roles with direct involvement in designing, managing, or researching components of biointelligent value creation systems. This enabled a multi-perspective view across technical, biological, and informatic aspects. Interviews were semi-structured, transcribed, and coded in MaxQDA. The analysis followed Mayring's approach to inductive category development [13], using meaning-based segments and an iteratively refined coding guide. Categories for challenges and requirements were developed and expanded during analysis; best practices were extracted when mentioned in relation to specific problems. Thematic saturation was reached, and consistency ensured through cross-interview comparisons and intercoder checks. These results form the basis for the categorization of requirements, challenges, and solution strategies in the following sections (Fig. 1).

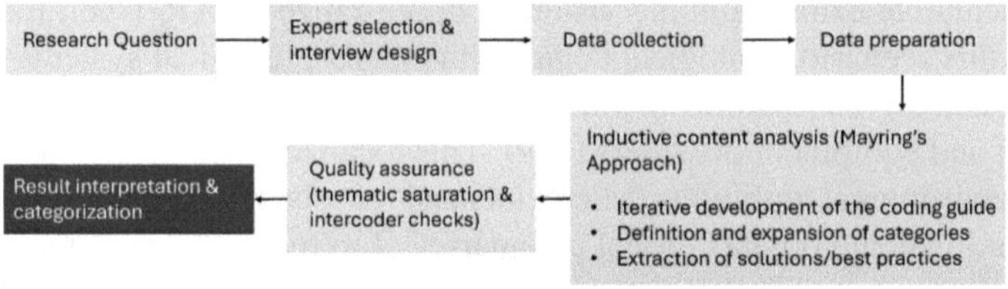

Fig. 1. Steps of inductive category development.

3 Results

3.1 Identified Challenges

The interviews reveal a set of recurring challenges that reflect the socio-technical complexity of IL in BVC.

A fundamental issue is the unpredictability and variability of biological inputs like cells, bacteria, and biomass. Unlike industrial components, these materials fluctuate naturally in quality and availability, often without clear early detection indicators. Existing sensor technologies are frequently too limited or slow to provide real-time insights into critical biological conditions such as cell vitality, which significantly limits the controllability of BIS. Furthermore, experts highlighted technical difficulties in system integration and interoperability, as diverse software systems, devices, and platforms hinder continuous information flow.

Beyond technical aspects, BVCs involve stakeholders from vastly different disciplines, including biotechnology, medicine, agriculture, manufacturing, and IT. This heterogeneity leads to misunderstandings, cultural frictions, and incompatible expectations. Communication is further hampered by inconsistent terminology, a lack of shared ontologies, and diverging levels of digital literacy. Many actors, particularly non-industrial ones, are unfamiliar with complex IT systems and uncertain about their information needs or interpretation. Consequently, information flow often becomes fragmented, inconsistent, or overly complex, either overwhelming users or depriving them of critical inputs.

The absence of clear responsibilities and governance mechanisms further compounds this fragmentation. Within these networks, data ownership, sharing authorization, and liability for misuse remain unclear. Data protection and privacy are especially sensitive in fields like healthcare, where ethical and regulatory standards are high. Without legally secure frameworks and mutual trust, data exchange is often avoided, hindering cooperation.

Economic and strategic issues further complicate the landscape. Data is often viewed as a competitive resource or private good, making participants reluctant to share it without compensation or control, particularly for personalized data. Concurrently, existing business models rarely incentivize transparency or information sharing. This creates competitive barriers, a lack of motivation, and underutilized data potential, especially in regional and cross-sectoral settings where collaborative value creation relies on mutual data access.

3.2 Derived Challenges for Information Logistic Systems

To address the identified challenges, robust and flexible Information Logistics Systems (ILS) tailored for Biointelligent Value Creation Systems (BVC) require several key features. First, standardized interfaces are essential for interoperability across heterogeneous devices and platforms, enabling seamless modular and decentralized system integration. Second, semantic interoperability, supported by shared vocabularies, protocols, and ontologies, must be established for meaningful and consistent communication among industrial and non-industrial actors. Additionally, ILS must support role-based access control, considering users' expertise, responsibilities, and data protection. This ensures experts have full data access while non-experts (e.g., patients, farmers) receive simplified, context-relevant information, enhancing usability, data sovereignty, and preventing misuse. Transparency and traceability emerged as recurring themes; stakeholders need to know data origin, generator, and processing, which is crucial for building trust, ensuring ethical compliance, and reconstructing processes. To facilitate this, governance structures must clearly define data ownership, responsibilities, and rights, particularly in asymmetric networks, ensuring legal certainty and collaboration. Furthermore, establishing incentive mechanisms for data sharing is vital, as many actors are reluctant to share sensitive information without concrete benefits like service improvements, reduced costs, or access to shared infrastructures.

Given the dynamic nature of BVC, scalability is a critical requirement; ILS must grow or shrink with the network, quickly onboard new partners, and support modular extension. Push/pull mechanisms, adapted to biological processes, are necessary to respond to event-driven situations like process deviations or quality changes along the supply chain, allowing stakeholders timely adjustments. To accommodate varying roles and knowledge levels, the system should support flexible role assignment and information tailoring, providing users with the precise level of detail needed while ensuring data integrity. Differentiated data protection levels must also be implemented for sensitive (e.g., patient health) versus non-sensitive data. As actors and frameworks evolve, ILS must be adaptable and reconfigurable. Finally, essential support for user guidance and data literacy is needed through intuitive, visual, and domain-specific tools, enabling even non-technical users to interact meaningfully with complex systems.

3.3 Current State of Solutions and Best Practices

While the requirements for ILS in BVC are conceptually established, practical and transferable solutions remain scarce. Despite growing interest in biological transformation in manufacturing [3, 5], the adaptation of information coordination systems to the complexities of BVC remains underexplored. Existing frameworks from Industry 4.0 and cyber-physical systems [14] primarily target deterministic processes and often overlook the variability introduced by living systems.

In related domains like healthcare logistics and biomanufacturing [15], tools such as LIMS or hospital IT systems show partial applicability. However, their fit for decentralized, heterogeneous networks like BVC is limited. Interfaces like OPC UA are well-developed for traditional automation but lack counterparts for biological-hybrid systems. Semantic interoperability, though widely acknowledged as critical, remains underdeveloped. Domain-specific ontologies exist but are seldom integrated across the disciplinary boundaries found in BVC.

Technologies for transparency and traceability—e.g., blockchain and audit trails—face challenges in multi-actor, biologically variable environments. Governance concepts like GAIA-X provide frameworks for data ownership and access, but tailored applications for adaptive BVC systems are rare. Incentive mechanisms for data sharing are especially underdeveloped; few digital platforms compensate voluntary data contributions, particularly from non-industrial actors.

Similarly, push/pull architectures are standard in IT but rarely adapted to real-time, event-driven needs in bioprocessing—e.g., when reacting to cell viability changes. Scalable, modular software designs like microservices are theoretically suitable but hindered by high integration costs, compatibility issues, and the lack of BVC-aligned reference architectures.

User-oriented dashboards and decision tools are also limited. While some exist, they are rarely adapted to the knowledge levels of BVC users, who may lack technical expertise.

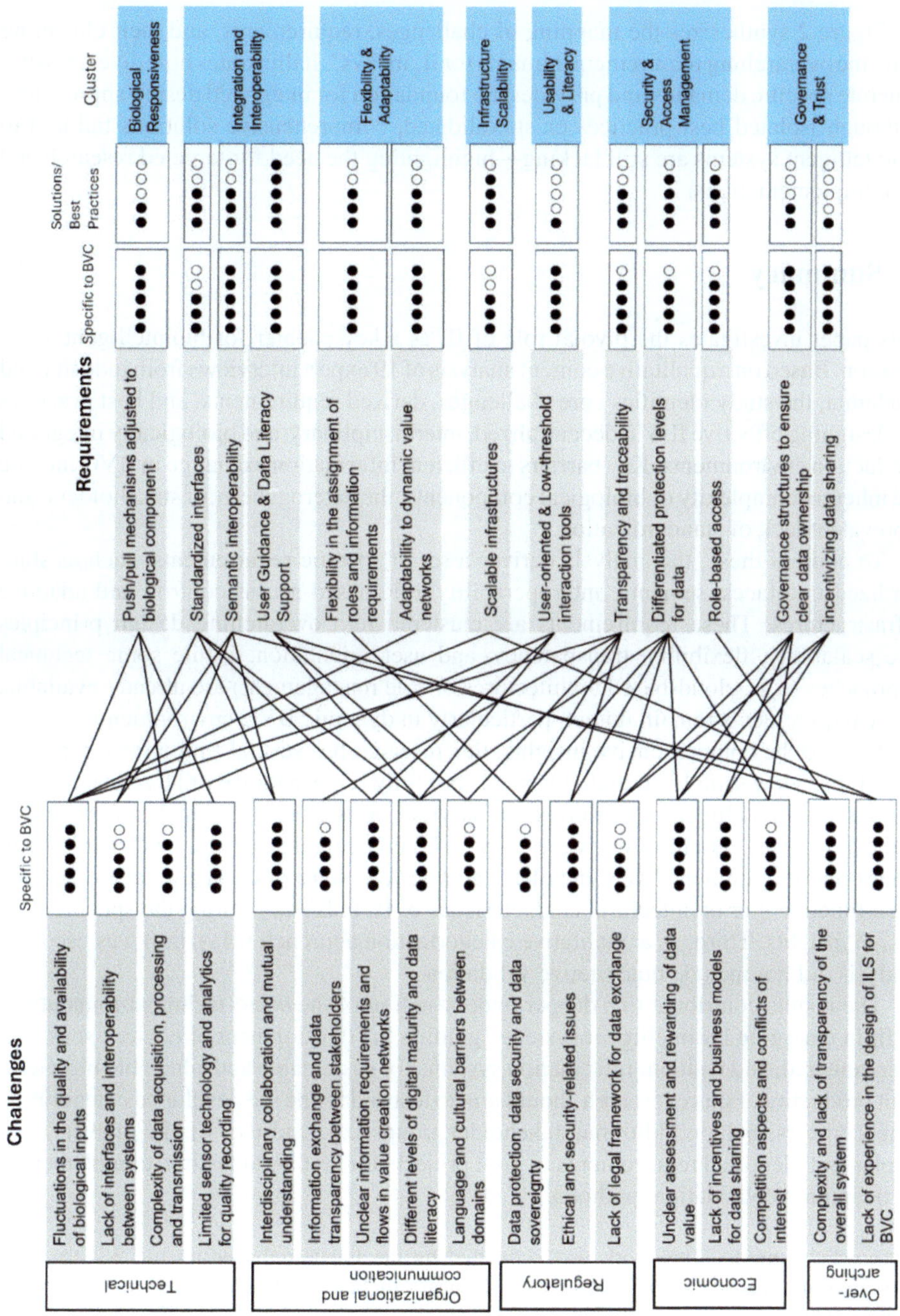

Fig. 2. Mapping of challenges, requirements, and requirement clusters in BVC.

In addition, the lack of shared tools for specifying information needs—such as a simplified, domain-adapted UML—continues to impede coordination.

Figure 2 synthesizes the mapping of challenges, requirements, and their clustering into nine overarching requirement domains for IL in BVC. It illustrates how diverse issues generate specific demands and provides the foundation for integrated design approaches. Although isolated best practices exist, validated, comprehensive solutions tailored to biointelligent systems are still lacking—highlighting the need for targeted research and pilot implementations.

4 Summary

This paper investigates the pivotal role of IL as a key enabler for biointelligent value creation. Based on a qualitative content analysis of 19 expert interviews from industry and academia, the study identifies core challenges, derived requirements, and best practices for designing effective ILS in decentralized, interdisciplinary, and biologically integrated production environments. Key barriers to efficient information exchange in BVC include the inherent complexity of biological components, the heterogeneity of stakeholders, and a prevalent lack of standardization.

To address these, the analysis derives a set of specific requirements, such as standardized interfaces, semantic interoperability, role-based access control, and adaptive infrastructures. These requirements are clustered into overarching design principles like scalability, flexibility, transparency, and user orientation. While some technical approaches (e.g., cloud-based architectures or role management) are already available, many requirements remain unmet, particularly in dynamic BVC environments.

Despite the comprehensive insights, this research has several limitations. First, the expert interview sample, though diverse, is not fully representative of all relevant stakeholders in biointelligent value creation, as viewpoints from policymakers, regulatory bodies, and various end-users were only indirectly represented. Second, while challenges and requirements were detailed, interviewees were often vague regarding concrete functional or architectural characteristics of IL in BVC, limiting the specificity of design insights. Third, as a qualitative categorization approach (Mayring) was used, the analysis did not include quantitative validation.

The results contribute to a deeper understanding of the socio-technical foundations of IL in emerging biointelligent systems, guiding the development of context-sensitive, user-centered, and resilient information systems for sustainable and adaptable biointelligent production. Future research should aim to test and refine the developed categories in other contexts, include additional stakeholder groups (e.g., end-users, policymakers), and implement identified requirements in pilot projects for empirical testing, complemented by quantitative validation methods.

Acknowledgements. This work was supported by the Landesministerium für Wissenschaft, Forschung und Kunst Baden-Württemberg (Ministry of Science, Research and the Arts of the State of Baden-Wurttemberg) within the Nachhaltigkeitsförderung (sustainability support) of the projects of the Exzellenzinitiative II.

References

1. Miehe, R., Bauernhansl, T., Beckett, M., Brecher, C., Demmer, A., Drossel, W.-G., et al.: The biological transformation of industrial manufacturing – Technologies, status and scenarios for a sustainable future of the German manufacturing industry. J. Manuf. Syst. **54**, 50–61 (2020). https://doi.org/10.1016/j.jmsy.2019.11.006
2. Miehe, R., Horbelt, J., Baumgarten, Y., Bauernhansl, T.: Basic considerations for a digital twin of biointelligent systems: applying technical design patterns to biological systems. CIRP J. Manuf. Sci. Technol. **31**, 548–560 (2020). https://doi.org/10.1016/j.cirpj.2020.08.006
3. Wegener, K., Damm, O., Harst, S., Ihlenfeldt, S., Monostori, L., Teti, R., et al.: Biologicalisation in manufacturing – Current state and future trends. CIRP Ann. **72**, 781–807 (2023). https://doi.org/10.1016/j.cirp.2023.05.005
4. Miehe, R., Bauernhansl, T., Schwarz, O., Traube, A., Lorenzoni, A., Waltersmann, L., et al.: The biological transformation of the manufacturing industry – envisioning biointelligent value adding. Procedia CIRP. **72**, 739–743 (2018). https://doi.org/10.1016/j.procir.2018.04.085
5. Full, J., Shoshi, A., Gamero, E., Baumgarten, Y., Protte, K., Kiemel, S., et al.: Biointelligent Waste-to-X systems: A novel concept for sustainable, decentralized and interconnected value creation. Procedia CIRP. **116**, 576–581 (2023). https://doi.org/10.1016/j.procir.2023.02.097
6. Miehe, R., Waltersmann, L., Sauer, A., Bauernhansl, T.: Sustainable production and the role of digital twins–Basic reflections and perspectives. J Adv Manuf. Process (2021). https://doi.org/10.1002/amp2.10078
7. Fastermann, P.: 3D-Drucken. Berlin, Heidelberg: Springer Berlin Heidelberg (2014)
8. Fraunhofer Institute for Manufacturing Engineering and Automation IPA. Solid CAR-T: Modular mini-factories for autonomous CAR-T cell production. https://www.ipa.fraunhofer.de/en/expertise/laboratory-automation-and-biomanufacturing-engineering/adapt/solid-cart.html, Accessed 28 Mar 2025
9. Shoshi, A., Xia, Y., Fieschi, A., Ackermann, T., Reimann, P., Weyrich, M., et al.: A flexible digital twin framework for ATMP production – towards an efficient car t cell manufacturing. Procedia CIRP. **125**, 124–129 (2024). https://doi.org/10.1016/j.procir.2024.08.022
10. Hartleif, S.M.: Gestaltung einer schlanken Informationslogistik im komplexen Produktionsumfeld: Universität Stuttgart (2022). https://doi.org/10.18419/opus-12438
11. Krcmar, H.: Informationsmanagement. Berlin, Heidelberg: Springer Berlin Heidelberg; (2015)
12. Miehe, R., Buckreus, L., Kiemel, S., Sauer, A., Bauernhansl, T.: A conceptual framework for biointelligent production—calling for systemic life cycle thinking in cellular units. Clean Technol. **3**, 844–857 (2021). https://doi.org/10.3390/cleantechnol3040049
13. Mayring, P.: Qualitative inhaltsanalyse. In: Mey, G., Mruck, K., (eds.) Hand-buch Qualitative Forschung in der Psychologie. Wiesbaden: VS Verlag für Sozialwissenschaften; 2010. p. 601–613. https://doi.org/10.1007/978-3-531-92052-8_42
14. Schuh, G., Kelzenberg, C., Wiese, J., Kessler, N.: Creation of digital production twins for the optimization of value creation in single and small batch production. Procedia CIRP. **93**, 222–227 (2020). https://doi.org/10.1016/j.procir.2020.04.125
15. Smith, B.K., Nachtmann, H., Pohl, E.A.: Improving healthcare supply chain processes via data standardization. Eng. Manag. J. **24**, 3 (2012). https://doi.org/10.1080/10429247.2012.11431924

Deriving Circular Economy Principles for Industrial Implementation - Operational Implications and Responsibilities Within Company Boundaries

Dominik Saubke(✉), Pascal Krenz, and Tobias Redlich

Helmut Schmidt University, Hamburg, Germany
dominiksaubke@hsu-hh.de

Abstract. In economics, new theoretical approaches have often emerged when our traditional paradigms surpassed their limits. Heterodox economic theories have gained increasing attention, responding to shifting contextual factors and challenging traditional, infinite, and unidirectional models of resource use. Since the seminal works of STAHEL and REDAY (1976), the *Circular Economy (CE)*, has emerged as a central economic concept for a sustainable economic transition in academic and political discussions, gaining significant traction over the past decade. However, much of the outcome of its discussions remains theoretical, with policymakers and institutions often overlooking the fact that successful implementation must occur at the company level. To address this gap, a social empirical research approach was conducted to identify and analyse 18 national and international strategy papers from governmental bodies, industry associations (e.g., ACATECH, VDI), and standardization organizations (e.g., DIN). Across over 1,000 pages of documentation, 14 *Action Areas* for implementing CE principles were condensed. The *Action Areas* are linked to 55 *Operational Implications* for industrial implementation, which were further assigned to nine corporate responsibilities to ensure pragmatic and actionable recommendations.

Keywords: Manufacturing · Sustainability · Circular Economy · Management

1 Introduction and Motivation

Few terms enjoy as much ubiquity across scientific, political, and economic discourse as the sustainable transformation [1]. However, it is becoming increasingly apparent that the invocation of sustainability alone is insufficient to address the structural transformations required in contemporary industrial value creation systems to achieve corporate sustainability [2]. Instead, the concept of a circular economy has emerged as a new guiding paradigm [3]. It integrates ecological objectives while simultaneously serving as a strategic framework for reconfiguring production and consumption structures and needs to be integrated into future corporate sustainability strategies [4, 5]. With Schuh (2023) introducing the concept of *Upgrade Circular Economy* [6], a benchmark was set that

H. Kohl et al. (Eds.): GCSM 2025, LNME, pp. 381–391, 2026.
https://doi.org/10.1007/978-3-032-21154-5_42

illustrates the potential of circular value creation. Through the deliberate extension of product life cycles and the reuse of high-quality components, it is possible to achieve cost savings in the double-digit percentage range. This is accompanied by the conservation of natural resources and significant cost advantages, enabling industrial organisations to gain a substantial competitive advantage through the implementation of circular business practices [7]. What is required is a shift away from linear process logic towards value creation systems based on the principles of circularity and situated at the point of demand. The consistent closing of industrial material loops, along with the integration of higher-order R-strategies (e.g. remanufacturing) into the product life cycle, brings formerly ambitious objectives, such as reducing CO_2 emissions and improving resource efficiency, within practical reach.

2 Objective of Research

The internal implementation of circularity principles within companies requires a deep understanding of the interdisciplinary interdependencies that significantly influence operational and strategic industrial processes. Although numerous approaches to integrating circular principles have emerged in recent years, many of these concepts remain abstract or generic, failing to sufficiently address the operational implications at the necessary level of granularity.

Against this backdrop, the present study addresses the research question of which core fields of action can be identified for the practical implementation of circular economy principles in industrial manufacturing companies and which concrete operational measures must be anchored within specific areas of responsibility. To answer this question, a total of 18 strategic documents, issued by national and international governmental institutions, industrial consortia such as ACATECH and VDI, as well as standardisation bodies including DIN, were systematically identified and subjected to structured analysis. The analysis was conducted using a qualitative empirical approach based on established principles of empirical social research. This provides a structured overview for orientation, offering both strategic and operational points of reference for the internal organisational implementation of circular value creation.

3 Methodological Approach

The methodological approach is based on a multi-stage mixed-method design, combining qualitative and theory-driven procedures to systematically derive concrete measures within a structure of defined *Action Areas*, subsequently transferred into a coherent responsibility framework. The primary objective was to identify functional responsibilities on a corporate level to introduce and maintain circular principles. The process initiated with an unsystematic literature review, aiming to capture a broad spectrum of relevant terminologies, role descriptions, and organisational patterns within the industrial sector. This explorative phase laid the groundwork for a subsequent intensive text analysis, in which content focusing on operation implications from the identified literature was extracted and categorised according to its thematic and conceptual relevance.

This analytical coding process is oriented towards the principles of *Grounded Theory* [8]. Through iterative coding cycles, recurring patterns and thematic clusters were identified, leading to the definition of *Action Areas*. These were subsequently mapped to a preliminary system of functional responsibilities within organizational boundaries (Fig. 1).

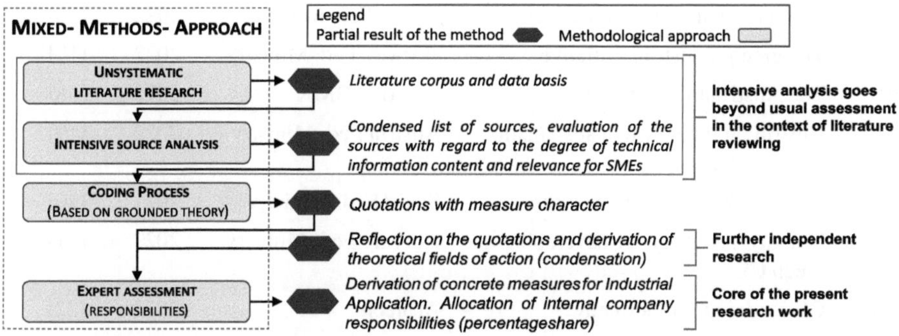

Fig. 1. Condensed overview of the selected mixed method approach.

In the final step, an expert-based validation of the *Action Areas* and linked *Operational Implications* was carried out. The expert assessment was conducted by a panel of 7 academics and industry professionals. Drawing on an interdisciplinary approach, the panel included social scientists, industrial engineers, mechanical engineers, and economic geographers. The previously identified roles and areas of responsibility were reflected upon, contextualised, and refined in close consultation with professionals from both industry and academia. The literature corpus was identified through a selective process. The selected publications originate from reputable institutions, organisations, and governmental bodies within the European Union, thereby ensuring a high degree of credibility and authority. All sources (listed Table 1.) were published within the past five years to preserve relevance and reflect current developments.

Table 1. Considered Sources for the Analysis.

	Name	Author	Year	Source
Q1	CE challenges and paths to transformation	German Fed. Gover	2024	[9]
Q2	CE at a glance	IHK Karlsruhe	2024	[10]
Q3	Circular markets for industrial value creation	BDI	2024	[11]
Q4	Draft national CE strategy	BMUV	2024	[12]
Q5	A comprehensive CE for Germany 2045	WWF	2023	[13]

(*continued*)

Table 1. (*continued*)

	Name	Author	Year	Source
Q6	German standardisation roadmap – CE	DIN	2023	[14]
Q7	CE and ressource efficient – CE	VDI	2022	[15]
Q8	Ce sectoral strategic guidelines	Italian Fed. Ministry	2022	[16]
Q9	The green industrial initiative	Norway Fed. Ministry	2022	[17]
Q10	The Austrian CE strategy	Austrian BMK	2022	[18]
Q11	Austria on the way to a sustainable and circular society	Austria Fed. Ministry	2022	[19]
Q12	Circular 2030 executive summary	Spain Fed. Ministry	2021	[20]
Q13	Norway's strategy for developing a green CE	Norway Fed. Ministry	2021	[21]
Q14	CE roadmap for Germany	Acatech	2021	[22]
Q15	A new action plan for the circular economy	EU Commission	2020	[23]
Q16	Action plan for the CE	EU Commission	2020	[24]
Q17	Roadmap towards the transition to CE	Poland Fed. Ministry	2019	[25]
Q18	A CE in the Netherlands by 2050	Nether. Fed. Ministry	2016	[26]

4 Responsibilities on a Corporate Level

Within the context of technology-driven manufacturing enterprises, the corporate level assumes a central role in governance and strategic orientation, as clearly illustrated by the presented framework. **Corporate Management [VI]** is conceived as an overarching entity that bears strategic, operational, and cultural responsibility across all organisational divisions. It defines the long-term strategic direction of the company, including market expansion, innovation strategies, and production objectives. It coordinates key functions such as **Product Development [VII]**, to ensure technological competitiveness, as well as **Marketing [II]**, which is responsible for market-oriented positioning and for communicating corporate philosophy. Additionally, it is primarily responsible for integrating and governing the entire value chain. This encompasses not only operational core areas such as **Quality Management [I]** and **Production [V]**, but also economically pivotal functions, including **Controlling [IX]**, **Logistics and Supply Chain Management [VIII]**, and **Sales [III]**. Here, Corporate Management acts as a connective interface, allocating resources, defining performance indicators, and facilitating the cross-functional exchange of knowledge and information. Sustainability is gaining increasing strategic relevance, as reflected in the establishment of the **Sustainability Division [IV]** as a new distinct domain of responsibility, but to establish true sustainability within the company, not all tasks can be delegated to this new division (Fig. 2).

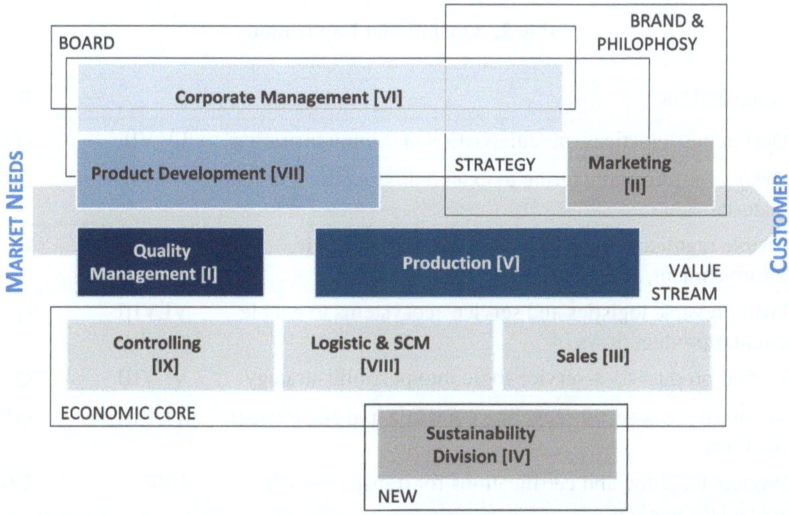

Fig. 2. Framing conditions for corporate responsibility in an industrial manufacturing company.

5 Core Results: Operational Implications

A comprehensive analysis was conducted, involving an in-depth review and coding of hundreds of pages to establish a foundational understanding of the topic, relevant stakeholders, and associated objectives. This process led to the identification of 194 key passages, of which 189 provided actionable insights aimed at enhancing circularity. These insights were further synthesised and consolidated into 14 overarching.

Action Areas for circular implementation. These are *[A] CE Requirements and Adaption in the Public Sector*, *[B] Product Portfolio Strategies concern CE Principles*, *[C] Financial Incentives and Circular Investment for New Business-Models*, *[D] Consumer Education and Engagement*, *[E] Entrepreneur and Management Coaching*, *[F] Transparent Monitoring and Reporting followed by Continuous Improvement*, *[G] Resource Efficiency, Best Practices, and Waste Minimization*, *[H] On-Site Implementation for Digitalization and Smart Technologies*, *[i] Quality Assurance for Circular Products through Obligations*, *[J] Trust Building and Consumer Orientation*, *[K] Local Collaboration and Networking*, *[L] Circularity in Product Design*, *[M] Forming of Production Networks* and *[N] New Norms and Standardisatio*n. Eleven *Action Areas* were identified as highly relevant for companies aiming to integrate circular principles at the corporate level, while Areas A, D, I, and N were excluded due to the limited influence companies can exert in these domains. These areas are directly linked to 55 operational measures outlined in Table 2.

Table 2. Operational Implications.

	Implementations	R	Source
B	1) Design best practices: modularisation and miniaturisation	I,V,VII	Q7, S19
	2) Design for remanufacturing to extend product life and retain material value	I,V,VII	Q7, S19
	3) Enable regulatory approval of products after repair, refurbishment, or ex.use	II,III,VI	Q6, S46
	4) Foster reverse logistics and service ecosystems to enable circular product flows	VI,VIII	Q6, S46
	5) Embed product-as-a-service logic into portfolio strategy	VI,VIII	Q6, S47
C	6) Establish cross-sector recycling consortia and stakeholder platforms	VI,VIII,IX	Q1, S69
	7) Product labelling and certifications for transparency in secondary markets	I,VI	Q6, S36
E E	8) Leverage entry points such as climate action and packaging management	IV,VI,VIII	Q7, S13f
	9) Use digitalisation for synergies between resource efficiency and circularity	VI	Q7, S13f
	10) Monitor consumpt. of machinery and equipment during operation and standby	VI	Q7, S13f
	11) Circular Horizon as a strategy to align innovation, investment, and operations	VI	Q6, S36
	12) Develop a company-specific structure for circular business management processes incorporating R-strategies and performance KPIs	VI	Q6, S37
	13) Skillsets and training guidelines to enable effective implementation of CE	VI	Q6, S45
	14) Conduct assessment against circ. business models to identify trans. potential	VI,IX	Q6, S45
	15) Develop standards that integrate digitalisation, BM, and managm. practices	VI,IX	Q6, S64
	16) Competency-based training concepts for the operatio. of CE principles	VI	Q6, S36
F	17) Increase transparency and trust in second-hand product quality and functionality through certification and clear declarations of conformity	I,II,III,VI	Q6, S36
	18) Develop a quality assurance system with standardised quality indices (e.g. grading) and conformity assessments for products in later life stages	I,IV,VII,VIII	Q6, S37
	19) Integrate reference metrics and reporting formats for CE data into reporting	IV,VI	Q6, S45

(*continued*)

Table 2. (*continued*)

	Implementations	R	Source
	20) Ensure compliance with data spec. requirements (accuracy, quality, format)	IV,VI,IX	Q6, S45
	21) Quantify the environmental impacts of materials used within the company supply chain and value creation process	IV,VII, VIII	Q6, S64
G	22) Foster cross-industry collaboration and leverage best-practice cases	IV,VI,IX	Q1, S13f
	23) Apply generic criterias tailored to the characteristics of the s product portfolio	VI,IX	Q6, S35
	24) Product-group-specific indicator sets for resource consumption in use phase	IV,V,VI	Q6, S35
	25) Continuous monitoring and periodic review to enable gradual optimisation	VI,VII,IX	Q6, S35
	26) Implement tech. processes to meet defined criteria and indicator benchmarks	V,VI	Q6, S35
G	27) Maintain up-to-date standards to support reporting frameworks like EFRS E5	IV,VI,VIII	Q6, S38
	28) Articulate the company's business model in the context of the circular economy	IV,VI	Q6, S46
	29) Submit and check data in databases like SCIP or EPREL	IV,VI,IX	Q6, S55
	30) Improve the utilisation efficiency of resources used within the supply chain	IV,VI,IX	Q14, S7
	31) Assess the economic and material potential of waste-designated resources	VI,VIII,IX	Q6, S37
H	32) Availability and utilisation through practical measures such as sensor deployment and integration with ERP systems	V,VIII	Q7, S13f
	33) Drive organisational transformation through change management	VI,IX	Q1, S10
	34) Assess cost-saving potential of circular measures through quantitative analysis	VI,IX	Q7, S10
	35) Support the mapping of dynamically evolving supply chains	VI,VIII,IX	Q7, S10
	36) Prepare the organisation for strategic and comprehensive CE measures	VI	Q1, S12
	37) Focus on digitalisation efforts to progressively enable the implementation of a Resource-efficient circular economy within the company	VI	Q1, S15
J	38) Quality assurance and conformity assessment for products with extended life	I,VI	Q6, S36

(*continued*)

Table 2. (*continued*)

	Implementations	R	Source
	39) Accelerate the diffusion of circular-ready products to maximise system-wide impact and resource efficiency	II,III,VII, VIII	Q14, S9
	40) Communicate corporate CE goals with transparency and accountability	VI	Q6, S45
K	41) Establish regional circular hubs to foster targeted support and collaboration across value networks	VI,VIII, IX	Q1, S13f
	42) Establish frameworks for the delegation of responsibility in post-use phases, including repair, remanufacturing, reuse, and recycled content application	VI	Q6, S46
	43) Define responsibility transfer mechanisms for product repair, refurbishment, reuse, as well as for the utilisation of recycled materials and components	VI,IX	Q6, S46
L	44) Define product-category-specific circularity metrics to steer design processes and reduce both resource consumption and environmental impact	I,VI,VII	Q6, S35
	45) Assess the funct. durability of products to support longevity / circular design	I,VI,VII	Q6, S60
	46) Guidelines that ensure non-destructive removal and facilitate reuse through secondary installation pathways	I,VI,VII	Q6, S61
	47) Utilise open interfaces for data exchange (integration with material databases)	VI,VIII	Q6, S46
	48) Publish tech. interfaces to accelerate the development of circ.supply chains	VI,VIII	Q6, S46
	49) Maximise the lifespan of packaging while maintaining opt. quality standards	VIII	Q6, S9
	50) Harmonise tech. standards for waste classification to promote resource efficiency and limit the use of primary materials in waste reduction strategies	VI,VIII	Q6, S37
	51) Develop a company-specific guideline document defining key concepts of the CE to align organisational understanding and strategies	IV,VI	Q6, S45
	52) Identify and implement open data exchange in compliance with standards	VI	Q6, S46
M	53) Develop normative guidelines for social standards to support the implementation of circular business models	VI	Q6, S49
	54) Product group-specific standards for durability, reparability, reusability, re-manufacturability, and recyclability based on the DIN EN 4555x series	VI,VIII	Q6, S35
	55) Implement and apply Material Flow Analysis (MFA) according to ISO/TC 323 Circular Economy standards	VI,VII	Q6, S55

The analysis revealed several *Action Areas* in which industrial enterprises need to engage in order to successfully implement the principles of a sustainable circular economy. In addition to identifying these primary areas, the analysis also enabled the derivation of specific *Operational Implications*. These identified implications were systematically aligned with the initially defined areas of responsibility within corporate boundaries. This structured alignment enabled a distribution of the sustainability transformation workload across multiple organisational levels, thereby preventing potential overload of management level or a single sustainability division. The 55 *Operational Implications* identified through this analysis exhibit diverse characteristics but uniformly represent actionable directives derived directly from the analysed texts. These explicitly defined *Operational Implications* position companies as responsible entities to increase organizational impact. Moreover, assigning responsibilities demonstrates that implementing circular economy principles cannot be limited to just one area of responsibility. Indeed, it becomes evident that many operational measures inherently represent concrete cross-functional tasks, requiring active collaboration among multiple stakeholders across different organisational units. Furthermore, the application-oriented nature of the identified *Operational Implications* reflects the inherently pragmatic approach for concrete industrial implementation. Thus, the conciseness and clarity of the results provide a valuable executive summary, effectively assisting companies in incorporating new aspects into their circular strategy.

6 Critical Analysis and Outlook

The present investigation comprehensively identifies 14 *Action Areas* essential for implementing circular economy principles within the industrial manufacturing context. It further specifies 55 practical *Operational Implications*, clearly assigning them to nine distinct organisational responsibilities. Therefore, the research objective has been successfully achieved. Nevertheless, the approach has certain limitations, including the exclusive use of documents written in English and the limited applicability due to overly generic suggestions in most of the literature used, which lacks practical relevance at the corporate level. Going forward, continuous updating and maintenance of this overview will be necessary to preserve its relevance and accuracy. Additionally, future research should include a broader set of documents and enhance findings through the active involvement of industry experts.

Acknowledgements. This research is funded by dtec.bw - Digitalization and Technology Research Center of the Bundeswehr [Project: Production Next Door]. Dtec.bw is funded by the European Union – NextGenerationEU.

References

1. Wirtschaft und Klimaschutz, Bundesministerium für (2022) Nachhaltigkeit in der Wirtschaft. https://www.bmwk.de/Redaktion/DE/Dossier/nachhaltigkeit, Access 23 Nov 2022

2. Saubke, D., Ihle, P., Krenz, P. et al.: Enhancing sustainability – strategic choices from and for small and medium-sized enterprises. In: Kohl, Seliger et al. (Hg.) Manufacturing Driving Circular Economy, Sustainable Manufacturing 2024; Proceedings GCSM 2024 (2025)
3. Koppe, J.C., Saubke, D., Redlich, T.: Towards circular cities - the dynamics of supply chains and local manufacturing. In: Proceedings CPSL2024 Hannover (2024). https://doi.org/10.15488/17722
4. Kurz, R., Wild, W.: Nachhaltigkeit und Unternehmen. uwf **23**, 323–328 (2015). https://doi.org/10.1007/s00550-015-0379-y
5. Eccles, R.G., Ioannou, I., Serafeim, G.: The impact of corporate sustainability on organizational processes and performance. Manage. Sci. **60**, 2835–2857 (2014). https://doi.org/10.1287/mnsc.2014.1984
6. Schuh, G., Mauß, W., Potente, T., et al.: Green Re-Assembly Upgrade Factory. Fraunhofer-Gesellschaft Bundesregierung (2022) Agenda 2030: Unsere Nachhaltigkeitsziele. Veröffentlicht (2023). https://www.bundesregierung.de/breg-de/themen/nachhaltigkeitspolitik/nachhaltigkeitsziele-erklaert-232174, Accessed 23 Nov 2022
7. Kuckartz, U.: Qualitative Inhaltsanalyse: Methoden, Praxis, Computerunterstützung, 3. überarbeitete Version. In: Grundlagentexte Methoden, Beltz Juventa, Weinheim (2016)
8. Bundesregierung Deutschland. Kreislaufwirtschaft: Herausforderungen und Wege der Transformation (2024)
9. Industrie- und Handelskammer Karlsruhe. Kreislaufwirtschaft im Überblick Oehlmann C (2024) Zirkuläre Märkte für industrielle Wertschöpfung "Made in Europa". Forderungen der BDI-Initiative Circular Economy für die Europawahl 2024 (2024)
10. Bundesministerium für Umwelt, Naturschutz, (…) und Verbraucherschutz. Nationale Kreislaufwirtschaftsstrategie: Stoffe vermeiden, wiederverwenden, verwerten (2024)
11. Tauer, R., Aechtner, J., Köberich, T., et al.: Eine umfassende Circular Economy für Deutschland 2045 zum Schutz von Klima und Biodiversität (2023)
12. Verein Deutscher Ingenieure e. V. und Deutsche Kommission Elektrotechnik Elektronik und Informationstechnik. Deutsche Normungsroadmap Circular Economy (2023)
13. Lange, U.: Circular Economy und ressourceneffiziente Kreislaufwirtschaft – Der Weg ins zirkuläre Wirtschaften (2022)
14. Italian Ministry for Technological Innovation, Transition D. Circular Economy Sectoral Strategic Guidelines (2022)
15. Norwegian Ministry of Trade, Industry and Fisheries. The Green Industrial Initiative (2022). https://www.regjeringen.no
16. Bundesministerium für Klimaschutz, Umwelt, Energie, Mobilität, Innovation und Technologie. Österreich auf dem Weg zu einer nachhaltigen und zirkulären Gesellschaft: Die österreichische Kreislaufwirtschaftsstrategie (2022)
17. Federal Ministry for Climate Action, Environment, Energy, Mobility, Innovation and Technology. Austrian Circular Economy Strategy (CES) (2022)
18. Government of Spain. España Circular 2030: Circular Economy Spanish Strategy – Executive Summary (2021)
19. Norwegian Ministry of Environment and Climate. Norway's Strategy for Developing a Green and Circular Economy (2021)
20. Kadner, S., Kobus, J., Hansen EG et al. (2021) Circular Economy Roadmap for Germany. acatech - Deutsche Akademie der Technikwissenschaften
21. Europäische Kommission. Aktionsplan für die Kreislaufwirtschaft – Für ein saubereres und wettbewerbsfähigeres Europa: Mitteilung der Kommission an das EU Parlament, den Rat, den Wirtschafts- und Sozialausschuss und den Ausschuss der Regionen (2020)
22. Europäische Kommission. Circular Economy Action Plan – For a Cleaner and More Competitive Europe (2024)

23. Ministry of Development, Republic of Poland. Roadmap Towards the Transition to Circular Economy (2019)
24. Netherlands Ministry of Infrastructure and Environment. A Circular Economy in the Netherlands by 2050 (2016). https://www.government.nl/documents/policy-notes/2016/09/14/a-circular-economy-in-the-netherlands-by-2050

Implementation Concept for Sustainable Empathic Technical Systems

Eckart Uhlmann[1,2], Julian Polte[1,2], and Philipp Lelidis[1(✉)]

[1] Fraunhofer Institute for Production Systems and Design Technology IPK, Pascalstr. 8-9, 10587 Berlin, Germany
philipp.lelidis@ipk.fraunhofer.de

[2] Institute for Machine Tools and Factory Management IWF, Technische Universität Berlin, Pascalstr. 8-9, 10587 Berlin, Germany

Abstract. This paper presents an implementation concept for a sustainable Empathic Technical Systems developed in the Fraunhofer flagship project "EMOTION," in which a novel approach to cooperative and decentralized production planning and control is being explored. By embedding empathic capabilities into traditional cognitive systems, ETS aim to address coordination challenges and optimize cooperation in increasingly complex industrial environments. The implementation of ETS enables the optimization of the entire production system in alignment with global goals such as sustainability, promoting environmentally friendly practices and responsible resource management. The concept of ETS ultimately contributes to the evolution of intelligent manufacturing systems by integrating affective and situational awareness, fostering enhanced collaboration between humans and machines. The concept demonstrates the implementation of the key requirements for successful ETS deployment, including perception through multimodal sensors, autonomous decision-making that aligns individual and collective goals, and continuous learning from interactions. The practical implementation of ETS is illustrated through the production of bipolar plates for fuel cells, highlighting the integration of advanced communication protocols and real-time data processing. By prioritizing human-centric approaches and sustainability, this research aims to pave the way for more resilient, adaptive, and efficient manufacturing systems that not only meet economic objectives but also promote social and environmental responsibility.

Keywords: Empathic Technical Systems · Distributed Manufacturing · Cooperation · Digital Twin · EMOTION

1 Introduction

The increasing complexity and dynamic of manufacturing processes necessitate production systems that are both flexible and adaptive. While cognitive systems have significantly advanced distributed control through automated decision-making, challenges persist in resolving conflicting goals such as lead time, resource conservation, and

H. Kohl et al. (Eds.): GCSM 2025, LNME, pp. 392–400, 2026.
https://doi.org/10.1007/978-3-032-21154-5_43

economic efficiency in cooperative settings. The notion of Empathic Technical Systems (ETS) extends beyond traditional cognitive paradigms by integrating affective and situational awareness, thereby fostering enhanced collaboration between humans and machines and machines themselves. This paradigm shift is in alignment with the principles of "Industry 5.0," emphasizing human-centric and sustainable manufacturing. By embedding empathic capabilities, ETS aim to address coordination challenges and optimize cooperation in increasingly complex industrial environments, contributing to the evolution of intelligent manufacturing systems.

This paper puts forward a proposal for the implementation of ETS. The proposed concept is capable of comprehending the states and intentions of all entities, leveraging this understanding to inform decision-making within production systems. Its aim is to enhance the alignment of economic, environmental, and social objectives throughout the production system by providing a framework for effective decision-making.

2 Background

2.1 Empathic Technical Systems

ETS are defined by [1] as systems designed to enhance cooperation within distributed production environments between humans and machines, as well as between machines themselves. These systems are characterized by their ability to understand the affective and cognitive states of other entities, which influences their decision-making processes. Key requirements for ETS include.

1. **Perception**: Utilizing multimodal sensors to collect physiological and psychological data from users and machines, enabling nuanced situational awareness.
2. **Communication**: Providing multimodal interfaces for effective human-machine interaction and secure protocols for machine-machine communication.
3. **Decision-making**: Autonomous decision-making that balances individual and collective goals through utility functions that integrate the interests of all.
4. **Learning**: The ability to learn from interactions and adapt to preferences and contexts, enhancing responsiveness.

By incorporating these features, ETS aim to create more adaptive, resilient, and sustainable production systems that foster collaborative environments. The **human-centered approach** of ETS emphasizes the importance of empathy in interactions, enabling better cooperation and conflict resolution by considering the needs and intentions of all stakeholders involved.

2.2 Collaborative Industrial Data Exchange

To facilitate collaborative environments founded upon the exchange of industrial data in a decentralized manner, [2] proposes an architecture integrating *Asset Administration Shells* (AAS), *Eclipse Dataspace Connector* (EDSC), and *Open Platform Communication Unified Architecture* (OPC UA). This architecture enhances industrial resilience, sustainability, and competitiveness by ensuring interoperability across different entities and company boundaries through a standardized data model based on AAS, enabling secure inter-organizational data exchange through EDSC, and facilitating internal factory data communication using OPC UA, supporting CO2-aware decision-making.

Eclipse BaSyx [3], an open-source middleware for Digital Twins based on AAS, facilitates scalable, cost-effective deployment across company boundaries, thereby disrupting data silos and optimizing production through digital testing. Its architecture fosters a unified data exchange framework for interoperability in smart manufacturing.

2.3 Adaptive, Decentralized Production Planning and Control

The communication technologies described in Sect. 2.2 are used by [4–7] for dynamic, order-based production system management by decoupling production processes from resources and integrating machine-level capabilities and real-time data. In each case a production order is managed by an algorithm, which interprets a process description and controls production steps in real-time. Resource allocation is performed just-in-time, based on the capabilities and the current state of the available resources, allowing the system to flexibly respond to machine failures, delays, or other unforeseen events. However, the systems lack key components of ETS, including multimodal sensing of human states, balancing individual and collective goals such as sustainable production, adapting to preferences as well as unforeseen events, and mechanisms for empathic collaboration. For example, while the system can dynamically reassign tasks when machines fail, they do not anticipate or automatically initiate maintenance processes for defective equipment. Thus, the systems remain reactive rather than adaptive and cooperative, missing the human-centricity that defines ETS.

3 Material

The implementation concept of an ETS is discussed with reference to the production of bipolar plates for fuel cells.

3.1 Product Bipolar Plate for Fuel Cells

The bipolar plates of the fuel cell are made of the graphite alloy FU 4369 HT. A flow field is milled into a bipolar plate to evenly distribute reactants such as hydrogen evenly across the electrode surface of a fuel cell. It helps maximize the reaction surface area and minimize pressure drops. The design of the flow field can significantly affect the efficiency of the fuel cell by optimizing gas distribution and controlling the formation of water or other byproducts. Additionally, a QR code is engraved for identification.

3.2 Production System

The production system for the bipolar plates is comprised of the nine actors in Table 1. Their overall states, namely *in operation, ready for operation, out of operation,* and *fault in operation*, can be derived from their current condition information.

Table 1. Actors of the Production system.

Actor type	Actors	Tasks
Warehouse	• Smart Warehouse	Material storage and warehouse management
Mobile robot with industrial robot arm	• AGV *Tend-o-Bot* [* MERGEFORMAT 8] • AGV *Adabot*	Material transport and machine loading
Worker	• Worker with exoskeleton *ErgoJack* [* MERGEFORMAT 9] • Worker with network-compatible screwdriver	Assembly, material transport and machine loading
Machining center	• Exeron HSC 600 • map LPZ 900 • Mikromat 10V-HSC	Flow field milling
Laser engraving system	• FOBA M2000-P	QR code engraving

To achieve the objective of a *smart warehouse*, it is essential to facilitate the exchange of information, the automation of equipment, the integration of processes, and the promotion of environmental sustainability [10]. For the former, communication is enabled for all actors via a network connection. Combined with the transport actors, the system facilitates equipment automation and process integration.

The transportation of the bipolar plates can be accomplished by one of the two analogous AGVs, namely *Tend-o-Bot* and *Adabot*, outlined in [8]. Alternatively, the task can be undertaken by the worker. The worker is equipped with the exoskeleton *ErgoJack* [9], which contains multiple sensors designed to ascertain the current state.

The purpose of the three machining centers, *Exeron HSC 600*, *map LPZ 900*, and *Mikromat 10V-HSC*, is to mill the flow fields of the bipolar plates. Subsequently, the laser engraving system *FOBA M2000-P* is utilized to engrave a QR code. A worker with a network-compatible screwdriver finally assembles the fuel cell.

4 Concept

This section articulates the implementation concept for ETS shown in **Fig. 1**.

4.1 Sustainability

Sustainability is a core principle guiding the optimization of the production system within ETS. By aligning production processes with global sustainability goals, ETS aim to minimize environmental impact while maximizing resource efficiency. This involves the integration of eco-friendly practices, such as reducing waste, saving energy, and reducing

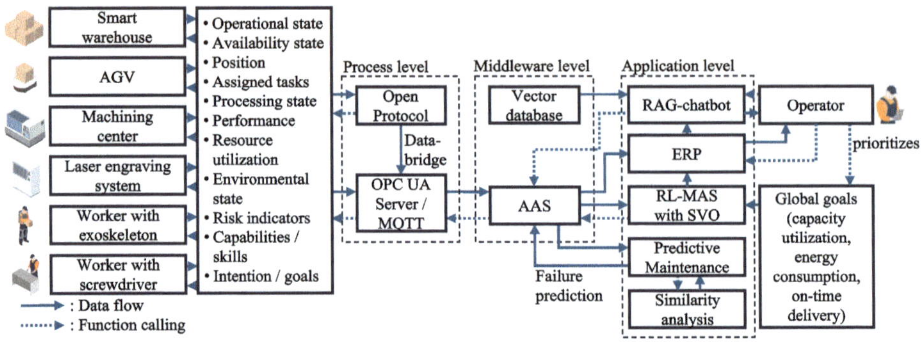

Fig. 1. Concept architecture of the ETS for a production system of fuel cells

material consumption throughout the production cycle. Additionally, the system incorporates real-time **monitoring of environmental metrics**, enabling proactive adjustments to operations that enhance sustainability. By leveraging data analytics, ETS can identify opportunities for improvement and implement strategies that not only advance economic objectives but also contribute to social equity and environmental stewardship. Ultimately, this commitment to sustainability fosters a holistic approach to manufacturing that meets the needs of the present without compromising the ability of future generations to meet their own needs, thereby promoting a responsible and resilient industrial ecosystem.

4.2 Perception

Depending on the direction of communication, a distinction is made between a digital model, a digital shadow, and a digital twin (DT), with the latter being characterized by automated, bi-directional communication between the physical or non-physical object and its digital representation [11]. The DTs are based on **AAS** and the submodel templates of the *Industrial Digital Twin Association* (IDTA), which, in combination with the communication described in Sect. 4.3, allow the real-time state of the object to be digitally represented. An empathic digital twin aims to optimize the interaction between the actors by promoting their perception, understanding, resonance and anticipation. The values of these submodels are utilized for anomaly detection and predictive maintenance. By analyzing the data generated by these models, it is possible to identify unusual patterns or deviations from expected performance, enabling early detection of potential problems. This proactive approach helps maintain the operational efficiency of systems and equipment while minimizing downtime and maintenance costs.

4.3 Communication

Real-time communication with **edge computing** enables efficient data processing and transmission at the source of data generation, minimizing latency and bandwidth usage. The use of network protocols such as **OPC UA**, **MQTT**, and **Open Protocol** enables seamless interoperability between different devices and systems. OPC UA ensures secure and reliable data exchange in industrial environments, while MQTT provides a lightweight messaging protocol ideal for IoT applications. Open Protocol further enhances flexibility by enabling integration across diverse platforms. Together, these technologies improve the responsiveness and efficiency of applications in real-time scenarios, supporting smarter decision-making and improved operational performance.

4.4 Human-Centricity

The transparency about the production system is key in enabling Operator 4.0, allowing to make informed decisions. The information from an **Enterprise Resource Planning** (ERP) system, the automatically assigned tasks from the MAS in Sect. 4.5, and the current state of the production system can be easily checked through the UI of the **AAS** deployed in Eclipse BaSyx and a **chatbot**. This chatbot [12] serves as an interactive interface, allowing users to inquire about their current tasks and receive real-time updates. Furthermore, it can provide more detailed information on each task, such as deadlines, priority levels, and associated resources from a vector database, thereby enhancing user engagement and ensuring transparency in the production planning process. By including **assist-as-needed**, ETS empower workers, enhance their productivity and well-being, reduce physical strain, and facilitate smoother interactions.

The state of the worker can be evaluated using an exoskeleton and a network-compatible screwdriver equipped with sensors. The **real-time monitoring** [13] allows the system to consider the physical state of the worker when allocating tasks, ensuring that workloads are manageable, ergonomic and capability-aligned. By fostering a collaborative environment where human operators are at the center of the production process, ETS not only improve operational efficiency but also promote a more satisfying and engaging workplace, leading to better outcomes for individuals and the organization.

4.5 Decision-Making

A flexible production planning and control system based on a **multi-agent system** (MAS) allows for dynamic task allocation while incorporating social value orientation. This approach enables the system to adapt to changing production requirements and optimize for the prioritized global goals [14]. Agents can communicate and collaborate to adjust production processes in real-time, ensuring that product adjustments are made efficiently. By considering **social value orientation** (SVO) [15], the system can also prioritize tasks that align with sustainable practices and social responsibility. This flexibility not only enhances productivity but also fosters a more responsible and responsive manufacturing environment.

Predictive maintenance facilitates forecasting potential equipment failures based on historical data. By treating maintenance as a dynamic task, the system can allocate resources proactively, minimizing downtime. Using **similarity analysis** with past cases, the estimated time for maintenance can be predicted, allowing for better automatic scheduling and reduced disruptions in the production process.

4.6 Learning

The learning capability of ETS is crucial for their adaptability and efficiency. In this context, a **socially aware Reinforcement Learning** (RL) approach is employed for empathic agents to optimize interactions between humans and machines as well as between machines themselves. By analyzing behavioral patterns, global goals and feedback from the environment, the agents can continuously refine and adapt their strategies to meet the needs of users and the demands of the production system.

Additionally, the concept of **fine-tuning** is implemented in conjunction with a **Retrieval-Augmented Generation** (RAG) approach to enhance the chatbots' performance and enable more accurate and context-aware decision-making. Through these learning mechanisms, ETS can ensure that they respond not only reactively but also proactively to changes in the production environment, significantly improving efficiency and collaboration in manufacturing.

5 Expected Results

The expected results of implementing ETS include a significant enhancement in production flexibility and efficiency through improved interaction and collaboration between human and machine actors. Building on the promising results of an MAS in a simulated production environment [16], given the integration of multimodal sensors, real-time communication protocols, and empathic digital twins based on AAS, ETS will enable a more precise understanding of the needs and intentions of all stakeholders, leading to optimized decision-making processes and more effective resource utilization. Furthermore, with reinforcement learning tailored to social interaction contexts, ETS progressively refine their decisions, ensuring that improvements compound over time. This ongoing optimization contributes to a self-improving system architecture, further justifying the expectation of resilient and sustainable performance while incorporating sustainable practices to minimize environmental impact. Ultimately, ETS aim to foster a resilient, adaptive, and economically successful production system that prioritizes social and ecological responsibility.

6 Conclusion and Future Work

Through the integration of empathic capabilities, ETS facilitate enhanced collaboration leading to improved decision-making and resource utilization. The continuous learning mechanisms embedded in ETS ensure adaptability to evolving production needs, ultimately contributing to a resilient and efficient manufacturing system that aligns with economic, social, and environmental objectives. Key technologies of the ETS implementation concept include the multimodal sensors for perception for machines and workers, real-time communication protocols such as OPC UA, MQTT and Open Protocol for seamless interaction, the digital twins based on AAS, and an MAS based on a Social Value Orientation for dynamic decision-making. By embedding these features, ETS can effectively interpret the actors' needs and intentions, fostering a collaborative environment that prioritizes sustainability and efficiency. The focus on altruism in the interactions between these actors further enhances cooperation, enabling a system that not only meets production goals but also considers the well-being of all actors involved. Future work will focus on refining the implementation of ETS to enhance decision-making capabilities, with particular attention to the role of altruism in actor interactions and validating the system as it is still undergoing evaluation.

Acknowledgements. This work was supported as a Fraunhofer FLAGSHIP PROJECT.

References

1. Rentzsch, M., Häberer, S.: Approaching self-organizing production: unwrapping the potential of empathy. Procedia CIRP **130**, 1687–1694 (2024)
2. Neubauer, M., et al.: Architecture for manufacturing-X: bringing asset administration shell, eclipse dataspace connector and OPC UA together. Manufact. Lett. **37**, 1–6 (2023)
3. Kuhn, T., Antonino, P.O., Schnicke, F.: Eclipse BaSyx – Industrie 4.0 einfach machen! - Das Open-Source-Betriebssystem für die Industrie 4.0. atp 63 10, pp. 54–61 (2021)
4. Ebner, A., Duewel, F., Pfrommer, J.: A OPC UA Based execution engine for production control in decentral organized manufacturing. In: The 22nd IEEE International Conference on Industrial Informatics (INDIN), pp. 1 – 8 (2024)
5. Fan, Z., Shi, D., Meyer, O., Seidelmann, J., Wang, H.: Asset adminstration shell-based flexible manufacturing system. In: IEEE 6th International Conference on Industrial Cyber-Physical Systems (ICPS), pp. 1 – 6 (2023)
6. Jirsa, J., et al.: Use of multi-agent system for industrial production control. IFAC-PapersOnLine **58**(9), 205–210 (2024)
7. Wagner, A., Volkmann, M., Hermann, J., Ruskowski, M.: Machining of Indiviudalized Milled Parts in a Skill-Based Production Environment. In: Silva, F.J.G., Pereira, A.B., Campilho, R.D.S.G. (eds.) Flexible Automation and Intelligent Manufacturing: Establishing Bridges for More Sustainable Manufacturing Systems. FAIM 2023, vol. 1, pp. 283–292. Springer, Cham (2023)
8. Lelidis, P., Vick, A.: Planning of time-efficient trajectories for mobile robots with differential-drives. In: 2024 IEEE 29th International Conference on Emerging Technologies and Factory Automation (ETFA), pp. 1 – 6 (2024)
9. Kuschan, J., Krüger, J.: Fatigue recognition in overhead assembly based on a soft robotic exosuit for worker assistance. CIRP Ann. Manuf. Technol. **70**(1), 9–12 (2021)
10. Zhen, L., Li, H.: A literature review of smart warehouse operations management. Front. Eng. Manag. **9**(1), 31–55 (2022)
11. Bergs, T., Gierlings, S., Auerbach, T., Klink, A., Schraknepper, D., Augspurger, T.: The concept of digital twin and digital shadow in manufacturing. Procedia CIRP **101**, 81–84 (2021)
12. Kohl, L., Eschenbacher, S., Besinger, P., Ansari, F.: Large language model-based Chatbot for improving human-centricity in maintenance planning and operations. In: 8th European Conference of the Prognostics and Health Management Society, pp. 1 – 12 (2024)
13. Lück, M., Hornung, T., Teklezgi, J.: Der Mensch in der intelligenten Fabrik: human-in-the-Loop: ein menschzentrierter Ansatz für die Wissenserweiterung mit maschinellem Lernen. Zeitschrift für wirtschaftlichen Fabrikbetrieb **119**(6), 456–459 (2024)
14. Uhlmann, E., Polte, J., Mühlich C.: Steigerung der Ressourceneffizienz durch agentenbasierte Produktionssteuerung. In: Nachhaltigkeit in der Produktion. Fertigungstechnik; Band 712. Pub.: Brecher, C.; Krüger, J.; Uhlmann, E.; Verl, A.; Riedel, O.; Lechler, A. Düsseldorf: VDI-Verlag, 2024, p. 1 – 15 (2024)
15. Crosato, L., Shum, H.P.H., Ho, E.S.L., Wei, C.: Interaction-aware decision-making for automated vehicles using social value orientation. IEEE Trans. Intell. Veh. **8**(2), 1339–1349 (2023)
16. Uhlmann, E., Polte, J., Mühlich, C.: Distributed real-time production control for resilient manufacturing systems. Procedia CIRP **120**, 1274–1279 (2023)

An Innovative AI-Based Operational Framework for Industry 5.0 Towards Sustainable Manufacturing

G. Di Francesco[1(✉)], A. Bianchini[2], F. Manuzzi[3], and A. Pompili[1]

[1] Opseed Srl, Via Col Di Lana 8, 40131 Bologna, BO, Italy
gianluca.difrancesco@opseed.com
[2] DIN, University of Bologna, Via Fontanelle 40, 47121 Forlì, FC, Italy
augusto.bianchini@unibo.it
[3] Turtle Srl, Via Sante Vincenzi, 46, 40138 Bologna, BO, Italy

Abstract. Manufacturing transformation is driven by integration of performance improvement and sustainability. To this end, Artificial Intelligence (AI) acts as a key enabler, accelerating the transition from Industry 4.0 to 5.0. This study introduces an innovative AI-driven operational framework: Sustainable Operational Intelligence (SOI) developed by Opseed.

The framework integrates human-machine collaboration, combining machine data, human activities (Work Process) and system elaborations (System Process) to build a corporate knowledge base. AI-driven processing enables natural language interaction with a manufacturing AI agent, able to extract insights to reduce losses and improve performance (Overall Equipment Effectiveness) and sustainability KPIs (CO2 equivalent emission (CO2e)). The framework was lab-tested on Collaborative Robot (Cobot) and Automatic Guided Vehicle (AGV) losses with two groups of not-expert operators, one using SOI and one without. Performance was evaluated based on Unplanned Downtime (UPDT) and daily CO2e per functional unit.

The results show significant reduction in UPDT and CO2e per functional unit. Aligned with autonomous maintenance principles, SOI enhances operator autonomy with continuous and targeted loss support. Moreover, integrating sustainability indices into the framework enables automatic mapping and identification of improvement actions.

Future work will extend to more failures and operators and be tested in automated packaging lines, addressing lab limitations.

Keywords: Industry 5.0 · Operational Framework · Sustainable Manufacturing · ESG Compliance · Artificial Intelligence · Operational Intelligence

1 Introduction

The manufacturing sector faces several key challenges: labour shortages due to demographic decline [1], inadequate skills to address increasing automation and digitalization [1], environmental sustainability institutional commitments [2] and the need to improve

H. Kohl et al. (Eds.): GCSM 2025, LNME, pp. 401–409, 2026.
https://doi.org/10.1007/978-3-032-21154-5_44

performances to meet growing market demands [3]. These factors heavily influence Environmental, Social and Corporate Governance (ESG) criteria, supporting the achievement of the United Nations' Sustainable Development Goals (SDGs) [4], particularly in three dimensions of sustainability: social, environmental, and economic.

Numerous studies have underlined the central role of sustainability in manufacturing. Traditional financial sustainability focuses on reducing the six major losses [5] to minimize downtime and waste, thereby improving Overall Equipment Effectiveness (OEE) and productivity. Environmental sustainability is increasingly crucial not only from a market perspective [6] and in reducing energy costs [7], but also from a regulatory standpoint, particularly with recent European directives [2]. It is therefore essential to integrate conventional metrics (like OEE) with new indicators such as Car-bon Footprint [8], life cycle impact measures, and Operational Environmental Equipment Effectiveness (OeEE) [9].

Industry 4.0 technologies and Artificial Intelligence (AI) can transform labor-reliant systems, but adults face challenges adapting, especially with traditional training formats [10–12].

Although Industry 4.0 has enabled the use of Manufacturing Execution Systems (MES) [13] and the Industrial Internet of Things (IIoT) [14], the resulting data often remain difficult to interpret, limiting their impact on decision-making. In addition, sustainability considerations are still rarely integrated into these systems.

More recently, AI has entered manufacturing operations in areas such as production planning, demand forecasting, and supply chain management [10, 16]. Yet, its adoption remains focused on specific applications like predictive maintenance, quality control [10, 16] and process optimization using machine learning [10, 17].

The gap this study seeks to address lies in developing an AI-based operational framework capable of governing manufacturing and Continuous Improvement (CI) processes, actively supporting operators in routine tasks and problem-solving. This includes integrating sustainability KPIs and human knowledge and machine data to produce more accessible and actionable data for improved decision-making.

The aim of this work is to introduce Sustainable Operational Intelligence (SOI), an innovative operational framework developed by Opseed, which places strong emphasis on skills development, operator engagement, and the continuous improvement cycle, impacting both traditional performance metrics (OEE, MTBF) and new environmental ones such as Carbon Footprint (CFP). The proposed framework does not aim to develop new technologies but rather to integrate existing ones in a structured way to support operational processes through the application of AI algorithms and systematic knowledge classification. The framework combines existing technologies (MES,IIot) with operator knowledge to support operational processes such as root cause analysis and technical interventions. Powered by an AI-based system, it enhances data utilization and knowledge sharing among operators, technicians, and machines, fostering human-machine interaction.

This approach facilitates the transition from Industry 4.0 to Industry 5.0, supporting the evolution towards a more human-centric, sustainable, and resilient manufacturing ecosystem [18].

2 Methodology

2.1 Sustainable Operational Intelligence (SOI) Framework

The proposed Operational Framework (SOI) is designed to support organizational processes in automated production and continuous improvement by assisting technical and operational personnel in daily activities and loss resolution. It integrates machine data, human activities (Work Process), and system processes (System Process) into a collaborative logic referred to as Human-System Interactions, enabling the creation of a shared digital knowledge base supported by an AI-powered information system, particularly through Natural Language Processing (NLP).

SOI is an evolving model capable of learning from human and machine inputs, progressing based on organizational maturity and processed KPIs. The methodological approach includes three phases: Assessment, Development, and Deployment (Fig. 1).

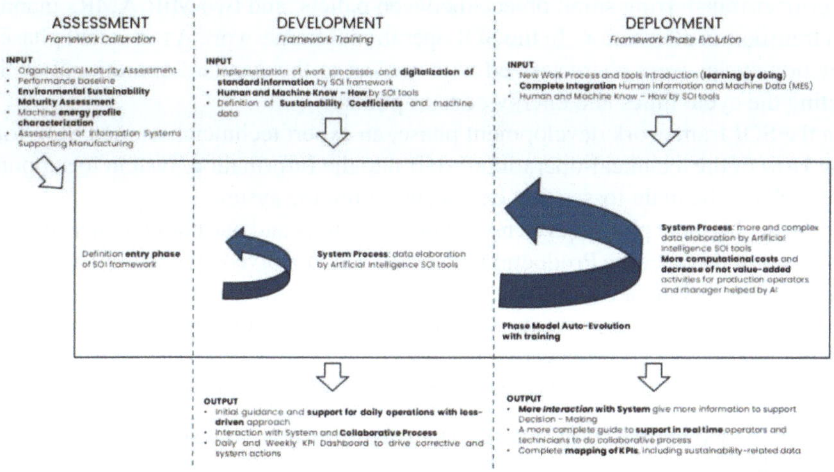

Fig. 1. Methodological approach for the application of the SOI operational framework

Assessment (Framework Calibration)

This phase includes an assessment of production area, with emphasis on continuous improvement and environmental sustainability. Key activities include the analysis of actual workflows, IT infrastructures, collection of performance and baseline indicators (OEE, MTBF, scrap, main losses) and energy profiling and daily emissions CO2 equivalent emission (CO_2e) per functional unit quantification.

This phase defines system calibration parameters.

Development (Framework Training)

This phase focuses on building the knowledge base to train the SOI framework: definition and digital structuring of corporate knowledge and mapping emissions per activity and associating them with environmental KPIs.

The SOI information system included operator interfaces with NLP input to document know-how and was trained through supervised learning on technical datasets.

Deployment (Framework Phase Evolution)

The final phase involves the full release of SOI to operators and the sharing of knowledge, along with the integration of human-machine data (MES), to facilitate data-driven decision-making and support Continuous Improvement activities.

The framework's information system focuses on AI agent capable of supporting operators during daily operational processes, via NLP. In addition, SOI enables automated performance mapping for both productivity and sustainability metrics, fostering a data-driven evolution of operations.

2.2 Validation SOI in a Lab Test

To validate the framework and its supporting AI-based Information System, a lab-scale experiment was conducted on simplified manufacturing systems involving a collaborative robot (Cobot) with pick-and-place functionality and Autonomous Mo-bile Robots (AMRs). The setup consisted of a Cobot (Universal Robot, model UR10e) with a vacuum gripper transferring small objects between pallets, and two MiR AMRs managing pallet transport in alternation. In the SOI operational frame-work Assessment phase: the production cycles were characterized with respect to the standard operational cycle by detecting the cycle times and energy consumptions.

In the SOI framework development phase, an expert technician trained SOI, loaded Know-How of the technical-operational staff into the Information System to support the framework for the main losses that can occur within the system.

In the deployment phase, seven non-expert operators and one maintenance technician (non-expert of this specific Production Systems) were involved, divided into two groups: Group 1 (with technician) faced the same stop events once without SOI sup-port and Group 2 addressed the same events with support from the SOI framework and AI agent. All participants received prior verbal training.

Performance was measured by comparing the following measures for two groups:

- Unplanned Downtime (UPDT): here defined as the time to restart the system and fix the crash
- Daily electricity consumption per functional unit (i.e., Specific energy consump-tion, SEC) [kWh/functional unit]
- Daily emissions rom electricity consumption per functional unit [gCO2e/functional unit]

The impact on OEE, SEC and Daily Emissions per unit was estimated considering a 24-h workday and using the references of the percentage of Unplanned Down-time for similar production systems in the literature.

The environmental impact assessment of the work cycles of Production System, based on the comparison between Group 1 and Group 2, began with the calculation of the energy consumption of the cobots. Due to limits of Power Analyzer, it was measured a second Cobot (Fanuc CRX-10iA) during Pick-Place operating cycle with simi-lar power consumption (Universal Robot 0,35 kW [19]; Fanuc Robot 0,30 kW [20]). The methodology is summarized as follows.

The first step was to identify the typical operational states of the cobot: OFF: Cobot powered off, IDLE: Cobot powered on but in standby and PROCESSING: Cobot actively performing tasks. For each state, the corresponding average active power was determined based on real-time power measurements of the Fanuc cobot cycle, recorded in 10-s intervals using a power analyzer (IIoT sensor Astreo KeraOne, 0.1 Hz sampling rate for three-phase and single-phase current absorption, with a measurement range of 0 to 200 A, 1% relative error, and a precision of 0.01 A). With power values defined for each state, the work cycles executed by operators in Group 1 and Group 2 were analyzed. The time spent in each operational state (OFF, IDLE, PROCESSING) during a typical cycle was determined for both systems. Subsequently, a daily time distribution was estimated by extrapolating the duration of each state over an operational day. This included assumed downtime and daily failure rates based on values reported in the literature for similar industrial systems [23, 24]. This allowed estimation of the daily hours spent in each operational state for the cobots. By multiplying the daily duration in each state with the corresponding average power, the daily energy consumption was calculated for each cobot. The next step involved the definition of functional unit, in line with the ISO 14067 [21] approach, which expresses carbon footprint results as mass of gCO_2e per functional unit, defined as a cycle of work. This unit was also used to express Specific Energy Consumption. Using these definitions, the Specific Energy Consumption was computed by dividing the daily energy use of each system by the number of functional units processed per day, separately for Group 1 and Group 2. Finally, the daily carbon emission for both systems was calculated by multiplying the measured daily energy consumption by the 2022 EU-27 Greenhouse Gas (GHG) emission intensity factor for electricity generation, as reported by the European Environment Agency [22]. The resulting emissions were normalized to the respective functional unit, yielding the Daily emissions from electricity consumption per functional unit. As a final step, the percentage variation in daily emissions per functional unit between Group 1 and Group 2 was calculated for each system.

3 Results and Discussion

The results of energy characterization of production system and validation test are reported in Fig. 2 and Table 1 (performance and sustainability index), respectively.

Figure 2 highlights the characterization of the production system both in terms of the production cycle, with the measurement of cycle times and expected production, and of energy profiling. This characterization is essential within the SOI Framework in the Assessment phase to integrate sustainability KPIs, such as daily emissions from electricity per functional unit, into the Continuous Improvement Cycle, supporting corrective and preventive actions. Traditionally, decisions are based on indicators like Safe-ty, Quality, Performance, and Cost, often neglecting sustainability indexes such as Carbon Footprint and Specific Energy Consumption, which are increasingly important for both environmental and financial sustainability. This case presents only emissions and SEC associated with Unplanned Downtime, but SOI also accounts for emissions from scrap and defective raw materials. Even with similar consumption during Stand-by and Run phases, failures still lead to CO_2e emissions due to unproduced parts, increasing SEC.

The Development phase, supported by AI, enables the creation of a Knowledge Management Database directly in the production environment, using a Loss-Driven

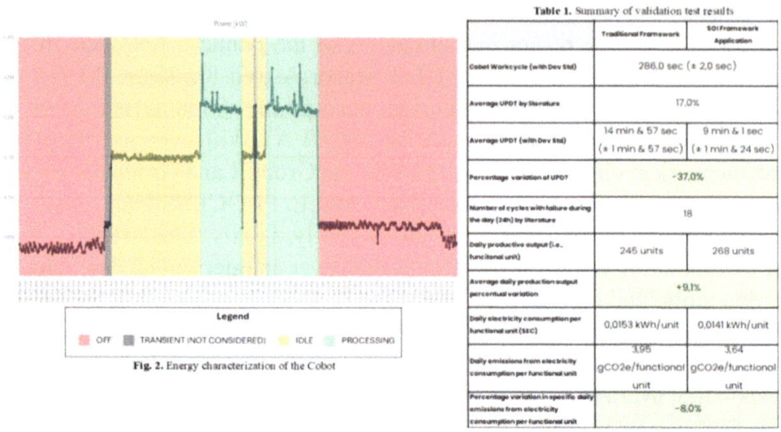

Fig. 2. Energy Characteization of the Cobot.

Table 1. Summary of validation test results

	Traditional Framework	SOI Framework Application
Cobot Workcycle (with Dev Std)	286,0 sec (± 2,0 sec)	
Average UPDT by literature	17,0%	
Average UPDT (with Dev Std)	14 min & 57 sec (± 1 min & 57 sec)	9 min & 1 sec (± 1 min & 24 sec)
Percentage variation of UPDT	-37,0%	
Number of cycles with failure during the day (24h) by literature	18	
Daily productive output (i.e., funcitonal unit)	245 units	268 units
Average daily production output percentual variation	+9,1%	
Daily electricity consumption per functional unit (SEC)	0,0153 kWh/unit	0,0141 kWh/unit
Daily emissions from electricity consumption per functional unit	3,95 gCO2e/functional unit	3,64 gCO2e/functional unit
Percentage variation in specific daily emissions from electricity consumption per functional unit	-8,0%	

approach, with a short development time (approx. 3 months). Procedural knowledge already present in the company is made accessible through AI, transforming human processes into collaborative workflows. This increases operator autonomy, reducing reliance on experts. In the validation experiment, the system was trained on typical losses in a Cobot–MIR setup and later supported operators with resolution suggestions. Operator-entered data, validated by technical experts, proved essential to interpreting MES data and supporting decisions.

Considering 17% unplanned downtime as reported in the literature [23, 24], the SOI application led to an 8.0% reduction in daily emissions per functional unit (Table 1). Although the absolute variations were small due to the low-consumption setup, this confirms the potential of SOI even in controlled contexts. Greater impact is expected in high-volume lines with lower Availability (60–70%) [25], where planned downtimes are also relevant but were excluded here.

Regarding OEE, the SOI framework led to improved effectiveness and a 37% reduction in UPDT, leading to a 9.1% increase in production output (Table 1). This was achieved thanks to guided loss resolution by the system, trained with company-specific knowledge. After a brief contextual interaction with the AI (approx. 2 min of fixed time), operators solved issues more autonomously, reducing the need for expert intervention, which could otherwise cause long delays. Although partial support was still needed, SOI minimized dependency. The framework also supports planned activities like cleaning and setup, contributing to planned downtime reduction. Technician from group without SOI took longer (15 min 13 s) than SOI group (max 10 min 1 s) due to unfamiliarity with this specific system, while SOI generated targeted system-specific support. The best operator performance (7 min 7 s) occurred during the final attempt, showing SOI's learning ability through operator feedback, streamlining future resolution steps. Moreover, SOI also enhances training by enabling a learning-by-doing approach through AI guidance, overcoming the limits of theoretical courses, traditional shadowing and written procedures difficult to ac-cess. In this test, despite recent training and voice instruction, operators struggled without SOI support, demonstrating its added value.

Limitations of the study relate to its lab environment and simplified system. The users were non-experts and the setup was small-scale. The test setup was intentionally simplified to limit external influences (such as process variability and reliance on expert personnel) and to better highlight the system's potential. This also made it possible to involve non-technical users without requiring extensive training. Real industrial cases will be addressed in future work, as they demand significantly more time (approximately three months) for proper implementation and validation.

4 Conclusion

The SOI framework, supported by an AI-based information system, embeds sustainability as a core component of manufacturing by integrating relevant indicators directly into decision-making workflows. By combining MES data with contextual insights from frontline operators on the shop floor, SOI enhances the interpretability and relevance of operational data.

Moreover, the SOI framework enables company knowledge to be captured, organized, and made easily accessible across the organization, regardless of expertise level, through NLP. This facilitates practical, on-the-job learning and supports the transition toward higher levels of autonomous maintenance, operational performance, and sustainability. In particular, the integration of sustainability into decision-making enables the generation of targeted corrective actions, contributing to measurable reductions in CO_2 emissions per functional unit.

Finally, the SOI framework consolidates technical and operational dimensions into an unified system that provides real-time support for daily activities and loss resolution. Through AI-driven interactions and collaborative processes, SOI promotes a continuous improvement cycle and enhances performance indicators (OEE, UPDT).

In future applications, SOI should be tested in real industrial contexts with multiple machines, auxiliary systems, and full-plant energy consumption, also considering the impact of scrap and raw materials, to assess broader environmental and performance.

References

1. Draghi, M.: Report on the future of European competitiveness. European Commission, Brussels (2024). https://commission.europa.eu
2. Union, E.: Directive (EU) 2022/2464 of the European parliament and of the council of 14 december 2022. Off. J. Eur. Union L **322**, 15–80 (2022)
3. World Economic Forum: The Future of Growth Report 2024. World Economic Forum, Geneva (2024). https://www.weforum.org/publications/the-future-of-growth-report/
4. United Nations Department of Economic and Social Affairs: Global Sustainable Development Report 2023: Times of Crisis, Times of Change – Science for Accelerating Trans-formations to Sustainable Development. United Nations, New York (2023). https://sdgs.un.org/gsdr/gsdr2023
5. Rusman, M., et al.: The overall equipment effectiveness (OEE) analysis in minimizing the Six Big Losses: An effort to green manufacturing in a wood processing company. IOP Conf. Ser. Earth Environ. Sci. **343**(1), 012034 (2019)
6. Asikis, T., et al.: How value-sensitive design can empower sustainable consumption. Royal Soc. Open Sci. **8**(1), 201418 (2021)
7. Centro Studi Confindustria: The Italian Economy at the Test of the Conflict in Ukraine: Summary and Main Conclusions. Confindustria, Rome (2022)
8. Nguyen, L.G., Hansuebsai, A.: Profitability consideration of corrugated paperboard production based on carbon footprint reduction and the improvement of overall equipment effectiveness. Information Modelling and Knowledge Bases XXIX, pp. 188–200. IOS Press (2018)
9. Laurensia, J., et al.: Integration of operational performance and eco-indicators for assessing environmental impacts of manufacturing processes in an automotive component company. J. Sustain. Sci. Manag. **18**(9), 184–197 (2023)
10. Ejjami, R., Boussalham, K.: Industry 5.0 in manufacturing: Enhancing resilience and responsibility through AI-driven predictive maintenance, quality control, and supply chain optimization. Int. J. Multidisc. Res. **6**(4) (2024)
11. Barnard, Y., et al.: Learning to use new technologies by older adults: Perceived difficulties, experimentation behaviour and usability. Comput. Hum. Behav. **29**(4), 1715–1724 (2013)
12. Davidson, M.: Foundry management & technology. Foundry Manag. Technol. **144**(9), 120 (2016)
13. Almada-Lobo, F.: The Industry 4.0 revolution and the future of Manufacturing Execution Systems (MES). J. Innov. Manag. **3**(4), 16–21 (2015)
14. Mantravadi, S., et al.: Application of MES/MOM for Industry 4.0 supply chains: A cross-case analysis. Comput. Ind. **148**, 103907 (2023)
15. O'Neill, M., et al.: Process visualization of manufacturing execution system (MES) data. In: IEEE SmartWorld 2021, pp. 1–8. IEEE, New York (2021)
16. Kehayov, M., et al.: Application of artificial intelligence technology in the manufacturing process and purchasing and supply management. Proc. Comput. Sci. **200**, 1209–1217 (2022)
17. Braghirolli, L.F., et al.: Improving production and maintenance planning with meta-learning-based failure prediction. J. Manuf. Syst. **75**, 42–55 (2024)
18. Leng, J., et al.: Industry 5.0: Prospect and retrospect. J. Manuf. Syst. **65**, 279–295 (2022)

19. Universal Robots A/S: UR10e e-Series Datasheet. Updated Dec 2024. https://www.universal-robots.com, Accessed 30 Apr 2025
20. FANUC Europe Corporation: CRX-10iA Collaborative Robot Datasheet. Rev. MDS-04018, 2023. https://www.fanuc.eu, Accessed 30 Apr 2025
21. International Organization for Standardization: ISO 14067:2018 – Greenhouse gases – Carbon footprint of products – Requirements and guidelines for quantification and communication. ISO, Geneva (2018)
22. European Environment Agency: Greenhouse gas emission intensity of electricity generation in Europe. EEA, Copenhagen (2024)
23. Fourlas, P.H., Tsarouhas, G.K.: Reliability and maintainability analysis of a robotic sys-tem for industrial applications: a case study. Int. J. Performability Eng. **11**(5), 453 (2015)
24. Jabrane, Z., Borgemo, E.: Evaluating usage of automated guided vehicles with respect to warehouse layout changes: a case study at Haldex Brake Products (2018)
25. Tonny, F.A., et al.: Optimization of overall equipment effectiveness (OEE) factors: case study of a vegetable oil manufacturing company. Manag. Sci. Lett. **13**(2), 124–135 (2023)

Industry 5.0-oriented Training Interventions: Sustainability and Pillars in the Case of Electronics

Alexios Papacharalampopoulos[1], Olga Maria Karagianni[1], Dovilė Eitmantytė[2], Dinas Vaitkaitis[3], Paulius Dobožinskas[3], and Panagiotis Stavropoulos[1](✉)

[1] Laboratory for Manuf. Systems and Automation (LMS), Mechanical Engineering and Aeronautics Department, University of Patras, 26504 Patras, Greece
pstavr@lms.mech.upatras.gr
[2] Kitron UAB, Taikos Pr. 151, 52119 Kaunas, Lithuania
[3] Kriziu Tyrimo Centras UAB, Prancuzu 62-23, 44457 Kaunas, Lithuania

Abstract. Industry 5.0 represents a paradigm shift in manufacturing, emphasizing human-centricity, sustainability, and resilience alongside technological advancement. This study presents a targeted training intervention in the electronics industry designed to introduce and operationalize the pillars of Industry 5.0, specifically implementing in training. The intervention addressed internal company incentives to enhance human-centric practices for operators' upskilling and wellbeing, as well as to embed new and procedures (such as green and resilient processes). Developed internally with support from training technology providers and monitored by a stakeholder responsible for overseeing Industry 5.0 adoption, the program combined theoretical instruction with experiential learning in a Learning Factory environment. Evaluation was conducted at both trainees' and company levels – using self-assessed confidence linked to self-efficacy, and a pillar-based assessment framework. Results demonstrated improvements in perceived upskilling and Industry 5.0 indicators post-intervention. Additionally, statistical analysis revealed strong correlations between sustainability and the other two pillars. The findings highlight the role of structured, hands-on training as a foundational step toward the operationalization of Industry 5.0.

Keywords: Industry 5.0 · Sustainability · Training

1 Introduction

Industry 5.0 is a forward-looking industrial paradigm that emphasizes the integration of advanced technologies with human collaboration, sustainability, and resilience [1]. It envisions a production ecosystem where efficiency is redefined, integrating technological progress, environmental stewardship, and social responsibility.

More concretely, training for Industry 5.0 should focus on developing a workforce that can effectively collaborate with advanced digital and cyber-physical systems while maintaining human-centric and sustainable production processes [2]. Such interventions

H. Kohl et al. (Eds.): GCSM 2025, LNME, pp. 410–418, 2026.
https://doi.org/10.1007/978-3-032-21154-5_45

emphasize adaptive learning, integrating emerging technologies with practical, hands-on training that enhances human decision-making and creativity [1, 3]. A key aspect is knowledge transfer, ensuring workers not only acquire technical skills but also develop competencies in problem-solving, decision-making, and sustainable resource management [4]. The importance of interdisciplinary education is often highlighted, blending engineering expertise with environmental and social aspects [5], fostering a workforce that can drive resilient and intelligent manufacturing solutions. It is noted that the term "intervention" functions as an external framework for monitoring the transformation of training; the added value lies in studying the feasibility of integrating Industry 5.0 in training systems and similar interventions have been implemented in the recent past [6], with different context.

At the same time, under the framework Industry 5.0 correlation between the three pillars has already been commented in literature. The pillars of Industry 5.0 -sustainability, resilience, and human-centric innovation – are deeply interrelated and form a cohesive framework aimed at enhancing industrial and societal well-being. It has been emphasized that while sustainability and resilience have often been developed as separate research domains, their integration is crucial for supply chain performance [7]. Also, three frameworks linking these concepts have been identified [8]: resilience as part of sustainability, sustainability as part of resilience, and both as co-equal goals, suggesting synergy when well-aligned but potential conflict if not. Also, in [9], they further reinforce the need for unified performance frameworks that concurrently measure sustainability and resilience in operational supply chains. Finally, a metrics-based methodology has been provided [10] to evaluate sustainable manufacturing that inherently supports resilient operations by addressing environmental, social, and economic dimensions across system levels.

Herein, the added value is acquiring proof for Industry 5.0 integration in training systems of advanced electronics industry and additionally, indicating also the coupling between sustainability and the other two pillars. In particular, this paper presents the design of the training intervention, the evaluation application, and the results at both the trainees' and company levels, followed by a discussion on the interplay of Industry 5.0 pillars and implications for sustainable transformation.

2 Method

2.1 Training Structure and Integration of Experiential Learning

The study employed a structured approach to assess the feasibility of the Industry 5.0 integration in training. The relevant industrial sector has been that of electronics manufacturing, and with respect to the selection of sixteen trainees, convenience sampling was applied. The design of the training followed a specific structure driven by internal (to the company) incentives. This led to a specific syllabus, which includes several socio-technical aspects. Table 1 summarizes indicative relevant training topics, while it is noted that a Learning Factory (equipment/testbed) [11] was also included in the training, engaging also experiential learning. Furthermore, a specific template for the Industry 5.0 integration was adopted [1], facilitating the procedure of adding specific skills and/or strategies. More specifically, the training topics addressed a range of technical, digital, and behavioral skills. The delivery combined theoretical instruction with hands-on practice, including structured sessions in a Learning Factory (LF)-like environment, where trainees applied concepts in simulated production settings. This experiential component played a central role in reinforcing learning outcomes, particularly for content related to sustainability and resilience. The LF setup included existing production equipment as well as newly introduced devices for documenting workflows and facilitating participant communication. Unfortunately, further details cannot be disclosed due to the sensitive nature of the company's production processes.

Table 1. Indicative topics included in the training.

Training topic	Desired skills
Introduction about company (e.g. company values, history, structure, leadership encouragement initiatives)	Behavioural/Social skills
ESD (Electrostatic Discharge) Safety/Requirements	Technical knowledge
Environmental protection, LEAN (5S), occupational safety	Behavioural/Social skills
Electromechanical basics (e.g. Terms and classification of components)	Technical knowledge
Manufacturing Execution System usage skills	Digital skills
IT usage skills (e.g. Intranet, cyber security, email, IT help desk etc.)	Digital skills
Quality management system (e.g. ISO standards)	Technical knowledge
Documentation reading	Technical knowledge
ERP system usage skills	Digital skills

2.2 Evaluation Procedure

Regarding the evaluation procedure, a multifold questionnaire was used [3]; its first part focused on the jobs and it was answered by the trainees and its second part targeted the company. There was also a third part, aiming at characterizing the intervention itself. To begin with, Fig. 1 summarizes the topics of the company-level questionnaire into the three pillars. Various aspects that play or can play role, besides the training, are taken into consideration.

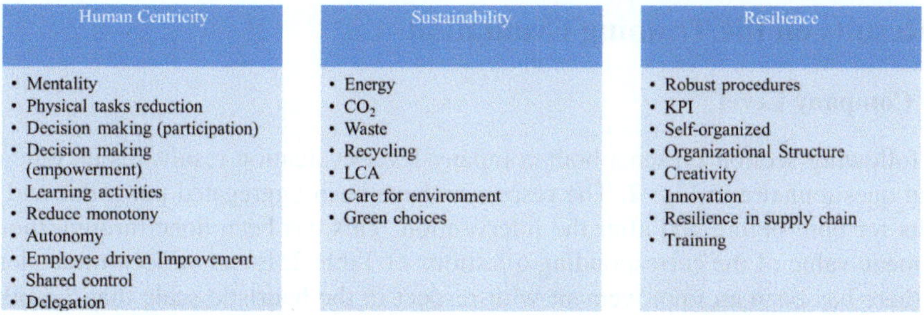

Fig. 1. Evaluation aspects from the perspective of company.

With respect to studying the training efficiency, the scale in which the trainees responded was a 5-scale Likert (1 strongly disagree, 2 rather disagree, 3 neutral, 4 rather agree, 5 strongly agree) and the questions of the evaluation are summarized in Table 2. As a matter of fact, the questions statement was linked to the participants' self-assessed confidence levels, aligning with established constructs of self-efficacy as an indicator of perceived skill development [12, 13]. At this point, there were no quantified tests; thus, the results can only be elaborated statistically at group level. While this approach captures subjective growth, a predictor for upskilling, it is acknowledged that future studies should incorporate objective assessments to capture the quantified impact with respect to upskilling. This all has been documented as part of a wider heuristic evaluation system [3] adopted herein, and in other studies as well.

Table 2. Upskilling-related questions and Industry 5.0 pillars (HC, R, S).

Questions utilized at trainees' level
I am familiar with 3 main values of the Company [Internal]
I know how to behave in the workplace in compliance with ESD safety [HC]
I understand the principles of lean methodology (5S principles) [S]
I know how to sort hazardous production waste [S]
I understand how to identify the direction of an electronic component [HC, Internal]
I know what information system "X" is and how to use it in practice [R]
I know "IT security standards" in the company and understand their importance [R]
I understand the meaning and importance of production documentation (product assembly instruction) [R]
I know what information system "Y" is and how to use it in practice [Internal]
I feel confident and I know how to sort waste in my daily routine [S]
I feel confident when working in different working areas in the company [R,HC]
I feel confident when communicating problems to other colleagues [HC]

3 Results on the Training Evaluation

3.1 Company Level

The following section presents both company-level evaluation results, as an outcome of the questionnaire in Fig. 2. The responses have been aggregated per pillar and the results for both before and after the intervention. This has been done through the use of a mean value of the corresponding questions of Table 2. It can be seen there clearly that there has been an improvement with respect to the heuristic scale that the pillars are measured in. This results in a percentage increase of 5% in the overall Industry 5.0 metric. This particular metric has been estimated as the mean value of all the responses in the upskilling-related questionnaire.

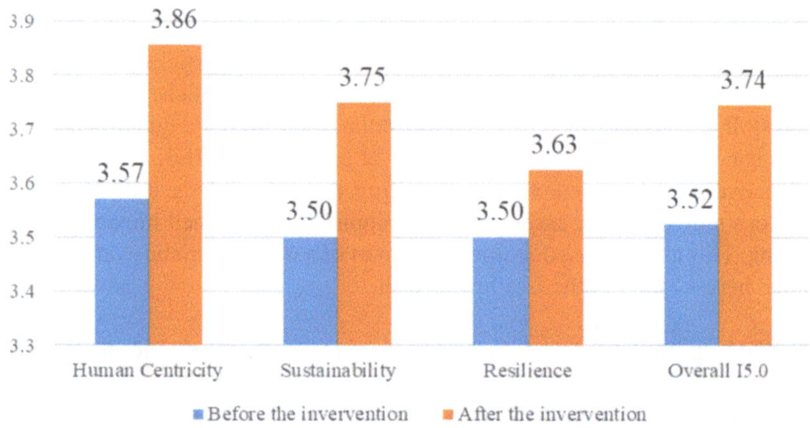

Fig. 2. Evaluation from the perspective of company. A 5-Likert scale has been used based on template questionnaires [3].

3.2 Trainees' Level

Next, it is examined how participants perceived their upskilling following the intervention. To this end, the box plots for the aggregated metrics (opinions on upskilling) with respect to the three pillars are given hereafter in Fig. 3. They regard the mean value of the difference after and before the intervention.

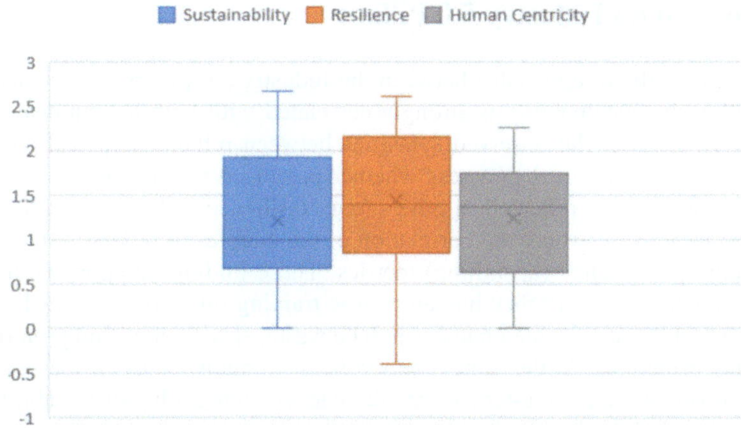

Fig. 3. Evaluation from the perspective of upskilling: Box plots of the aggregated responses. The scale is differential Likert scale (After – Before).

To partially summarize the results so far, the intervention has been proved to be successful, since the averages are positive, from both perspectives of trainees and company. This is of course based on averages, since the individual values may have absolutely no significance in this pre-post experiment. However, it would be quite interesting to try to generalize the results for future similar interventions. To this end, a one sample t-test of the differences are considered to compare the sample mean of the differential Likert-scale responses to a hypothesized population mean of 0. Table 3 summarizes these results and is indicative of the generalizability, through the results of the t-tests; the lower limit of the confidence intervals is positive in all cases. Likert scales were treated as interval-level data, since composite scores, such as sustainability, are considered [14]. Additionally, the normality was checked (Shapiro-Walk test).

Table 3. Generalization of training efficiency with respect to pillars [14, 15].

Metric	Test	Normality	Confidence Interval Lower limit
Individual questions Q_n	$\mu > 0$	Not necessarily	> 0
Sustainability		Yes	> 0.8
Resilience		Yes	> 1
Human Centricity		Yes	> 0.9

Also, even if it is checked against one of the questions with low score, with the help a paired t-test, the outcome of the generalizability does not change; only the normality and the type of test. This can partially act as a control group; probably due to the fact that the subject in training was only in theory and not in the Learning factory and can potentially be used to balance the statistical variations in opinions (i.e. due to misunderstandings and mischaracterizing) at group level. The generalizability is the case even if the average of two questions is used. This is a good indication that the Learning Factory was a really helpful tool, boosting the efficiency of training.

4 Discussion on Industry 5.0 Pillars

Shifting focus to the relationships between the Industry 5.0 pillars, the results suggest that upskilling in sustainability is strongly correlated with both human-centricity and resilience. To validate the observed synergies between pillars, statistical correlations were calculated. This aimed to assess whether perceived improvements in one pillar (e.g., sustainability) are reflected in others (e.g., resilience or human-centricity). This is indicated in Table 4, where the correlation of the couples of pillars is shown, utilizing the training efficiency (upskilling) metrics. These findings support the triple bottom line (People, Planet, Profit): human-centric training improves job satisfaction and engagement, while sustainable practices reduce waste, save energy, and promote circular economy principles. At the same time, profit is strengthened as businesses improve efficiency, resilience, and long-term competitiveness by integrating sustainability-driven innovation into their operations.

To further explore these relationships, a Principal Component Analysis (PCA) was conducted [16], taking into account the responses to the questionnaires as sequences of data $Q_n[k]$, where Q_n is the response to the question n and k is the ID of the trainee. Through grouping these questions, three components were identified, with Q3 and Q4 loading on separate ones. These three components could be considered as sustainability-related skills, but their exact definition exceeds the purposes of the current work. It is also noted that, through Barlett's test of sphericity, a significant p-value ($p < 0.05$) indicates that there are sufficient correlations among your items to justify PCA.

Table 4. Practical Correlation Industry 5.0 pillars.

Investigated couple	Pearson's R	p-value
Sustainability – Resilience	0.772	< 0.001
Sustainability – Human Centricity	0.727	0.001
Resilience – Human Centricity	0.769	< 0.01

Overall, the interplay of the Industry 5.0 pillars observed through this training intervention provides a compelling case for integrated training design. These insights inform both academic understanding and practical deployment of human-centered, sustainable manufacturing practices.

5 Conclusions and Future Outlook

The intervention was shown to be successful, with positive outcomes observed in training-related efficiency and an enhanced characterization of the company's capabilities. Both of these facts constitute a proof of concept towards successfully integrating Industry 5.0 into training systems. In this procedure, sustainability emerged as positively correlated with the other two pillars, likely due to shared workflows and technological foundations across these areas. These interconnections suggest a mutually reinforcing relationship between the pillars of the intervention.

Looking ahead, for practical reasons, a potential reduction of the questionnaire and a speed-up of the training procedure seem probable in the near future. Also, a roadmap for future skills could be shaped due to the correlation of the pillars. Furthermore, future evaluations should consider incorporating additional assessment methods, including standardized tests, to enable more precise quantification of the outcomes and eventually, full operationalization of Industry 5.0. This could be done with both knowledge retention questions as well as workflow checking procedures.

Acknowledgements. This has been supported by the EC project BRIDGES 5.0 (contract No. 101069651).

References

1. Papacharalampopoulos, A., Stavropoulos, P., Ziarsolo, U., Karagianni, O.M.: Teaching learning factories 5.0: shaping training, skilling and reskilling for the future. In: International Assoc. for the Management of Technology Conference (pp. 159–167). Springer, Cham (2024)
2. Gamberini, L., Pluchino, P.: Industry 5.0: a comprehensive insight into the future of work, social sustainability, sustainable development, and career. Austral. J. Career Dev. **33**(1), 5–14 (2024)
3. Papacharalampopoulos, A., Karagianni, O.M., Stavropoulos, P., et al.: On a heuristic evaluation system for Industry 5.0 with respect to interventions: the case of training in businesses. Procedia CIRP 132, 122–128 (2025)
4. Narkhede, G., Chinchanikar, S., Narkhede, R., Chaudhari, T.: Role of Industry 5.0 for driving sustainability in the manufacturing sector: an emerging research agenda. J. Strategy Manage. (ahead-of-print) (2024)
5. Supriya, Y., Bhulakshmi, D., Bhattacharya, S., Gadekallu, T.R., et al.: Industry 5.0 in smart education: concepts, applications, challenges, opportunities, and future directions. IEEE Access (2024)
6. Papacharalampopoulos, A., Stavropoulos, A., Karagianni, O.M., Ippolito, M.: Integrating successfully Industry 5.0 in industry: technological perspective and experiential training at process level. In: 19th CIRP Conference on Intelligent Computation in Manufacturing Engineering
7. Negri, M., Cagno, E., Colicchia, C., Sarkis, J.: Integrating sustainability and resilience in the supply chain: a systematic literature review and a research agenda. Bus. Strateg. Environ. **30**(7), 2858–2886 (2021)
8. Marchese, D., Reynolds, E., Bates, M.E., Morgan, H., Clark, S.S., Linkov, I.: Resilience and sustainability: similarities and differences in environmental management applications. Sci. Total. Environ. **613–614**, 1275–1283 (2018)
9. Ramezankhani, M.J., Torabi, S.A., Vahidi, F.: Supply chain performance measurement and evaluation: a mixed sustainability and resilience approach. Comput. Ind. Eng. **126**, 531–548 (2018)
10. Huang, A., Badurdeen, F.: Metrics-based approach to evaluate sustainable manufacturing performance at the production line and plant levels. J. Clean. Prod. **192**, 462–476 (2018)
11. Abele, E., Metternich, J., Tisch, M., Chryssolouris, G., et al.: Learning factories for research, education, and training. Procedia CIRP **32**, 1–6 (2015)
12. Hadi, P.: The influence of self-efficacy on employee performance mediated by work motivation and work engagement. Int. J. Res. Bus. Soc. Sci. **12**(2) (2023)

13. Landrum, B.: Examining students' confidence to learn online, self-regulation skills and perceptions of satisfaction and usefulness of online classes. Online Learn. **24**(3), 128–146 (2020)
14. Tanujaya, B., Prahmana, R.C.I., Mumu, J.: Likert scale in social sciences research: problems and difficulties. FWU J. Soc. Sci. **16**(4), 89–101 (2022)
15. The jamovi project: jamovi. (Version 2.3) [Computer Software] (2022). Retrieved from https://www.jamovi.org
16. R Core Team:. R: A Language and environment for statistical computing. (Version 4.1) [Computer software] (2021). Retrieved from https://cran.r-project.org
17. Revelle, W.: psych: Procedures for Psychological, Psychometric, and Personality Research. [R package] (2019). Retrieved from https://cran.r-project.org/package=psych

Education and Training

Virtual Remotes Teaching Laboratories in the Progress of a Sustainable Education

José Alejandro González-Medina(✉), Paula González-Suárez, and Pedro Manuel Hernández-Castellano

University of Las Palmas de Gran Canaria, Las Palmas de Gran Canaria, Spain
{jose.gonzalez225,paula.gonzalez131}@alu.ulpgc.es, pedro.hernandez@ulpgc.es

Abstract. Education plays a crucial role in sustainable development. The correct transmission of knowledge, creativity, and critical thinking to the next generations is essential to guarantee the health of modern society. However, data shows that absenteeism and failure among students remain significant issues. Particularly in engineering, a fundamental field for the progress of communities and the quality of life, there seems to be a lack of enthusiasm among new generations. Additionally, in this age of Industry 4.0, the rapid pace of technological advancement means that educational content often fails to align with the realities of the industry. For these reasons, the University of Las Palmas de Gran Canaria (ULPGC) has initiated a project aimed at incorporating virtual remote teaching laboratories (VRTL) to enhance engineering education. The goal is to facilitate practical training for students through digital twins that emulate the educational use of university equipment. This approach is intended to foster easier and more effective competency development, keeping students motivated while reducing material resource consumption. This work presents an analysis of the prototype and testing of a VRTL focused on a laser cutting machine, demonstrating the effectiveness and potential future development of this concept.

Keywords: Remote laboratories · engineering education · digital twins · Industry 4.0

1 Introduction

School absenteeism and early dropouts continue to hinder student achievement and long-term educational outcomes, underscoring the urgent need for more engaging and adaptable learning models. In response to the limitations of traditional methods, Learning Experience Design (LXD) has emerged as a promising approach to deliver personalized and meaningful education [1, 2]. This need for innovation is particularly critical in Engineering Education, where the rapid pace of technological advancement under Industry 4.0 demands updated and resource-efficient training methods. However, many institutions face challenges due to limited resources, which compromise the quality of hands-on learning experiences [1, 3, 4]. Remote Laboratories (RLs) offer a practical solution by

H. Kohl et al. (Eds.): GCSM 2025, LNME, pp. 421–429, 2026.
https://doi.org/10.1007/978-3-032-21154-5_46

enabling students to access real laboratory equipment remotely, thus supporting cost-effective, scalable, and industry-aligned learning [1, 4]. When integrated with Teaching Laboratories (TLs), RLs contribute to the goals of Education 4.0 by leveraging digital technologies to enhance educational delivery [5]. The adoption of such technologies has been steadily increasing since 1995, driven by expanding digital infrastructure and capabilities [4, 6].

A key component of this transformation is the Digital Twin (DT), a virtual model that replicates the behavior and structure of a real system, enabling real-time simulation of industrial processes [7]. DTs support the integration of cyber-physical systems and enhance engineering education by offering safe, cost-effective, and scalable learning environments [1, 8]. However, a complete disconnection between the DT and physical reality would represent a mistake [6]. In engineering, this disconnection could lead to professionals with limited understanding of material constraints and sustainability. Nevertheless, as it is shown in recent studies, strong digital and cognitive skills can significantly benefit students in their professional futures [9].

Therefore, the development of Virtual Remote Teaching Laboratories (VRTLs) is a crucial area of research in both Engineering and Sustainable Education. These tools must be designed to maximize educational value while minimizing material costs, without losing the essence of human-centered teaching. Often, the success of digital learning depends more on the strategy than the technology itself [10, 11]. This requires a solid understanding of educational methodologies and LXD, followed by prototype implementation and evaluation.

2 Theoretical Framework

This section outlines the foundational concepts relevant to the study, focusing on contemporary educational methodologies and the intersection between design and education through the lens of LXD.

2.1 Educational Methodologies

Education is fundamentally structured around three core pillars: learning, assessment, and teaching. The educator's role is to facilitate the interaction of these elements by adapting materials, stimulating interest, encouraging independent learning, and tracking student progress. These interactions form another triadic relationship between the teacher, the student, and the content, which serves as the foundation for various educational methodologies [1, 12].

The nature of this triad shapes the structure and philosophy of different pedagogical approaches. Two dominant paradigms (behaviorism and constructivism) offer contrasting views. Behaviorism emphasizes a unidirectional flow of knowledge, where students passively receive information. In contrast, constructivism promotes a dialogic and participatory model, positioning the teacher as a facilitator and the student as an active agent in constructing knowledge [13].

Constructivist approaches underpin many innovative educational methodologies, often referred to as active methodologies, which prioritize student engagement and autonomy. These include widely adopted strategies such as flipped classroom, project-based learning, cooperative learning, gamification, problem-based learning, thinking-based learning, and service learning [14].

2.2 Learning Experience Design

LXD is a design-driven discipline that focuses exclusively on improving the quality of educational experiences. Like other branches of design, LXD is rooted in a holistic, interdisciplinary, and learner(user)-centered philosophy, with the primary aim of making the learning process engaging, relevant, enjoyable, and informative. This creative approach is particularly effective in contexts characterized by uncertainty, where it helps transform ambiguous challenges into structured and meaningful outcomes [2].

At its core, LXD prioritizes the learner's perspective, emphasizing their goals, preferences, and the environments in which they learn. It calls for the active involvement of students in a co-creative process, ensuring their voices are integrated throughout the design and development of the learning experience [2]. As with many creative methodologies, LXD can be nonlinear and unpredictable. However, its flexibility makes it highly adaptable across diverse academic programs.

The LXD process typically unfolds through six iterative phases [15]: assess the need for the problem, conduct user research, design the experience, prototype the experience, test the experience, and implement the experience.

2.3 Sustainable Education

The relationship between education and sustainability is becoming increasingly undeniable. The Sustainable Development Goals, created by the United Nations, have a dedicated section focused on quality education [16]. There, we can see that it is not only referring to equal access to education without discrimination based on ethnicity, gender, disabilities, and other vulnerable reasons, which is highly important in itself. It does not refer only to the teaching of knowledge to promote the sustainable development of society, either. Rather, quality education also encompasses "the number of youth and adults who have relevant skills, including technical and vocational skills, for employment, decent jobs, and entrepreneurship" [16].

In this way, sustainable education can be defined as a method of teaching knowledge and skills that ensures the present and future development of society in harmony with the world's ecology, economy, and individual growth. As it was said in the introduction, there is a general problem of educational absenteeism and a particular problem of competence teaching in engineering, compromising the sustainability of the educational system (particularly in Spain). In this way, the object of this piece of work is to show the early implementation and the results of trying to reach a more sustainable educational methodology in engineering through VRTLs.

This technology has the potential to give access to educational resources in remotes parts of the world which could not be reached any other way. At the same time, VRTLs can give the opportunity to experience environments of high risk and danger with security, enriching the formation of students. Both aspects also have a high relation with sustainability education, as was previously explained.

3 Implementation

In this context, the School of Industrial and Civil Engineering (EIIC) at the University of Las Palmas de Gran Canaria (ULPGC) has launched an educational innovation project. This project aims to foster greater student responsibility and engagement in their learning process through the implementation of VRTL in courses across three undergraduate engineering degrees. This contribution focuses on the VRTL created for the course "Product Development Technologies" in the fourth year of the Bachelor's Degree in Industrial Design Engineering and Product Development (GIDIDP) [1].

3.1 Project Scoop

The project aims to advance engineering education through several strategic objectives. It promotes educational innovation by integrating emerging technologies that enhance teaching quality and support the digital transformation of learning environments. By eliminating physical constraints, the VRTL increases accessibility and flexibility, enabling students to conduct hands-on practice remotely. It also improves resource efficiency by allowing remote use of costly lab equipment, reducing wear and operational expenses. Aligned with Industry 4.0 requirements, the project helps students develop essential digital and technical skills. Initially focused on laser cutting and engraving, the VRTL serves as a proof of concept at ULPGC, laying the groundwork for future expansion.

3.2 User Experience Research and Development Guidelines

A preliminary study was conducted to understand the experiences of students in the IDIDP program regarding laboratory practices. It involved a focus group with 5 participants and a closed-ended questionnaire completed by 20 students. The focus group provided qualitative insights into students' motivations and expectations, while the questionnaire helped validate these findings across a broader sample.

Students expressed dissatisfaction with current lab practices, citing a lack of support and unclear objectives. They called for more practical, hands-on projects involving real equipment to relate theory to practice and enhance their professional portfolios. While lab sessions were seen as helpful for understanding theoretical concepts, students noted a disconnection between theory and application. Despite these issues, many showed initiative by using online resources and developing personal interests in the topics covered.

In response, a set of design and development criteria was established to guide the implementation of the VRTL. These include creating a highly interactive and intuitive platform with action-based tasks and spatial navigation, ensuring accessibility across devices, supporting collaborative use, and integrating seamlessly with existing course materials and the institution's educational platform. These criteria aim to improve the learning experience and better align practical training with academic instruction.

3.3 Description of the VRTL Designed

As part of a university course, a pilot VRTL was developed to simulate the operation of an industrial laser cutting and engraving machine. This simulation was built using a digital twin, created through CAD modeling software, which accurately replicates the machine's physical structure and key components—such as the cutting head, motion system, and control screen—while omitting non-visible systems like cooling units and power electronics to streamline the model. The digital twin was integrated into a virtual environment using the Unity Game Engine. It was chosen for its accessibility and the wealth of online learning resources.

Unity Game Engine uses C# as its main code language. Nevertheless, it offers many helpful user-interface characteristics for new developers. Through both coding and interface-helped programming, the different actions and movements of the DT were created. Thanks to its easy model integration, the different machine pieces were imported into the game engine, modeled previously, as it was said, in CAD and 3D mesh modeling programs, in this case, Autodesk Fusion 360 and Blender. Also, Unity enables us to integrate a user interface, creating interaction features and graphical design content as well.

A major feature of the VRTL is its user-friendly interface, which resembles a simplified computer-aided manufacturing (CAM) tool. After logging in through a secure portal, students enter a virtual workspace where they can view a sample piece, configure settings, and operate the machine. The CAM imitation interface was modeled after the software used with the actual machine, allowing students to configure two sets of parameters on different geometric elements. The virtual machine includes a realistic control system, enabling users to perform actions such as opening and closing the workspace door, moving the machine's axes, setting the origin point, and adjusting the focal distance. Navigation within the 3D environment mimics standard 3D modeling software, making it intuitive for students to explore and interact with the system. They can then use a virtual control screen to move the cutting head, set the origin, and start or pause the cutting process. The ability to adjust the focal distance adds another layer of interactivity and experimentation (see Fig. 1).

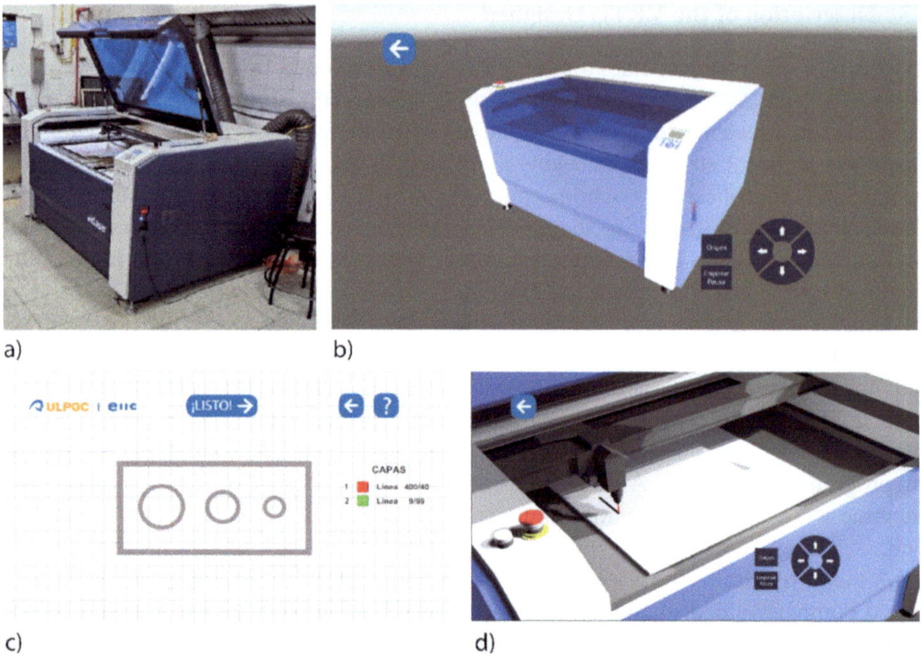

Fig. 1. a) Image of the real machine; b) Image of the DT of the machine, at the bottom-right of the image it can be seen the head movement panel with the set origin and the start-stop button; c) Image of the CAM interface, at the right side it can be seen the layers configurations, at the top the button that let access to the students to the DT, and al the top-right the go-back button and the help button; d) Image of the DT doing an operation.

4 Results and Discussion

The VRTL was tested during the 2024/25 academic year with three student groups. Two groups (GI and GII), totaling 18 students, used the VRTL after completing in-person lab sessions, while a third group (GIII) of 11 students used it beforehand. All sessions were supervised by a member of the research team, ensuring consistent guidance and support.

The evaluation aimed to assess the acceptance and effectiveness of the VRTL as a learning tool in comparison to traditional methods. Students in groups GI and GII completed one questionnaire after their in-person lab session (Q1) and another following their use of the VRTL (Q2), resulting in 18 and 7 responses, respectively. Meanwhile, group GIII completed one questionnaire before using the VRTL (Q3) and another after the lab session (Q4), yielding 11 and 9 responses (see Table. 1). All surveys were completed independently and included both closed and open-ended questions. These focused on key aspects such as accessibility, interface design, experiment quality, learning outcomes, and the integration of the VRTL into the course. The summarized findings are presented below.

Table 1. Structure of the questionaries and groups

Group	Before VRTL	After VRTL
GI y GII	Q1	Q2
GIII	Q3	Q4

A majority of students reported feeling anxious when using physical lab equipment due to fear of causing damage or injury, with 83% indicating this in Q1 and 91% in Q3. Most students believed that using a virtual model beforehand could help reduce this anxiety and better prepare them for lab sessions, with up to 94% supporting this view in Q1. After using the VRTL, 86% of students in Q2 and 78% in Q4 agreed that it accurately reflected the behavior of the physical machine. However, opinions were more divided on whether it improved their understanding of the machine's operation, with only 33–43% affirming this. Regarding the timing of use, most students preferred accessing the VRTL before the in-person lab, as indicated by 57% in Q2 and 78% in Q4, suggesting its value as a preparatory tool. Additionally, around 56–57% of students reported feeling more confident using the physical machine after interacting with the VRTL, although some cited technical issues such as loading delays and system errors that limited its effectiveness. Students also noted limitations in the virtual model, including a lack of movement precision and restricted customization options. Despite these issues, all respondents expressed interest in continuing to use the VRTL in future lab sessions.

The analysis of questionnaire responses reveals consistent student perceptions of the VRTL's purpose and value, with many indicating that using it before in-person lab sessions enhanced their confidence in handling real equipment. While no significant differences were found between groups who used the VRTL at different times, some technical issues affected the experience. Low participation in the post-use surveys (likely due to their voluntary and independent nature) limited the overall response rate.

5 Conclusions and Future Work

This work explores how emerging technologies, particularly those linked to Industry 4.0, can enhance engineering education by improving practical training and overall student experiences. A central achievement is the development of a VRTL prototype. Although it is still in its early stages, pilot testing has shown promising results, with students reporting positive learning outcomes. Future improvements will incorporate student feedback, integrate the VRTL fully into course curricula, and explore strategies to boost participation. Looking ahead, incorporating AI and machine learning into VRTLs could enable greater personalization, support continuous learning, and extend their use beyond formal education into areas like interactive manuals and training simulations.

References

1. González-Suárez, P., González-Medina, J.A., Hernández-Castellano, P.M., Araña Suárez, R.E.: Propuesta de implementación de Laboratorios Docentes Remotos Virtuales para la

mejora de la Educación en Ingeniería de los Procesos de Fabricación, en Realidad extendida: Nuevas tecnologías en la mejora de la docencia, en Via Docendi, no. 5. , Universidad de Las Palmas de Gran Canaria. Servicio de Publicaciones y Difusión Científica (2024). https://doi.org/10.20420/1786.2024.745

2. Tawfik, A.A., Gatewood, J., Gish-Lieberman, J.J., Hampton, A.J.: Toward a definition of learning experience design. Technol. Knowl. Learn. **27**(1), 309–334 (2022). https://doi.org/10.1007/s10758-020-09482-2
3. Lee, J., Davari, H., Singh, J., Pandhare, V.: Industrial artificial intelligence for industry 4.0-based manufacturing systems. Manuf. Lett. **18** (2018). https://doi.org/10.1016/j.mfglet.2018.09.002
4. Silva, I.N.D., García-Zubía, J., Hernández-Jayo, U., Alves, J.B.D.M.: Extended remote laboratories: a systematic review of the literature from 2000 to 2022. IEEE Access **11**, 94780–94804 (2023). https://doi.org/10.1109/ACCESS.2023.3271524
5. Terkowsky, C., Frye, S., May, D.: Online engineering education for manufacturing technology: is a remote experiment a suitable tool to teach competences for "Working 4.0"?, Eur. J. Educ. **54** (2019). https://doi.org/10.1111/ejed.12368
6. Heradio, R., De La Torre, L., Galan, D., Cabrerizo, F.J., Herrera-Viedma, E., Dormido, S.: Virtual and remote labs in education: a bibliometric analysis. Comput. Educ. **98**, 14–38 (2016). https://doi.org/10.1016/j.compedu.2016.03.010
7. Wu, J., Yang, Y., Cheng, X., Zuo, H., Cheng, Z.: The Development of Digital Twin Technology Review, pp. 4901–4906 (2020). https://doi.org/10.1109/CAC51589.2020.9327756
8. Guinaldo, M., De La Torre, L., Heradio, R., Dormido, S.: A virtual and remote control laboratory in moodle: the ball and beam system. In: IFAC Proc. vol. 46, no 17, pp. 72–77, (2013). https://doi.org/10.3182/20130828-3-UK-2039.00021
9. Zhao, B., Zhou, J.: Research hotspots and trends in digitalization in higher education: a bibliometric analysis. Heliyon **10**(21), e39806 (2024). https://doi.org/10.1016/j.heliyon.2024.e39806
10. Csernoch, M.: Lean digital education to resolve the paradox of the illusion of digital prosperity. J. Innov. Knowl. **10**(2), 100676, (2025). https://doi.org/10.1016/j.jik.2025.100676
11. Rahman, Md.M., Hossain, Md.E.: Digital dynamics in technology adoption: exploring socio-economic development through technology and education. Technol. Soc. **82**, 102906 (2025). https://doi.org/10.1016/j.techsoc.2025.102906
12. Wankat, P.C., Oreovicz, F.S.: Teaching Engineering, Second Edition. Purdue University Press (2015).https://doi.org/10.2307/j.ctv15wxqn9
13. Boghossian, P.: Behaviorism, constructivism, and socratic pedagogy. Educ. Philos. Theory **38**(6), 713–722 (2006), https://doi.org/10.1111/j.1469-5812.2006.00226.x
14. Solórzano, D.A.N.: Active methodologies in the classrooms. Int. J. Psychosoc. Rehabil. **24**(5), 5918–5928 (2020). https://doi.org/10.37200/IJPR/V24I5/PR2020566
15. Glaser, N., AlZoubi, D., Earnshaw, Y., Shaffer, E.L., Yang, M.: Formative design and development of a three-dimensional collaborative virtual learning environment through learning experience design methods. J. Form. Des. Learn. **6**(2), 63–76 (2022). https://doi.org/10.1007/s41686-022-00072-2
16. Goal 4 | Department of Economic and Social Affairs. Acceded: 18 of June of 2025. [Online]. Available in: https://sdgs.un.org/goals/goal4#targets_and_indicators

Learning Factory Remanufacturing Challenge: A Hands-On Training for Circular Economy

Kai Rüdele(✉) and Matthias Wolf

Institute of Innovation and Industrial Management (IIM), Graz University of Technology, Kopernikusgasse 24, 8010 Graz, Austria
kai.ruedele@tugraz.at

Abstract. Learning factories (LFs) have become an integral element of engineering education, especially when it comes to making new and on-trend subjects tangible. This also includes the abstract matters of sustainability and circular economy (CE), which are influenced by the design of products and production, and therefore closely linked to engineering expertise. This article presents how a new learning module was used to implement CE principles within a specific LF to foster sustainable practices and education. As a starting point, engineering students were familiarized with fundamental knowledge of CE. Following this, they were tasked with developing their own concepts for a disassembly line that reprocesses returned product components in the most ecological and economic way. These student-generated designs were discussed, evaluated, and have proven to be basically operable and effective. By integrating theoretical inputs with direct applications, this initiative not only enhances students' understanding of CE, but also enables them to implement it in actual operational practice. Additionally, our research is a first step towards the institutionalization of CE practices within a realistic learning environment. The findings and recommendations contribute to the broader discourse on education for sustainability and the role of LFs in promoting environmentally conscious industrial processes.

Keywords: circular economy · learning factory · engineering education · R-strategies

1 Introduction

In the last two decades, the concept of circular economy (CE) has gained momentum in scientific research [1, 2]. Also, governments and the public are urging industrial companies to improve their ecological sustainability [3, 4]. It is necessary to promote processes with low adverse environmental impacts, characterized by a conservation of resources and low carbon footprints [3]. Consequently, it is desirable to increase awareness, knowledge, and skills regarding circular economy and value preservation to response to the destruction of the environment [5, 6], but also to qualify for future job opportunities [7]. One way to tackle this issue at an early stage is to incorporate the basics of CE into engineering education [8, 9], and, if possible, in an interactive and practical manner [10]. Learning factories (LFs) as close-to-reality learning environments appear suitable for this purpose [11, 12].

H. Kohl et al. (Eds.): GCSM 2025, LNME, pp. 430–438, 2026.
https://doi.org/10.1007/978-3-032-21154-5_47

In this context, a learning module was designed for an existing LF to teach basic concepts of CE. The related practical exercise focuses on an organised disassembly and treatment of returned products. By transforming the related task into a playful group activity, both the predefined challenge and the devised solutions become tangible. This article outlines the implementation, outcomes, and lessons learned of this module and affirms the suitability of LFs in teaching future engineers in CE topics.

2 State of the Art

LFs are physical learning environments where processes and technologies are similar to industrial sites, which allows a direct approach to production while emphasizing experimentation and problem-based learning; they are also fostering continuous improvement through their own actions and interactive involvement of the participants [13]. Operations are replicated in detail and only abstracted where necessary or unavoidable, so that learners can work under authentic but risk-free conditions [14]. The acquisition of competencies is usually achieved by means of structured self-learning processes supported by different teaching methods (with the instructor remaining passive). A LF's infrastructure needs to be flexible and adaptable to fulfill technical requirements and its didactic purpose.

LFs promote and integrate tacit and explicit knowledge, as well as practical problem-solving skills. Moreover, they spark curiosity and self-guided learning, which makes them open platforms for ideation and conception. Additionally, they encourage critical and independent thinking, which inculcates leadership qualities among learners [15]. LFs also have the ability and capacity to convert abstract content into simulated learning, which can then be converted into concrete learning situations. As a result, they have not only become established in engineering education in recent decades [14, 15], but are also increasingly being used to teach ecological sustainability in general and CE in particular [12, 16].

Recently introduced courses and exemplary use cases dealing with CE are an indication of the efforts being conducted in the LF community and the growing popularity of this subject [16]. For instance, Dvorak et al. developed a four-part LF training that encompasses the wide spectrum of CE, from appropriate product design to reverse logistics, as well as the management of disassembly and cost models [17]. The LF-based value creation system proposed by Petrusch et al. does not end with manufacturing, but also comprises stations for material collection and sorting. This setup puts participants into the role of production employees who are challenged to find the required information and procedure to keep fictitious components of a product in the loop [18]. Koller et al. describes the process chain of their LF from diagnosis of unsorted batches to disassembly, cleaning, inspection, reconditioning (incl. Replenishment), and reassembly of motors with commercially available technologies, as well as the underlying didactic concept [19]; additionally, this regard, increases in the efficiency of remanufacturing through digital technologies and automation are discussed.

The focal point of all these studies often is in the various R-frameworks [20], which classify the different stages of resource use and waste treatment in a CE. The 10 Rs as outlined by Potting et al. [21], which are used in higher education [22] and in the context of LFs [23], seem to be the most common approach now [20, 24]. Particularly strategies that extend the lifespan of a product and its parts (i.e., repair, refurbish, remanufacture, repurpose) appear to be relevant for LFs [23, 25]. However, unlike reuse (= discarded products directly used by others for the same purpose) and recycling (= converting waste into raw material), these strategies are not clearly defined. Table 1 provides a brief overview of terminology delimitations.

Table 1. Distinctive characteristics of reconditioning discarded products

	Repair	Refurbish	Remanufacture	Repurpose
Pottering et al. [21]	Servicing of a defective product so that it can be used with its original function	Restoring an old product to bring it up to date	Using parts of discarded product in a new product with the same function	Use a discarded product or its parts in a new product with a different function
other sources	Individual and hand-crafted [26] restoring of working order on product [27]	Replacing failed components only to fix issues and get a cosmetically close-enough-to-new state [28]; usually on module level [27]	Bringing used goods up to the quality level of a new product [29]; Replacing components to meet new specification [28]	Open-loop reuse with a new purpose or in another context, often due to deteriorating performance [30]

Apart from the fact that the terms discussed above are often used interchangeably, there may also be cases where these types of reconditioning are combined: thus, remanufacturing can also describe the process of rebuilding a product to meet or exceed the original manufacturer's specifications by repairing worn-out components and replacing others with new ones [28]. Meanwhile, there are also decision support tools to identify the most economically and environmentally sound R-strategies, which minimizes the risk for inefficiencies [31].

3 Learning Module on Circular Economy

The LEAD Factory (an acronym for lean, energy efficiency, agility, and digitalization) is operated by the Graz University of Technology and used in academic and industrial contexts for education, workshops, and research [32]. This LF demonstrates an assembly line for the three variants of the TUG Scooter consisting of more than 60 parts (Fig. 1 a).

The goal of the new module is that participants not only grasp the principles of CE (i.e., explicit knowledge), but also be capable of putting them into practice (i.e., skills and tacit knowledge). The learning module is organized as described in Table 2, building on the previous attempts and concepts [18, 25, 33]. As usual for LFs, the traditional instructor-centered teaching at the beginning just provides the theoretical basis needed for the subsequent problem-based practical exercise. The centerpiece of the module is hands-on group work. There, students are exposed to a situation where discarded scooters with various signs of wear and tear (scratches up to broken or missing parts) are returned to the LEAD Factory for further treatment. Initially, the groups have to focus on designing a disassembly line (Fig. 1 b), followed by creating a concept that outlines the handling of individual parts. This business model should primarily clarify technical aspects such as sorting and processing. It is also crucial to consider various uncertainties, such as the condition, quantity, and timing of returns. Finally, the groups' ideas should be compiled in a short concluding presentation.

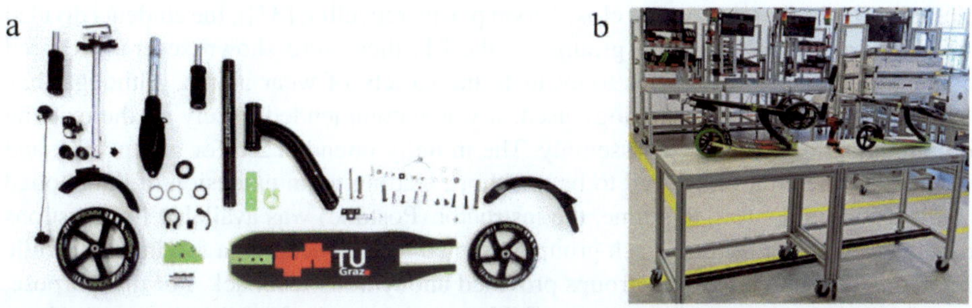

Fig. 1. (a) Disassembled TUG Scooter; (b) Starting point for the disassembly line (preexisting assembly line in the background).

Table 2. Didactic concept of the CE learning module (inspired by Schroth et al. [34])

Phase	Skill level	Content/activity	Objective
Theory fundamentals	Awareness/ Remember	Interactive lecture, also with video material that showcase real-world examples of CE	Familiarization with R-strategies [21] and the state of research [24]
Demonstration	Understand	Exemplary application of CE for a known process in LF [33]	Knowledge improvement
Hands-on training	Apply/ Analyze	Group work on practical and close-to-reality use cases	Application of acquired knowledge; enhancing problem-solving skills
Exposition	Evaluate	Presentation of results and discussion	Judging information, ideas, and concepts

The assignment was deliberately kept open by not imposing any directive concerning the R-strategy to be implemented. Any approach that results in an extended or new life cycle of the product and/or its parts, thereby reducing the consumption of natural resources and minimizing waste, is permissible. Nonetheless, since in mechanical engineering, reconditioning is often associated with (machine) processing, this task is internally referred to as the "remanufacturing challenge".

In December 2024, the new learning module was taught for the first time and on the last day of a four-day course to ensure that participants were already familiar with the product and its parts, as well as with the assembly process. On the first three days, various topics such as lean production, quality management, and digitalization were covered. Any bias or self-selection in the course enrollment can be ruled out, as the new content on CE had not been announced nor taught before. In total, 17 students, of which most were mechanical engineering students in different Master's programs, participated.

Following a theory session of about 30 min (25 min presentation with short video clips, plus 5 min for showcasing closed-loop plastic recycling [33]), the students divided into three roughly equal-sized groups. In the LF, they were shown several prepared scooters as examples of returns to evaluate the variety of wear marks. Although they were free to choose the technology used, it was recommended to rely on the existing equipment, at least for the disassembly. The initially intended 2 h for group work and presentation preparation proved to be too short, leading to an extension of the allotted time to 150 min. During this time, the instructor (Postdoc) was available for questions from the groups. At the end, each group presented their concept in a 5-min pitch, while the members of the other two groups provided anonymous feedback. For this purpose, an evaluation sheet containing two text fields to comment on the (dis)advantages plus a 5-point Likert scale to rate four aspects was distributed: in the category of *Ecological impact*, the improvements over disposal had to be assessed, while *Circularity* relates to the fulfillment of the R-strategies; *Creativity* was included due to its relevance as outlined by Pinzone & Taisch [6]. The characteristics and ratings of the proposed concepts can be found in Table 3.

All the concepts presented have in common that recycling is carried out by external entities outside the LEAD Factory. Apart from this, the groups' solutions and pitches emphasized different aspects. Group 1 estimated costs and emission savings, while group 2 determined the number of reprocessed scooters needed to cover costs (break-even analysis). In contrast, group 3 was the only one that utilized the existing assembly line; members of the other groups noted that a separation of reworked parts from new ones must be ensured if quality differences are to be expected. The discussions and submitted evaluation forms showed that both instructors and students acknowledged the differences.

The overall rating of the course was 4.8 out of 5, indicating the same high level as usual [32]. Many of the students indicated that they liked the (combination of) practical exercises, group work and reflection; all key elements of experiential learning that enforce tacit knowledge. Two out of the 17 participants stated that CE should have been covered in more detail.

4 Conclusion and Outlook

As the circular economy emerges as an inevitable component of resource security, the demand for proficient engineers and skilled manufacturing experts able to keep materials in circulation for as long as possible grows. At the same time, learning factories which

Table 3. Comparison of the concepts developed in group work

		Group 1	Group 2	Group 3
Description		Refurbishing on sub-assembly level; limited repair capabilities; remanufacturing via externals	Categorizations of returned and processed scooters incl. Needed processes	Internal rework station feeds existing warehouse/assembly line
Assumptions		4,000 returns p.a.; takt time of 27 min	5,000 returns p.a.; takt time of 18 min	10 reworked scooter per day
# stations		2	1 + external processing	2 (sorting + rework)
# workers		2	1	< 1
R-strategies [24]		R4, R5	R3–6, R8	R3, R4, R6, R8
Advantages		Functional & visual check lists; Routine replacement of safety-critical parts	Flexible price model; Batches where useful	Comprehensive flow-chart
Shortcomings		Returns to suppliers doesn't ensure CE	Too optimistic assumptions for external processes	No economic considerations
Rating (5 = best)	Implementability	4.18	3.82	3.33
	Ecological impact	4.18	4.00	4.00
	Circularity	4.45	4.11	4.17
	Creativity	3.72	3.82	3.33

mimic industrial situations have proven to be an effective tool in engineering education, especially for new and complex subjects. Based on the principle of learning by doing, participants gather new and long-lasting knowledge and can transfer these capabilities much easier to real-life situations.

This paper presented how circular economy is addressed in a specific learning factory. The emphasis is on the design, initial conduct, and outcomes of a one-day learning module for master's students in mechanical engineering, focusing on remanufacturing challenges of a predefined product. After a theory input and with circularity in mind, students had to develop approaches to address the remanufacturing challenge in groups. They disassembled used scooters and decided which R-strategy could be pursued for each component. As a subsequent task, the concepts were presented and discussed.

The results demonstrate that students successfully identified multiple solutions to the posed challenge, showcasing their ability to conceptualize ideas with minimal initial guidance. Participant feedback and the active class contribution indicate a strong appreciation for both the topic of circular economy and the specific task, underscoring the value of experiential learning in engineering curricula.

For the next classes, the proposed groups' concept will not only be presented but also demonstrated in the LEAD Factory in the best possible way, so that the participants can experience the strengths and weaknesses. Also, a more distinct quantification of the concepts by means of a score is desirable.

References

1. Schöggl, J.-P., Stumpf, L., Baumgartner, R.: The narrative of sustainability and circular economy - a longitudinal review of two decades of research. Resour. Conserv. Recycl. **163**, 105073 (2020). https://doi.org/10.1016/j.resconrec.2020.105073
2. Reike, D., Vermeulen, W., Witjes, S.: The circular economy: new or refurbished as CE 3.0? - exploring controversies in the conceptualization of the circular economy through a focus on history and resource value retention options. Res. Conserv. Recycl. **135**, 246–264 (2018). https://doi.org/10.1016/j.resconrec.2017.08.027
3. European Commission: The clean industrial deal: a joint roadmap for competitiveness and decarbonization. https://commission.europa.eu/document/download/9db1c5c8-9e82-467b-ab6a-905feeb4b6b0_en, Accessed 03 Apr 2025
4. Xu, Y., Yang, L., Hossain, M., et al.: Unveiling the trajectory of corporate green innovation: the roles of the public attention and government. J. Clean. Prod. **444**, 141119 (2024). https://doi.org/10.1016/j.jclepro.2024.141119
5. van Langen, S., Vassillo, C., Ghisellini, P., et al.: Promoting circular economy transition: a study about perceptions and awareness by different stakeholders' groups. J. Clean. Prod. **316**, 128166 (2021). https://doi.org/10.1016/j.jclepro.2021.128166
6. Pinzone, M., Taisch, M.: Key competencies for circular manufacturing. Human Aspects Ad. Manuf. **80**, 120–125 (2023). https://doi.org/10.54941/ahfe1003514
7. International Labour Organization: Skills for a greener future: a global view based on 32 country studies. https://www.ilo.org/wcmsp5/groups/public/---ed_emp/documents/publication/wcms_732214.pdf, Accessed 202504 Apr
8. Sanchez-Romaguera, V., Dobson, H., Tomkinson, C.: Educating engineers for the circular economy. In: 8th International Conference in Engineering Education for Sustainable Development (2016)
9. O'Born, R., Heimdal, A.: Experiences from teaching circular economy concepts to engineering students. In: Proceedings of the 24th International Conference on Engineering & Product Design Education (2022). https://doi.org/10.35199/EPDE.2022.76
10. Rodríguez-Chueca, J., Molina-García, A., García-Aranda, C., et al.: Understanding sustainability and the circular economy through flipped classroom and challenge-based learning: an innovative experience in engineering education in Spain. Environ. Educ. Res. **26**(2), 238–252 (2020). https://doi.org/10.1080/13504622.2019.1705965
11. Gento, Á., Pimentel, C., Pascual, J.: Teaching circular economy and lean management in a learning factory. In: Training engineering students for modern technological advancement, pp. 183–201 (2022). https://doi.org/10.4018/978-1-7998-8816-1.ch009
12. Weyand, A., Thiede, S., Mangers, J., et al.: Sustainability and circular economy in learning factories - case studies. In: Proceedings of the 12th Conference on Learning Factories (2022). https://doi.org/10.2139/ssrn.4080162
13. International Association of Learning Factories: Definition of "learning factory". https://ialf-online.net/, Accessed 04 Apr 2025

14. Abele, E., Eichhorn, N.: Process Learning Factory - training students and management for excellent production processes. In: Advanced manufacturing systems and technology. 63–73 (2008)
15. Jing, Z., Turi, J., Lu, S., et al.: Sustainability through factory-based learning in higher education. Sustainability **15**(6), 5376 (2023). https://doi.org/10.3390/su15065376
16. Barth, J., Ruzicic, N., Mennenga, M., et al.: Development of an IALF overarching learning module for circular economy. In: Learning factories of the future, pp. 366–373 (2024). https://doi.org/10.1007/978-3-031-65411-4_43
17. Dvorak, J., Hörsting, R., Gleich, et al.: Circular production in learning factories: a teaching concept. In: Learning Factories of the Future, pp. 358–365 (2024). https://doi.org/10.1007/978-3-031-65411-4_42
18. Petrusch, N., Briese, C., Lange, A., et al.: Learning factory for circular economy and the corporate environment. In: Proceedings of the 13th Conference on Learning Factories (2023). https://doi.org/10.2139/ssrn.4456416
19. Koller, J., Grosse Erdmann, J., Herold, M., et al.: RemanLab - conceptualization and realization of a learning factory for remanufacturing. In: Proceedings of the 13th Conference on Learning Factories (2023). https://doi.org/10.2139/ssrn.4469183
20. Kirchherr, J., Reike, D., Hekkert, M.: Conceptualizing the circular economy: an analysis of 114 definitions. Resour. Conserv. Recycl. **127**, 221–232 (2017). https://doi.org/10.1016/j.resconrec.2017.09.005
21. Potting, J., Hekkert, M., Worrell, E., et al.: Circular economy: measuring innovation in the product chain. Policy Report of Netherlands Environmental Assessment Agency (2017)
22. Renfors, S.: Education for the circular economy in higher education: an overview of the current state. Int. J. Sustain. High. Educ. **25**(9), 111–127 (2024). https://doi.org/10.1108/IJSHE-07-2023-0270
23. Barth, J., Metternich, J.: Remodeling learning factories for circular economy. In: Proceedings of the 15th Conference on Learning Factories (2025). In press
24. Morseletto, P.: Targets for a circular economy. Resour. Conserv. Recycl. **153**, 104553 (2020). https://doi.org/10.1016/j.resconrec.2019.104553
25. Rüdele, K., Raml, A., Ramsauer, C.: Improving subject-related content of learning factories through incorporating students' ideas to address circular economy. In: Proceedings of the 15th Conference on Learning Factories (2025). In press
26. Alexiadis, I.: Remanufacturing und refurbishing, https://www.wissensdemokratie.de/remanufacturing-und-refurbishing, Accessed 04 Apr 2025
27. Thierry, M., Salomon, M., Van Nunen, J., et al.: Strategic issues in product recovery management. Calif. Manage. Rev. **37**(2), 114–135 (1995). https://doi.org/10.2307/41165792
28. Elalj, S.: Remanufactured vs. refurbished: what are the differences? https://www.refurb.me/blog/remanufactured-vs-refurbished, Accessed 04 Apr 2025
29. Döpper, F., Koller, J.: Remanufacturing in the learning factory. Industry 4.0 Science **40**(4), 85–89 (2024)
30. Willskytt, S., Böckin, D., André, H., et al.: Framework for analysing resource-efficient solutions. In: 12th Biennial International Conference on EcoBalance (2016)
31. Mügge, J., Hahn, I., Riedelsheimer, T., et al.: Digital twins for circular economy - enabling decision support for R-strategies. Industrie 4.0 Manage. **38**(6), 42–46 (2022). https://doi.org/10.30844/IM_22-6_42-46
32. Ramsauer, C., Rüdele, K., Wolf, M., et al.: Experiences from 10 years LEAD Factory and a roadmap for the next decade. In: Learning factories of the future, pp. 3–12 (2024). https://doi.org/10.1007/978-3-031-65400-8_1
33. Rüdele, K., Streßler, T., Ramsauer, C.: Circular economy in learning factories using the example of transfer molding. In: Learning factories of the future, pp. 349–357 (2024). https://doi.org/10.1007/978-3-031-65411-4_41

34. Schroth, T., Hummel, V., Schuhmacher, J., et al.: Investigation of AI algorithms for the clustering and combination of pick and stow operations in warehouses and development of a learning module for undergraduates. In: Learning Factories of the Future, pp. 221–229 (2024). https://doi.org/10.1007/978-3-031-65411-4_27

Empowering SMEs for Circularity: A Competency-Based Learning Approach

Benjamin Gellert(✉), Katrin Singer-Coudoux, Ayla Antonia Eser, and Ronald Orth

Fraunhofer Institute for Production Systems and Design Technology, Pascalstr. 8-9, 10587 Berlin, Germany
benjamin.gellert@ipk.fraunhofer.de

Abstract. To achieve true sustainability, manufacturing companies must transition from a linear to a circular economy. This requires well-trained professionals, particularly in product design, where crucial decisions about a product's circularity are made. New or adapted circular competencies are therefore essential. Small and medium-sized enterprises (SMEs), however, often lack the resources to systematically develop and implement training strategies for such socio-ecological transformations. The research project SustainAble addresses this challenge by proposing a competency catalog that enables SMEs to identify the circular economy competencies relevant to their workforce, with an additional focus on circular product design. This paper outlines the development of the catalog. The methodology combines a systematic literature review of circular economy competency models with AI-assisted profiling using JobCraftAI, a generative AI tool developed by the project IT partner. The identified competencies were systematically synthesized, categorized, validated in expert workshops and compiled in the SustainAble Competency Catalog. The result comprises 81 general circular economy competencies and 53 product design-specific competencies grouped into 20 competency categories. This research shows a practical framework for competency development and supports future applications such as adaptive learning and targeted training strategies.

Keywords: circular economy · competency · competency development · product design · SMEs · organizational learning

1 Introduction

The ongoing ecological crisis, most notably the intensifying effects of global warming, has placed environmental, social, and governance considerations at the forefront of corporate responsibility. In this context, transitioning from a linear to a Circular Economy (CE) has emerged as a critical strategy for reducing raw material consumption, minimizing waste generation, and promoting sustainable development [1, 2]. However, this transformation is not merely a technical shift. It demands a workforce equipped with both the motivation and the appropriate competencies to implement circular practices effectively [3].

H. Kohl et al. (Eds.): GCSM 2025, LNME, pp. 439–447, 2026.
https://doi.org/10.1007/978-3-032-21154-5_48

For manufacturing companies, and particularly for small and medium-sized enterprises (SMEs), the transition to circularity presents specific challenges. They often operate with limited resources and lack the necessary strategic frameworks to systematically address training and competency development [4]. As a result, many SMEs struggle to identify the skills and knowledge their workforce needs to support the circular transformation. This applies especially in product design, a phase that has a significant impact on a product's environmental performance throughout its life cycle. Choices there influence which materials are used, how long the product will last, how easily it can be maintained or repaired, and what happens at the end of its use phase [5, 6].

Out of these challenges the following research question emerges: *How can relevant CE competencies be structured to support competency development in SMEs?*

In response, the research project SustainAble proposes a competency catalog SMEs can use to identify the relevant CE competencies for their workforce. Achieving this objective requires the systematic identification and classification of key competencies essential for the implementation of CE practices, both in general and with specific emphasis on the domain of product design. The result is the SustainAble Competency Catalog – a structured framework that consolidates CE competencies from a systematic literature review and AI-generated insights.

2 Foundations of Competency-Based Learning for CE

Competence development is widely understood as the process of expanding the knowledge, skills and capabilities of individuals and teams within a specific context [7]. In contrast to isolated knowledge or skill acquisition, competence refers to an integrated ability to act effectively in complex, real-world situations. According to Boyatzis [8] and Mulder [9], competence includes not only cognitive and practical elements but also attitudes, values and the capacity to apply these in a professional context. The European Commission [10] further distinguishes between knowledge as the theoretical foundation, skill as the ability to apply knowledge, and competence as the proven ability to perform effectively across a variety of personal and professional contexts. While skills are often task-specific, competence involves a broader capacity to act autonomously, adaptively and responsibly, especially when facing novel or unpredictable situations.

To translate this understanding into practical application, organisations increasingly rely on competency models. These structured frameworks define the essential competencies required for success in specific roles or functions [11]. To support comparability and transparency in learning outcomes, many systems adopt formalised competency frameworks. In Germany, the Deutscher Qualifikationsrahmen (DQR) provides a structured reference that links qualifications to eight defined levels based on learning outcomes [12]. At the European level, frameworks such as ESCO (European Skills, Competences, Qualifications and Occupations) provide cross-sectoral competence descriptors that support labour mobility, curriculum design and policy alignment. For projects like SustainAble, they offer a valuable reference for describing competencies in a way that is recognised and comparable across organisational and national boundaries.

Another concept in competency-based learning is the use of defined job roles. Roles serve as anchors that link real-world tasks and responsibilities to specific competencies.

A well-defined role captures what a person is expected to do and makes it possible to identify the knowledge, skills and attitudes needed for effective performance. The role concept also allows for more effective assessment of competence by tying it directly to observable job-related behaviours. As Boyatzis [8] points out, competence becomes meaningful only when applied in the context of specific performance expectations. A practical example of this approach is the role of the circular product designer.

3 Methodology

A state-of-the-art analysis was conducted to explore existing expertise on CE competencies (Fig. 1). The analysis followed a two-way approach:

(1) Systematic Literature Review – This involved a structured analysis of existing CE competency models, focusing both on general frameworks and on those specifically targeting product design. The goal was to assess the current state-of-the-art in literature and synthesize relevant competencies from academic and applied sources.
(2) AI-Generated Competency Profiling – JobcraftAI, developed by project partner Peers Solutions, was used to generate circular competencies for the specific role of a product designer. JobcraftAI is a generative AI-based tool that extracts the necessary skills for any role across industries. It is specifically trained to derive role-specific skillsets and thus offers a novel, data-driven complement to the literature-based analysis.

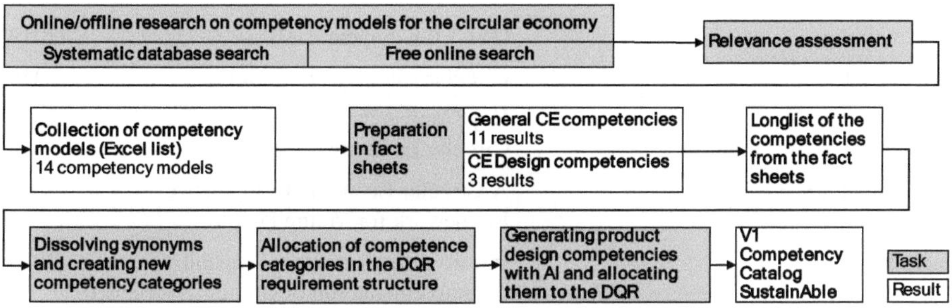

Fig. 1. Literature review process

Following the literature review, expert validation workshops were conducted to refine the results.

3.1 Systematic Literature Review

A comprehensive overview of existing CE competency models was obtained through an extensive review of online and offline sources. The research was divided into a systematic search in Scopus with an additional free online search. The systematic database search identified 178 papers using the search query TITLE-ABS-KEY("circular economy"

AND "skill" OR "circular economy" AND "competence" OR "Competence Model Circular Economy") which have been reduced to 27 following a relevance assessment of the title and abstract according to the research question. For the free online research, the search string "Competence Model Circular Economy" was used both in English and German, which resulted in another 8 papers.

To ensure relevance and quality, the full texts of the collected literature was reviewed according to the inclusion criteria, topicality (2016 and onwards), general description of CE competencies and language (German/English), resulting in a list of 14 competency models (Table 1).

Table 1. Relevant competency models from the literature

Author	Title	Year
Consoli et al	Do green jobs differ from non-green jobs in terms of skills and human capital?	2016
De Wolf et al	ThreeC-Creating competencies for a circular economy	2016
Bsi group	Executive Briefing: BS 8001 – a Guide	2017
los Rios, Irel Carolina de; Charnley, Fiona J.S	Skills and capabilities for a sustainable and circular economy: The changing role of design	2017
Burger et al	The heterogeneous skill-base of circular economy employment	2019
EU Comission	CYCLE Competencies	2019
Zwiers et al	Circular Literacy Framework	2020
Sumter et al	Key Competencies for Design in a Circular Economy	2021
Andorlini et al	A Circular Economy Competence Framework for Young People	2022
Dagilienė et al	Impact of organizational capabilities on the implementation of circularity-oriented activities	2024
Ebner von Eschenbach et al	Bildung für nachhaltige Entwicklung	2024
Lagorio et al	The future of green skills for the manufacturing sector	2024
Lagorio et al	The future of green skills for the manufacturing sector – Product-process Design	2024
Maer-Matei et al	Skills and Jobs in Demand for the Circular Economy	2024

Subsequently, the 14 competency models were compiled into fact sheets. They contain: a description, competencies and/or competency categories used, R-strategies (Refuse, Rethink, Repair, ...) addressed, and the focus in the value creation cycle (design, procurement, ..., end of life). During the research process, it became evident that a significant number of competency models focused specifically on product design. Consequently, the reviewed papers were categorized into two groups: general circular economy competencies and circular product design competencies. In conclusion, 11 competency models were assigned to general circular competencies and 3 to circular product design competencies. Nevertheless, none of the papers offered a comprehensive framework. Each had shortcomings in different areas, such as not integrating general circular and circular product design competencies in one framework, targeting the wrong group or lacking an SME focus, which highlighted the need for a new model.

Coming from the 14 models, a longlist of all competencies mentioned was created, resulting in 117 general circular competencies and 26 competencies for circular product design. By grouping together similar competencies under one synonym, the longlist was shortened and separated into general circular competencies and circular product design competencies. The synonyms, used as headings for the grouped competencies, were then dissolved to create new competency categories. Subsequently, competencies for product design were generated with AI and sorted into the competency categories.

To structure the competencies within the newly developed competency categories, the German Qualification Framework (DQR) was applied. The DQR distinguishes between professional competencies, subdivided into "knowledge" and "skills", and personal competencies, subdivided into "social skills" and "independence" [13]. It is a widely recognized framework for defining competencies and has been used in the context of knowledge and competence management at the IPK. Following the classification of competencies according to the DQR, an initial version (V1) of the SustainAble Competency Catalog was developed, including 93 general circular competencies and 49 competencies for circular product design.

3.2 Validation Workshops

Based on the literature review, a list of competency categories and associated competencies was developed. To validate and refine this preliminary set, three expert workshops were conducted. Each workshop included three senior and one junior researcher with extensive experience in CE and competency management, and the project partner with experience in learning systems and training architecture. In these sessions, each competency was systematically reviewed within its respective category. The list was streamlined by eliminating redundancies, merging overlapping categories, reallocating competencies to more suitable categories, and refining the competency descriptions. The result was the consolidated version (V2) of the SustainAble catalog.

The competencies derived from the literature were initially aligned with the DQR dimensions knowledge, skills, social skills, and independence. However, during the expert discussions, it was proposed to revise these dimensions to better reflect the needs of CE competency development. As a result, an adapted four-dimensional structure was introduced:

- Knowledge: refers to facts, principles and practices acquired through learning and understanding.
- Practical skills: the ability to apply knowledge and use know-how to solve problems, including both cognitive and practical abilities.
- Social skills: the ability to work effectively with others, understand their perspectives, and contribute to shaping social and work environments.
- Personal skills: the ability to act independently and responsibly, reflecting on actions and improving one's capabilities

4 SustainAble Competency Catalog

The SustainAble Competency Catalog contains in total 134 competencies allocated in 20 categories, resulting from the consolidation of literature-based findings and the integration of AI generated product design competencies. The catalog is divided into general CE and circular product design competencies.

4.1 General CE Competencies

The general CE section of the catalog contains 11 competency categories (Fundamentals of Sustainability, Fundamentals of CE, Systems Thinking, Strategic Thinking, Innovative and Creative Thinking, Resilience, Resource Management, Waste Management, Data Management, Regulatory, Business Models), encompassing a total of 81 competencies. All competencies include concise definitions. Table an excerpt from the catalog, showcasing the two exemplary competency categories "Fundamentals of Circular Economy" and "Innovative and Creative Thinking", along with their corresponding specific competencies (Table 2).

Table 2. Excerpt from SustainAble Competency Catalog – General Competencies.

Dimensions (adapted from DQR)	Competency categories		
	Fundamentals of Circular Economy	Innovative and Creative Thinking	…
Knowledge	Principles, Procedures, and Regulations of the CE R-strategies Product life cycle management	Visions of the CE Visions of sustainability	…
Practical Skills	Assessment of product circularity Reduction of product's negative environmental impacts	Design thinking for challenges Innovation and idea management	…
Social Skills	Promoting principles of CE	Spreading motivation for the CE Encouraging creativity	…

(*continued*)

Table 2. *(continued)*

Dimensions (adapted from DQR)	Competency categories		
	Fundamentals of Circular Economy	Innovative and Creative Thinking	…
Personal Skills	Ethics in CE	Identifying opportunities Innovative thinking Solution-oriented mindset Lateral thinking Embracing new leadership	…

4.2 Circular Product Design Competencies

Building on the general framework from Sect. 4.1, this part focuses on competencies essential for enabling CE through product design. As mentioned in Sect. 3, JobCraftAI was utilized to generate additional role-specific competencies. For this purpose, the job profile *"Product Designer Circular Economy"* was created.

The results of JobCraftAI were then compared with literature-based findings. Relevant AI-generated skills not found in the literature were reviewed and added to the catalog. The final product design version of the catalog contains nine categories (Fundamentals of Circular Economy and Sustainability, Product Design and Development, Innovative and Creative Thinking, Usage and Product Life Cycle, Materials, Logistics, Communication, Regulatory Compliance, Business Models), encompassing a total of 53 competencies with concise definitions. Analogous to Sect. 4.1, each competency was classified into the DQR-based dimensions.

5 Conclusion and Outlook

This paper introduces the *SustainAble Competency Catalog*, a structured framework that consolidates and organizes CE competencies for SMEs, with a particular focus on product design. The catalog was developed through a two-way approach: a comprehensive systematic literature review and AI-generated competency profiling using JobCraftAI. This methodology allowed for both the integration of established scientific knowledge and the incorporation of industry-specific insights.

The literature review revealed a broad and diverse landscape of existing competency models in the CE context. The models emphasize not only technical and environmental aspects of CE but also encompass areas such as regulatory knowledge, business models, data management, and key soft skills including communication, resilience, systemic thinking, or creativity. Particularly notable is the presence of dedicated product design competency models in literature, underscoring the central role of design decisions in shaping a product's circularity potential. To provide a comprehensible and practical structure, the catalog uses an adapted version of the DQR, categorizing competencies into knowledge, practical skills, social skills, and personal skills categories. By integrating JobCraftAI into the process, the catalog benefits from a dynamic, role-specific perspective. This AI-generated input enriches the academic findings with current, practice-oriented skill sets relevant to the evolving demands of the CE.

Future steps should consider a comprehensive review with industry professionals to examine whether the relevant practical competencies are sufficiently covered. This process would help to further align the identified competencies with real-world requirements. Additionally, the catalog shall serve as a foundation for mapping existing learning resources to the respective competencies. Over the long term, the catalog is intended to serve as a foundational element within an adaptive learning path generator. When integrated with a self-assessment tool, this system would enable users to evaluate their existing competencies and receive personalized learning pathways accordingly.

6 Disclaimer

The SustainAble Project is funded by: German Federal Ministry for Economic Affairs and Climate Action based on a decision of the German Bundestag.

References

1. Ellen MacArthur Foundation: Towards the circular economy Vol. 1: an economic and business rationale for an accelerated transition (2013). https://www.ellenmacarthurfoundation.org/towards-the-circular-economy-vol-1-an-economic-and-business-rationale-for-an, Accessed 16 Apr 2025
2. Cantzler, J., Creutzig, F., Ayargarnchanakul, E., et al.: Saving resources and the climate? A systematic review of the circular economy and its mitigation potential. Environ. Res. Lett. **15**, 123001 (2020). https://doi.org/10.1088/1748-9326/abbeb7
3. Guerreschi, A., Piras, L., Heck, F.: Barriers to efficient knowledge transfer for a holistic circular economy: insights towards green job developments and training for young professionals. Youth **3**, 553–578 (2023). https://doi.org/10.3390/youth3020038
4. Sharma, N.K., Govindan, K., Lai, K.K., et al.: The transition from linear economy to circular economy for sustainability among SMEs: a study on prospects, impediments, and prerequisites. Bus Strat Env **30**, 1803–1822 (2021). https://doi.org/10.1002/bse.2717
5. los Rios, I.C., de Charnley, F.J.: Skills and capabilities for a sustainable and circular economy: the changing role of design. J. Cleaner Prod. **160**, 109–122 (2017). https://doi.org/10.1016/j.jclepro.2016.10.130
6. den Hollander, M.C., Bakker, C.A., Hultink, E.J.: Product design in a circular economy: development of a typology of key concepts and terms. J. Ind. Ecol. **21**, 517–525 (2017). https://doi.org/10.1111/jiec.12610
7. Becker, G.S.: Human capital: a theoretical and empirical analysis, with special reference to education, 2. ed. Human behavior and social institutions, vol 5. Columbia Univ. Press, New York, NY (1975)
8. Boyatzis, R.: The Competent Manager. A Model For Effective Performance (1982)
9. Mulder, M.: Conceptions of Professional Competence. In: Billett, S., Harteis, C., Gruber, H. (eds.) International Handbook of Research in Professional and Practice-based Learning, pp. 107–137. Springer, Netherlands, Dordrecht (2014)
10. Commission, E., Cedefop, I.C.F., Foundation, E.T.: European Skills, Competences, Qualifications and Occupations (ESCO): Handbook. Publications Office of the European Union, Luxembourg (2019)

11. Gangani, N., McLean, G.N., Braden, R.A.: A competency-based human resource development strategy. Perform. Improv. Q. **19**, 127–139 (2006). https://doi.org/10.1111/j.1937-8327.2006.tb00361.x
12. Bundesministerium für Bildung und Forschung Der DQR. https://www.dqr.de/dqr/de/der-dqr/der-dqr_node.html, Accessed 24 Apr 2025
13. Bundesministerium für Bildung und Forschung DQR-Niveaus. https://www.dqr.de/dqr/de/der-dqr/dqr-niveaus/dqr-niveaus_node.html, Accessed 24 Apr 2025

Student Sessions

Municipal Waste Management in a Circular Economy

Peter Thornton(✉) and Asela K. Kulatunga

University of Exeter, Exeter, UK
pat206@exeter.ac.uk

Abstract. Municipal waste management has become a growing issue worldwide, contributing to pollution and climate change. It's been increasingly affected by a rising population, urbanisation, an increasing consumption of goods and increased resource scarcity. While Municipalities have inadequate and outdated waste infrastructure, and other barriers to contend with. Over the last few years, national Governments have started to introduce plans, strategies and regulations to change the current linear system into a circular system. With targets from the UK Government approaching in 2035, there are some Local Authorities in the UK who are far from the target, including authorities in Devon.

The report aims to identify inefficiencies and barriers in the current linear system at a local level, in Devon, looking at how they could be addressed, and proposing solutions which could be applied locally and holistically to the UK, making the system more circular. It has collected publicly available secondary data from Governments and other sources, using it to analysis and compare the UK to a range of other countries. With the outcomes of this report being recommendations to local authorities and the UK government, with potential solutions which could be implemented to improve the circular economy.

Keywords: Waste Management · Circular Economy · Sustainability · Municipal Solid Waste

1 Introduction and Background

This reports on municipal waste management in a circular economy (CE) and how it could be improved across the UK using a holistic approach. Focusing on the waste processes in Devon County Council (DCC) and its Local Authorities (LAs), having some of the best and worst recycling rates in the UK [1].

The aim is looking at how municipal waste can be recycled and reintegrated into a CE better, across the UK and the LAs. The objectives are to collect and analyse data of the current system in DCC and its LAs; look at current and future governmental policies,

This paper was prepared for presentation during the student sessions of GCSM 2025 and underwent an expedited peer review process, coordinated by all project supervisors from four collaborating universities. The project supervisor was Dr Asela K Kulatunga.

H. Kohl et al. (Eds.): GCSM 2025, LNME, pp. 451–458, 2026.
https://doi.org/10.1007/978-3-032-21154-5_49

identify the existing barriers, the impact of community engagement and technology, and find solutions able to be implemented into LAs current systems.

The goal is to provide insight and recommendations that support the transition, help to improve resource efficiency, reduce impact and align practices with targets.

Global Waste Generation. Currently the world is producing municipal solid waste (MSW), with more 'being created every year, contributing to the triple planetary crisis of climate change, pollution and biodiversity loss.' [2]. Global and governmental organisations, started realising 'actions must be taken urgently to halt waste growth and to shift towards zero waste and circular economy models.' [2]. The world generates over 2 billion tonnes of municipal waste, with it expected to rise by 70% by 2050, with most of the increase coming from developing countries [3].

Waste in UK. Waste in the UK is split up by nation, where they set their own waste policy and strategy. Currently 2+ types of recycled waste is to be collected by the Waste Collection Authority (WCA) under the Environmental Protection Act 1990 [4]. The Waste (England and Wales) Regulations 2011, set out that paper, metal, plastic or glass is collected separately, when certain conditions are applied [4]. Under the Environment Act 2021, waste collections will be standardised across all LAs, however it's not due to take place until 31st March 2026 [4, 5]. Theres other changes which are going to be made [4, 5] and will help England move towards its CE strategy [5].

Waste in Devon. Devon is currently a 2-tiered authority, where the LAs are WCAs and DCC is the Waste Disposal Authority (WDA). There are 2 Unitary Authorities which deliver waste collection and disposal services [6] and are separate from DCC. There is the South West Devon Waste Partnership with the purpose 'to provide a disposal route for residual waste, from the partnership area that avoided landfill and reduced our carbon impact, whilst being reliable and economical.', the solution to this was to use an Energy from Waste (EfW) plant [7].

2 Literature Review

MSW is a growing issue globally, coupled with finite resources and the need to shift from the linear economy to a more sustainable approach. The CE is a system where resources don't become waste and are regenerated, expanding the lifespan of resources, helping to mitigate environmental and other impacts [8]. MSW has varying definitions internationally, however, is typically waste generated by community activities [9].

Principles in literature are around minimising waste, resource use, and creating value out of them at the end of their life [10]. With frameworks being discussed, such as the R strategy and the Ellen MacAuthur Foundation model ('Butterfly model'). The R strategy minimises waste and maximises resources efficiency, however there can be lots of R's able to develop CE framework, increasing the complexity. The 'Butterfly model' aims eliminate waste and pollution, the circularise resources and regenerate nature [11]. The CE has no consensus on its definition and needs objectives to be defined to enable transition [12], having core ideas of reducing waste and maximising resources remain.

Technological advancements could help aid MSW systems, like using Artificial Intelligence (AI), Machine Learning (ML) and Internet of Things (IoT), allowing for improvements in sorting, collection and logistics [13]. Potentially improving the efficiency of waste collections and recycling rates and reduce contamination [13].

The feasibility of implementing the CE, has the potential for job creation, innovation and cost savings [13, 14], allowing for new business models and markets [14]. EfW technologies benefit from converting waste to energy and heat, however, have barriers around costs. CE practices can reduce management costs and earn from recovered resources, allowing for innovation and job creation [13]. The environmental impact of the CE conserves resources, improves soil quality, reduces waste and pollution [13]. Life cycle assessment studies comparing the impact, are difficult to analysis and need surrounding context, with potential to receive different results from different models [15].

Implementing the CE brings many benefits, solutions, barriers and challenges. Current limitations in infrastructure, having been designed for a linear system, need to be reconfigured [16] with some lacking capacity and experts, affecting the efficiency and effectiveness of local strategies [13]. LAs financially struggle, especially with initial costs of EfW plants and recycling facilities, due to limited income [13]. The current linear economy runs at a low cost compared to the CE, making it difficult for business to compete [16]. Behaviour and culture of the public, often lack awareness, understanding and knowledge [13], which has a big effect on reuse and recycling [13]. Policy barriers, such as conflict regulations can have fragmentation and inconsistency hindering the CE [13, 17]. Extended Producer Responsibility (EPR) enforcement has issues around monitoring compliance [17]. Intellectual Property restricts the sharing of information and slows the development of CE [17].

3 Methodology

The methodology uses a mixed method due to combine quantitative analysis of waste management costs and statistics, and qualitive analysis for understanding the policies and community factors. Quantitative analysis identifies trends across Devon, UK and internationally, while qualitive analysis explores how policy, technology and engagement affect the CE. Secondary data was sourced from Governmental reports, DCC, LAs, Eurostat and the World Bank. Some datasets from 2022, may be influenced by factors like COVID-19 and the cost-of-living crisis.

4 Results

Waste in Devon. Across the LAs in Devon, most have a recycling rate above the national average of 42.3% for England, apart from Exeter (29.1%), Plymouth and Torbay [18]. Spatial analysis indicates higher waste generation in the south east of Devon, correlating with the population density, while the composition reveals that non-recycled waste dominates, then composted and recycled [19].

The waste flow in Devon, shows that most waste ends up being either recycled (26.9%), composted (27.6%) or sent to an EfW plant (44%), with 1.4% going to landfill. This suggests a strong reliance on EfW as a primary treatment method, supporting diversion from landfill but limiting the material recovery potential. The majority of this waste processed, stays within the UK, however some does leave going into either Romania, Ireland, Spain, Germany and Malaysia, according to the LAs websites, indicating limited international dependence for waste treatment.

Waste in UK. All nations in the UK according to UK Government data, have had an increasing recycling rate since 2010, with Wales consistently having the best for many years, due to food waste being kept separate from all other types and being sent to anaerobic digestors [20]. With Wales being found 2nd best out of 48 countries in the report [21]. With England, Scotland and Northern Ireland in the top 15 [21]. When completing a comparison between Germany, Devon, European Union (EU)-27, UK, Australia, England, Norway, Sweden, Japan and New Zealand, using 2022 data. It was found that most countries with a higher recycling rate had longstanding policies and strategies, also found in [21]. Germany had the highest recycling rate at 69.2%, with only 0.46% landfill rate (Eurostat). It found that New Zealand had the lowest recycling rate at 10.2% with about 90% going to landfill and Japan had the highest incineration rate just under 80%. The EU and England were found to have similar targets for 2035, and that Australia and New Zealand have similar targets to each other, looking to reduce the amount of organic waste and total waste generated. With all the countries having a form of CE policy or strategy.

There are a couple of overarching policies in the UK and others which have helped improve the recycling rate more directly by encouraging sustainable practices like the Landfill tax [22]. The Landfill Tax was found to have reduced landfill by around 38 million tonnes over 13 years. Some do not provide enough of an incentive to repair, refurbish or reuse products, like the Waste Electrical and Electronic Equipment (WEEE) Regulations 2013. New policies and changes are coming in or recently implemented, like the Simpler Recycling policy and plastic film recycling [23].

Barriers. With many barriers affecting recycling rate and waste management. Policies and regulations are a barrier due inconsistent approaches to recycling [28], with it being addressed by the 'Simpler Recycling' policy [23, 24]. Public participation and behaviour, often lacks awareness and understanding of local waste policies, and contaminated waste [25]. Financial barriers are mostly around limited funds and high operating costs. LAs in the UK are facing increasing financial pressure [26]. Infrastructure barriers are in terms of capacity of recycling and waste facilities. There are technological barriers in terms of the adoption of advanced waste treatment, the increasing complexity, funding required and data challenges [27].

The Cost of Waste. Taking the mean gate fee from [28, 29] per material in 2024 and the waste produced in Devon [19], it was found that recycled materials are 25.2% of the waste but cost 49.5%, and composting is 28.3% of the waste and 6.1% of the cost. This shows that there is more value for money in composting and EfW than recycling and landfill. With LAs spending more on social care, its meant other services get less, such as waste collections [30]. There are ways of making waste economically viable by closing landfill and implementing EfW plants, which have other benefits [31, 32]. There is opportunity for systems like ML to help improve and optimize processes [33]. A rise in product-like-a-service business models, helping improve longevity and recyclability of products, reducing company expenses [34].

Circularity of Waste. There are ways of improving the circularity of waste, especially to waste that's been landfilled or discarded, like WEEE. Done through urban mining, which extracts materials from waste, allowing salvage of finite materials and limits the impact on the environment [35]. Logistics can impact the circularity of waste by creating closed loop supply chains, having forward and reverse flows between producers and consumers, allowing for less waste. Using reverse logistics, can allow for businesses to make additional revenue by reducing raw material consumption, increasing the value of recovered products and reducing waste [36]. Technology can affect the waste in the CE by allowing for improvement in efficiency, quality of the service and sustainability [37]. Combining systems like IoT, AI and ML, allows for better analysis and optimisation of routes and schedules by using algorithms [37]. The impact of the community can have drastic effects on waste management, like in decision making, the implementation of circular infrastructure and education of the community. Policy has an impact on the circularity of waste; however, this area of literature is sparse [38, 39], it's found that behavioural and regulatory policies are favoured to help foster the CE.

Proposed Solutions. The proposed solutions have advantages and disadvantages and may not all suit existing systems. Material processing solutions include advanced sorting and recovery systems, including ways to separate waste more precisely and which cleans contaminated waste, like cleaning off food or removing outer packaging. Collection process solutions could use smart bins to monitor the amount of waste or optimize waste collection and could use bins which separate the waste for users. Others include policy changes, like standardising policies, or providing higher incentive for recycling. Implementing waste partnerships, providing financial incentives to the LAs. Another approach is to create regional companies who manage waste, like the water and sewage companies in the UK.

5 Discussion and Conclusion

The report has looked at waste in Devon, across the UK and internationally, highlighting both the barriers to and opportunities for achieving a CE. The findings reveal wide variation in recycling rates and infrastructure between LAs, mainly due to fragmented policies and the freedom for councils to decide what to recycle. Financial constraints further constrain LAs, limiting their investment in infrastructure and technology. Upcoming reforms, like the Simpler Recycling policy, may help reduce inconsistencies, however the effectiveness will only come clear when fully implemented.

Technological innovations like smart bins, route optimisation and EfW, can offer real potential to reduce costs and environmental impact. To improve the lasting progress this relies upon behavioural change and community engagement, which is needed for building trust and normalising practices. Transitioning to a more CE, requires the shared responsibility of stakeholders and for them to work together.

This report has focused mainly on policies in England, and only partially explored aspects like gate fees and WEEE recycling. Highlighting the need for further research and case studies to identify proven solutions. The UK's transition to more of a CE offers economic and environmental benefits, though this requires public engagement, investment and coherent policy to reduce inefficiencies and lost opportunities.

References

1. Yurday, E.: Which English councils are best and worst at recycling? Top 10 and bottom 10 revealed (2024). https://www.nimblefins.co.uk/. https://www.nimblefins.co.uk/best-cheap-ukhome-insurance/best-worst-councilsengland-recycling. Accessed 5 Dec 2024
2. UNEP. Global Waste Management Outlook 2024 - Beyond an age of waste: Turning rubbish into a resource. [online] United Nations Environment Programme, p.3,8 (2024). https://doi.org/10.59117/20.500.11822/44939
3. Lama, N.: The World has a Waste Problem. Here's How to Fix It. [online] International Finance Corporation (2024). https://www.ifc.org/en/blogs/2024/the-world-has-a-waste-problem. Accessed 27 Mar 2025
4. Smith, L.: Household waste collection in England and Wales (2019). https://commonslibrary.parliament.uk/household-wastecollection-in-england-and-wales/. Accessed 25 Mar 2025
5. DEFRA. Simpler Recycling in England: policy update (2024). https://www.gov.uk/government/publications/simpler-recycling-inengland-policy-update/simpler-recycling-in-england-policy-update#timelines. Accessed 26 Mar 2025
6. DCC. How is local government in the county of Devon currently structured? - Devon County Council (2025). https://www.devon.gov.uk/devolution-and-lgr/how-is-localgovernment-in-the-county-of-devon-currently-structured/. Accessed 27 Mar 2025
7. Plymouth City Council. South West Devon Waste Partnership | PLYMOUTH.GOV.UK (2015). https://www.plymouth.gov.uk/south-west-devon-waste-partnership. Accessed 27 Mar 2025
8. Chareonvong, C., et al.: Implementation and education of circular economy in community solid waste management: a systematic literature review. J. Educ. Learn. **14**(3), 97–97 (2025). https://doi.org/10.5539/jel.v14n3p97
9. Reis, W.F., Barreto, C.G., Guilherme, M.: Circular economy and solid waste management: connections from a bibliometric analysis. Sustainability **15**(22), 15715 (2023). https://doi.org/10.3390/su152215715
10. Daphne, T.: R-strategies for a circular economy (2023). https://www.circularise.com/. https://www.circularise.com/blogs/r-strategies-for-a-circular-economy. Accessed 30 Mar 2025
11. Negrete-Cardoso, M., Rosano-Ortega, G., Álvarez-Aros, E.L., Tavera-Cortés, M.E., Vega-Lebrún, C.A., Sánchez-Ruíz, F.J.: Circular economy strategy and waste management: a bibliometric analysis in its contribution to sustainable development, toward a post-COVID-19 era. Environ. Sci. Pollut. Res. **29**(41), 61729–61746 (2022). https://doi.org/10.1007/s11356-022-18703-3
12. Edmond, R., Fang, P.: Circular economy in municipal solid waste management: innovations and challenges for urban sustainability. J. Environ. Prot. **16**(02), 35–65 (2025). https://doi.org/10.4236/jep.2025.162003
13. Fayal, E., Carmo, V., Fernandes, R.M., Melo, S., de, V.: Applications of digital technologies for overcoming challenges in municipal solid waste reverse logistics: a systematic literature review. Engenharia Sanitaria e Ambiental **29**(3) (2024). https://doi.org/10.1590/s1413-415220240048

14. Brunner, P.H., Morf, L.S.: Waste to energy, indispensable cornerstone for circular economy: a mini-review. Waste Manag. Res. **43**(1) (2024). https://doi.org/10.1177/0734242x241227376
15. Chenavaz, R.Y., Dimitrov, S.: From waste to wealth: policies to promote the circular economy. J. Clean. Prod. **443**(141086), 141086 (2024). https://doi.org/10.1016/j.jclepro.2024.141086
16. WRAP. Extended Producer Responsibility (EPR) (2024). https://www.wrap.ngo/what-we-do/our-services/policy-and-insights/extended-producer-responsibility-EPR. Accessed 2 Apr 2025
17. Akomea-Frimpong, I., et al.: A review of barriers to circular economy implementation in solid waste management. Res. Square (Research Square) (2023). https://doi.org/10.21203/rs.3.rs-2936217/v1
18. DEFRA. Local authority collected waste generation annual results 2023/24 (England and regions) and local authority data annual results 2023/24 (2025). https://assets.publishing.service.gov.uk/media/67ffbbdced87b816085467bf/LA_and_Regional_Spreadsheet_2023-24.ods. Accessed 5 Apr 2025
19. Booth, M.: Waste Performance Statistics 2023–24. [online] Devon Authorities Strategic Waste Committee (2024). https://democracy.devon.gov.uk/documents/s50065/Waste%20Performance%20Statistics%202023-24.pdf. Accessed 5 Apr 2025
20. Welsh Government. New stats: Why is Wales an outlier in UK recycling rates? (2022). https://www.gov.wales/new-stats-why-wales-outlier-uk-recycling-rates. Accessed 9 Apr 2025
21. Eunomia. Global Recycling League Table – Phase One Report – Eunomia Research and Consulting (2024). https://eunomia.eco/reports/global-recycling-league-tablephase-one-report/. Accessed 9 Apr 2025
22. HMRC. Landfill Tax: increase in rates (2024). https://www.gov.uk/government/publications/landfill-tax-rates-for-2025-to-2026/landfill-tax-increase-inrates. Accessed 12 Apr 2025
23. Mottram, L.: Exeter EFW. [PowerPoint presentation]. Exeter EFW. 22 October 2024 (2024)
24. DEFRA. Simpler Recycling in England: policy update (2024). https://www.gov.uk/government/publications/simpler-recycling-inengland-policy-update/simpler-recycling-in-england-policy-update. Accessed 12 Apr 2025
25. Oluwadipe, S., Garelick, H., McCarthy, S., Purchase, D.: A critical review of household recycling barriers in the United Kingdom. Waste Manag. Res.: J. Sustainable Circular Econ. **40**(7) (2021). https://doi.org/10.1177/0734242x211060619
26. Sandford, M., Brien, P.: Why are local authorities going 'bankrupt'? [online] House of Commons Library (2024). https://commonslibrary.parliament.uk/why-are-local-authorities-goingbankrupt/. Accessed 12 Apr 2025
27. Hariyani, D., Hariyani, P., Mishra, S., Sharma, M.K.: Leveraging digital technologies for advancing circular economy practices and enhancing life cycle analysis: a systematic literature review. Waste Manag. Bull. **2**(3) (2024). https://doi.org/10.1016/j.wmb.2024.06.007
28. Zest Recycle. Waste Trading Prices | Essex & London (2024). https://www.zestrecycle.co.uk/prices/. Accessed 6 Mar 2025
29. letsrecycle.com. (n.d.). Prices. https://www.letsrecycle.com/prices/. Accessed 6 Mar 2025
30. Local Government Association. Save local services: How is £1 of council funding spent? | Local Government Association (2024). www.local.gov.uk. https://www.local.gov.uk/about/campaigns/save-local-services/save-local-services-how-ps1-councilfunding-spent. Accessed 13 Apr 2025
31. Kalkanis, K., et al.: Transforming waste to wealth, achieving circular economy. Circular Econ. Sustain. **2** (2022). https://doi.org/10.1007/s43615-022-00225-2
32. Islam, N.F., Gogoi, B., Saikia, R., Yousaf, B., Narayan, M., Sarma, H.: Encouraging circular economy and sustainable environmental practices by addressing waste management and biomass energy production. Reg. Sustain. **5**(4), 100174 (2024). https://doi.org/10.1016/j.regsus.2024.100174

33. Alsabt, R., Alkhaldi, W., Adenle, Y.A., Alshuwaikhat, H.M.: Optimizing waste management strategies through artificial intelligence and machine learning - an economic and environmental impact study. Clean. Waste Syst. **8**, 100158–100158 (2024). https://doi.org/10.1016/j.clwas.2024.100158
34. Aiguobarueghian, I., Adanma, U.M., Ogunbiyi, E.O. Solomon, N.O.: Waste management and circular economy: a review of sustainable practices and economic benefits. World J. Adv. Res. Rev. **22**(2), 1708–1719 (2024). https://doi.org/10.30574/wjarr.2024.22.2.1517
35. Edmond, C.: What is urban mining and how does it help the planet? [online] World Economic Forum (2023). https://www.weforum.org/stories/2023/11/urban-mining-circular-economy-ewaste/. Accessed 13 Feb 2025
36. Berberoglu, Y., Kazancoglu, Y., Sagnak, M.: Circularity assessment of logistics activities for green business performance management. Bus. Strategy Environ. **32**(7) (2023). https://doi.org/10.1002/bse.3390
37. Hess, C., Dragomir, A.G., Doerner, K.F., Vigo, D.: Waste collection routing: a survey on problems and methods. CEJOR **32**(2), 399–434 (2023). https://doi.org/10.1007/s10100-023-00892-y
38. Malek, W., Mortazavi, R., Cialani, C., Nordström, J.: How have waste management policies impacted the flow of municipal waste? An empirical analysis of 14 European countries. Waste Manage. **164**, 84–93 (2023). https://doi.org/10.1016/j.wasman.2023.03.040
39. Losa, R.: Public policies on circular economy: a systematic review. Ecol. Econ. **228**, 108452 (2025). https://doi.org/10.1016/j.ecolecon.2024.108452

Development of a Chatbot for ESG Reporting for Micro, Small, and Medium-Scale Enterprises

Tejas Suresh Bane, Teja Raghu Varma Alluri(✉), and Maxim Mintchev

Technische Universität Berlin, Pascalstr. 8-9, 10587 Berlin, Germany
tejraghuvarma.a@gmail.com

Abstract. With the Implementation of the Corporate Sustainability Reporting directives and the European Sustainability Reporting Standards (ESRS), micro, small, and medium-scale (SME) industries face a dilemma. Increasing demands from partners or clients, who are subject to mandatory reporting standards, have heightened indirect pressure on these businesses to disclose their sustainability practices. However, SMEs find it challenging because they often lack expertise, resources and clarity to what to report and how. In response to this gap, our study presents the development of a user-friendly Chatbot to assist the MSME's in navigating through the Voluntary Sustainability Reporting Standards for the Non-Listed SME's (VSME) metrics, issued by EFRAG. This study examines the importance of sustainability practices and disclosures and explores how enterprises can position themselves competitively by highlighting their sustainable initiatives. The chat-bot, built using the Botpress platform and driven by a natural language processing (NLP) framework, simplifies the complex reporting requirements into conversational guidance. It offers step-by-step support in understanding key performance indicators (KPIs) and aligning with ESG expectations. This paper outlines the regulatory background, identifies the practical obstacles of the SME's face, and how this intelligent conversational tool simplifies access to sustainability reporting and practices. The result reveals a promising out-come for digital sustainability reporting and emphasizes scalable ESG adoption across Europe's MSME's sectors.

Keywords: Chatbot · Voluntary standard for non-listed small- and medium-sized enterprises (VSME) · Environmental Social & Governance (ESG) · CSRD · ESRS · Natural Language Processing (NLP)

1 Introduction

With the urgency of climate change, social responsibility, and corporate accountability, the EU has advanced sustainability policy through the European Green Deal and the Corporate Sustainability Reporting Directive (CSRD), adopted in 2022. The CSRD

This paper was prepared for presentation during the student sessions of GCSM 2025 and underwent an expedited peer review process, coordinated by all project supervisors from four collaborating universities. The project supervisors was Maxim Mintchev.

H. Kohl et al. (Eds.): GCSM 2025, LNME, pp. 459–465, 2026.
https://doi.org/10.1007/978-3-032-21154-5_50

expands mandatory sustainability disclosures to around 10,000 companies, requiring alignment with the European Sustainability Reporting Standards (ESRS) [1], which cover climate, labor, circular economy, biodiversity, and anti-corruption topics. While directly binding for large and listed companies, the directive indirectly impacts micro-, small-, and medium-sized enterprises (SMEs) that operate as suppliers or partners in CSRD-compliant value chains. These SMEs are increasingly required to provide ESG-related data for scope 3 emissions, human rights due diligence, and sustainable procurement, creating a trickle-down compliance effect despite no direct legal obligation [2].

The CSRD directly applies to large companies and listed SMEs in the EU, but its effects extend far beyond this group. Many smaller, non-listed SMEs are pulled in through supply chain requirements, as larger partners need ESG data to meet their own reporting obligations [3]. This may include details on emissions, labor practices, and sourcing, information that many SMEs have never been required to track. For these businesses, the challenge is less about willingness than understanding what is needed and how to gather it.

This core concern goes far beyond simple regulatory deficiencies and includes lack of available resources and specific expertise. Commissioning ESG consultants or creating internal compliance units put a heavy burden on pockets of small business. Thus, a significant digital and information gap remains, preventing widespread ESG integration of SME levels.

To address these issues, a conversational chatbot tool that has been developed to guide SMEs through sustainability reporting filtering complex key performance indicators (KPI's) into sequential prompts and guidance and context aware guidance, thus offering a low-cost, user-friendly solution to improve value chain transparency in line with ESG expectations.

2 Background and Current Approaches to ESG Reporting

In the last two decades, the scope of sustainability reporting has evolved from being voluntary to mandatory. Early frameworks like the Global Reporting Initiative (GRI) set a universally recognized standardized method of reporting impacts on environmental, social, and economic aspects. Investor-oriented standards like the Sustainability Accounting Standards Board (SASB) and the Task Force on Climate-related Financial Disclosures (TCFD) were launched aiming to cover the effect of sustainability factors on financial results [4]. With a view to consolidating this diverse framework, the International Sustainability Standards Board (ISSB) was established in 2021 and later published IFRS S1 and IFRS S2 in 2023 with the goal of preparing an international standard for sustainability and climate-related reporting [5].

In Europe, significant progress has been seen in the regulatory environment. The Corporate Sustainability Reporting Directive (CSRD) [6] requires large companies and publicly traded small and medium-sized companies (SMEs) to report information in line with the European Sustainability Reporting Standards (ESRS), which have been developed by the European Financial Reporting Advisory Group (EFRAG) [7]. The ESRS is based on the concept of double materiality, including the impact of sustainability concerns on the organization and the impact of the organization on society and the environment. The standards are classified under general requirements, general disclosures, and specific guidelines on environmental (E1–E5), social (S1–S4), and governance (G1) considerations.

Although non-listed SMEs are not directly in scope, their involvement in the supply chains of CSRD-reporting companies means they are often asked to provide ESG-related information. To support these businesses, EFRAG released the Voluntary Sustainability Reporting Standard for non-listed SMEs (VSME) in 2024 [8]. VSME offers three modules: the Basic Module, which includes 12 simplified KPIs that can be reported without a formal materiality assessment, the Narrative Module (PAT) which also includes disclosures on policies, actions, and targets, and the Business Partners Module (BP), with additional data requested by clients, lenders, or other stakeholders.

Other EU measures also influence ESG reporting, including the EU Taxonomy Regulation [9], which defines environmentally sustainable activities, and the Sustainable Finance Disclosure Regulation (SFDR) [10], which mandates ESG transparency for financial market participants. Together, these frameworks form a complex regulatory environment in which SMEs must increasingly understand and communicate their sustainability performance.

A large number of small and medium-sized businesses lack the capacity, funds, or knowledge to manage ESG needs on an individual basis. Larger companies often recruit consultants, join associations, or engage in workshops, but these options are expensive and not always accessible to smaller businesses. Normative and EcoVadis are two examples of online ESG platforms that offer scalable solutions [11]. However, they are usually aimed at bigger, more established businesses and may prove prohibitively expensive for small and medium-sized businesses to manage. This gap creates an opportunity for alternative approaches, such as interactive guidance systems. For example, a chatbot trained on the VSME framework could provide real-time, low-barrier support, answering questions, explaining concepts like "Scope 1 emissions" in plain language, and acting as an onboarding companion that bridges EU-level ESG ambitions with SME-level capabilities.

3 Methodology

A chatbot was selected to give SMEs a simple, low-cost way to start ESG reporting, directly addressing the gap of limited expertise and time. A conversational interface reduces complexity by asking one question at a time, explaining terms when needed, and skipping sections that do not apply, helping users reach a usable first draft aligned with VSME. Botpress was chosen instead of other platforms or building a custom system because it provides a ready-made chatbot platform that can be fed with VSME content and configured to follow a defined conversation logic. Its visual flows, nodes, and cards, together with built-in memory and Natural Language Understanding (NLU) engines, allow fast, reliable setup with lower development and maintenance effort. The master flow is organized into modular workflows made of nodes that either collect data or give guidance; simple logic rules route users based on their answers. All inputs are then compiled into a VSME-aligned draft ESG report for further refinement outside the chatbot.

3.1 Requirements Definition (Data Types)

To generate an ESG draft report aligned with the Voluntary Sustainability Reporting Standard for non-listed Micro, Small, and Medium-sized Enterprises (VSME) framework, the chatbot collects a defined set of qualitative and quantitative data. This includes company information, environmental indicators (energy, water, waste, emissions), social indicators (workforce composition and ethics), and governance indicators (policies, structures, compliance) for SMEs as a whole. The data is gathered through structured conversational interactions, ensuring that SMEs can provide relevant inputs in an accessible format [5].

In addition to core ESG categories, the chatbot incorporates two VSME-specific modules: the Policies, Actions, and Targets (PAT) module, which records sustainability policies, implemented actions, and defined goals; and the Business Narrative module, which captures the company's ESG story, strategies, achievements, and future vision. This structured data is compiled into a draft ESG report that SMEs can further refine to meet specific reporting requirements, providing both a foundation for disclosure and a clear framework for continued reporting development.

3.2 Data Structure

Nodes are the basic units of conversational logic in Botpress. A flow begins at a Start Node and ends at an End Node. Between them, the conversation moves through nodes via defined transitions. Each node represents a step that can perform actions and then route to the next step. Nodes have a name and two execution moments, On Enter (what happens when the node is reached) and On Receive (what happens after user input). When no further transition is available, the flow ends between Start and End, three node types structure the dialogue

A Standard node executes an ordered set of cards, such as showing a message, asking a question, setting a variable, running a small code snippet, or checking a condition, and then following its configured transition. An Autonomous node encapsulates a reusable sub-flow (with its own Start/End) and may also run its own cards before or after that sub-flow, which supports modular design for larger conversations. A Trigger node performs immediate navigation to a target node or flow without executing cards. Together, these nodes and their cards enable sequential, transparent logic while keeping the overall conversation modular and easy to adapt to different reporting paths.

The conversation moves along transitions between nodes. Some transitions are unconditional and always taken; others depend on what the company has already shared. Inputs such as energy use, headcount, and industry are saved as scoped variables and reused to tailor later questions, adjust the path, personalize messages, and finally assemble a single draft ESG report. Each module (General Reporting, PAT, Business Narrative) contributes to its own structured variables to that final output.

The built-in Knowledge Base (KB) acts like a quick reference shelf for the bot. It indexes curated materials, including VSME guidance and relevant EU and German regulations, so the bot can answer free form questions like "What is Scope 2?" Without breaking the flow. Lookups can be enabled at specific steps to keep data collection focused while still giving clear, on-demand explanations when needed.

3.3 Conversational Flow Structure

The ESG chatbot's conversation flow, which can be seen in Fig. 1, begins at the Start node (Trigger node), which serves as the entry point to the interaction. From there, the user is directed to the "Greeting" stage (Autonomous node), which not only welcomes the user but also determines the initial direction of the conversation. At this point, the user can choose between two autonomous flows: "Assistance with ESG" or "Overview of ESG". The "Overview of ESG" node provides a concise explanation of ESG concepts and the VSME framework. After delivering the overview, the bot immediately prompts the user to decide whether they are ready to proceed with ESG reporting support. If the user agrees, the flow connects back to the "Assistance with ESG" autonomous module, ensuring the conversation does not end prematurely and maintains a seamless transition from general information to actionable guidance within the "Assistance with ESG" autonomous node, the chatbot guides the user through a structured process. This includes obtaining consent for storing data using Standard node, collecting basic company information using Autonomous node, and assessing the company's ideal reporting framework using Autonomous node. Once this assessment is complete, a Standard node is used to confirm whether the user wishes to proceed with VSME-based reporting. If confirmed, the conversation moves to the VSME Module Selection (Standard node), from which the user can enter any of the three main autonomous modules, Basic Modules, PAT (Policies, Actions, and Targets) Modules, and Business Narrative Modules. Each of these modules runs as an independent sub-flow, collecting and validating relevant inputs before returning to the module selector to allow other modules to be completed in any sequence. Once all required modules are finished, the flow proceeds to the Draft Report Generation autonomous node, where all collected data is compiled into a VSME-aligned ESG report. The conversation concludes at the End node (Trigger node), marking the completion of the reporting process.

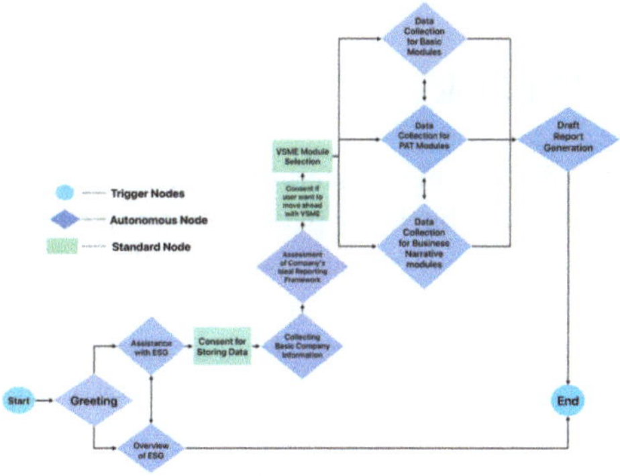

Fig. 1. Conversational Logic Map

4 Results and Discussion

To evaluate the chatbot's performance, simulated sessions were conducted with four fictional SMEs representing the food processing, textile manufacturing, metal fabrication, and electronics assembly sectors. The company profiles and data were based on publicly available SME benchmarks to create realistic testing scenarios. The sessions were performed by the development team, one acting as a trained evaluator to test functionality and another as a first-time user to assess usability. Each company followed the same master conversational flow, with the chatbot adapting its questions to the company's size, sector, and provided responses.

During the simulations, the chatbot successfully collected and organized environmental, social, and governance (ESG) indicators following the VSME framework. Environmental inputs included electricity, water, emissions, and waste; social indicators covered employee numbers, gender ratios, and contract types; and governance data focused on sustainability policies and compliance mechanisms. The PAT and Business Narrative modules captured each company's policies, actions, and sustainability goals, producing a structured and editable draft report for each simulated SME.

Performance was assessed qualitatively based on data accuracy, contextual understanding, report completeness, and user experience. Across all sessions, the chatbot accurately processed all relevant inputs and successfully filtered out non-essential information such as user remarks or unrelated text. When vague responses were entered, the chatbot prompted clarifications instead of assuming data, maintaining data integrity. Manual comparison of chatbot-generated reports with the VSME checklist showed over 95% alignment in content and structure, confirming the system's ability to generate coherent and comprehensive draft reports.

Data security and confidentiality were also considered. The chatbot currently operates on the Botpress Cloud environment, where all collected data are stored temporarily and automatically deleted after report generation. No personal or financial data are retained. For future deployments, options for local hosting or on-premises operations will be explored to address SME concerns regarding sensitive information. Overall, the chatbot demonstrated strong reliability, adaptability, and clarity in guiding SMEs through ESG data collection and report generation, lowering barriers to sustainability reporting and enhancing user understanding of the VSME framework.

5 Conclusion and Outlook

The research addressed the challenges faced by SMEs in meeting growing ESG data demands from CSRD-compliant business partners without the means for complicated compliance procedures. The chatbot created, based on the VSME model, proved effective in simulated SME situations, walking users through applicable KPIs. Clarifying terminology and generating a structured draft ESG summary report. The conversational nature of the tool reduced barriers to entry and enhanced user comprehension, benefits often missing from static templates or long-form guidelines.

Thus, the chatbot provides a practical solution for these companies to get started with ESG reporting, offering a basis for subsequent refinement or incorporation into formal disclosures. Future enhancements would involve automated data uploads, multilingual capability, industry-specific guidance, and optional connections to verification services. As VSME is refined, ongoing updates to the chatbot will ensure its continued relevance. In total, this is a scalable, flexible solution to facilitate SMEs in making their initial organized advances towards sustainability reporting.

References

1. European Commission. Corporate Sustainability Reporting Directive (CSRD). Official Journal of the European Union (2021)
2. European Financial Reporting Advisory Group. Amended ESRS exposure draft 2025 (2024)
3. SME United, ESG data tracking system (2023)
4. Eccles, R., Krzus, M.: Implementing the task force on climate-related financial disclosures recommendations: an assessment of corporate readiness. Schmalenbach Bus. Rev. **71**, 287–293 (2018)
5. European Commission. An SME Strategy for a Sustainable and Digital Europe. European Commission, Brussels (2020)
6. Federal Ministry of Labour and Social Affairs. The German Act on Corporate Due Diligence in Supply Chains (LkSG). BMAS (2021)
7. Hartmann, J., Perego, P.: Mapping the CSR reporting landscape. J. Bus. Ethics **182**(2), 431–448 (2023)
8. The Voluntary standard for non-listed micro-, small- and medium-sized undertakings (VSME), EFRAG (2025)
9. Revell, A., Blackburn, R.: SMEs and sustainability: exploring barriers and motivations. Bus. Strateg. Environ. **31**(2), 729–740 (2022)
10. Schaltegger, S., Hörisch, J., Freeman, R.E.: Business cases for sustainability: a critical analysis. Organ. Environ. **35**(2), 242–269 (2022)
11. How EcoVadis Helps Our Requesting Customers Drive Sustainability With Their Suppliers. EcoVadis (2023). https://resources.ecovadis.com/buyers/how-ecovadis-helps-our-requesting-customers-drive-sustainability-with-their-suppliers. Accessed 03 Sep 2025

Circular Economy Gamification for Industrial Contexts

Caitlin Vinopal, Hasret Sümeyye Uyan, Mohammed Aslam Hajanajubudeen, Felix von Amelunxen(✉), and Valentin Eingartner

Technische Universität Berlin, Pascalstr. 8-9, 10587 Berlin, Germany
felix.von.amelunxen@tu-berlin.de

Abstract. Current linear take-make-waste manufacturing models have been shown to be unsustainable long term [1], leading to significant waste, depletion of water bodies, and reduction in ecosystems. This has led to the adoption of strategies, such as Circular Economy (CE), to ensure both the availability of resources and financial viability for the future. The CE-Hub Berlin aims to address the discrepancy in CE competency, particularly within Berlin's industrial infrastructure and its predominantly Small and Medium-sized Enterprises (SMEs), where implementation is constrained by limited stakeholder expertise. This project, developed in collaboration with CE-Hub Berlin, addresses these barriers by creating an interactive, web-based game designed to transfer knowledge from an industrial and manufacturing perspective. Increasing interest and knowledge in circularity will increase implementation, creating more sustainable and robust manufacturing for the future.

Keywords: Circular Economy · Competence Building · Gamification

1 Introduction

To ensure the availability of resources for future generations while maintaining current living standards, a significant reduction in resource consumption is required. Industrial activity plays a central role in environmental degradation, resource depletion, and waste generation. The Circular Economy (CE) offers a promising pathway for sustainable transformation by decoupling economic growth from resource use. It aims to extend product and service lifecycles, minimize waste and pollution, and regenerate natural systems [2]. Despite technological progress and growing societal awareness, the implementation of CE strategies remains challenging. Common barriers include structural incompatibilities with existing systems, stakeholder resistance, and limited competencies in CE-related practices [3]. Gamification – the use of game elements in non-game contexts – has emerged as an effective tool to address these challenges by fostering engagement and facilitating competency development.

This paper was prepared for presentation during the student sessions of GCSM 2025 and underwent an expedited peer review process, coordinated by all project supervisors from four collaborating universities. The project supervisors were Felix von Amelunxen and Valentin Eingartner.

H. Kohl et al. (Eds.): GCSM 2025, LNME, pp. 466–473, 2026.
https://doi.org/10.1007/978-3-032-21154-5_51

This academic project seeks to close the CE competence gap among current and future professionals by promoting systems thinking and active learning through gamified approaches.

2 Theoretical Background

Within CE there are many R-frameworks discussed and one of the extended versions called the 9R-framework [4] describes key strategic and technical processes. The first 3 strategies (R0 Refuse, R1 Rethink, R2 Reduce) are focused on smarter product use and optimization of resources in manufacturing; refusing features, products or materials, rethinking product use, and reducing waste in manufacturing. Next are the 5 strategies focused on extending the lifespan of the products. Those are R3 Reuse a product for its original function, R4 Repair and maintenance or broken part, R5 Refurbish an old product, R6 Remanufacture old components to as-new condition to be used in new products, and R7 Repurpose old products for a new function. Finally, the last 2 strategies relate to useful application of materials; R8 Recycle materials from old products for new products and R9 Recover energy from old materials. Closed-loop systems, like those using the R-frameworks, are a fundamental step toward circularity [5]. They require both systematic approaches as well as technological support, and increasing CE competence is a first step to understanding the necessary steps for transition [6].

2.1 CE Competency Development in Industrial Contexts

Especially for SMEs, competency development is an important factor in implementing CE. One approach adopted in industrial contexts is a structured program with financial incentives to assist SMEs [7]. The program analyzes current capabilities, generates projects to address real-world problems and facilitates peer networks for collaboration. Knowledge management is necessary within the organization to ensure knowledge creation, transfer and retention processes optimize CE knowledge and practices [8]. To guarantee skill acquisition, it is important to integrate hands-on training modules, for example in eco-design, remanufacturing and life cycle assessment, with technical and vocational training [9].

2.2 Gamification Supported Learning

Gamification influences psychological mechanisms of learners to deliver effective learning and behavioral changes. Gamified elements activate intrinsic and extrinsic motivation to increase commitment to learning and knowledge retention [10]. Intrinsically motivating elements include visually appealing graphics, storylines, and overcoming challenges, while extrinsic motivators include points, badges, leaderboards, competitions and external social factors. Learning can be greatly affected and motivated by the influence of peer users [10]; [11]. A game context, particularly simulations, also enables critical thinking and problem-solving skills by engaging users both emotionally and cognitively [3]. Including elements improves knowledge creation, retention, and recall, increases learner motivation and keeps learners engaged.

Some games were of particular inspiration for the project. Risk and Race [3] is a business simulation game with decision making and resource allocation that leads to branching outcomes. Recyclebank [11] is an eco-action game to reward learning and real-world recycling actions with points and feedback. CodeCombat [12] is a coding teaching game in which typing actions of a video game character in computer code (python, javascript, etc.) leads to improved knowledge mastery compared to traditional methods. Finally, Quizlet [13] teaches material through flash cards, timed matching and other competitive games, leading to improved motivation and knowledge retention. Although various gamification applications exist, they do not focus on CE from a manufacturing point of view. To fill this gap another option is required.

3 Development of CE Learning Game

3.1 Objectives

This project targets professionals working in industrial and manufacturing contexts, particularly those seeking accessible learning opportunities during short breaks or in between tasks. Accordingly, the game is designed to be engaging, intuitive, and completable within 5–10 min, while still delivering meaningful educational value. The primary learning objectives are as follows:

1. **Understanding CE Principles and the 9R Framework.** Introducing foundational concepts of circularity and demonstrating their practical relevance.
2. **Recognizing Circular Value Chains.** Visualizing material flows across different value chain stages to promote systems-level awareness.
3. **Developing Systemic Thinking.** Encouraging players to consider interdependencies, rebound effects, and long-term impacts of their decisions.
4. **Fostering Interactive Learning.** Enhancing knowledge retention through active, decision-based participation.

In addition to these pedagogical aims, several design objectives were defined based on gamification literature, feasibility constraints, and initial user needs. These include ensuring scalability, encouraging replayability, minimizing development time, and maintaining an engaging balance between play duration and educational content.

3.2 Methodological Analysis

Following the definition of learning and design objectives, the development team generated a list of 11 potential game concepts aimed at fostering CE competencies. These ideas varied in complexity and format, ranging from crossword puzzles on CE principles and technologies, to video-based modules with integrated quizzes, and long-term scenario games simulating strategic decisions over a 5–30 year period in a factory setting. For each concept, the team outlined core game mechanics, potential gamification elements, and assessed development feasibility within the available time frame. The concepts were then evaluated against the predefined objectives and scored accordingly. The four highest-ranking ideas were selected for further development, including comparative analysis of similar games and refinement of core gameplay mechanics. These four shortlisted concepts included: A matching game linking R-strategies to their definitions and applications, a decision-based scenario game requiring selection of the most appropriate R-strategy, a competitive multiple-choice quiz, and a word-guessing game (e.g., similar to Hangman) using CE-related terminology and clues.

The final game design emerged from a combination of these elements, resulting in a scenario-based game structured around multiple-choice decisions. Selected choices are explained through pop-up definitions, and word-guessing mini-games are integrated into key lifecycle decisions. The game also includes a scoring system and leaderboard to enhance engagement and replayability.

3.3 Building the Game

To structure the gameplay, a decision tree was developed (see Fig. 1) outlining the ten planned stages and the corresponding decision pathways. This structure enables interactive mechanics where earlier choices influence the availability of later options. In the interface, unlocked choices are highlighted in green, while options restricted by previous decisions appear in orange, illustrating dependencies across the product lifecycle. Each decision was assigned both a sustainability score and a profitability score, highlighting the interplay between ecological and economic considerations and reinforcing that these aspects are not mutually exclusive. The scoring system is based on the following aggregated metrics:

$$\begin{aligned} \textit{Sustainability} &= \textit{Material Usage} + \textit{Energy Usage} + \textit{Water Usage} + \\ &\textit{Waste Created} + \textit{Product Lifetime} + \textit{Carbon Footprint} \end{aligned} \tag{1}$$

$$\textit{Profitability} = \textit{Waste Created} + \textit{Processing Time} + \textit{Cost} + \textit{Profit} \tag{2}$$

Scores were assigned on a -5 to + 5 scale for each metric with the aim to collect more points. Minigame scores were added to a bonus score based on the number of attempts left out of 3. The scoring system is created with respect to both the balancing of the game and reality, which makes the assigned specific scores for every decision out of scale. Updates will continue to be made to provide more founded values.

For example, selecting *Planned Obsolescence* results in a negative sustainability score due to increased material use, waste generation, and reduced product lifetime. Despite quicker processing time and potentially higher profits through repeat purchases, these gains are offset by higher resource consumption, leading to a neutral profitability score overall. In contrast, the decision to *Use Remanufactured Components*, which can preserve up to 65% of original car parts [14], yields clear sustainability benefits through reductions in material, energy, and water use, as well as lower emissions. However, it requires more time for sorting, repair, and testing, reflecting a moderate trade-off in processing efficiency.

To support development, a range of digital tools was used. Version control and collaboration were managed via GitHub, while AI tools such as Claude AI and DeepSeek R1 supported the generation of code snippets and integration of gameplay mechanics. AI assistance was particularly useful for implementing game logic and visuals; however, all outputs were reviewed and adjusted to ensure consistency with the project's educational structure and scoring framework.

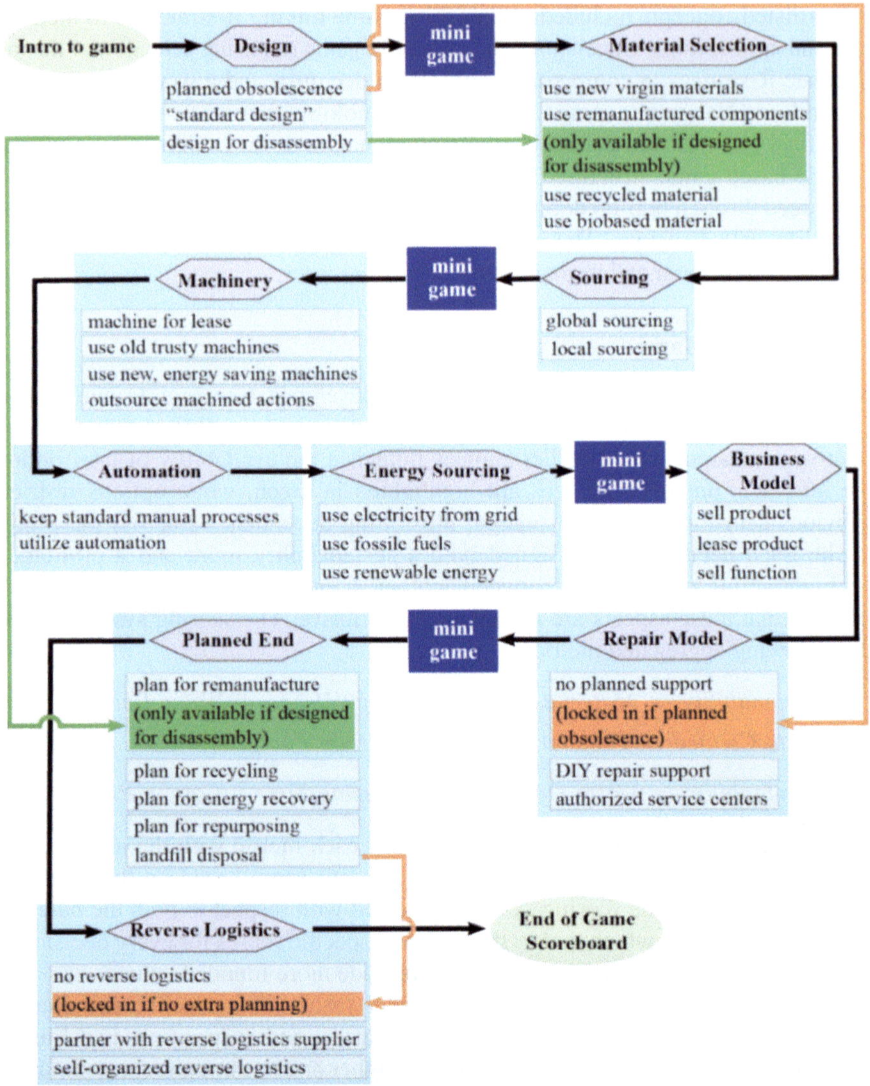

Fig. 1. Decision Tree of Game

4 Results and Discussion

The project resulted in the successful development of a web-based gamified learning tool designed to educate users on CE principles within an industrial context. Players assume the role of an automotive manufacturer based in Berlin and are tasked with making 10 strategic decisions across a product's lifecycle - from design, production, use, and end of life phases (see Fig. 1). Each decision influences real-time sustainability and profitability scores, visualized through a live tracker and animated feedback. To support the learning objectives, the game integrates several key features and gamification elements:

Decision-Based Scenarios. At each of the ten stages, users are presented with predefined decision paths that impact the storyline and score. This structure simulates real-world trade-offs and maintains player engagement through narrative continuity.

Real-Time Feedback. A dynamic score tracker displays changes in sustainability, profitability, and bonus points, offering immediate feedback to reinforce learning and highlight the consequences of each decision.

Option Disabling Logic. Certain choices influence the availability of future options, illustrating lifecycle dependencies. For instance, selecting "Planned Obsolescence" during product design disables later repair choices, while opting for "Landfill Disposal" eliminates reverse logistics options.

Mini-Games. At the end of each stage, word-guessing mini-games provide bonus points based on the number of attempts taken. These interactive elements reinforce specific CE concepts (e.g., renewable energy use) while enhancing engagement, especially around lesser-known topics.

Badging System. Upon completing the game, users receive a badge and title—such as "Circular Champion" or "Balanced Strategist"—based on their cumulative performance. This encourages reflection on their strategic approach and learning outcomes.

Leaderboard. Players can compare their final scores across different playthroughs via a leaderboard, fostering a sense of competition and promoting replayability.

Decision Summary. A comprehensive summary screen allows users to review all choices made during the game along with their respective impacts, supporting post-play reflection and consolidation of knowledge.

4.1 Feedback and Limitations

Initial feedback from academic advisors emphasized the need for greater clarity, including the definition of key terms, improved decision explanations, and an increased number of decision points within the mini-games. These suggestions were implemented to enhance the game's educational impact.

Beta testers appreciated the informative content and found the game engaging, though some perceived it as too long. Suggestions included interface improvements for mobile compatibility and enhanced visual design. However, confusion persisted regarding the consequences of certain decisions, with testers seeking clearer feedback and comparative explanations. While some ambiguity was introduced intentionally to encourage replayability, additional refinements are needed to address these concerns.

A key limitation of this project is that the game has not yet been tested with its intended target group – industrial SMEs – due to time constraints. While gamification effectively supports the communication of CE fundamentals, its educational impact may diminish with increasing task complexity and required expertise. As motivation and learning outcomes are highly context-dependent, further research is needed to evaluate long-term effects. Future iterations should therefore include pilot studies with the target audience to validate effectiveness and usability.

5 Conclusion and Future Work

The continued dominance of linear "take-make-waste" production models underscores the need for accessible and engaging tools that promote CE thinking. This project responds by developing a gamified learning tool designed to raise awareness and build CE competencies, particularly within industrial contexts.

Future development will focus on improving clarity around decision outcomes, integrating optional in-depth learning features, and validating the tool with SME employees. Sector-specific adaptations (e.g., textiles, electronics), mobile compatibility, and integration into the CE-Hub Berlin platform will further enhance its accessibility and relevance. By refining this approach, the project contributes to fostering CE expertise and supporting the broader adoption of sustainable production practices.

References

1. Ellen MacArthur Foundation, What is the linear economy? https://www.ellenmacarthurfoundation.org/what-is-the-linear-economy. Accessed 17 Aug 2025
2. Ellen Macarthur Foundation. Circular economy introduction. https://www.ellenmacarthurfoundation.org/topics/circular-economy-introduction/overview
3. Manshoven, S., Gillabel, J.: Learning through play: a serious game as a tool to support circular economy education and business model innovation. Sustainability **13**(23), 13277 (2021). https://doi.org/10.3390/su132313277
4. José, P., Hekkert, M.P., Ernst, W., Aldert, H.: Circular economy: measuring innovation in the product chain (2017)
5. Kara, S., Hauschild, M., Sutherland, J., McAloone, T.: Closed-loop systems to circular economy: a pathway to environmental sustainability? CIRP J. Manuf. Sci. Technol., Advance online publication (2022). https://doi.org/10.1016/j.cirp.2022.05.008
6. Sakao, T., Bocken, N., Nasr, N., Umeda, Y.: Implementing circular economy activities in manufacturing for environmental sustainability. CIRP J. Manuf. Sci. Technol. (n.d.). https://doi.org/10.1016/j.cirp.2024.06.002
7. Maher, R., Yarnold, J., Pushpamali, N.N.C.: Circular economy 4 business: a program and framework for small-to-medium enterprises (SMEs) with three case studies. J. Clean. Prod., 137114 (2023). https://doi.org/10.1016/j.jclepro.2023.137114
8. Scipioni, S., Russ, M., Niccolini, F.: From barriers to enablers: the role of organizational learning in transitioning SMEs into the circular economy. Sustainability **13**(3), 1021 (2021). https://doi.org/10.3390/su13031021
9. Ab Hamid, N., Mohd Kamaruzaman, F., Rasul, M.S.: Concept of circular economy in technical and vocational education: a systematic literature review. Int. J. Educ. Res. Rev. **13**(4) (2023). https://doi.org/10.11591/ijere.v13i4.28125
10. Cabot, A.G., Lopez, E.G., Alvaro, S.C., Martinez, J.M.G., de-Marcos, L.: Measuring the effects on learning performance and engagement with a gamified social platform in an MSc program. Comput. Appl. Eng. Educ. (2019). https://doi.org/10.1002/cae.22186
11. Ouariachi, T., Li, C., Elving, W.J.L.: Gamification approaches for education and engagement on pro-environmental behaviors: searching for best practices. Sustainability **12**(11), 4565 (2020). https://doi.org/10.3390/su12114565
12. CodeCombat: Learn to code by playing a game. (n.d.). CodeCombat. https://codecombat.com/

13. Quizlet: Study tools & learning resources for students and teachers | Quizlet. (n.d.). Quizlet. https://quizlet.com/
14. Geist, H., Balle, F.: Remanufactured products, components, and their materials: a circularity engineering focused empirical status quo analysis. Sustain. Prod. Consumption **45**, 525–537 (2024). https://doi.org/10.1016/j.spc.2024.02.003

Closing the Loop: Recovering Value from End-of-Use Polyurethane Screens for Circular Economy

Tamm Omar[1,2(✉)], Dhanraj Kharal[1,2], Mahsen Salah Al-Ani[1,2], Fazleena Badurdeen[1,2], and I. S. Jawahir[1,2]

[1] Department of Mechanical and Aerospace Engineering, University of Kentucky, Lexington, KY, USA
tamm.omar@uky.edu
[2] Institute for Sustainable Manufacturing (ISM), University of Kentucky, Lexington, KY, USA

Abstract. Polyurethane (PU) is a widely used polymer in mining and aggregation industries due to its excellent mechanical and chemical properties. In the United States alone, thousands of kilotons of PU enter waste streams each year, yet only a small percentage is recycled and returned to the market, raising critical environmental and human health concerns. One US-based mining company, for example, discards approximately 6,000 end-of-use (EoU) PU screening panels annually, totaling around 50 tons of landfill waste. Despite its functional efficacy, these panels are discarded at the EoU generating enormous industrial waste as PU cannot be remelted or recast. To address this issue, this paper explores circular strategies for repurposing these discarded PU panels to extend their life cycle and reduce waste. It offers insight into how EoU products from one industry can be effectively adapted for applications in another sector, supported by relevant analysis. Ultimately, this work supports the circular economy (CE) by mitigating environmental impacts, preserving resources, and proposing an economically viable repurposing business model.

Keywords: Polyurethane · Circular Economy · End-of-Use (EoU) · Value Recovery

1 Introduction

The mining and manufacturing sectors are among the leading contributors to global industrial waste, with a substantial portion of end-of-use (EoU) products ending up in landfills. One notable example is Methylene diphenyl diisocyanate (MDI) based Polyurethane (PU) screening panels used to line mining shafts/chutes routinely discarded at EoU despite retaining many of their high-value mechanical properties. The

This paper was prepared for presentation during the student sessions of GCSM 2025 and underwent an expedited peer review process, coordinated by all project supervisors from four collaborating universities. The project supervisor was Fazleena Badurdeen.

H. Kohl et al. (Eds.): GCSM 2025, LNME, pp. 474–482, 2026.
https://doi.org/10.1007/978-3-032-21154-5_52

US alone consumes around 3700 kilo tons of PU annually for manufacturing different products [1]. One manufacturer report selling over 125,000 panels/year, all of which ultimately end up in landfills, amounting to over 1100 tons/yr of waste (based on the manufacturer's statement; name withheld due to confidentiality) (see Fig. 1). Globally, the same company collaborates with other brands to produce more than 1 million panels/yr, landfilling over 10,000 tons of panels that may have other potential applications. This represents not only a substantial economic loss but also an unsustainable waste stream.

(a) New Panels (b) Use of Panels in Mines (c) Landfilling Panels

Fig. 1. Panels with current life cycle stages

While recycling is often the most common solution companies pursue, it is important to recognize other higher value recovery strategies. The circular economy (CE) provides a broader framework aimed at transforming the traditional linear "take-make-dispose" model into a more sustainable circular system. It is grounded in three core principles: eliminating waste and pollution, circulating products and materials, and regenerating natural systems [2]. To support these principles, CE emphasizes different R's strategies such as reuse (with/without repair for same/other applications), remanufacture and recycle applied in a hierarchical manner based on their resources and energy efficiency, and economic value creation [3]. This approach ultimately enables more sustainable value retention across product life cycles. PU panels, while not biodegradable, retain their structural integrity far beyond its operational use. Reusing it in secondary applications, rather than landfilling, offers an opportunity to reduce environmental burdens and extend material value [3].

This paper investigates the EoU value recovery options for PU panels considering the case of a copper mine where panels are routinely discarded. To explore a sustainable reuse pathway, a pilot project was conducted, where the discarded panels were utilized as ground reinforcement elements. The economic and environmental performance, functionality, and long-term impact of the reuse application was evaluated to demonstrate how they can find renewed purpose in different industrial sectors. The study reveals a scalable model for EoU PU panel waste reduction that benefits both industries and the environment.

2 Background

PU panels are extensively used in mining operations for separation and filtration of raw materials. The panels examined in this study are produced through open casting, a process in which pre-polymer mixtures are poured into heated molds, then cured to form highly abrasion-resistant screens [4]. The panels also have steel inserts to enhance stiffness and enable modular installation. They are engineered for high wear resistance to withstand ore abrasiveness, vibration, and throughput.

Traditional EoU strategies for plastics typically focus on recycling and energy recovery. The two main mechanical recycling pathways, shredding and grinding, are inefficient for the PU panels studied due to the material's crosslinked structure and the presence of embedded materials like steel reinforcements [5]. While these approaches can reclaim material in powder form, according to the manufacturer less than 10% can be reused in new panels as a filler, with limited mechanical benefit. Recent research in polymer sustainability emphasizes the need for higher-order recovery strategies, which preserves more of the material's embodied energy and functionality [6]. The panels cannot be remanufactured or refurbished, as doing so would compromise the material integrity required; they can also no longer be reused for the original application due to the loss of functionality. However, they could potentially be repurposed for alternative uses after evaluating feasibility and other relevant factors. Deploying such strategies is central for CE, which aims to "close the loop" by designing out waste and recovering value from EoU products [2].

Despite the extensive use of PU panels in mining, there is limited published literature on repurposing. Most scholarly work on PU waste focuses on foams, automotive parts, or construction waste [7], highlighting a critical gap this study seeks to address.

3 Research Methodology

The three-phase methodology followed (Fig. 2) focused on panel EoU research, evaluation, and selection of potential reuse applications and validation.

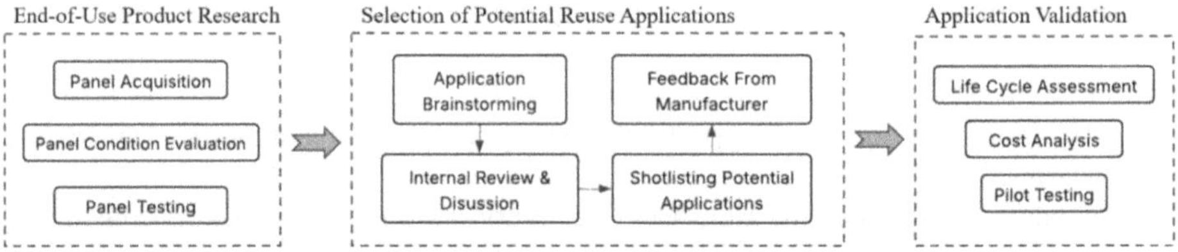

Fig. 2. Research Methodology

3.1 End-of-Use Product Research

EoU PU panels sourced from a copper mine located in Arizona were evaluated for size, shape, and physical condition by visual inspection for any signs of surface damage, deformation or excessive wear. A Shamir Advanced Technological Engineering Company (SATEC) Universal Testing Machine (UTM) was then used to conduct a three-point bending test to measure load-bearing capacity. This test evaluates the panel's structural performance to identify repurposing applications feasible within its load-carrying capacity. It also indicates the rate of deterioration between the new and old panels. In addition, several other material properties, such as its deterioration under prolonged exposure to sunlight, corrosion resistance, hardness, surface slipperiness when wet, weight, aesthetic appearance, and non-biodegradability, were also considered.

3.2 Selection of Potential Reuse Applications

The EoU panels cannot be reused for the same application (i.e., in mines) for reasons described previously. As such, this phase focused on identifying other applications in which they can be used (i.e., for a different purpose, sometimes referred to as repurposing). During the initial phase of selecting an application for panel reuse, fourteen concepts were generated through brainstorming and preliminary market research. After evaluating each idea for feasibility, practicality, and demand, the list was refined to nine promising options. The shortlisted potential applications were evaluated based on required criteria such as material properties, mechanical strength, cost, etc., and a multi-criteria decision matrix was then used to eliminate non-feasible applications. Finally, two applications emerged as most viable: geogrids and anti-fatigue mats. Reusing as geogrids was chosen due to ease of implementation, and strong potential for successful adoption.

Geogrids are widely used in construction, agriculture, and landscaping to reinforce soil, improve load distribution, and prevent erosion. Made from polymers like high density polyethylene (HDPE) or polypropylene, geogrids are designed with a grid-like structure for soil interlock, enhancing the structural integrity of surfaces such as roads, paddocks, retaining walls, and foundations. The PU panels share many key mechanical qualities of traditional geogrids, including high compressive and tensile strength, abrasion and chemical resistance, and grid-like geometry. The rigidity, modularity and resilience of PU panels make them a compelling substitute in applications where traditional geogrids are used.

3.3 Application Validation

EoU PU panels can serve as an alternative to conventional HDPE geogrids on farms. For example, on horse farms, they can help prevent soil displacement resulting from horses digging and trampling. Horse farms regularly face infrastructure challenges, such as keeping paddocks dry under heavy use, preventing mud and erosion in high-traffic zones, and ensuring stable footing that protects horse health. These issues often necessitate durable ground reinforcement solutions that are both cost-effective and sustainable. To assess the feasibility of the reuse application, two discarded panels were installed as geogrids in a high horse traffic area of the farm (see Fig. 3).

A comparative Life Cycle Analysis (LCA), a standards-based methodology for evaluating the environmental impacts throughout the entire life cycle of a product, process, or system [8], of EoU PU panel reuse vs. using new geogrids in horse farms was conducted to assess the environmental impacts of the two applications. Additionally, a cost analysis was conducted to evaluate the potential for establishing a viable business model.

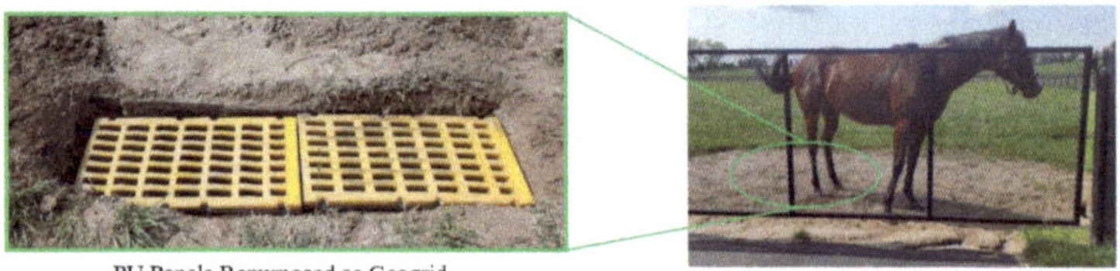

Fig. 3. EoU screening panels installed as geogrid mats

4 Results and Discussion

The results of the deflection testing demonstrate that the EoU panel retains a substantial portion of its original stiffness and load-bearing capacity (see Fig. 4). The initial slope of the EoU panel (orange) is slightly lower which suggests a minor reduction in stiffness or modulus. However, this difference is minimal and may also be influenced by experimental variability. Both panels exhibit comparable performance in the bending test, reaching approximately 1050 N at the yield point. This further confirms that the structural integrity of the used panel remains largely intact. These findings support the decision to pursue repurposing over recycling, as the EoU panels maintain sufficient functionality for reuse without requiring reprocessing.

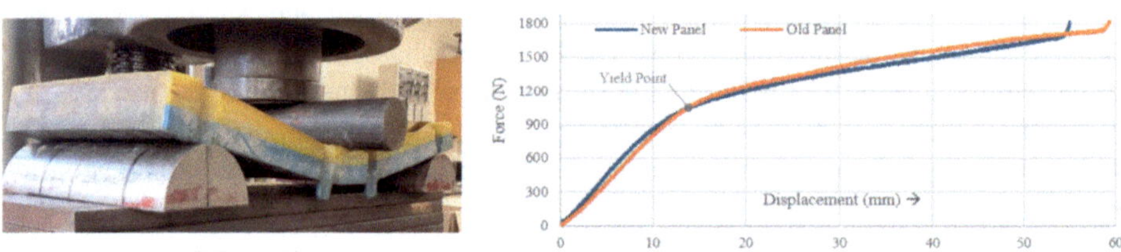

(a) SATEC Machine (b) Results

Fig. 4. Deflection testing results for a new and a used panel

The primary factors that were taken into consideration in establishing the cost model were the expense of loading/unloading EoU panels onto a truck, transportation, and long-term storage. It was assumed that no initial investment cost would be incurred. Equation-1 below illustrates the cost model for repurposed EoU PU panels.

$$\frac{\$Cost}{panel} = \frac{4\left(\frac{PT}{60}\right)L + MD + \left(\frac{2P}{V}\right)C}{P} \tag{1}$$

where, P = number of panels, T = loading/unloading time per panel (min), L = truck loaders rate ($/hr), M = transport rate ($/mile), D = transportation distance (miles), V = number of panels that can be stored vertically in a warehouse, and C = storage cost ($/sq ft per month).

This cost model (Table 1) can be used to identify the minimum selling price of EoU PU panels for use in horse farms as a replacement for commercially available new geogrids. It was assumed that panels are transported using a standard semi-truck

(TF1 Dry Van Trailer) [9] from Arizona mine to Kentucky horse farm, that can carry around 1500 panels. The geogrids currently being used on the farm cost $21.92 per 2 sq ft. covering a 50 m^2 would require about 269 units and cost approximately $5900. If the EoU panels, each 2 sq ft., are sold at $3.91, the total cost drops to about $1050, saving nearly $4850. Even at a mid-range price of $10 per unit (i.e. $2690), farms can still save more than half the cost compared to using conventional geogrids. This pricing ensures manufacturer profitability while offering a cost effective and competitive solution, especially at large scale.

Table 1. Cost analysis for panel reuse [10, 11]

Description		$/panel
Direct	EOU panels loading at mine (@$18/hr for 1 min)	$0.30
	Transportation to manufacturer (@$2.19/mile for 1500 panels, 1790 miles)	$2.61
	Unloading (storage) & Reloading (upon order) (@$18/hr for 2 min)	$0.60
	Transportation to customer (@$2.19/mile for 1500 panels, 10.7 miles)	$0.02
	Final unloading at site (@$18/hr for 1 min)	$0.30
Indirect	Panels stocking cost (@$126/month to store 1500 panels)	$0.08
Total cost:		$3.91

The comparative LCA between EoU PU panels and geogrids currently used by the farm was conducted to estimate the Global Warming Potential (GWP), ozone depletion (OZ), human and freshwater toxicity, and acidification. openLCA 2.4.0 with the Ecoinvent database was used to conduct the LCA. Material and production data for each system were extracted from Ecoinvent. For PU panels, product physical information and transportation data were provided by its manufacturer, while, for the geogrid, standard product data (material, size, shape, weight, etc.,) were obtained by publicly available information provided by its manufacturer [12]. The LCA was performed considering appropriate assumptions where necessary and special attention was given to avoid biasing in the comparison between the two product systems. Reinforced land area of 50 m^2 was chosen as the functional unit for the study, to ensure consistency in the analysis. This requires the use of about 200 geogrid panels and 269 PU panels. The study was conducted only considering the United States markets, assuming a 15-year product life cycle for both systems. The system boundaries used are shown in Table 2.

Since this study only considers the EoU PU panels, it does not account for the environmental impact of producing the PU panels for the first life. Instead, it focuses only on the impacts of reusing these panels for the new application, comparing the environmental impacts avoided by replacing new geogrids. The LCA results are summarized in Table 3.

Two case scenarios were examined to assess potential reuse applications, with the aim of determining the circumstances under which reusing EoU panels offers environmental benefits. As seen from the results, reusing EoU PU panels as geogrid panels offer substantial environmental benefits. However, it is highly influenced by the impacts

Table 2. System Boundary for LCA

EoU PU Panels		Geogrid	
EoU to Gate	Recovery of discarded PU panels	**Cradle to Gate**	Recycled material recovery and processing
	Transportation to distribution center		Manufacturing and packaging production
	Machining and cleaning		Packaging and transportation
	Delivery to end user		Delivery to end user

Table 3. LCA Comparison Results

Impact categories	Unit	Geogrid Panels	Reuse of EOU PU Panels	
			(Arizona - Kentucky)	(Arizona - California)
Climate change	kg CO2-Eq	**670.44**	**1369.43**	**428.77**
Ozone depletion	kg CFC-11-Eq	8.26E-06	2.09E-05	6.40E-06
Ecotoxicity: freshwater	CTUe	1.46E + 04	1.40E + 04	4.35E + 03
Human toxicity: carcinogenic	CTUh	2.30E-04	3.50E-04	1.10E-04
Acidification	kg SO2-Eq	2.342	2.70842	0.87148

due to transporting EoU panels from the mine to location of reuse. When the EoU panels are transported over long-distances (Arizona – Kentucky vs. Arizona - California), the environmental impact advantage diminishes due to the higher emissions associated, potentially outweighing the benefits of reuse. This LCA case study heavily relies on Ecoinvent data due to unavailability of detailed material and production information, and it ignores the resource depletion and social factors which could have strengthened the assessment and support broader applicability of the reuse strategy.

5 Conclusions and Future Work

This study demonstrates the potential of reusing EoU PU panels from the mining industry as a viable strategy to advance CE practices. Through deflection testing, LCA, and cost modeling, this study confirms that discarded PU panels are a viable replacement for geogrid used in agricultural settings. The pilot installation provided real-world validation of their utility in ground stabilization. Compared to manufacturing new geogrids, reusing EoU PU panels demonstrated clear and quantifiable CE benefits as this approach eliminates material demand entirely, delivers at least 82% greater cost efficiency, improves environmental performance by over 36%, and completely eliminates the landfilling of the waste, particularly when transportation distances are minimized. Collectively, these outcomes highlight the environmental responsibility, economic viability, and societal relevance of the proposed solution, demonstrating its potential as a scalable pathway for value recovery.

Future efforts will focus on expanding the reuse of EoU PU panels in non-critical load bearing applications such as a subsurface layer beneath artificial turf and modular sports flooring, as well as land reinforcement in areas prone to soil erosion. Additional testing (e.g., panel contamination from mine debris) will also be conducted in future studies.

Acknowledgment. This work is funded by the PU panel manufacturer to investigate potential circular and sustainable strategies for managing EoU waste.

References

1. Liang, C., et al.: Material flows of polyurethane in the United States. Environ. Sci. Technol. **55**(20), 14215–14224 (2021). https://doi.org/10.1021/acs.est.1c03654
2. Ellen MacArthur Foundation.: Circular Economy (CE) principles (2024). https://www.ellenmacarthurfoundation.org/topics/circular-economy-introduction/overview. Accessed 28 July 2025
3. Kirchherr, J., Yang, N.H.N., Schulze-Spüntrup, F., Heerink, M.J., Hartley, K.: Conceptualizing the circular economy (Revisited): an analysis of 221 definitions. In: Resources, Conservation and Recycling, vol. 194 (2023). https://doi.org/10.1016/j.resconrec.2023.107001
4. Hearon, K., Gall, K., Ware, T., Maitland, D.J., Bearinger, J.P., Wilson, T.S.: Post-polymerization crosslinked polyurethane shape-memory polymers. J. Appl. Polym. Sci. **121**(1), 144 (2011). https://doi.org/10.1002/APP.33428
5. Jefferson,H., et al.: Plastics recycling: challenges and opportunities. Philos. Trans. Roy. Soc. B: Biol. Sci. **364** (1526), 2115–2126 (2009). www.ncbi.nlm.nih.gov/pmc/articles/PMC2873020/
6. Beena Unni, A., Muringayil Joseph, T.: Enhancing polymer sustainability: eco-conscious strategies. Polymers **16**(13), 1769 (2024). https://doi.org/10.3390/POLYM16131769
7. Sonnenschein, M.F.: Polyurethanes: Science, Technology, Markets, and Trends. Wiley, Hoboken (2021)
8. Samani, P.: Synergies and gaps between circularity assessment and life cycle assessment (LCA). In: Science of the Total Environment, vol. 903 (2023). https://doi.org/10.1016/j.scitotenv.2023.166611
9. Strick Trailers: Sheet and Post Semi Trailers 2019. www.stricktrailers.com/products_/sheet-and-post-semi-trailers. Accessed 30 July 2025
10. ZipRecruiter: Truck Loader (2025). www.ziprecruiter.com/Salaries/Truck-Loader-Salary. Accessed 29 July 2025
11. Chris,H.: What is total cost per mile for truckload carriers? FreightWaves (2020). www.freightwaves.com/news/understanding-total-operating-cost-per-mile
12. Premier Equestrian: Premier Geo Grid | Mud Control Mats & Grids (2025) https://premierequestrian.com/product/premier-geo-grid-mats-ground-stabilization/#geo-contact. Accessed 20 July 2025

Solid-State Upcycling of Recovered Aluminum Using Friction Extrusion and AFSD Technologies

Alan Hensley(✉), Naser Alqseer, Lillian Melton, R. Sarvesha, James Caudill, Fazleena Badurdeen, and I. S. Jawahir

Department of Mechanical and Aerospace Engineering and Institute for Sustainable Manufacturing, University of Kentucky, Lexington, KY 40506, USA
nal326@uky.edu

Abstract. The increasing demand for sustainable manufacturing has prompted the development of innovative approaches to upcycle and more effectively utilize recovered materials for sustainable value creation. This study involves two solid-state processes, Friction Extrusion and Additive Friction Stir Deposition (AFSD), for producing deposits from post-consumer aluminum waste, which can subsequently be utilized in the creation of new aluminum products. The mechanical performance of this upcycling method is evaluated to establish a foundation for future research to facilitate Circular Economy (CE) with sustainability benefits. Recovered material is processed through this hybrid approach to demonstrate its practical feasibility. Also, metal powders are incorporated to assess the alloying capability of this method. The properties of both the as-recovered and processed materials are investigated to evaluate the change in material characteristics throughout the method. The goal of this method is to improve material efficiency and reduce energy consumption by avoiding remelting recovered materials and addressing efficiency losses associated with existing solid-state recycling processes. This approach has the potential to lower energy usage and minimize waste throughout the recycling pathway, thereby aligning with CE principles by enabling reintegration of scrap/waste and recovered materials from end-of-life (EoL) products into the value chain as higher-value products.

Keywords: Sustainable Manufacturing · Upcycling · Circular Economy

1 Introduction

Sustainability has become an increasingly important focus in manufacturing and is commonly addressed through one of two strategies. One approach is to optimize the use of input materials; the other is to maximize value recovery from EoL products for reuse,

---

This paper was prepared for presentation during the student sessions of GCSM 2025 and underwent an expedited peer review process, coordinated by all project supervisors from four collaborating universities. The project supervisor was I.S. Jawahir.

H. Kohl et al. (Eds.): GCSM 2025, LNME, pp. 483–490, 2026.
https://doi.org/10.1007/978-3-032-21154-5_53

remanufacture or recycling. This study focuses on the latter, investigating an alternate approach for aluminum recovery. Aluminum is traditionally recycled through melting, a process that consumes a significant amount of energy and has harmful environmental impacts. Solid-state processes have emerged as a promising alternative to conventional recycling methods, offering the potential for lower energy consumption, reduced environmental impact and superior material properties facilitating upcycling. However, these processes remain relatively uncommon, especially at industrial scales. This project explores a solid-state method for reintroducing discarded aluminum into the production cycle, spanning from material sorting to stock material creation, thus enabling the creation of new input materials with better properties and without excessive energy use. Specifically, it utilizes the solid-state processes of Additive Friction Stir Deposition (AFSD) and Friction Extrusion in a hybrid manufacturing method to create aluminum deposits from post-consumer scrap. These deposits may then be used to manufacture new aluminum products. AFSD and friction extrusion both utilize a rotating tool to generate frictional heat and induce localized plastic flow. In the case of AFSD, the mechanism is applied in a layer-by-layer fashion, enabling the incremental build up of material typical of additive manufacturing (AM) techniques. In both cases, tool features such as surface protrusions enhance mixing and promote uniform material flow during processing. The combination of elevated temperatures and high strain rates results in dynamic recrystallization and metallurgical bonding. As these processes operate entirely below the melting point of the material, they consume significantly less energy than conventional melting-based recycling methods, while still producing fully consolidated deposits.

2 Literature Review and Background

The deterioration of aluminum from the traditional recycling process has persisted as a major issue. Much of this degradation is due to the melting process, during which oxidation and contamination introduce impurities and result in the loss of material (Jerina et al., 2018). This results in the loss of aluminum, along with the other elements in the mixture, which can alter the material's composition. This is countered by the addition of more elements to replace those lost and correct the composition, forcing the life cycle loop of the material to remain open. This process also alters the microstructure, requiring post-processing work on the recycled material. Further, the energy required to melt the material is substantial, estimated between 11 and 20 GJ per ton (Güley et al., 2013). Solid-state processes offer reduced energy consumption and operate at significantly lower temperatures. Despite their theoretical advantages, these processes still require further development to achieve industrial viability (Shamsudin et al., 2016). Beyond basic material reuse, solid-state techniques have also shown promise for alloying. Wang et al. (2024) demonstrated that aluminum alloy composition can be altered by incorporating alloying powders during solid-sate processing. Among the emerging techniques, AFSD has gained attention as a promising method for aluminum upcycling. Conventionally, AFSD is used for the creation of near-net components from feedstock rods of input material, typically metals. Instead of the typically solid feedstock rods, Yoder et al. (2022) compacted aluminum chips directly into feedstock rods for AFSD. Although their approach yielded high-quality deposition, it introduced inefficiencies within the initial feedstock production step. Specifically, the compacted feedstock had a reduced density (~70%), leading to low deposition rates and an energy consumption of over 4.5kWh/kg, significantly more than the energy consumption achievable with the higher deposition rates of solid feedstock, which can approach 1 kWh/kg. Improving feedstock density could therefore enhance the energy efficiency of AFSD-based upcycling. Achieving higher density through mechanical compaction, however, becomes increasingly difficult due to the exponential rise in required pressing force at higher levels of compaction (Njjar et al., 2024). One potential solution to this limitation is friction extrusion, a solid-state process which converts materials from a variety of input forms into fully consolidated, uniform bars and can be adapted to produce various cross-sectional geometries. With the appropriate die configurations, friction extrusion can be used to directly convert partially compacted scrap into dense AFSD feedstock (Wang et al., 2024). This study investigates a new solid-state upcycling method in which aluminum scrap is first compacted and then processed through friction extrusion to produce feedstock suitable for AFSD. The resulting feedstock is subsequently deposited via AFSD to form billets that can serve as stock material for further manufacturing. Additionally, alloying elements are introduced to evaluate their diffusion and distribution across both processes.

3 Methodology

To investigate the solid-state upcycling of aluminum scrap, a methodology comprised of five main steps was developed. This approach, illustrated in Fig. 1, includes the following stages: collection and processing, alloying, compaction, friction extrusion, and AFSD. Material property evaluations, including mechanical testing and microstructural analysis, were performed following both friction extrusion and AFSD processes. In this study, the collection and processing phase consists of recovering, cleaning, and shredding discarded single-use aluminum trays and lids. Prior to alloying, the composition of the recovered material was determined using electron dispersion spectroscopy (EDS), which identified the alloy as AA 3003. To enhance mechanical properties, the material was modified toward the composition of AA 6061. This was achieved by dry-mixing copper and magnesium powders with shredded aluminum in weight fractions corresponding to the AA 6061 specification. For compaction, a gradual pressing workflow was adopted to maximize the material loaded into the die prior to the final compaction. Three presses were employed in sequence: an arbor press, a manual hydraulic press, and a full-capacity hydraulic press. This sequence allowed for incremental increase in compaction. Although this multi-step approach could be consolidated into a single pressing operation in an industrial setting, it was adopted in this study to accommodate limitations related to equipment availably and scale. Following compaction, the billets were loaded into the friction extrusion machine and processed to produce an alloyed, densified feedstock. This process facilitated the distribution of the added elements and further consolidated the compacted, mechanically bonded scrap into a unified structure. These rods are then deposited by the AFSD process. Finally, samples are extracted from both the feedstock rod produced via friction extrusion and the AFSD deposit for subsequent property evaluation.

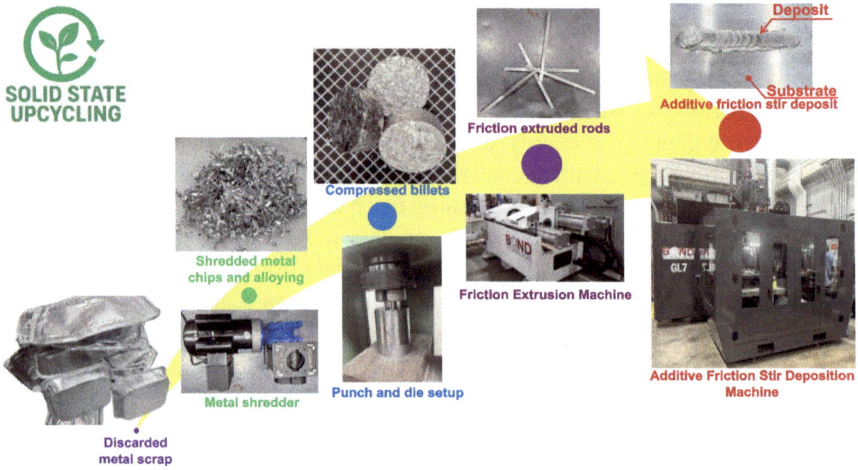

Fig. 1. The adopted methodology

4 Results and Discussion

A total of 2.1 kg of AA 3003 was used for this study. The shredded material (typical fragments were approximately 30 × 150 mm) was evenly divided into two metal containers of approximately 1 kg each, and the necessary amounts of copper powder (~30 μm) and magnesium powder (~300 μm) were added to match the target composition of AA 6061. Despite the relatively low volume, caution was taken to avoid static discharge and other possible ignition sources for the powders, and the dry mixing was performed in proximity to a class D fire extinguisher. The blended material was compacted using the previously described multi-step pressing workflow, forming 21 billets (60 mm diameter, ~25 mm height) with an average density of 65%, which required approximately 95 kN each. These billets were then processed by the friction extrusion system, to produce 13 mm square cross-section rods with an approximate length of 150 mm. However, this process did not yield ideal results. Due to suboptimal processing parameters, the extruded material deviated from the intended geometry. Despite this, the resulting feedstock was still usable and was subsequently deposited using a BOND GL7 AFSD machine. Processing parameters were chosen based on typical AA 6061 feedstock; however, these proved to be unsuitable for the upcycled material, leading to poor bonding and a rough final deposit.

Cross-sectional analysis of both the feedstock rod and AFSD deposit was conducted to evaluate hardness and microstructure. The feedstock rod (seen in Fig. 2a) shows the spiral pattern generated by the friction extrusion process. Lower shear conditions near the center region resulted in visible porosity and retained boundaries between fragmented scrap pieces. This is also shown by the hardness map (Fig. 2a), with the material near the center exhibiting similar hardness to the recovered scrap material, which had an average hardness of 39 0.1 HV. Although this heterogeneity is not ideal, the density of the feedstock rod exceeded 95%, representing a substantial improvement over the initial compacted billets.

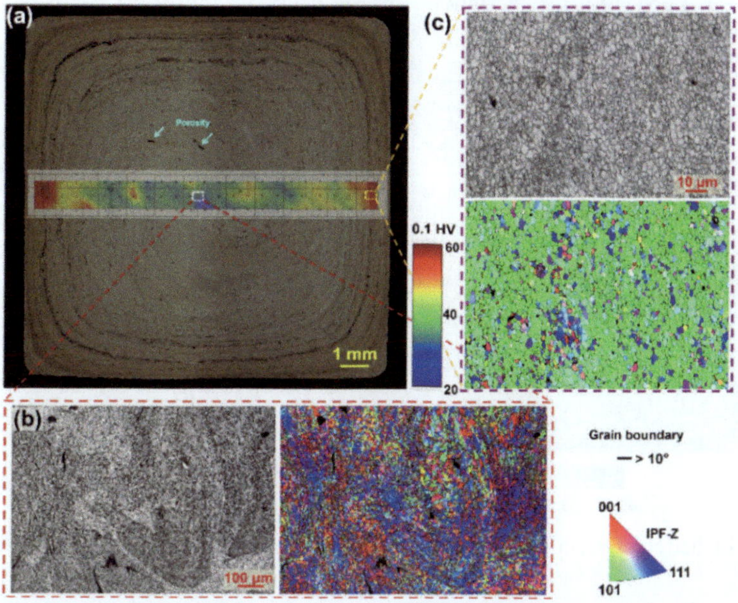

Fig. 2. a) Etched cross-section of a rod formed by friction extrusion with overlaid hardness map. b) Grain structure of the rod's center region. c) Grain structure of the rod's outer region.

The outer regions of the feedstock rod, which experienced more intense stirring and deformation, exhibited higher hardness and a more mixed structure. A series of indents were also performed across a cross-section of the deposit (Fig. 3a), mapping the region spanning the deposit/substrate boundary. Flow lines are also visible here from the stirring motion of the process, similar to those seen in the feedstock rod, though from a different angle. The hardness map reveals hardness becoming much more uniform within the deposited material, with elevated regions coinciding with the darker flow lines that likely experienced higher strain during deformation. On average, the hardness of the deposit was similar to the more processed outer regions of the feedstock rod. The increase in hardness is attributed to significant grain refinement and dynamic recrystallization induced by the stirring action during both AFSD and friction extrusion processes. These effects can be reversed by post-process heat treatment, depending on the final application requirements. The EBSD images shown in Figs. 2a, 2b, and 3b, clearly show the significant difference between the grain structures generated by the two processes.

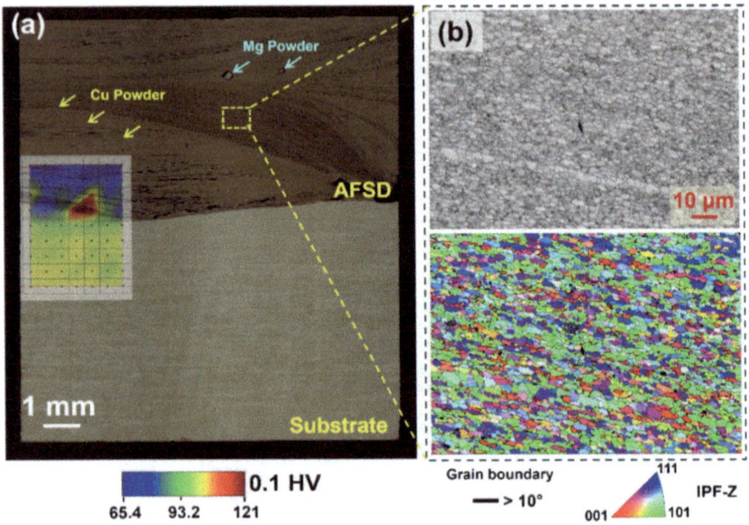

Fig. 3. a) Etched cross-section of the AFSD deposit with overlaid hardness map. b) Grain structure deposit.

Residual, unincorporated powder particles were also observed, likely due to non-optimized processing conditions. The traces of these elements are particularly visible in Fig. 4, where SEM imagery highlights them in both the feedstock rod and deposit. Similar to the hardness profiles, the outer regions of the feedstock rod appear more thoroughly processed, though some alloying elements remain unevenly distributed. In contrast, the AFSD deposit exhibits a more uniform composition, as these traces disappear almost completely. Although some image noise may account for the color variation in the images, the absence of clearly distinct particles suggests improved mixing and distribution of the alloying elements. Further improvements may be achieved by optimizing processing parameters and selecting powder particle sizes that enhance incorporation during both friction extrusion and AFSD. Notably, some intact particles were still visible within the deposit, particularly those corresponding to the smaller copper particles. Along with the partially absorbed elements these residual particles could lead to issues in further processes such as accelerated tool wear in machining processes caused by sudden changes in hardness within the material.

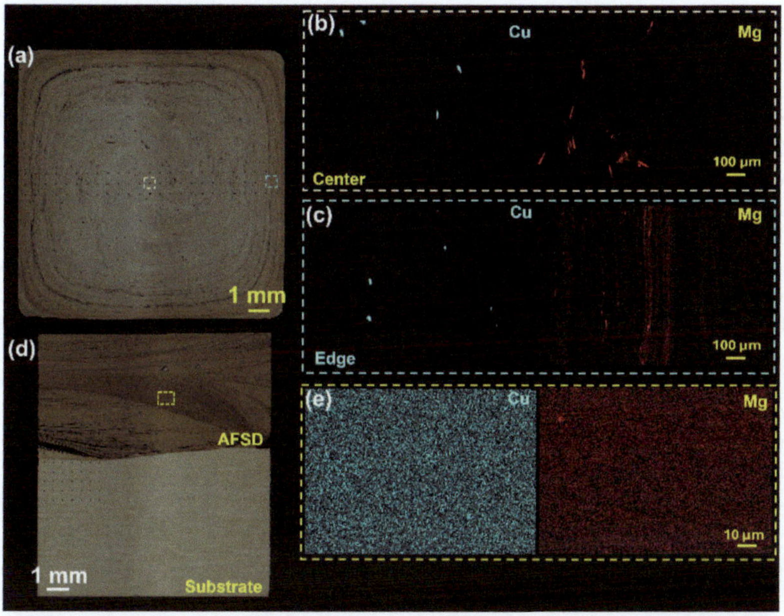

Fig. 4. Distribution of copper and magnesium in the feedstock rod and deposit.

5 Conclusions and Future Work

This study demonstrates a solid-state upcycling approach as an alternative to conventional aluminum recycling. Recovered and shredded aluminum scrap from single-use trays and lids was first consolidated into dense, metallurgically bonded feedstock through friction extrusion. The resulting material was then deposited using AFSD to produce a small-scale deposit, validating the feasibility of this upcycling pathway. Friction extrusion served as the primary densification step, converting scrap material into processable feedstock, while AFSD enabled material deposition through layer-by-layer thermomechanical processing. Importantly, these outcomes were achieved despite unoptimized processing parameters, indicating strong baseline potential for further refinement. Future work should evaluate the net energy efficiency of the full process, including compaction, extrusion, deposition, and post-deposition heat treatment. A key question is whether the energy consumed in these steps is offset by the savings from avoiding melting and from utilizing fully dense feedstock during AFSD. If the total energy footprint remains lower than that of conventional recycling or other solid-state routes, this method could represent a more sustainable and scalable pathway for aluminum upcycling. In doing so, it would directly support CE principles by keeping material recovered from EoL products in circulation and enabling its transformation into higher value forms. Additional studies without alloying could help isolate the effects of processing, and other solid-state consolidation techniques could also be explored to identify more efficient or industrially scalable alternatives for preparing AFSD-compatible feedstock from aluminum scrap.

References

Jerina, L., Medved, J., Godec, M., Vončina, M.: Influence of the specific surface area of secondary material on the solidification process and microstructure of aluminium alloy AA7075. J. Therm. Anal. Calorim. **134**(1), 455–462 (2018)

Güley, V., Güzel, A., Jäger, A., Khalifa, N.B., Tekkaya, A.E., Misiolek, W.Z.: Effect of die design on the welding quality during solid state recycling of AA6060 chips by hot extrusion. Mater. Sci. Eng., A **574**, 163–175 (2013)

Shamsudin, S., Lajis, M.A., Zhong, Z.W.: Solid-state recycling of light metals: a review. Adv. Mech. Eng. **8**(8), 1687814016661921 (2016)

Wang, T., et al.: Upcycled high-strength aluminum alloys from scrap through solid-phase alloying. Nat. Commun. **15**(1), 10664 (2024)

Yoder, J., Hahn G., Zhao N., Brennan, R., Cho, K., Yu, H.: Additive friction stir deposition-enabled upcycling of automotive cast aluminum chips. Add. Manuf. Lett. (2022)

Al Njjar, A., Mazloum, K., Sata, A.: Comparison of single action die and floating die for compaction of aluminum alloy AA7075 using simulation and experimental study. Eng. Res. Express **6**(1), 015412 (2024)

Ecodesign Approaches for Mechanical Component Development at the University of Bologna Motorsport Team

Noemi Gurieri(✉), Emanuele Morelli, and Matteo Garulli

University of Bologna, Viale del Risorgimento 2, 40136 Bologna, Italy
{noemi.gurieri,emanuele.morelli5,matteo.garulli}@studio.unibo.it

Abstract. UniBo Motorsport, the motorsport team of the University of Bologna, is currently addressing the transition toward sustainable manufacturing, a key challenge that requires balancing high performance with environmental responsibility. In collaboration with one of its industrial sponsors, the team conducted a detailed analysis based on ISO 14067, focusing on the carbon footprint assessment of steering system components and adopting the Life Cycle Assessment (LCA) methodology in accordance with ISO 14040 and ISO 14044. These standards enabled the identification of the main requirements for an integrated Environmental Management System (EMS), primarily through the evaluation of manufacturing processes, the selection of alternative materials, the optimization of machining operations, and the definition of improved, sustainability-oriented design solutions for mechanical components.

Keywords: Ecodesign · LCA · Formula SAE · Carbon Footprint · Sustainable Manufacturing

1 Introduction

1.1 Formula Student and University of Bologna Motorsport

Formula Student is an international engineering design competition aimed at university students, offering a practical environment to apply theoretical knowledge through the design, manufacturing, and testing of single-seater race cars. UniBo Motorsport, the official team of the University of Bologna, was founded in 2009 and has evolved from a loosely structured group into a professionally organized association. Since 2017, the team has experienced rapid growth, expanding to include electric and moto projects, all managed under a unified administration.

This paper was prepared for presentation during the student sessions of GCSM 2025 and underwent an expedited peer review process, coordinated by all project supervisors from four collaborating universities. The project supervisor was G. Campana.

H. Kohl et al. (Eds.): GCSM 2025, LNME, pp. 491–498, 2026.
https://doi.org/10.1007/978-3-032-21154-5_54

As of 2024–2025, the team exceeds 200 members and participates in major international events with three race vehicles. This article focuses exclusively on four-wheeled prototypes used in Formula SAE competitions, specifically analysing the Cost Event, a static event aimed at evaluating the economic and industrial feasibility of the vehicle. The team is required to submit a comprehensive cost report, detailing materials, assembly operations, and manufacturing processes in accordance with competition rules. Additionally, recent editions introduced the estimation of the Carbon Footprint (CFP), encouraging sustainable design. This year's sustainability report concentrates on the steering system, critically assessing current methodologies and proposing improvements to enhance environmental and industrial viability.

2 Method, Materials and Processes

UniBo Motorsport team's design process is based on a circular logic of continuous improvement, supported by a structured knowledge transfer between senior members and new hires, and guided by two fundamental objectives: performance optimization and environmental impact reduction. The Carbon Footprint (CFP) has been adopted as the key metric for measuring climate-altering emissions across the entire life cycle of components, in accordance with international standards ISO 14067, ISO 14040, and ISO 14044.

The analysis was conducted according to the four standard phases of the Life Cycle Assessment (LCA) methodology: definition of the objectives and system boundaries ("from cradle to gate"), data collection (Life Cycle Inventory), impact assessment using certified emission factors (Life Cycle Impact Assessment), and critical interpretation of the results. This approach allowed us to identify the phases with the greatest impact in terms of greenhouse gas emissions and develop concrete optimization strategies, such as adopting alternative materials, improving energy efficiency, and selecting certified and localized suppliers. Finding the optimal solution begins with an analysis of the assembly into which the component will be integrated, taking into account the interfacing components and the mechanical stresses to which it will be subjected.

Within the steering system, three commercial components have been considered: the quick release steering insert, the ball bearing, and the double universal joint. The intermediate component connecting these elements cannot be sourced as a commercial part, as no existing solution meets the specific interface and geometric requirements necessary to link all three components effectively.

To optimise emissions associated with the manufacturing of this section of the steering line, two main factors can be addressed: the sizing of the bearing and the production of the custom component. The first aspect can be easily minimised by defining the expected service life of the system, thus allowing for the smallest possible shaft diameter compatible with the bearing. Conversely, the manufacturing process of the custom component presents a more multifaceted challenge. Given that the geometry is largely constrained by the need to interface with fixed commercial components, significant modifications are not feasible.

Therefore, the analysis has primarily focused on the selection of materials and the choice of production methods, aiming to reduce the environmental impact while maintaining mechanical performance and functional integration.

The component under study is the final element of the steering assembly, linking the steering wheel's quick-release mechanism to the double universal joint that transmits motion to the steering column.

The custom shaft designed, features an end that connects directly to the double universal joint through a tongue-and-groove coupling. The central portion interfaces with the bearing, which allows the steering column to rotate relative to the fixed steering support. Finally, it is equipped with a flange for coupling to the quick release component, enabling rapid connection and disconnection of the steering wheel and dashboard. To ensure the passage of electrical wiring necessary for power supply and data exchange with the dashboard, dedicated holes have been machined in the shaft.

The cross-sectional dimensions were particularly constrained by the interfaces with commercial components. Once the bearing was selected and the minimum diameter defined as the diameter of the double universal joint bore, the shaft body diameter was fixed. This section was then diverged to expand at the circumference where the bolted connection holes for the quick release system of the steering wheel are located.

The bearing was sized by taking into account both the number of races scheduled and the hours of testing planned with the vehicle during the pre-race period. By analyzing typical track configurations for Formula Student competitions and estimating the forces acting on the bearing through simulations conducted on a virtual model of the vehicle, it was possible to evaluate not only the maximum loads but also the average loads to which the bearing would be subjected.

The bearing experiences both radial and axial loads due to load transfer occurring during high-speed cornering and the reaction forces that must be applied when the driver brakes or negotiates tight turns. For this reason, an angular contact bearing with double row of balls was selected. This choice was also motivated by the impossibility of installing two separate bearings due to space constraints and the particular configuration of the steering column.

After selecting the bearing, the diameter of the intermediate shaft section was established. The minimum required strength of the material for manufacturing the shaft was then evaluated. Given the large diameter section, the required strength is not particularly demanding, allowing prioritization of both lightness and durability throughout the component's life cycle.

The main options were twofold: either optimize for lightness by designing the shaft for a single season's service life, or favor reusability by ensuring operation for at least two seasons, potentially considering the use of high-performance steels. The study led to the development of the following geometry (see Fig. 1).

Once the geometry has been defined, it is still possible to adapt it with a view to simplifying the production of the part or to facilitating assembly within the assembly. Specifically, once the minimum section has been identified, it is possible to choose between different materials and machining processes with

Fig. 1. Steering System (left); Quick release assembly (centre) and shaft (right)

which to make the part, also considering the environmental impact related to production. In this way, year after year, it has been possible to reduce not only the weight of the component, improving its mechanical performance, but also the number of machine hours and the complexity of the processes, with a consequent reduction in costs and emissions.

As regards the choice of material, the selection was conducted considering the required characteristics in terms of mechanical strength, stiffness, workability, weight, and corrosion resistance. After a preliminary evaluation phase, three main materials were examined:

- **AA 7075-T6.** A high-strength aluminum alloy, with an excellent strength-to-weight ratio, excellent machinability. Widely used in motorsports and aerospace.
- **A356-T6.** An aluminum alloy characterised by good mechanical strength and light weight, suitable for casting processes. A valid alternative for low-cost production and high repeatability.
- **AISI 316.** An austenitic stainless steel. It is a very resistant and tenacious material, with excellent resistance to corrosion, but it has a higher density and requires more energy-intensive processes.

These three materials were analyzed in detail to compare both performance and environmental impact. The comparison between Ergal 7075-T6 and AISI 316 is particularly relevant.

The latter was initially considered due to its corrosion resistance, an important factor for components exposed to rain and humidity during races. However, for the same geometry, AISI 316 does not match aluminum in terms of strength-to-weight ratio or machinability. Although it is one of the easiest stainless steels to machine (being less martensitic), it still demands longer machining times and higher energy inputs, making it less suitable for this application.

Aluminum, on the other hand, does have a critical limitation: its lower resistance to cyclic fatigue, which is significant for components subject to repeated loads and vibrations, such as the steering column's connecting elements. Nonetheless, in the context of Formula Student, the actual stress conditions allow the component to remain functional for multiple seasons, despite aluminum's inherent fatigue limitations.

A potentially valuable comparison could have been made with composite materials, known for their exceptional strength-to-weight performance. However, the decision to avoid composites was taken deliberately based on environmental sustainability criteria. Recycling epoxy composites is possible, but not yet viable; the energy required for manufacturing and curing often outweighs the benefits of recovery. Within an eco-design framework, these factors made composites unsuitable for the current application.

The analysis focused primarily on the transformation of the raw material into the finished component, one of the most impactful phases in terms of energy consumption. To quantify the actual energy used and compare the environmental impact of different materials, CAM simulations were conducted using the software ISCAR, which provided the power absorbed by the machine tool and the chip removal rate for each material. Starting from this data it was possible to calculate:

- the volume of material removed to obtain the final geometry,
- the processing time needed,
- the total consumption of electricity from the machine tool, considering machines' efficiency,
- finally, the environmental impact in terms of CO_2 equivalent associated with the process.

The CFP values associated with the raw materials were obtained from the project database Idemat, which provide the emission coefficients in $kgCO_{2eq}$ per kg of material. The integration of these data with those relating to the energy consumed led to estimate with good precision the total emissions generated by the production of each component, considering the entire production phase, "from the rough to the finished piece". The analysis was further extended considering other emissions contributions throughout the life cycle, including:

- **Extraction of resources and raw materials**: emissions related to the extraction and transformation of natural resources into semi-finished products;
- **Transformation into finished product**: emissions from industrial processes, such as melting, machining and heat treatments;
- **Transportation**: contribution of logistics activities along the supply chain;
- **End of life**: emissions resulting from the disposal, recycling or recovery operations of the component once it has been decommissioned.

This approach allows an objective comparison of design and production choices not only in terms of functional performance, but also in terms of environmental sustainability, promoting a more informed and responsible decision-making process.

3 Discussion and Results

Compared to the other options analyzed (A356-T6 and AISI 316), Ergal allows the creation of complex geometries with an excellent compromise between processing time, energy consumption and emissions, as highlighted in the following Table 1:

Table 1. Comparison of the environmental impacts

Cathegory	AA7075	AISI 316	A356
Gross weight [kg]	1.48	4.12	1.00
Used material impact [kg CO_{2eq}]	10.75	21.67	7.04
Processing impact [g CO_{2eq}]	57.88	509.33	26.30
Transport [kg CO_{2eq}]	2.24	6.22	1.51
Molds, supports, fusion [kg CO_{2eq}]	–	–	6.94
Total [kg CO_{2eq}]	**13.06**	**28.40**	**15.52**

This choice is part of a broader ecodesign approach, a conscious and systematic design process focused, from the initial stages, on developing solutions that minimize environmental impact throughout the product's entire life cycle. It is not just about choosing less impactful materials, but about rethinking the entire process, from supplier selection to end-of-life strategies, with sustainability in mind. In this perspective, even starting from the best possible solution, the overall environmental impact can be further reduced through targeted actions along the value chain, as suggested in Chap. 2. One of the most effective strategies is the reduction of supply distances: by purchasing the material from local or national suppliers (e.g. Italian or European), it is possible to significantly reduce the emissions related to logistics. Table 2 clearly shows the emissions resulting from the transport of one tonne of aluminum from China, highlighting the importance of reducing transportation.

Table 2. Impacts for the transportation for AA7075-T6

Transport	Description	Eco-Intensity $[\frac{kgCO_{2eq}}{km/t}]$	Distance [km]	Impact [kgCO_{2eq}/t]
Ship	Ship + container	0.0048	19170	91.545
Train	Train + container, electric	0.0100	1050	10.340
Truck	Truck + container, Diesel, 28 t net	0.0300	136	4.237
Truck	Truck + trailer, Diesel, 24 t net	0.0284	15	0.426
Subtotal				**106.548**

An additional key consideration concerns the source of energy used in primary aluminum production: in China, it still relies heavily on fossil fuels with high emission intensity, such as coal, whereas in Italy and across Europe, the energy mix is cleaner and environmental regulations are more stringent. This results in a more efficient production process with a lower environmental impact.

In addition to geographical origin, ecodesign also involves a conscious choice between primary and secondary aluminum. The production of primary aluminum from bauxite is one of the most energy-intensive industrial processes, whereas the use of secondary (i.e. recycled) aluminum can reduce emissions by up to 95%. However, it is essential that the secondary aluminum is refined, obtained through controlled selection and remelting processes that ensure high chemical and mechanical quality, an essential requirement for use in the motorsport sector. Integrating these measures (high-performance materials, local sourcing, and quality recycling) is a direct expression of ecodesign principles and allows performance and environmental responsibility to be combined in a single virtuous process. The Table 3 shows the differences between primary and secondary aluminum.

Table 3. Primary vs Secondary aluminum Emissions

	Primary Aluminum	Secondary Aluminum
Energy consumption $[MWh/t]$	14	0.7
Emissions $[kgCO_{2eq}]$	6.36	0.69
Performance	Standard	−5%
Price [e/kg]	2.2–2.5	1.2–1.8

Final choice: based on these considerations, UniBo Motorsport chose to use secondary refined Ergal, sourced from Italian suppliers, explicitly specifying the environmental and quality requirements. This choice meets not only the team's technical standards but also the desire to fully integrate ecodesign principles into the development process of the team.

4 Conclusion

In the constant pursuit of speed, precision and performance, Formula Student is much more than a simple competition: it is an educational and experimental environment where technical innovation is integrated with design vision, methodological rigor, and environmental responsibility. When performance and sustainability are truly balanced, engineering transcends the dimension of pure technical expertise and becomes a conscious transformation tool, capable of driving progress without compromising environmental balance.

The analysis conducted on the steering component represented an opportunity to adopt this new perspective. What initially appeared to be a material selection process revealed itself to be a complex system of interconnected choices (mechanical, economic, and ethical) oriented towards a solution capable of combining technical excellence and environmental responsibility. Parameters such as material weight, energy consumption and associated emissions are no longer considered individually, but evaluated within a systemic vision which integrates environmental sustainability, production efficiency, and traceability along the entire supply chain.

What was initially an experimental project, developed in an informal context, is now a structured and dynamic environment in which engineering is applied in a concrete way to address the real challenges. In this scenario, sustainability is no longer considered a design constraint but a strategic lever to generate innovation.

Acknowledgement. The authors would like to sincerely thank Prof. Eng. PhD Giampaolo Campana for his guidance and support, acknowledge the UniBo Motorsport Team and the University of Bologna, and express their gratitude to the sponsors, whose support makes our Formula Student initiative possible and to all the GCSM Team.

References

1. European Commission: Ecodesign for Sustainable Products Regulation (ESPR). https://ec.europa.eu/environment/publications/proposal-ecodesign-sustainable-products-regulation_en, Accessed 27 Sept 2025
2. Santochi, M., Giusti, R.: Tecnologia meccanica e studi di fabbricazione. Hoepli, Milano (2003)
3. Campana, G., Mele, M.: Sistemi integrati di lavorazione. Pitagora Editrice, Bologna (2015)
4. Sandvik Coromant: https://www.sandvik.coromant.com/it-it, Accessed 27 Sept 2025
5. Iscar Italia: https://www.iscaritalia.it/index.aspx/countryid/9/lang/it, Accessed 27 Sept 2025
6. Delft University of Technology: Idemat Database – Environmental impact data for materials. https://www.ecocostsvalue.com, Accessed 27 Sept 2025

Metal Hydride Test Bench Development for Investigation of Sustainable and Efficient Hydrogen Storage

Clemens Biet(✉), Sanjeev Kumar Sundara Raju, Ayush Chaturvedi, and Parasuram Mahadevan

Technische Universität Berlin, Berlin, Germany
clemens.biet@tu-berlin.de
https://www.tu.berlin/imef

Abstract. This paper presents the development of a metal hydride test bench designed to investigate sustainable and efficient hydrogen storage technologies. The system utilizes LaNi5, a widely studied metal hydride with favorable absorption/desorption kinetics and stability. A thermodynamic simulation model is implemented alongside the physical setup to predict system behavior and support experimental design. The test bench includes pressure and temperature control systems, safety testing, and precise measurement instrumentation. This platform enables future testing of various hydride materials, including recycled or industrial waste-based alloys, supporting broader research efforts toward safe, low-energy hydrogen storage solutions.

Keywords: metal hydride · $LaNi_5$ · hydrogen · tank design · modelling

1 Introduction and Motivation

As part of the transition from fossil energy sources towards renewables, hydrogen is seen as an important energy carrier of the future. While offering advantages such as high gravimetric energy density, direct bidirectional conversion with electrical power, and the potential for completely carbon-free energy conversion, the storage of molecular hydrogen still poses challenges due to the low density of hydrogen at normal conditions. Both established technologies for storing hydrogen at higher energy densities, specifically high-pressure hydrogen gas tanks and cryogenic liquid hydrogen storage, require large amounts of energy. Metal hydrides offer an efficient low low-energy alternative for hydrogen storage, providing low cost and high safety at moderate energy density.

This paper was prepared for presentation during the student sessions of GCSM 2025 and underwent an expedited peer review process, coordinated by all project supervisors from four collaborating universities. The project supervisor was C. Biet.

H. Kohl et al. (Eds.): GCSM 2025, LNME, pp. 499–507, 2026.
https://doi.org/10.1007/978-3-032-21154-5_55

1.1 Behavior of Metal Hydrides

When metals are exposed to gaseous hydrogen, the hydrogen molecules can be absorbed into the solid metal structure. This adsorption process is an exothermic reaction. Depending on the pressure of the hydrogen atmosphere, the process can also be reversed into a desorption of hydrogen from the metal, which in turn is an endothermic reaction. The general metal hydride reaction can be written as

$$\mathrm{M} + \frac{x}{2}\mathrm{H}_2 \Leftrightarrow \mathrm{MH}_x + \mathrm{Q} \tag{1}$$

where M is the metal and Q is the heat.

1.2 Potential for Sustainable Energy Storage

In addition to the low energy requirements for working with metal hydrides compared to high-pressure or cryogenic hydrogen storage, other aspects also strengthen its potential as a sustainable energy storage. The energy required for "charging" the metal hydride consists of the compression power, which is relatively low due to the low required pressure level, and the dissipation of heat, which is also relatively low if the environment is utilized efficiently. When "discharging" the metal hydride, a heat source is needed. Quite often, the hydrogen is used in a fuel cell or combustion process, which in turn provides excess heat, which can be used to power the desorption process.

Although specific materials are commonly used for metal hydride hydrogen storage, such as $LaNi_5H_6$ or MgH_2, in general, most metals do absorb hydrogen to some degree. As a result, even widely available materials can be used.

1.3 Goals and Integration in Subsequent Projects

The presented research project marks the beginning of a comprehensive investigation into metal hydrides. The goal for this initial step is to design, manufacture, and commission a metal hydride hydrogen storage tank that can be used in a lab environment to investigate metal hydride materials at moderate temperature and pressure levels. The developed test bench will provide the foundation for follow-up projects. First, established materials such as $LaNi_5$ /FeTi should be able to be tested for their performance properties. Later, alternative materials, including industrial waste and possible product recycling options, shall be investigated for their potential as sustainable and affordable metal hydride energy storage. Additionally, a numeric model of the metal hydride tank is implemented to support the experimental activities.

2 Methodology

The project is structured in five main work packages, each containing multiple tasks, as shown in the flowchart in Fig. 1. Following the initial theoretical research on metal hydrides, the work packages procurement, physical setup, and

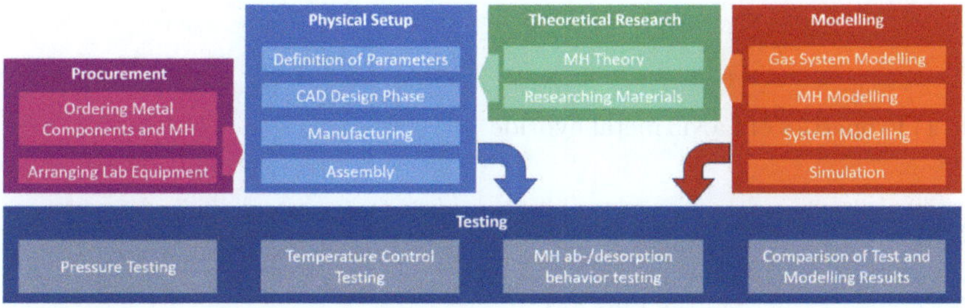

Fig. 1. Visualization of the project work packages and tasks.

numerical modelling have been worked on in parallel. These are followed by the experimental testing work package since it depends on the finished test bench as well as the simulation results. Even though the experimental setup is the main goal for this initial project towards investigation of metal hydrides as a sustainable and efficient hydrogen storage, the numerical modelling and simulation are an important addition. It allows for a deeper understanding of the thermodynamic behavior of the metal hydride test bench, provides parameters for the hardware configuration, and offers the possibility to conduct measurement simulation comparisons.

3 Selection of Metal Hydride Type

Figure 2 shows the operating pressures and temperatures for common metal hydride materials. In order to achieve low energy requirements for the metal hydride storage, a moderate pressure range between 1 bar and 20 bar as well as a temperature range between -35 °C and 100 ° C can be defined. Another important property is the hydrogen capacity. While most types are within the region of low single-digit weight percent, some high-temperature metal hydrides can reach hydrogen capacities of up to 13 wt%. Lastly, price and commercial availability pose another important aspect for the procurement of the metal hydride. In this project, $LaNi_5$ was chosen. It is a very commonly used alloy in low-temperature metal hydride storage systems. Some additional advantages of $LaNi_5$ are: fast absorption and desorption rates, very good cycling stability, lower sensitivity to gas purity (poisoning), easier activation [1]. The properties of the acquired material are summarized in Table 1.

4 Thermodynamic Modelling of the Metal Hydride Tank

The numeric simulation of the metal hydride tank is a powerful tool to support the design and testing activities in this otherwise experimental study. It helps to gain a better understanding of the physical processes and can provide valuable parameters for the design phase of the metal hydride tank, as well as serving as a cross-reference for measurement data.

Table 1. Properties of LaNi5 metal hydride powder.

property	value
grav. hydrogen capacity	1.5 wt%
vol. hydrogen density	$105\ ^{\mathrm{kgH_2}}/\mathrm{m}^3$
reaction enthalpy ΔH	$-31^{\mathrm{kJ}}/\mathrm{molH_2}$
reaction entropy ΔS	$-110^{\mathrm{J}}/\mathrm{KmolH_2}$
plateau pressure p_p	1.7 atm (20°C)

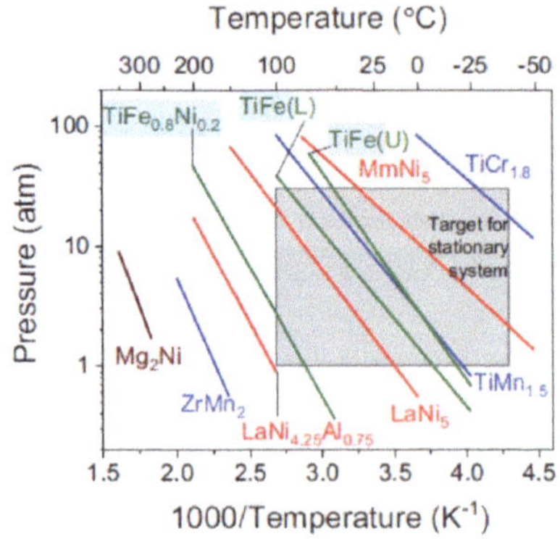

Fig. 2. Operating pressures and temperatures for common metal hydride materials [3].

The modeling approach for the metal hydride storage system is taken from Xiao et al. [2]. The model of the metal hydride tank is divided into the hydrogen gas phase, the solid metal hydride phase, and a transport model that resembles the interaction between these two phases. Following, the central equations for the numeric model are presented.

4.1 Mass Balance and Mass Flow

Mass balances are allied to the metal hydride, the gaseous hydrogen, and the solid materials in the metal hydride storage system.

$$\frac{dm_{\mathrm{MH}}}{dt} = \dot{S}_{\mathrm{MH}} \qquad \frac{dm_{\mathrm{H_2}}}{dt} = \dot{m}_{\mathrm{H_2}} + \dot{S}_{\mathrm{H_2}} \qquad \frac{dm_{\mathrm{s}}}{dt} = -\dot{S}_{\mathrm{H_2}} \tag{2}$$

The mass source term of metal hydride $\dot{S}_{\mathrm{MH}}$ is positive during absorption and negative during desorption. Accordingly, the mass source term for the gaseous hydrogen $\dot{S}_{\mathrm{H_2}}$ is negative during absorption and positive during desorption. The relation between these two source terms can be written as

$$\dot{S}_{\mathrm{MH}} = -\dot{S}_{\mathrm{H_2}} \frac{M_{\mathrm{MH}}}{\psi M_{\mathrm{H_2}}} \tag{3}$$

Regarding the external gaseous mass flow $\dot{m}_{\mathrm{H_2}}$, a throttle equation as in [4, p.21] is being used.

$$\dot{m}_{\mathrm{H_2}} = A_{\mathrm{eff}} \sqrt{\rho_0 p_0} \sqrt{\frac{2\kappa}{\kappa - 1} \left[\left(\frac{p_1}{p_0} \right)^{\frac{2}{\kappa}} - \left(\frac{p_1}{p_0} \right)^{\frac{\kappa+1}{\kappa}} \right]} \tag{4}$$

4.2 Reaction Kinetics and Equilibrium Pressure

The following equation describes the reaction mass mass flow of the metal hydride depending on the gas pressure in relation to the equilibrium pressure.

$$\dot{S}_{\mathrm{MH}} = C_{\mathrm{a}} e^{\frac{-E_{\mathrm{a}}}{RT}} \ln\left(\frac{p_{\mathrm{H_2}}}{p_{\mathrm{eq,a}}}\right)(m_{\mathrm{sat}} - m_{\mathrm{MH}}), \text{ for } p_{\mathrm{H_2}} > p_{\mathrm{eq,a}} \tag{5}$$

$$\dot{S}_{\mathrm{MH}} = C_{\mathrm{d}} e^{\frac{-E_{\mathrm{d}}}{RT}} \frac{p_{\mathrm{H_2}} - p_{\mathrm{eq,d}}}{p_{\mathrm{eq,d}}} m_{\mathrm{MH}}, \text{ for } p_{\mathrm{H_2}} < p_{\mathrm{eq,d}} \tag{6}$$

The Van't Hoff equation can be used to determine the equilibrium pressure based on the parameters of the metal hydride used.

$$\ln\left(\frac{p_{\mathrm{eq,a/d}}}{p_0}\right) = -\frac{\Delta H_{\mathrm{a/d}}}{RT} + \frac{\Delta S_{\mathrm{a/d}}}{R} - k_{\mathrm{p}}\left(\frac{m_{\mathrm{MH}}}{m_{\mathrm{sat}}} - \frac{1}{2}\right) \tag{7}$$

4.3 Energy Balance

The first law of thermodynamics can be used to derive an energy balance for the given system in order to determine a common tank temperature. In this approach, it is assumed that the temperatures of the hydrogen gas and the solid metal hydride in the system at similar due to a large shared contact area.

$$(m_{\mathrm{H_2}} c_v + m_s c_s)\frac{dT}{dt} = \frac{m_{\mathrm{H_2}}}{dt}(c_p T_\infty - c_v T) + \dot{Q} - \frac{dm_{\mathrm{a/d}}}{dt}\left[(c_v - c_s)T + \frac{\Delta H_{\mathrm{a/d}}}{M_{\mathrm{H_2}}}\right] \tag{8}$$

4.4 Equation of State

Due to low pressure and temperature levels, an ideal gas approach for the equation of state is used to calculate the gas pressure.

$$pV = mRT \tag{9}$$

4.5 Simulation Results

A simulation of the designed metal hydride test bench was modeled and run in MathWorks®Simulink. Results presented in Fig. 3 show a three-step process of filling, holding, and emptying the tank. The initial pressure is 1 atm. The mass-flow is the controlled variable with an ingoing rate of 0.00139 kg/h for the first 1000 s and an outgoing rate of 0.001 kg/s during the last 1000 s. The temperature is regulated by a convective cooling condition with a constant cooling medium temperature of 300 K. The temperature, pressure, and mass flow time signals align with the characteristic expected behavior of metal hydrides.

5 Design of Metal Hydride Test Bench

The test bench developed in this project is an open-loop configuration, as can be seen in Fig. 4. A high-pressure hydrogen source serves as a supply for the test bench. A pressure regulator reduces the supply pressure to 40 bar, which is the design pressure of the test bench. A magnetic valve connects the supply to the main metal hydride storage tank. A needle valve is positioned at the combined inlet/outlet of the main tank to limit the maximum volume flows. A second magnetic valve connects the main tank to an ambient pressure exhaust.

Regarding the temperature regulation, the main tank is a double-walled design that is connected to a temperature control unit using automotive cooling water. The device has a temperature range of 0 °C to 100 °C.

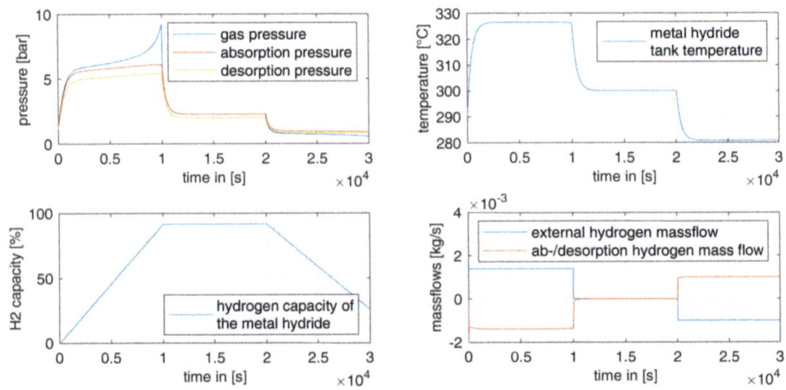

Fig. 3. Simulation results of the thermodynamic metal hydride model.

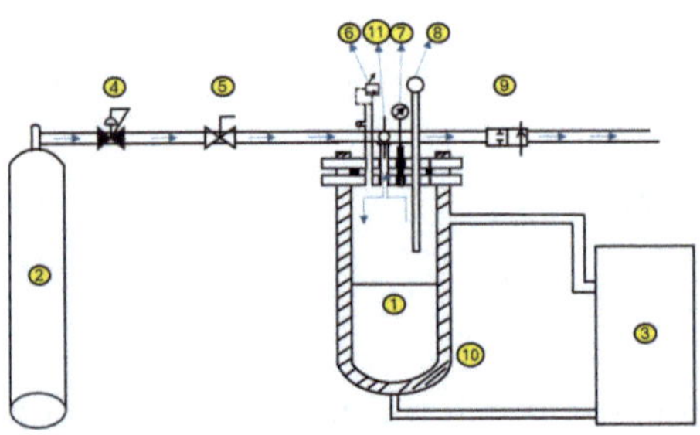

No.	Part Name
1	Metal Hydride Tank
2	HP Gas Supply Tank
3	Temperature Control Unit
4	Pressure Regulator (100 bar → 10 bar)
5	Inlet Control Valve
6	Safety Valve (Pressure Limit: 40 bar)
7	Pressure Sensor
8	Temperature Sensor
9	Outlet Control Valve
10	Cooling Water Jacket
11	Fine Tune Flow Control Valve

Fig. 4. Schematic visualization of the experimental setup.

The structure of the pressure vessel was designed based on CE compliance under EN-13445 PED. Different shapes and flange types have been considered. The final pressure vessel design is shown in Fig. 5. The minimum thickness e was determined using the following equations.

$$P_{\text{max}} = \frac{2fze}{D} \tag{10}$$

(a) MH-Tank.

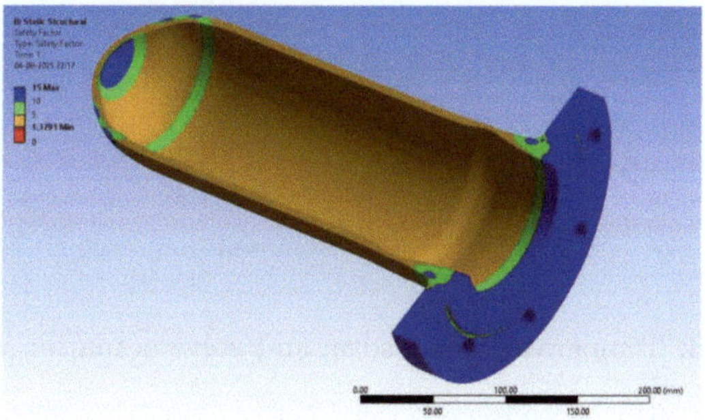

(b) FEM simulation of the pressure vessel.

Fig. 5. Pressure vessel CAD.

where D is the diameter, f is the nominal design stress and z is the joint coefficient. Following the analytical design approach, the final geometry was analysed using a numerical FEM approach, as shown in Fig. 5b

5.1 Measurement Equipment and Sensors

The sensors used include a Type K thermocouple for temperature measurement, a piezoelectric pressure sensor with a range of 100 bar, and a thermal hydrogen mass flow sensor. All sensors are connected to the data acquisition system cDAQ9178 from National Instruments, using the NI9215 voltage input module, the NI9214 thermocouple module, as well as an NI9401 digital input/output module. The electromagnetic valves, as well as the pressure sensor, are supported by a 12 V power supply. A resistor with $500\,\Omega \pm 0.1\%$ was used to convert the 4..20 mA current output of the pressure sensor to a 2..10 V voltage signal.

6 Commissioning of the Test Bench

Before exposing the test bench to pressurized hydrogen, a pressure and leakage test was conducted. First, the tank was filled with pressurized air, testing for correct operation of the valves and sensors as well as identifying leaks in the systems. Afterwards, a pressure test with nitrogen up to the maximum design operating pressure of 40 bar is conducted to verify the structural safety of the system. A pressure test with nitrogen up to 21 bar is shown in Fig. 6.

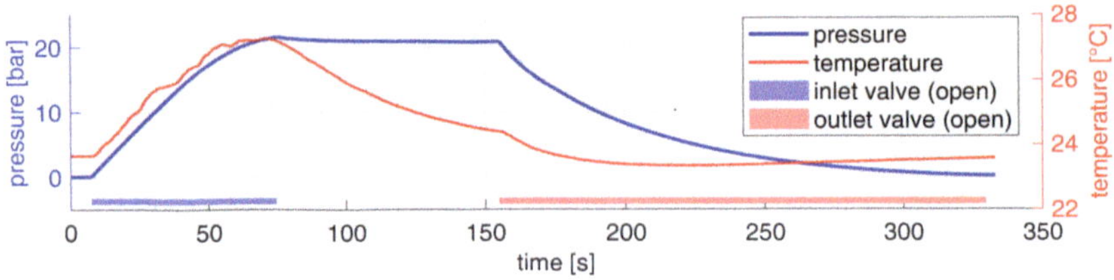

Fig. 6. Temperature, pressure, and valve actuation during nitrogen pressure tests.

6.1 Metal Hydride Filling and Activation Procedure

To avoid oxidation of the $LaNi_5$ powder, which would decrease its hydrogen storage performance, the metal hydride tank needs to be filled in a nitrogen environment. Afterwards, an initial activation procedure for $LaNi_5$ needs to be followed which includes a vacuum to less than 10^{-3} Pa at a temperature of 90 °C for 2 h, a cooling phase and a final phase with hydrogen at pressure between 3 and 5 MPa for 3 h.

7 Summary and Outlook

In the pursuit of sustainable energy storage technologies, metal hydrides offer an attractive alternative for hydrogen storage due to their inherent safety, moderate energy requirements, and potential for recyclability. This paper details the development of a laboratory-scale metal hydride test bench, including its design, construction, simulation, and commissioning.

The project utilizes $LaNi_5$ as the base hydride material, selected for its reliable performance, high cycling stability, and commercial availability. A thermodynamic model based on mass and energy balance equations, reaction kinetics, and the Van't Hoff relation is implemented to simulate absorption and desorption processes. The simulation aids in predicting system behavior and validating experimental data.

The experimental setup comprises a CE-compliant pressure vessel, thermal regulation system, high-precision sensors, and control electronics. The system undergoes comprehensive leak and pressure testing using air and nitrogen before hydrogen activation.

With the test bench operational, the platform enables systematic evaluation of established and novel hydride materials under controlled conditions. This includes exploring the use of recycled metals and industrial byproducts, aiming to improve sustainability and cost-efficiency. The integration of simulation and hardware establishes a robust foundation for ongoing research in metal hydride-based hydrogen storage.

References

1. Klopčič, N., Grimmer, I., Winkler, F., Sartory, M., Trattner, A.: A review on metal hydride materials for hydrogen storage. J. Energy Storage **72**, part B, 108456 (2023). https://doi.org/10.1016/j.est.2023.108456
2. Xiao, J., Tong, L., Bénard, P., Chahine, R.: Thermodynamic analysis for the hydriding-dehydriding cycle of metal hydride system. Energy **191**, 116535 (2020). https://doi.org/10.1016/j.energy.2019.116535
3. Liu, H., et al.: An overview of TiFe alloys for hydrogen storage: Structure, processes, properties, and applications. J. Energy Storage **68**, 107772 (2023). https://doi.org/10.1016/j.est.2023.107772
4. Merker, G. P., Teichmann, R.: Grundlagen Verbrennungsmotoren. ATZ/MTZ-Fachbuch, Springer Vieweg Wiesbaden (2014). https://doi.org/10.1007/978-3-658-03195-4

Gap Analysis of Documentation Template for Open Source Hardware and Adoption in Documentation Workflows with Large Language Models

Neeraj Chodankar(✉), Sagar Didaga Ramakrishna, Sejal More, Atharva Chitrakar, Julien Colomb, and Robert Mies

Technische Universität Berlin, Pascalstr. 8-9, 10587 Berlin, Germany
{neeraj.lahu.chodankar,sagar.didaga.ramakrishna, sejal.rajesh.more,atharva.kiran.chitrakar}@campus.tu-berlin.de

Abstract. High-quality documentation is essential for reproducible and scalable Open Source Hardware. However, many projects apply existing documentation templates inconsistently, resulting in missing or outdated information. This study uses a two-phase approach to evaluate documentation quality and explore automation support. First, we benchmarked a minimal Open Source Hardware documentation template against seven representative hardware repositories to identify recurring gaps in adoption. Second, we tested Large Language Models (ChatGPT and Gemini AI) as Natural Language Processing assistants to automate documentation tasks such as detecting changes, updating files, and highlighting missing information. Results show that while the template already covers essential documentation categories, projects often fail to apply it consistently. Large Language Model-based assistants can help maintainers by reducing manual effort, improving cross-file consistency, and prompting for incomplete assets. Together, these findings outline a practical path toward semi-automated, maintainable, and reproducible documentation workflows for the Open Source Hardware community.

Keywords: Open Source Hardware · Documentation Templates · Large Language Models · Natural Language Processing · Reproducibility

1 Introduction

Open Source Hardware (OSH) refers to physical objects whose designs are shared on the Internet and licensed for study, production, modification, and distribution by anyone. Its formal definition, drafted by the OSH community in 2010, is maintained by the Open Source Hardware Association (OSHWA) [1]. While inspired by free and open-source software, OSH faces unique challenges due to material, regulatory, and manufacturing complexities [2]. Moving from concept to reproducible hardware requires more than design files: Indeed, it depends on comprehensive documentation, which is essential for dissemination, replication, and sustainability [1].

H. Kohl et al. (Eds.): GCSM 2025, LNME, pp. 508–514, 2026.
https://doi.org/10.1007/978-3-032-21154-5_56

In practice, documentation quality enables reproducibility. The openness and complexity of an open-source hardware product are reflected in the structure and richness of its documentation. Despite its importance, OSH documentation often lacks critical elements such as editable source files, structured metadata, Bill of Materials, test results, and certification records [3]. These omissions reduce traceability, hinder maintenance, and limit scalability. Previous analyses further show that while basic information (e.g., licensing) is usually provided, version histories, assembly instructions, and validation data are frequently incomplete. This suggests that the gap lies less in defining what should be documented than in ensuring consistent and detailed adoption of existing practices.

Recent research indicates that LLMs can serve as NLP assistants to help address this gap by extracting metadata, generating structured changelogs, prompting for compliance information, and guiding maintainers in template use [4]. However, it is not yet clear how existing repositories reflect the strengths and weaknesses of available templates. A systematic analysis of real OSH projects can show which elements are consistently documented, which are often missing, and where templates may need clarification or support. These insights also provide the foundation for automation. By identifying which documentation assets are consistently missing or incomplete, the analysis highlights exactly where support tools are needed. This creates a direct link to our second focus: testing whether LLMs can act as assistants to fill these gaps, automate updates, and maintain consistency.

Building on this motivation, our study addresses two research questions. First, how can analysis of existing OSH repositories reveal gaps and inform the refinement of documentation template (RQ1)? This step provides evidence about which documentation assets are well adopted, which are frequently missing, and where maintainers need the most support. Second, how LLMs such as ChatGPT and Gemini AI help address these gaps by acting as NLP assistant in documentation workflows (RQ2)? By linking the identification of gaps (RQ1) with the testing of automation (RQ2), the study explores both the current state of OSH documentation and the potential for LLM-based tools to reduce manual effort, maintain consistency, and support more reliable application of templates. A two-phase design was applied: first, a comparative analysis between a minimal documentation template and representative OSH projects to identify recurring gaps; and second, a use case–driven methodology [5] to test LLMs' ability to detect repository changes and generate structured updates. Our findings suggest that while the template already covers essential categories, consistent and detailed adoption is the primary challenge. NLP assistants may offer a complementary role by automating updates and flagging missing assets, supporting more reliable application of existing templates. This can benefit OSH developers and maintainers, particularly newcomers by reducing the documentation burden, improving reproducibility, and ensuring that projects remain accessible and sustainable over time. In this study we are considering the LLMs as NLP assistants since there is no NLP assistant for development of OSH documentation. In this way we analyse the possibilities and verify our study which will eventually be used as inputs for the development of NLP assistant.

2 Methodology

2.1 Documentation Template Assessment

To address the first research question, we applied a comparative analysis to identify gaps between an OSH meta documentation template and actual practices observed in the community. The reference framework was the *Hardware Documentation Guide* by Colomb. et al. [6], which provides a comprehensive checklist of 88 documentation assets organised into chapters such as a *Readme as an entry door,* on *Community, Product Development and User Analysis, Function–Behaviour–Structure Models, Product Build, Project History,* and *User Guide.*

From an initial pool of 32 use cases, seven representative projects were selected based on completeness, stage of development, diversity of application domains, and public accessibility. The final set included the External Automated Defibrillator (prototype development), Precious Plastic Kit (prototype development), MakAir ventilator (testing phase), YAEMK Keyboard (replication), Open EVSE (replication), Analog Thing (market availability) and Lautsprecher Teufel (market availability). This mix ensured coverage of projects at different levels of maturity and across varied domains, (see Annex Table 1 in the supplementary file) [7].

First, each project was systematically reviewed across repository files (e.g., README, /docs, /CAD), linked resources (manuals, wikis), and version control histories. Each of the 88 assets was classified as Available (1), Partially Available (0.5), or Not Available (0) (see Annex Table 2 in the supplementary file for detailed results) [7]. In the second phase (use case–driven extraction), we examined whether projects contained documentation assets not covered by the template. All analyses were conducted manually using a predefined checklist, reviewed independently by four researchers for reliability, with findings recorded in a shared spreadsheet.

2.2 NLP Scenario Development

To address the second research question, we adopted a structured, use case–driven methodology inspired by Wang et al. (2015) [5], in which scenarios are formalised with identifiers, actors, goals, preconditions, main flows, alternate flows, exception flows, and postconditions to enable systematic execution. This framework was adapted to the context of OSH documentation by modelling scenarios that an NLP assistant could recognise and act upon, ensuring consistency with real documentation practices.

The process began with requirement elicitation, identifying documentation tasks from both the minimal template and observed repository practices. In the next phase, actors and goals were mapped, linking roles such as developers, maintainers, and the NLP assistant with their objectives. Each use case was then expressed in the standardised format, defining the main flow of actions along with alternate and exception flows to handle incomplete or inconsistent documentation.

In the mapping-to-system-logic phase, scenarios were translated into detection rules, triggers, expected outputs, and consistency checks. These were implemented in the automation and execution phase, where the assistant parsed repository inputs, applied

NLP pipelines, and generated structured outputs such as updated documentation sections or alerts. Finally, postconditions were verified to ensure that the assistant's outputs matched the scenario expectations.

This methodology provided a rigorous, reusable way to model documentation tasks, not to produce software test cases as in Wang et al. [5], but to define OSH-specific documentation scenarios for NLP assistants. These structured scenarios then served as the foundation for the practical case study described in Sect. 3.2 (see Annex Fig. 1 in the supplementary file for the flowchart) [7].

3 Results and Discussion

3.1 Documentation Template Assessment

Table 1. Results from Template Evaluation

Analysis Aspect	Findings	Results/Implications
1) Coverage of Template Assets in Projects	Assets fall into three consistency ranges: • **>5–7 (7 assets):** consistently present • **>2–5 (30 assets):** inconsistently documented • **0–2 (50 assets):** frequently missing	Documentation is **uneven**: basic assets strong, intermediate incomplete, advanced often absent. Confirms earlier observations of inconsistency
2) Assessment of Template w.r.t Project documentation	Review of project repositories found **no significant new assets** beyond the minimal template	Confirms template **already covers essential documentation categories**. The real issue is **inconsistent and incomplete application**, not missing elements
3) Overall Conclusion	–	The template is a **valid evaluation framework**, but maintainers need support to **consistently apply and update** documentation across project lifecycle

A structured evaluation of seven OSH projects was carried out by reviewing 88 documentation assets, each scored as *fully available* (1), *partially available* (0.5), or *not available* (0). The assets were grouped into three score ranges: >5–7 (consistently present, 7 assets), >2–5 (inconsistently documented, 30 assets), and 0–2 (frequently missing, 50 assets). Results show that some key assets such as community details, licenses, Product development and CAD files, etc., were consistently present, while a larger number of assets like Hardware Summary Overview, safety guides, maintenance documentation, etc., were often incomplete or missing.

The analysis shows clear patterns. Basic assets, such as project descriptions and licensing information, were almost always available, reflecting their practicality and established importance within the OSH community. Intermediate assets, such as Bill of Materials and CAD files, were generally present but often incomplete or outdated. Advanced assets, such as version histories, assembly instructions, and testing reports, were the least consistently documented. Even in projects selected for their maturity, these elements were often missing or insufficiently detailed. This uneven coverage confirms earlier observations [8] that documentation practices in OSH remain inconsistent, particularly for assets that require ongoing maintenance and validation.

Moreover, we analysed the assets that are present in the current documentation but not in the template and can be potentially added to minimal template. This evaluation was done in multiple iterations so as to ensure accuracy. After assessing the projects, no significant new documentation assets were identified that should be added to the template. This suggests that the template already covers the essential categories needed for OSH projects. However, many assets were incomplete or inconsistently applied across repositories. The real challenge therefore lies not in expanding the template but in ensuring its consistent and detailed adoption throughout the project lifecycle. This finding has two key implications: (i) the template can serve as a reliable reference and evaluation tool for OSH documentation, and (ii) maintainers often struggle with applying it comprehensively, especially when updates must be propagated across multiple files. These insights directly motivate the next stage of our study (Sect. 3.2), where we test whether LLMs such as ChatGPT and Gemini AI can act as assistants to support maintainers by detecting repository changes, prompting for missing information, and automating documentation updates.

3.2 NLP Use-Case Scenario

The methodology was applied to three case studies on open-source hardware repositories, showing how LLMs acting as NLP assistants can support documentation tasks while also exposing their current limitations. In the MakAir ventilator repository, the addition of a new CAD file (*Patient_Filter_Box_rev7.step*) was successfully detected by the assistant. It extracted the part name and revision and updated the Bill of Materials, assembly guide, and changelog. Missing items such as drawings and test reports were flagged in the pull request description. The assistant followed the defined main flow of detection, metadata extraction, consistency checks, and pull request creation, with only minor formatting corrections required. This case shows that assistants can reliably enforce cross-file consistency and reduce repetitive work, though the updates still required human review before integration(see Annex Table 3 for input specifications, Annex Fig. 3 and Annex Fig. 4 for output snapshots from ChatGPT and Gemini AI, Annex C.1 for ChatGPT results and Annex C.2 for Gemini AI results) [7].

The second case study, conducted on the YAEMK split keyboard repository, tested a material change from PLA to PETG. The assistant automatically revised the Bill of Materials, build guide, and changelog, and marked the material transition in the version history. It also produced PETG-specific printing notes and added a checklist of missing items such as slicing profiles or updated photographs. While the outputs were usable and consistent, the limitations became clearer: the assistant could not provide the engineering justification for the change, confirm that suggested print settings were accurate, or assess the effect on tolerances and durability. These tasks required human testing and validation, showing that the assistant can support consistency but cannot replace technical expertise (see Annex Table 4 for input specifications, Annex Fig. 5 and Annex Fig. 6 for output snapshots from ChatGPT and Gemini AI, Annex D.1 for ChatGPT results and Annex D.2 for Gemini AI results) [7].

The third case study used the Open Automated External Defibrillator repository to examine translation and localization, focusing on producing German versions of documentation sections. The assistant generated draft translations of the README, assembly guide, and user manual, maintained a glossary of technical terms, and flagged ambiguous phrases for review. This demonstrated how NLP tools may reduce the workload of multilingual documentation. However, the translations showed limitations: the assistant struggled with regulatory and domain-specific terminology, and it could not ensure that critical safety-related language was precise or compliant. This means outputs required close human checking before use (see Annex Table 5 for input specifications, Annex Fig. 7 and Annex Fig. 8 for output snapshots from ChatGPT and Gemini AI, Annex E.1 for ChatGPT results and Annex E.2 for Gemini AI results) [7]. Taken together, the three cases show that LLMs acting as NLP assistants can help automate documentation updates, translation, and consistency checks, but their role remains complementary. They reduce manual effort and highlight gaps, yet human maintainers are still essential for validation, interpretation, and ensuring technical and regulatory accuracy.

4 Conclusion

This study showed that while the existing OSH documentation template already covers essential categories, its consistent and detailed adoption across projects remains a major challenge. By benchmarking seven representative repositories, we identified recurring gaps in advanced assets such as version histories, assembly instructions, and testing reports. Scenario-based tests with ChatGPT and Gemini AI further demonstrated that LLMs can complement human maintainers by detecting repository changes, updating linked documentation, and flagging missing elements, thereby reducing manual effort and improving consistency. Together, these findings suggest a pathway toward semi-automated documentation pipelines in OSH, where templates provide the framework and NLP assistants help enforce adoption, ultimately lowering the documentation burden, supporting newcomers, and improving reproducibility.

Contributorship. Chodankar, Neeraj., Ramakrishna, Sagar Didaga., More, Sejal., & Chitrakar, Atharva (2025). *Annex to "Gap Analysis of Documentation Template for Open Source Hardware and Adoption in Documentation Workflows with Large Language Models"*. Zenodo. https://doi.org/10.5281/zenodo.17353556.

References

1. Open Source Hardware: The Turing way Community. https://book.the-turing-way.org/reproducible-research/open/open-hardware

2. Bonvoisin, J., et al.: What is the 'source' of open source hardware? J. Open Hardw. **1**(1), 5 (2017). https://doi.org/10.5334/joh.7
3. Bonvoisin, J., et al.: How participative is open source hardware? Insights from online repository mining. Des. Sci. **4**, e19 (2018). https://doi.org/10.1017/dsj.2018.15
4. Fu, W., et al.: LLM4SecHW: leveraging domain specific large language model for hardware debugging (2024). arXiv:2401.16448
5. Wang, C., et al.: UMTG: a toolset to automatically generate system test cases from use case specifications. In: Proceedings of the 2015 10th Joint Meeting on Foundations of Software Engineering, pp. 942–45. ACM (2015). https://doi.org/10.1145/2786805.2803187
6. Colomb, J., et al.: Hardware documentation guide. Version 1.0, Open-Make Project (Zenodo digital archive) (2025). https://open-make.github.io/RHardware-minimaltemplate/
7. Chodankar, N., Ramakrishna, S.D., More, S., Chitrakar, A.: Annex to "gap analysis of documentation template for open source hardware and adoption in documentation workflows with large language models. Zenodo (2025). https://doi.org/10.5281/zenodo.17353556
8. Bonvoisin, J., Mies, R.: Measuring openness in open source hardware with the open-o-meter. Procedia CIRP **78**, 388–393 (2018). https://doi.org/10.1016/j.procir.2018.08.306
9. Dai, J.X., et al.: Issues and challenges of knowledge management in online open source hardware communities. Des. Sci. **6**, e24 (2020). https://doi.org/10.1017/dsj.2020.18
10. Stirling, J., et al.: HardOps: utilising the software development toolchain for hardware design. Int. J. Comput. Integr. Manuf. **35**(12), 1297–309 (2022). https://doi.org/10.1080/0951192X.2022.2028188
11. Miljković, N., Trisovic, A., Peer, L.: Towards FAIR principles for open hardware (2021). arXiv: https://doi.org/10.48550/arXiv.2109.06045
12. Fu, W., et al.: LLM4SecHW: leveraging domain specific large language model for hardware debugging (2024). arXiv:2401.16448
13. Nematollahi, M., et al.: Application of triangulation approach in requirements engineering process: a study in analysis and design of the pediatrics epilepsy information system. Hormozgan Med. J. **21**(1), 65–76 (2017). https://doi.org/10.18869/acadpub.hmj.21.1.65

Author Index

H. Kohl et al. (Eds.): GCSM 2025, LNME, pp. 515–517, 2026.
https://doi.org/10.1007/978-3-032-21154-5

MIX
Papier aus verantwortungsvollen Quellen
Paper from responsible sources
FSC® C105338

If you have any concerns about our products,
you can contact us on
ProductSafety@springernature.com

In case Publisher is established outside the EU,
the EU authorized representative is:
Springer Nature Customer Service Center GmbH
Europaplatz 3, 69115 Heidelberg, Germany

Printed by Libri Plureos GmbH
in Hamburg, Germany